Supersymmetry and String Theory

Beyond the Standard Model

The past decade has witnessed dramatic developments in the fields of experimental and theoretical particle physics and cosmology. This Second Edition is a comprehensive introduction to these recent developments and brings this self-contained textbook right up to date. Brand-new material for this edition includes the ground-breaking Higgs discovery and results of the WMAP and Planck experiments. Extensive discussion of theories of dynamical electroweak symmetry breaking, metastable supersymmetry breaking, an expanded discussion of inflation and a new chapter on the landscape, as well as a completely rewritten coda on future directions, give readers a modern perspective on this developing field. A focus on three principal areas – supersymmetry, string theory and astrophysics and cosmology – provides the structure for this book, which will be of great interest to graduates and researchers in the fields of particle theory, string theory, astrophysics and cosmology. The book contains several problems, and password-protected solutions will be available to lecturers at www.cambridge.org/9781107048386.

Michael Dine is Professor of Physics at the University of California, Santa Cruz. He is an A. P. Sloan Foundation Fellow, a Fellow of the American Physical Society and a Fellow of the American Academy of Arts and Sciences. Prior to this, Professor Dine was a Research Associate at the Stanford Linear Accelerator Center, a long-term member of the Institute for Advanced Study and Henry Semat Professor at the City College of the City University of New York.

Reviews of the first edition

"An excellent and timely introduction to a wide range of topics concerning physics beyond the standard model, by one of the most dynamic researchers in the field. Dine has a gift for explaining difficult concepts in a transparent way. The book has wonderful insights to offer beginning graduate students and experienced researchers alike."

Nima Arkani-Hamed, Harvard University

"How many times did you need to find the answer to a basic question about the formalism and especially the phenomenology of general relativity, the Standard Model, its supersymmetric and grand unified extensions, and other serious models of new physics, as well as the most important experimental constraints and the realization of the key models within string theory? Dine's book will solve most of these problems for you and give you much more, namely the state-of-the-art picture of reality as seen by a leading superstring phenomenologist."

Lubos Motl, Harvard University

"This book gives a broad overview of most of the current issues in theoretical high energy physics. It introduces and discusses a wide range of topics from a pragmatic point of view. Although some of these topics are addressed in other books, this one gives a uniform and self-contained exposition of all of them. The book can be used as an excellent text in various advanced graduate courses. It is also an extremely useful reference book for researchers in the field, both for graduate students and established senior faculty. Dine's deep insights and broad perspective make this book an essential text. I am sure it will become a classic. Many physicists expect that with the advent of the LHC a revival of model building will take place. This book is the best tool kit a modern model builder will need."

Nathan Seiberg, Institute for Advanced Study, Princeton

Supersymmetry and String Theory, Second Edition

Beyond the Standard Model

MICHAEL DINE
University of California, Santa Cruz

Shaftesbury Road, Cambridge CB2 8EA, United Kingdom

One Liberty Plaza, 20th Floor, New York, NY 10006, USA

477 Williamstown Road, Port Melbourne, VIC 3207, Australia

314–321, 3rd Floor, Plot 3, Splendor Forum, Jasola District Centre, New Delhi – 110025, India

103 Penang Road, #05–06/07, Visioncrest Commercial, Singapore 238467

Cambridge University Press is part of Cambridge University Press & Assessment, a department of the University of Cambridge.

We share the University's mission to contribute to society through the pursuit of education, learning and research at the highest international levels of excellence.

www.cambridge.org
Information on this title: www.cambridge.org/9781107048386

First published 2015

A catalogue record for this publication is available from the British Library

Library of Congress Cataloging-in-Publication data
Dine, Michael, author.
Supersymmetry and string theory : beyond the standard model / Michael Dine,
University of California,
Santa Cruz. – Second edition.
pages cm.
Includes bibliographical references and index.
ISBN 978-1-107-04838-6 (Hardback)
1. Supersymmetry. 2. String models. I. Title.
QC174.17.S9D56 2015
539.7′258–dc23 2015013721

ISBN 978-1-107-04838-6 Hardback

Additional resources for this publication at www.cambridge.org/9781107048386

This book is dedicated to Mark and Esther Dine

Contents

Preface to the First Edition

As this is being written, particle physics stands on the threshold of a new era, with the commissioning of the Large Hadron Collider (LHC) not even two years away. In writing this book, I hope to help prepare graduate students and postdoctoral researchers for what will hopefully be a period rich in new data and surprising phenomena.

The Standard Model has reigned triumphant for three decades. For just as long, theorists and experimentalists have speculated about what might lie beyond. Many of these speculations point to a particular energy scale, the teraelectronvolt (TeV) scale, which will be probed for the first time at the LHC. The stimulus for these studies arises from the most mysterious – and still missing – piece of the Standard Model: the Higgs boson. Precision electroweak measurements strongly suggest that this particle is elementary (in that any structure is likely to be far smaller than its Compton wavelength), and that it should be in a mass range where it will be discovered at the LHC. But the existence of fundamental scalars is puzzling in quantum field theory, and strongly suggests new physics at the TeV scale. Among the most prominent proposals for this physics is a hypothetical new symmetry of nature, supersymmetry, which is the focus of much of this text. Others, such as technicolor, and large or warped extra dimensions, are also treated here.

Even as they await evidence for such new phenomena, physicists have become more ambitious, attacking fundamental problems of quantum gravity and speculating on possible final formulations of the laws of nature. This ambition has been fueled by *string theory*, which seems to provide a complete framework for the quantum mechanics of gauge theory and gravity. Such a structure is necessary to give a framework to many speculations about Beyond the Standard Model physics. Most models of supersymmetry breaking and theories of large extra dimensions or warped spaces cannot be discussed in a consistent way otherwise.

It seems, then, quite likely that a twenty-first-century particle physicist will require a working knowledge of supersymmetry and string theory, and in writing this text I hope to provide this. The first part of the text is a review of the Standard Model. It is meant to complement existing books, providing an introduction to perturbative and phenomenological aspects of the theory, but with a lengthy introduction to non-perturbative issues, especially in the strong interactions. The goal is to provide an understanding of chiral symmetry breaking, anomalies and instantons that is suitable for thinking about possible strong dynamics and about dynamical issues in supersymmetric theories. The first part of the book also introduces grand unification and magnetic monopoles.

The second part of the book focuses on supersymmetry. In addition to global supersymmetry in superspace, there is a study of the supersymmetry currents, which are important for understanding dynamics and also for understanding the BPS conditions which play an important role in field theory and string theory dualities. The Minimal Supersymmetric Standard Model (MSSM) is developed in detail, as well as the basics of supergravity and supersymmetry breaking. Several chapters deal with supersymmetry dynamics, including dynamical supersymmetry breaking, Seiberg dualities and Seiberg–Witten theory. The goal is to introduce phenomenological issues (such as dynamical supersymmetry breaking in hidden sectors and its possible consequences), and also to illustrate the control that supersymmetry provides over dynamics.

I then turn to another critical element of Beyond the Standard Model physics: general relativity, cosmology and astrophysics. The chapter on general relativity is meant as a brief primer. The approach is more field theoretic than geometrical, and the uninitiated reader will learn the basics of curvature, the Einstein Lagrangian, the stress tensor and the equations of motion and will encounter the Schwarzschild solution and its features. The subsequent two chapters introduce the basic features of the Friedmann–Robertson–Walker (FRW) cosmology, and then very early universe cosmology: cosmic history, inflation, structure formation, dark matter and dark energy. Supersymmetric dark matter and axion dark matter, and mechanisms for baryogenesis, are all considered.

The third part of the book is an introduction to string theory. My hope, here, is to be reasonably comprehensive while not being excessively technical. These chapters introduce the various string theories, and quickly compute their spectra and basic features of their interactions. Heavy use is made of light cone methods. The full machinery of conformal and superconformal ghosts is described but not developed in detail, but conformal field theory techniques are used in the discussion of string interactions. Heavy use is also made of effective field theory techniques, both at weak and strong coupling. Here, the experience in the first half of the text with supersymmetry is invaluable; again supersymmetry provides a powerful tool to constrain and understand the underlying dynamics. Two lengthy chapters deal with string compactifications; one is devoted to toroidal and orbifold compactifications, which are described by essentially free strings; the other introduces the basics of Calabi–Yau compactification. Four appendices make up the final part of this book.

The emphasis in all of this discussion is on providing tools with which to consider how string theory might be related to observed phenomena. The obstacles are made clear, but promising directions are introduced and explored. I also attempt to stress how string theory can be used as a testing ground for theoretical speculations. I have not attempted a complete bibliography. The suggested reading in each chapter directs the reader to a sample of reviews and texts.

What I know in field theory and string theory is the result of many wonderful colleagues. It is impossible to name all of them, but Tom Appelquist, Nima Arkani-Hamed, Tom Banks, Savas Dimopoulos, Willy Fischler, Michael Green, David Gross, Howard Haber, Jeff Harvey, Shamit Kachru, Andre Linde, Lubos Motl, Ann Nelson, Yossi Nir, Michael Peskin, Joe Polchinski, Pierre Ramond, Lisa Randall, John Schwarz, Nathan Seiberg, Eva Silverstein, Bunji Sakita, Steve Shenker, Leonard Susskind, Scott Thomas, Steven Weinberg, Frank Wilczek, Mark Wise and Edward Witten have all profoundly influenced me, and this influence is reflected in this text. Several of them offered comments on the text or provided specific advice and explanations, for which I am grateful. I particularly wish to thank Lubos Motl for reading the entire manuscript and correcting numerous errors. Needless to say, none of them are responsible for the errors which have inevitably crept into this book.

Some of the material, especially on anomalies and aspects of supersymmetry phenomenology, has been adapted from lectures given at the Theoretical Advanced Study Institute, held in Boulder, Colorado. I am grateful to K. T. Manahathapa for his help during these schools, and to World Scientific for allowing me to publish these excerpts. The lectures "Supersymmetry phenomenology with a broad brush" appeared in *Fields, Strings and Duality*, eds. C. Efthimiou and B. Greene (Singapore: World Scientific, 1997), "TASI lectures on M theory phenomenology" appeared in *Strings, Branes and Duality*, eds. C. Efthimiou and B. Greene (Singapore: World Scientific, 2001) and "The strong CP problem" in *Flavor Physics for the Millennium: Proc. TASI 2000*, ed. J. L. Rosner (Singapore: World Scientific, 2000).

I have used much of the material in this book as the basis for courses, and I am also grateful to students and postdocs (especially Patrick Fox, Assaf Shomer, Sean Echols, Jeff Jones, John Mason, Alex Morisse, Deva O'Neil and Zheng Sun) at Santa Cruz, who have patiently suffered through much of this material as it was developed. They have made important comments on the text and in the lectures, often filling in missing details. As teachers, few of us have the luxury of devoting a full year to topics such as this. My intention is that the separate supersymmetry or string parts are suitable for a one-quarter or one-semester special topics course.

Finally, I wish to thank Aviva, Jeremy, Shifrah and Melanie for their love and support.

Preface to the Second Edition

Much has happened since the appearance of *Supersymmetry and String Theory: Beyond the Standard Model* in 2006. The LHC, after a somewhat bumpy start, has performed spectacularly, discovering what is almost certainly the Higgs particle of the simplest version of the Standard Model in 2012, reproducing and improving a broad range of other Standard Model measurements and excluding significant swathes of the parameter space of proposed ideas for Beyond the Standard Model (BSM) physics.

There have also been important observational and experimental developments in astrophysics and cosmology. The Wilkinson Microwave Anisotropy Probe (WMAP), the Planck satellite and a variety of other experiments have greatly improved our understanding of the cosmic microwave radiation background. We have more reliable measures of the dark matter and dark energy densities and a good measurement of the spectral index, n_s. It is likely that we will soon have some information on, and possibly a measurement of, the scale of inflation coming from studies of B-mode polarization. At the same time, direct and indirect searches for weakly interacting massive particle (WIMP) dark matter have significantly constrained the space of masses and couplings. However, there remain, as of the time of writing, some intriguing anomalies. Furthermore, axion searches have made significant progress and are probing significant parts of the plausible parameter space.

On the theoretical side there have been a number of developments. Within the study of the Standard Model, there has been enormous progress in QCD computations; indeed, these have played an important role in the Higgs discovery. Lattice gauge theorists have continued to make strides in computation of quantum chromodynamics (QCD) quantities, such as quark masses, while embarking on the study of theories relevant to issues in BSM physics. Within supersymmetric models, metastable dynamical supersymmetry breaking has emerged as both an interesting feature of supersymmetric dynamics and a possible mechanism for supersymmetry realization in nature. Other important new ideas include general gauge mediation.

But perhaps the most important theoretical development has been the response to the Higgs discovery, as well as BSM (particularly supersymmetry) exclusions. The observed Higgs mass is compatible with supersymmetry only if the superpartners are quite heavy (tens of TeV) or under special circumstances. Many other BSM ideas face similar challenges. This has sparked a search for alternatives and also a rethinking of notions of naturalness. The big questions are:

1. Is there some form of new physics that accounts for the hierarchy between the weak and other scales, which is perhaps difficult to see or which occurs at a scale somewhat above the current LHC reach?

2. Are our ideas about naturalness somehow misguided? Would a more refined viewpoint point to some energy scale slightly higher than a TeV, which might be accessible to future LHC experiments or some higher-energy accelerator? This has focused renewed attention on ideas such as little Higgs models and Randall–Sundrum models, as well as the possibility that the scale of supersymmetry breaking is simply higher.
3. The possibility that simple-minded notions of naturalness may not be correct has increased interest in the landscape hypothesis.

In this present edition of this book I have attempted to incorporate these developments and to provide some possible directions for investigations of BSM physics. Additions include:

1. new sections on the Higgs discovery;
2. discussion of developments in perturbative QCD computations;
3. expanded discussion of lattice gauge theory, with an emphasis on results of the simulations for quantities such as quark masses;
4. updated discussion of dark matter experiments;
5. updated discussion of the neutrino mass matrix;
6. updated discussion of inflation in light of WMAP, Planck and other experiments;
7. more extensive discussion of solutions to the hierarchy problem outside supersymmetry, especially the little Higgs and Randall–Sundrum models;
8. sections on metastable dynamical supersymmetry breaking that include the Intriligator, Shih and Seiberg models but treat the issue quite generally;
9. an introduction to general gauge mediation;
10. more extensive discussion of the landscape, hypothesis and its connection to and possible implications for notions of naturalness;
11. replacement of the previous "Coda" by a discussion of possible future directions in light of the first four years of LHC, dark matter searches, cosmological observations and theoretical developments.

I have also taken the opportunity to correct many errors in the first edition. I am grateful to the many readers who have pointed these out. I am sure that errors will remain, and I have only myself to blame for these.

Michael Dine

Santa Cruz, California

A note on the choice of metric

There are two popular choices for the metric of flat Minkowski space. One, often referred to as the West Coast metric, is particularly convenient for particle physics applications. Here

$$ds^2 = dt^2 - d\vec{x}^2 = \eta_{\mu\nu} dx^\mu dx^\nu. \tag{0.1}$$

This has the virtue that $p^2 = E^2 - \vec{p}^{\,2} = m^2$. It is the metric of many standard texts in quantum field theory. But it has the annoying feature that ordinary space-like intervals – conventional lengths – acquire a minus sign. So, in most general relativity textbooks as well as string theory textbooks, the East Coast metric is standard:

$$ds^2 = -dt^2 + d\vec{x}^{\,2}. \tag{0.2}$$

Many physicists, especially theorists, become so wedded to one form or another that they resist – or even have difficulty – switching back and forth. This is a text, however, that is intended to deal with particle physics, general relativity and string theory. So, in the first half of the book, which deals mostly with particle physics and quantum field theory, we will use the West Coast convention (0.1). In the second half, dealing principally with general relativity and string theory, we will switch to the East Coast convention (0.2). For both author and readers this may be somewhat disconcerting. While I have endeavored to avoid errors from this somewhat schizophrenic approach, some will have surely slipped in. But I believe that this freedom to move back and forth between the two conventions will be both convenient and healthy. If nothing else, this may be the first textbook in physics in which the author has deliberately used both conventions (many have done so inadvertently).

At a serious level, in computations the researcher must always be careful to be consistent. It is particularly important to be careful when borrowing formulas from papers and texts, and especially when downloading computer programs, to make sure that one has adequate checks on such matters as signs. I will appreciate being informed of any such inconsistencies, as well as of other errors both serious and minor, which have crept into this text.

Text website

Even as this book was going to press, there were important developments in a number of these subjects. The website http://scipp.ucsc.edu/~dine/book/book.html contains updates, errata, solutions of selected problems and additional selected reading.

PART 1

EFFECTIVE FIELD THEORY: THE STANDARD MODEL, SUPERSYMMETRY, UNIFICATION

1 Before the Standard Model

Two of the most profound scientific discoveries of the early twentieth century were special relativity and quantum mechanics. With special (and general) relativity came the notion that physics should be local. Interactions should be carried by dynamical fields in space–time. Quantum mechanics altered the questions which physicists ask about phenomena; the rules governing microscopic (and some macroscopic) phenomena were not those of classical mechanics. When these ideas were combined they took on their full force, in the form of *quantum field theory*: particles themselves are localized, finite-energy, excitations of fields. Otherwise mysterious phenomena, such as the connection of spin and statistics, were immediate consequences of this marriage. But quantum field theory posed serious challenges for its early practitioners. The Schrödinger equation seems to single out time, making a manifestly relativistic description difficult. More seriously, but closely related, in quantum field theory the number of degrees of freedom is infinite, in contrast with the quantum mechanics of atomic systems. In the 1920s and 1930s, physicists performed conventional perturbation theory calculations in the quantum theory of electrodynamics, namely quantum electrodynamics (QED), and obtained expressions which were neither Lorentz invariant nor finite. Until the late 1940s these problems stymied any quantitative progress, and there was serious doubt whether quantum field theory was a sensible framework for physics.

Despite these concerns, quantum field theory proved a valuable tool with which to consider problems of fundamental interactions. Yukawa proposed a field theory of the nuclear force in which the basic quanta were mesons. The corresponding particle was discovered shortly after the Second World War. Fermi was aware of Yukawa's theory and proposed that weak interactions arose through the exchange of some massive particle – essentially the $W^{\pm}$ bosons, which were finally discovered in the 1980s. The large mass of these particles accounted for both the short range and the strength of the weak force. Because of its very short range, one could describe it in terms of four fields interacting at a point. In the early days of the theory, these were the proton, neutron, electron and neutrino. Viewed as a theory of four-fermion interactions Fermi's theory was very successful, accounting for all experimental weak interaction results until well into the 1970s. Yet the theory raised even more severe conceptual problems than QED. At high energies the amplitudes computed in the leading approximation violated unitarity, and the higher-order terms in perturbation theory were very divergent.

The difficulties of QED were overcome in the late 1940s, by Bethe, Dyson, Feynman, Schwinger, Tomanaga and others, as experiments in atomic physics demanded high-precision QED calculations. As a result of their work, it was now possible to perform perturbative calculations in a manifestly Lorentz-invariant fashion. Exploiting covariance the infinities could be controlled and, over time, their significance came to be understood. Quantum electrodynamics achieved enormous successes, explaining the magnetic moment of the electron to extraordinary precision as well as the Lamb shift in hydrogen and other phenomena. One now, for the first time, had an example of a system of physical law that was consistent both with Einstein's principles of relativity and with quantum mechanics.

There were, however, many obstacles to extending this understanding to the strong and weak interactions, and at times it seemed that some other framework might be required. The difficulties came in various types. The infinities of Fermi's theory of weak interactions could not be controlled as in electrodynamics. Even postulating the existence of massive particles to mediate the force did not solve the problems. But the most severe difficulties came in the case of the strong interactions. The 1950s and 1960s witnessed the discovery of hundreds of hadronic resonances. It was hard to imagine that each should be described by still another fundamental field. Some theorists pronounced field theory dead and sought alternative formulations (among the outgrowths explorations was string theory, which has emerged as the most promising setting for a quantum theory of gravitation). But Gell-Mann and Zweig realized that *quarks* could serve as an organizing principle. Originally, there were only three, u, d and s, with baryon number $1/3$ and charges $2/3$, $-1/3$ and $-1/3$ (in units of the electric charge) respectively. All the known hadrons could be understood as bound states of objects with these quantum numbers. Still, there remained difficulties. First, quarks were strongly interacting and there were no successful ideas for treating strongly interacting fields. Second, those searching for quarks came up empty handed.

In the late 1960s a dramatic series of experiments at SLAC, and a set of theoretical ideas due to Feynman and Bjorken, changed the situation again. Feynman had argued that one should take seriously the idea of quarks as dynamical entities (for a variety of reasons he hesitated to call them quarks, referring to them as *partons*). He conjectured that these partons would behave as nearly free particles in situations where momentum transfers were large. He and Bjorken realized that this picture implied a scaling in deep inelastic scattering phenomena. The experiments at SLAC exhibited just this phenomenon and showed that the partons carried the electric charges of the u and d quarks.

But this situation was still puzzling. Known field theories did not behave in the fashion conjectured by Feynman and Bjorken. The interactions of particles typically became *stronger* as the energies and momentum transfers grew. This is the case, for example, in quantum electrodynamics and a simple quantum mechanical argument, based on unitarity and relativity, would seem to suggest it is true in general. But there turned out to be an important class of theories with the opposite property.

In 1954 Yang and Mills wrote down a generalization of electrodynamics where the $U(1)$ symmetry group is enlarged to a non-Abelian group, with massless gauge bosons transforming in the adjoint representation of the group. While mathematically quite beautiful, these *non-Abelian gauge theories* remained oddities for some time. First, their possible place in the scheme of things was not known (Yang and Mills themselves suggested that perhaps their vector particles were the ρ mesons). Moreover, their quantization was significantly more challenging than that of electrodynamics. It was not at all clear that these theories really made sense at the quantum level, that is, that they respected the principles of both Lorentz invariance and unitarity. The first serious effort to quantize

Yang–Mills theories was probably due to Schwinger, who chose a non-covariant but manifestly unitary gauge and carefully verified that the Poincaré algebra was satisfied. The non-covariant gauge, however, was exceptionally awkward. Real progress in formulating a covariant perturbation expansion was made by Feynman, who noted that naive Feynman rules for these theories were not unitary but that this difficulty could be removed, at least in low orders, by adding a set of fictitious fields ("ghosts"). A general formulation was provided by Faddeev and Popov, who derived Feynman's covariant rules in a path integral formulation and showed their formal equivalence to Schwinger's manifestly unitary formulation. A convincing demonstration that these theories are unitary, covariant and *renormalizable* was finally given in the early 1970s by 't Hooft and Veltman, who developed elegant and powerful techniques for performing real calculations as well as formal proofs.

In the original Yang–Mills theories the vector bosons were massless and their possible connections to known phenomena were obscure. However, Carl R. Hagen, Francois Englert, Gerald S. Guralnik, Peter W. Higgs, Robert Brout, and T. W. B. Kibble discovered a mechanism by which these particles could become massive. In 1967, Weinberg and Salam wrote down a Yang–Mills theory of weak interactions based on what has come to be referred to as the "Higgs mechanism". This finally realized Fermi's idea that weak interactions arise from the exchange of a very massive particle. To a large degree this work was ignored until 't Hooft and Veltman proved the unitarity and renormalizability of these theories. At this point the race to find precisely the correct theory and study its experimental consequences was on; Weinberg's and Salam's first guess turned out to be correct.

The possible role of Yang–Mills fields in strong interactions was, at first sight, even more obscure. To complete the story required another important fact of hadronic physics. While the quark model was very successful, it was also puzzling. The quarks were spin-$1/2$ particles, yet models of the hadrons seemed to require that the hadronic wave functions were symmetric under the interchange of quark quantum numbers. A possible resolution, suggested by Greenberg, was that the quarks carried an additional quantum number, called color, coming in three possible types. The statistics puzzle was solved if the hadron wave functions were totally antisymmetric in color. This hypothesis required that the color symmetry, unlike, say, isospin, should be exact and thus special. While seemingly contrived, it explained two other facts: the width of the π^0 meson and the value of the e^+e^- cross section to hadrons, each of which was otherwise was too large by a factor three.

To a number of researchers the exactness of this color symmetry suggested a possible role for Yang–Mills theory. So, in retrospect there was an obvious question: could it be that an $SU(3)$ Yang–Mills theory, describing the interactions of quarks, would exhibit the property required to explain Bjorken scaling, i.e. that the interactions become weak at short distances? Of course, things were not quite so obvious at the time.The requisite calculation had already been done by 't Hooft but the result seems not to have been widely known nor its significance appreciated. David Gross and his student Frank Wilczek set out to prove that no field theory had the required scaling property, while Sidney Coleman, apparently without any particular prejudice, assigned the problem to his graduate student David Politzer. All soon realized that Yang–Mills theories do have the property of asymptotic freedom: the interactions become weak at high momentum transfers or at short distances.

Experiment and theory now entered a period of remarkable convergence. Alternatives to the Weinberg–Salam theory were quickly ruled out. The predictions of quantum chromodynamics (QCD) were difficult, at first, to verify in detail. The theory predicted small violations of Bjorken scaling, depending logarithmically on energy, and it took many years to measure them convincingly. But there was another critical experimental development which clinched the picture. The existence of a heavy quark beyond the u, d and s had been predicted by Glashow, Iliopoulos and Maiani and was a crucial part of the developing Standard Model. The mass of this *charm* quark had been estimated by Gaillard and Lee. Appelquist and Politzer predicted, almost immediately after the discovery of asymptotic freedom, that heavy quarks would be bound in narrow vector resonances. In 1974 a narrow resonance was discovered in e^+e^- annihilation, the J/ψ particle, which was quickly identified as a bound state of a charm quark and its antiparticle.

Over the next 25 years, this Standard Model was subjected to more and more refined tests. One feature absent from the original Standard Model was CP(T) violation. Kobiyashi and Maskawa pointed out that if there were a third generation of quarks and leptons, then the theory could accommodate the observed CP violation in the K meson system. Two more quarks and a lepton were discovered, and their interactions and behavior were as expected within the Standard Model. Jets of particles which could be associated with *gluons* were seen in the late 1970s. The W and Z particles were produced in accelerators in the early 1980s. At CERN and SLAC, precision measurements of the Z mass and width provided stringent tests of the weak-interaction part of the theory. Detailed measurements in deep inelastic scattering and in jets provided precise confirmation of the logarithmic scaling violations predicted by QCD. The Standard Model passed every test.

At the time at which the first edition of this book went to press, the Standard Model had triumphed in almost every realm. The low-energy weak interactions were completely described by the Weinberg–Salam theory with corrections from the strong interactions, many well understood. At high energies the W and Z particles had been produced in great numbers in accelerators, and their properties – i.e. production rates and decays – compared with the theory, including the effects of QCD, at the one part per mil level. The Tevatron had performed precise studies of jet production in excellent agreement with QCD and lattice gauge theory had witnessed an enormous leap in reliability and precision, reproducing features of the hadron spectrum and yielding quantities of importance for the study of the weak decays of B mesons, for example. The only missing piece was the Higgs particle, or whatever entity was responsible for the breaking of the electroweak symmetry. In 2012, that changed. The 5σ discovery of a scalar particle was announced at CERN on July 4. By the end of the first run of the LHC at the end of the year, a good deal of circumstantial evidence had accumulated that this particle was indeed the Higgs scalar of the simplest Standard Model. 't Hooft and Veltman had received the Nobel Prize for their work on non-Abelian gauge theories in 1999. During the first 14 years of the new millennium, these successes have been recognized by several Nobel Prizes: Gross, Politzer and Wilczek for the understanding of strong interactions (2004); Nambu for his work on spontaneous symmetry breaking; Kobayashi and Maskawa for the mechanism of

CP violation in the Standard Model (2008); and Englert and Higgs for the proposal of the Higgs particle (2013). Since the publication of the first edition of this book, a Nobel Prize has been awarded for the discovery of dark energy (Perlmutter, Reiss and Schmidt, 2011).

So the question which I raised in 2006, Why write a book about Beyond the Standard Model physics?, is all the sharper now. It is still true that, for all its simplicity and success in reproducing the interactions of elementary particles, the Standard Model cannot represent a complete description of nature. In the first few chapters of this book we will review the Standard Model and its successes, including the recent discovery of the Higgs particle, which is a triumph not only for our understanding of the electroweak theory but of QCD as well. Then we will discuss some of the Standard Model's limitations. These include the *hierarchy problem*, which, at its most primitive level, represents a failure of dimensional analysis; the presence of a large number of parameters; the strong CP problem, i.e. the presence of a very small dimensionless number which violates CP. We will confront the incompatibility of quantum mechanics with Einstein's theory of general relativity, the inability of the Standard Model to account for the small but non-zero value of the cosmological constant (an even more colossal failure of dimensional analysis) and its failure to account for basic features of our universe, the excess of baryons over antibaryons, dark matter and structure. Then we will set out on an exploration of possible phenomena which might address these questions. These include: supersymmetry, technicolor and large or warped extra dimensions as possible solutions to the hierarchy problem; grand unification as a partial solution to the overabundance of parameters; and the axion for the strong CP problem. Still more ambitious is superstring theory, as a possible solution to the problem of quantizing gravity, which incorporates many features of these other proposals. We will consider the experimental constraints on new physics, which have become more severe with the first LHC run, and discuss the prospects for the future at the LHC and beyond. Finally, we will acknowledge the possibility that the resolution of some of these puzzles might involve a *landscape* or *multiverse*.

Suggested reading

A complete bibliography of the Standard Model would require a book by itself. A good deal of the history of special relativity, quantum mechanics and quantum field theory can be found in *Inward Bound*, by Abraham Pais (1986), which also includes an extensive bibliography. The development of the Standard Model is also documented in this very readable book. As a minor historical note I would add that the earliest reference in which I came across the observation that a Yang–Mills theory might underlie the strong interactions is due to Feynman, in about 1963 (Roger Dashen, personal communication, 1981), who pointed out that in an $SU(3)$ Yang–Mills theory three quarks would be bound together, as would quark–antiquark pairs.

2 The Standard Model

The interactions of the Standard Model give rise to the phenomena of our day to day experience. They explain virtually all the particles and interactions which have been observed in accelerators. Yet the underlying laws can be summarized in a few lines. In this chapter we describe the ingredients of this theory and some of its important features. Many dynamical questions will be studied in subsequent chapters. For detailed comparisons of theory and experiment there are a number of excellent texts, described in the suggested reading at the end of the chapter.

2.1 Yang–Mills theory

By the early 1950s physicists were familiar with approximate global symmetries such as isospin. Yang and Mills argued that the lesson of Einstein's general theory was that symmetries, if exact, should be local. In ordinary electrodynamics the gauge symmetry is a local Abelian symmetry. Yang and Mills explained how to generalize this to a non-Abelian symmetry group. Let's first review the case of electrodynamics. The electron field $\psi(x)$ transforms under a gauge transformation as follows:

$$\psi(x) \to e^{i\alpha(x)}\psi(x) = g_\alpha(x)\psi(x). \tag{2.1}$$

We can think of $g_\alpha(x) = e^{i\alpha(x)}$ as a group element in the group $U(1)$. The group is Abelian: $g_\alpha g_\beta = g_\beta g_\alpha = g_{\alpha+\beta}$. Quantities such as $\bar\psi\psi$ are gauge invariant, but derivative terms such as $i\bar\psi\, \partial\!\!\!/ \psi$, are not. In order to write down the derivative terms in an action or equation of motion, one needs to introduce a gauge field A_μ transforming under the symmetry transformation as

$$\begin{aligned} A_\mu &\to A_\mu + \partial_\mu\alpha \\ &= A_\mu + ig(x)\partial_\mu g^{-1}(x). \end{aligned} \tag{2.2}$$

This second form allows more immediate generalization to the non-Abelian case. Given A_μ and its transformation properties, we can define a covariant derivative,

$$D_\mu\psi = (\partial_\mu - iA_\mu)\psi. \tag{2.3}$$

This derivative has the property that it transforms like ψ itself under the gauge symmetry:

$$D_\mu\psi \to g(x)D_\mu\psi. \tag{2.4}$$

We can also form a gauge-invariant object from the gauge fields A_μ themselves. A simple way to do this is to construct the commutator of two covariant derivatives,

$$F_{\mu\nu} = i[D_\mu, D_\nu] = \partial_\mu A_\nu - \partial_\nu A_\mu. \tag{2.5}$$

This form of the gauge transformations may be somewhat unfamiliar. Note in particular that the charge of the electron, e (the gauge coupling) does not appear in the transformation laws. Instead, the gauge coupling appears when we write down a gauge-invariant Lagrangian:

$$\mathcal{L} = i\bar{\psi}\, \not{D}\psi - m\bar{\psi}\psi - \frac{1}{4e^2}F_{\mu\nu}^2, \tag{2.6}$$

where the "slash" notation is defined by $\not{a} = a^\mu \gamma_\mu$. The more familiar formulation is obtained if we make the replacement

$$A_\mu \to eA_\mu. \tag{2.7}$$

In terms of this new field the gauge transformation law is

$$A_\mu \to A_\mu + \frac{1}{e}\partial_\mu \alpha \tag{2.8}$$

and the covariant derivative is

$$D_\mu \psi = (\partial_\mu - ieA_\mu)\psi. \tag{2.9}$$

We can generalize this to a non-Abelian group, $\mathcal{G}$, by taking ψ to be a field (fermion or boson) in some representation of the group; $g(x)$ is then a matrix which describes a group transformation acting in this representation. Formally, the transformation law is the same as before,

$$\psi \to g(x)\psi(x), \tag{2.10}$$

but the group composition law is more complicated:

$$g_\alpha g_\beta \neq g_\beta g_\alpha. \tag{2.11}$$

The gauge field A_μ is now a matrix-valued field, transforming in the adjoint representation of the gauge group:

$$A_\mu \to gA_\mu g^{-1} + ig(x)\partial_\mu g^{-1}(x). \tag{2.12}$$

Formally, the covariant derivative also looks exactly as before:

$$D_\mu \psi = (\partial_\mu - iA_\mu)\psi, \quad D_\mu \psi \to g(x) D_\mu \psi. \tag{2.13}$$

Like A_μ, the field strength is a matrix-valued field:

$$F_{\mu\nu} = i[D_\mu, D_\nu] = \partial_\mu A_\nu - \partial_\nu A_\mu - i[A_\mu, A_\nu]. \tag{2.14}$$

Note that $F_{\mu\nu}$ is not gauge *invariant* but, rather, covariant:

$$F_{\mu\nu} \to gF_{\mu\nu}g^{-1}, \tag{2.15}$$

i.e. it transforms like a field in the adjoint representation, with no inhomogeneous term.

The gauge-invariant action $\mathcal{L}$ is formally almost identical to that of the $U(1)$ theory:

$$\mathcal{L} = i\bar{\psi}\, \not{D}\psi - m\bar{\psi}\psi - \frac{1}{2g^2}\,\mathrm{Tr}\,F_{\mu\nu}^2. \tag{2.16}$$

Here we have changed the letter we use to denote the coupling constant: we will usually reserve e for the electron charge and use g for a generic gauge coupling. Note also that it is necessary to take the trace of F^2 to obtain a gauge-invariant expression.

The matrix form for the fields may be unfamiliar, but it is very powerful. One can recover expressions in terms of more conventional fields by defining

$$A_\mu = A_\mu^a T_a, \tag{2.17}$$

where T_a are the group generators in the representation appropriate to ψ. Then, for $SU(N)$, for example, if the T_as are in the fundamental representation, we have

$$\mathrm{Tr}(T_a T_b) = \frac{1}{2}\delta_{ab}, \quad [T^a, T^b] = if^{abc}T^c, \tag{2.18}$$

where f^{abc} are the structure constants of the group and

$$A_\mu^a = 2\,\mathrm{Tr}(T_a A^\mu), \quad F_{\mu\nu}^a = \partial_\mu A_\nu^a - \partial_\nu A_\mu^a + f_{abc}A_\mu^a A_\nu^b. \tag{2.19}$$

While they are formally almost identical, there are great differences between the Abelian and non-Abelian theories. Perhaps the most striking is that the equations of motion for the A_μs are non-linear in non-Abelian theories. This behavior means that, unlike the case of Abelian gauge fields, a theory of non-Abelian fields without matter is a non-trivial, interacting, theory with interesting properties. With and without matter fields, this will lead to much richer behavior even classically. For example, we will see that non-Abelian theories sometimes contain solitons, localized finite-energy solutions of the classical equations. The most interesting of these are the magnetic monopoles. At the quantum level these non-linearities lead to properties such as asymptotic freedom and confinement.

Using the form in which we have written the action, the matter fields ψ can appear in any representation of the group; one just needs to choose appropriate matrices T^a. We can also consider scalars, as well as fermions. For a scalar field ϕ, we define the covariant derivative $D_\mu\phi$ as before and add to the action a term $|D_\mu\phi|^2$ for a complex field or $(D_\mu\phi)^2/2$ for a real field.

2.2 Realizations of symmetry in quantum field theory

The most primitive exercise we can do with the Yang–Mills Lagrangian is to set $g = 0$ and examine the equations of motion for the fields A^μ. If we choose the gauge $\partial_\mu A^{\mu a} = 0$, all the gauge fields obey

$$\partial^2 A_\mu^a = 0. \tag{2.20}$$

So, like the photon, all the gauge fields A^a_μ of the Yang–Mills theory are massless. At first sight there is no obvious place for these fields in either the strong or the weak interactions. But it turns out that in non-Abelian theories the possible ways in which the symmetry may be realized are quite rich. First, the symmetry can be realized in terms of massless gauge bosons; this is known as the *Coulomb phase*. This possibility is not relevant to the Standard Model but will appear in some of our more theoretical considerations later. A second way is known as the *Higgs phase*. In this phase, the gauge bosons are massive. In the third, the *confinement phase*, there are no physical states with the quantum numbers of isolated quarks (particles in the fundamental representation), and the gauge bosons are also massive. The second phase is relevant to the weak interactions; the third, confinement, phase to the strong interactions.[1]

2.2.1 The Goldstone phenomenon

Before introducing the Higgs phase it is useful to discuss global symmetries. While we will frequently argue, like Yang and Mills, that global symmetries are less fundamental than local ones, they are important in nature. Examples are isospin, the chiral symmetries of the strong interactions and baryon number. We can represent the action of such a symmetry much as we represented the symmetry action in Yang–Mills theory:

$$\Phi \to g_\alpha \Phi, \tag{2.21}$$

where Φ is some set of fields and g is now a constant matrix, independent of spatial position. Such symmetries are typically accidents of the low-energy theory. Isospin, for example, as we will see arises because the masses of the u and d quarks are small compared with other scales of quantum chromodynamics. Then g is the matrix

$$g_{\vec{\alpha}} = e^{i\vec{\alpha}\cdot\vec{\sigma}/2} \tag{2.22}$$

acting on the u and d quark doublet. Note that $\vec{\alpha}$ is not a function of space but a continuous parameter, so we will refer to such symmetries as continuous global symmetries. In the case of isospin it is also important that the electromagnetic and weak interactions, which violate this symmetry, are small perturbations on the strong interactions.

The simplest model of a continuous global symmetry is provided by a complex field ϕ transforming under a $U(1)$ symmetry,

$$\phi \to e^{i\alpha}\phi. \tag{2.23}$$

We can take for the Lagrangian for this system

$$\mathcal{L} = |\partial_\mu \phi|^2 - m^2|\phi|^2 - \frac{1}{2}\lambda|\phi|^4. \tag{2.24}$$

If $m^2 > 0$ and λ is small, this is simply a theory of a weakly interacting, complex scalar. The states of the theory can be organized as states of definite $U(1)$ charge. This is the unbroken

[1] The differences between the confinement and Higgs phases are subtle, as was first stressed by Fradkin, Shenker and 't Hooft. But we now know that the Standard Model is well described by a weakly coupled field theory in the Higgs phase.

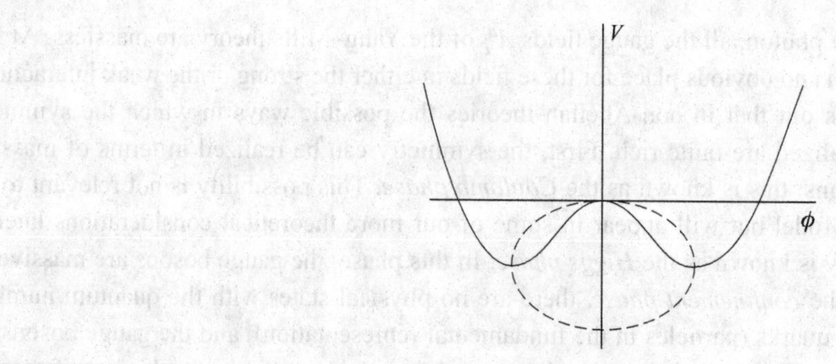

Fig. 2.1 Scalar potential with negative mass-squared. The stable minimum leads to broken symmetry.

phase. However, m^2 is just a parameter and we can ask what happens if $m^2 = -\mu^2 < 0$. In this case the potential,

$$V(\phi) = -\mu^2|\phi|^2 + \lambda|\phi|^4, \tag{2.25}$$

looks as in Fig. 2.1. There is a set of degenerate minima,

$$\langle\phi\rangle_\alpha = \frac{\mu}{\sqrt{2\lambda}}e^{i\alpha}. \tag{2.26}$$

These ground states are obtained from one another by symmetry transformations; in somewhat more mathematical language, we say that there is a manifold of vacuum states. Quantum mechanically it is necessary to choose a particular value of α. As will be explained in the next section, if one chooses α then no local operator, e.g. no small perturbation, will take the system into a state of different α. To simplify the writing, take $\alpha = 0$. Then we can parameterize the complex field ϕ in terms of real fields σ and π:

$$\phi = \frac{1}{\sqrt{2}}[v + \sigma(x)]e^{i\pi(x)/v} \approx \frac{1}{\sqrt{2}}[v + \sigma(x) + i\pi(x)]. \tag{2.27}$$

Here $v = \mu/\sqrt{\lambda}$ is known as the *vacuum expectation value* (vev) of the field ϕ. In terms of σ and π, the Lagrangian takes the form

$$\mathcal{L} = \frac{1}{2}[(\partial_\mu\sigma)^2 + (\partial_\mu\pi)^2 - 2\mu^2\sigma^2 + \mathcal{O}(\sigma,\pi)^3]. \tag{2.28}$$

So we see that σ is an ordinary real, scalar field of mass-squared $2\mu^2$, while the π field is massless. The fact that it is massless is not a surprise: the mass represents the energy cost of turning on a zero-momentum excitation of π, but such an excitation is just a symmetry transformation $v \to ve^{i\pi(0)}$ of ϕ. So there *is* no energy cost.

The appearance of massless particles when a symmetry is broken is quite general and is known as the Nambu–Goldstone phenomenon; π is called a Nambu–Goldstone boson. In any theory with scalars, the choice of a minimum may break some symmetry. This means that there is a manifold of vacuum states. The broken-symmetry generators are those which transform the system from one point on this manifold to another. Because there is no energy cost associated with such a transformation, there is a massless particle associated with each broken-symmetry generator. This result is very general. Symmetries can be broken not only by the expectation values of scalar fields but also by the expectation values of composite operators, and the theorem holds. A proof of this result is provided in Appendix B. In nature there are a number of excitations which can be identified as Goldstone or almost-Goldstone ("pseudo-Goldstone") bosons. These include spin waves in solids and the pi mesons. We will have much more to say about pions later.

2.2.2 Aside: choosing a vacuum

In quantum mechanics there is no notion of a spontaneously broken symmetry. If one has a set of degenerate classical configurations, the ground state will invariably involve a superposition of these configurations. If we took σ and π in Eq. (2.27) to be functions only of the time t then the σ–π system would just be an ordinary quantum mechanical system with two degrees of freedom. Here σ would correspond to an anharmonic oscillator of frequency $\omega = \sqrt{2}\mu$. Placing this particle in its ground state, one would be left with the coordinate π. Note that π, in Eq. (2.27), is an angle, like the azimuthal angle, in ordinary quantum mechanics. We could call its conjugate variable L_z. The lowest lying state would be the zero-angular-momentum state, a uniform superposition of all values of π. In field theory at finite volume, the situation is similar. The zero-momentum mode of π is again an angular variable, and the ground state is invariant under the symmetry. At infinite volume, however, the situation is different. One is forced to choose a value of π.

This issue is most easily understood by considering a different problem: rotational invariance in a magnet. Consider Fig. 2.2, which shows a ferromagnet with spins aligned at an angle θ. We can ask: what is the overlap of two states, one with $\theta = 0$, one at θ, i.e. what is $\langle\theta|0\rangle$? For a single site the overlap between the state $|+\rangle$ with $\theta = 0$ and the rotated state is

$$\left\langle + \left| e^{i\tau_1\theta/2} \right| + \right\rangle = \cos(\theta/2). \tag{2.29}$$

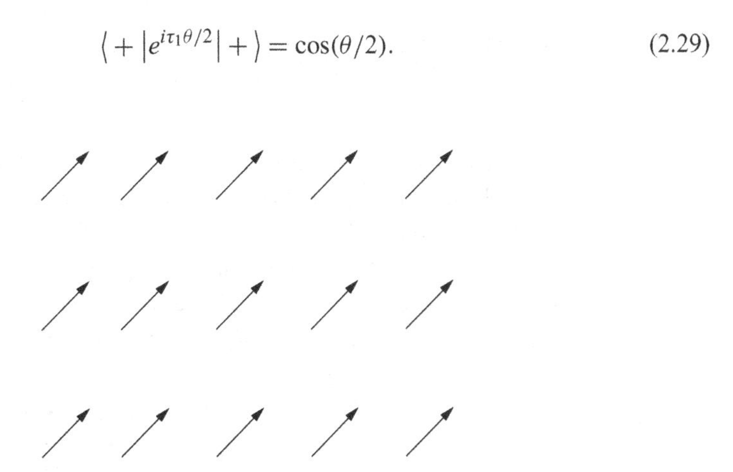

Fig. 2.2 In a ferromagnet the spins are aligned but their direction is arbitrary.

If there are N such sites, the overlap behaves as follows:

$$\langle\theta|0\rangle \sim [\cos(\theta/2)]^N, \tag{2.30}$$

i.e. it vanishes exponentially rapidly with the "volume", N.

For a continuum field theory, states with differing values of the order parameter v also have no overlap in the infinite-volume limit. This is illustrated by the theory of a scalar field ϕ with Lagrangian

$$\mathcal{L} = \frac{1}{2}(\partial_\mu \phi)^2. \tag{2.31}$$

For this system there is no potential, so the expectation value $\phi = v$ is not fixed. The Lagrangian has a symmetry, $\phi \to \phi + \delta$, for which the charge is just

$$Q = \int d^3x\, \Pi(\vec{x}) \tag{2.32}$$

where Π is the canonical momentum. So we want to study

$$\langle v|0\rangle = \langle 0|e^{iQ}|0\rangle. \tag{2.33}$$

We must be careful how we take the infinite-volume limit. We will insist that this be done in a smooth fashion, so we will define

$$\begin{aligned} Q &= \int d^3x\, \partial_0 \left(\phi e^{-\vec{x}^2/V^{2/3}}\right) \\ &= -i\int \frac{d^3k}{(2\pi)^3}\sqrt{\frac{\omega_k}{2}}\left(\frac{V^{1/3}}{\sqrt{\pi}}\right)^3 e^{-\vec{k}^2 V^{2/3}/4}[a(\vec{k}) - a^\dagger(\vec{k})]. \end{aligned} \tag{2.34}$$

Now one can evaluate the matrix element, using

$$e^{A+B} = e^A e^B e^{-[A,B]/2}$$

(provided that the commutator is a c-number), obtaining

$$\langle 0|e^{iQ}|0\rangle = e^{-cv^2 V^{2/3}}, \tag{2.35}$$

where c is a numerical constant. So the overlap vanishes with the volume. You can convince yourself that the same holds for matrix elements of local operators. This result does not hold in 0+1 and 1+1 dimensions, because of the severe infrared behavior of theories in low dimensions. This is known to particle physicists as Coleman's theorem, and to condensed matter theorists as the Mermin–Wagner theorem. This theorem will make an intriguing appearance in string theory, where it is the origin of energy–momentum conservation.

2.2.3 The Higgs mechanism

Suppose that the $U(1)$ symmetry of the previous section is local. In that case, even a spatially varying $\pi(x)$ represents a symmetry transformation and, by a suitable gauge choice, it can be eliminated. In other words, by a gauge transformation we can bring the field ϕ to the form

$$\phi = \frac{1}{\sqrt{2}}[v + \sigma(x)]. \tag{2.36}$$

In this gauge, the gauge-invariant kinetic term for ϕ takes the form

$$|D_\mu \phi|^2 = \frac{1}{2}(\partial_\mu \sigma)^2 + \frac{1}{2}A_\mu^2 v^2 + \cdots. \tag{2.37}$$

The second term is a mass term for the gauge field A_μ. To determine the actual value of the mass, we need to examine the kinetic term for the gauge fields,

$$-\frac{1}{2g^2}(\partial_\mu A^\nu)^2 + \cdots. \tag{2.38}$$

So the gauge field must have mass $m_A^2 = g^2 v^2$.

This phenomenon, that the gauge boson becomes massive when the gauge symmetry is spontaneously broken, is known as the Higgs mechanism. While formally quite similar to the Goldstone phenomenon, it is also quite different. The fact that there is no massless particle associated with motion along the manifold of ground states is not surprising – these states are all physically equivalent. Symmetry breaking, in fact, is a paradoxical notion in gauge theories, since gauge transformations describe entirely equivalent physics (gauge symmetry is often referred to as a redundancy in the description of a system). Perhaps the most important lesson here is that gauge invariance does not necessarily mean, as it does in electrodynamics, that the gauge bosons are massless.

2.2.4 Goldstone and Higgs phenomena for non-Abelian symmetries

Both the Goldstone and Higgs phenomena generalize to non-Abelian symmetries. In the case of global symmetries, for every generator of a broken global symmetry there is a massless particle. For local symmetries, each broken generator gives rise to a massive gauge boson.

As an example, relevant both to the strong and the weak interactions, consider a theory with a symmetry $SU(2)_L \times SU(2)_R$. Take M to be a Hermitian matrix field,

$$M = \sigma I + i\vec{\pi} \cdot \vec{\sigma}. \tag{2.39}$$

Under the above symmetry, which we first take to be global, M transforms as follows:

$$M \to g_L M g_R \tag{2.40}$$

with g_L and g_R $SU(2)$ matrices. We can take the Lagrangian to be

$$\mathcal{L} = \mathrm{Tr}\,(\partial_\mu M^\dagger \partial^\mu M) - V(\mathrm{Tr}(M^\dagger M)). \tag{2.41}$$

This Lagrangian respects the symmetry. If the curvature of the potential at the origin is negative, M will acquire an expectation value. If we take:

$$\langle M \rangle = \langle \sigma \rangle I \tag{2.42}$$

then some of the symmetry is broken. However, the expectation value of M is invariant under the subgroup of the full symmetry group with $g_L = g_R^\dagger$. In other words, the unbroken symmetry is $SU(2)$. Under this symmetry, the fields $\vec{\pi}$ transform as a vector. In the case of the strong interactions, this unbroken symmetry can be identified with isospin. In the case of the weak interactions, there is an approximate global symmetry reflected in the masses of the W and Z particles, as we will discuss later.

2.2.5 Confinement

There is still another possible realization of gauge symmetry: confinement. This is crucial to our understanding of strong interactions. As we will see, Yang–Mills theories, in the case where there is not too much matter, become weak at short distances and strong at large distances. This is just what is required to understand the qualitative features of the strong interactions: free-quark and free-gluon *behavior* at very large momentum transfers, but strong forces at larger distances so that there are in fact no free quarks or gluons. As is the case for the Higgs mechanism, there are no massless particles in the spectrum of hadrons: QCD is said to have a "mass gap." These features of strong interactions are supported by extensive numerical calculations, but they are hard to understand through simple analytical or qualitative arguments (indeed, if you can offer such an argument, you could win a Clay prize of $1 million). We will have more to say about the phenomenon of confinement when we discuss lattice gauge theories.

One might wonder: what is the difference between the Higgs mechanism and confinement? This question was first raised by Fradkin and Shenker and by 't Hooft, who also gave an answer: there is often no qualitative difference. The qualitative features of a theory without massless gauge fields as a result of the Higgs phenomenon can be reproduced by a confined strongly interacting theory. However, the detailed predictions of the weakly interacting Weinberg–Salaam theory are in close agreement with experiment but those of the strongly interacting theory are not.

2.3 The quantization of Yang–Mills theories

In this book we will encounter a number of interesting classical phenomena in Yang–Mills theory but, in most of the situations in nature on which we are focusing, we will be concerned with the quantum behavior of the weak and strong interactions. Abelian theories such as QED already present considerable challenges. One can perform canonical quantization in a gauge, such as the Coulomb gauge or a light cone gauge, in which unitarity is manifest – all the states have positive norm. But, in such a gauge the covariance of the theory is hard to see. Or one can choose a gauge where Lorentz invariance is manifest, but not unitarity. In QED it is not too difficult to show, at the level of Feynman diagrams, that these gauge choices are equivalent. In non-Abelian theories, canonical quantization is still more challenging. Path integral methods provide a much more powerful approach to the quantization of these theories than the canonical methods mentioned above.

A brief review of path integration appears in Appendix C. Here we discuss gauge fixing and derive the Feynman rules. We start with the gauge fields alone; adding the matter fields – scalars or fermions – is not difficult. The basic path integral is

$$\int [dA_\mu] e^{iS}. \tag{2.43}$$

The problem is that this integral includes a huge redundancy: the gauge transformations. To deal with this, we need to make a gauge choice, for example

$$G^a\left(A^a_\mu\right) = \partial_\mu A^{\mu a} = 0. \tag{2.44}$$

We insert unity in the form

$$1 = \int [dg]\delta\left(G\left(A^g_\mu\right)\right)\Delta[A]. \tag{2.45}$$

Here we have reverted to our matrix notation: G is a general gauge-fixing condition; A^g_μ denotes the gauge transform of A_μ by g. The quantity Δ is a functional determinant known as the Faddeev–Popov determinant. Note that Δ is gauge invariant: $\Delta[A^h] = \Delta[A]$. This follows from the definition

$$\int [dg]\delta\left(G\left(A^{hg'}_\mu\right)\right) = \int [dg]\delta\left(G\left(A^{g'}_\mu\right)\right), \tag{2.46}$$

where, in the last step, we have made the change of variables $g \to h^{-1}g$. We can write a more explicit expression for Δ as a determinant. To do this, we first need an expression for the variation of the As under an infinitesimal gauge transformation. Writing $g = 1 + i\omega$, and using the matrix form for the gauge field, we have

$$\delta A_\mu = \partial_\mu \omega + i[\omega, A_\mu]. \tag{2.47}$$

This can be written elegantly as a covariant derivative of ω, where ω can be thought of as a field in the adjoint representation:

$$\delta A_\mu = D_\mu \omega. \tag{2.48}$$

If we make the specific choice $G = \partial_\mu A^\mu$ then to evaluate Δ we need to expand G about the field A_μ for which $G = 0$:

$$G(A + \delta A) = \partial_\mu D^\mu \omega = \partial^2 \omega + i[A_\mu, \partial_\mu \omega] \tag{2.49}$$

or, in index form,

$$G\left(A^a_\mu\right) = (\partial^2 \delta^{ac} + f^{abc} A^{\mu\, b} \partial_\mu)\omega^c. \tag{2.50}$$

So

$$\Delta[A] = \det(\partial^2 \delta^{ac} + f^{abc} A^{\mu\, b} \partial_\mu)^{-1/2}. \tag{2.51}$$

We will discuss strategies to evaluate this determinant shortly.

At this stage, we have reduced the path integral to

$$Z = \int [dA_\mu]\delta(G(A))\Delta[A] e^{iS} \tag{2.52}$$

and we can write down the Feynman rules. The δ-function remains rather awkward to deal with, though, and this expression can be simplified through the following trick. Introduce a function ω (not to be confused with the ω of Eq. (2.48)) and average over ω with a Gaussian weight factor:

$$Z = \int [d\omega] e^{i\int d^4x(\omega^2/\xi)} \sum \int [dA_\mu]\delta(G(A) - \omega)\Delta[A]e^{iS}. \tag{2.53}$$

We can do the integral over the δ-function. The quadratic terms in the exponent are now given by

$$\int d^4x\, A^{\mu a}\left[-\partial^2\eta_{\mu\nu} + \partial_\mu\partial_\nu\left(1 - \frac{1}{\xi}\right)\right]A^{\nu a}. \tag{2.54}$$

We can invert this to find the propagator. In momentum space,

$$D_{\mu\nu} = \frac{\eta_{\mu\nu} + (\xi - 1)k_\mu k_\nu/k^2}{k^2 + i\epsilon}. \tag{2.55}$$

To write down explicit Feynman rules, we need also to deal with the Faddeev–Popov determinant. Feynman long ago guessed that the unitarity problems of Yang–Mills theories could be dealt with by introducing fictitious scalar fields with the wrong statistics. Our expression for Δ can be reproduced by a functional integral for such particles:

$$\Delta = \int [dc^a][dc^{a\dagger}]\exp\left(i\int d^4x[c^{a\dagger}(\partial^2\delta^{ab} + f^{abc}A^{\mu\, c}\partial_\mu)c^b]\right). \tag{2.56}$$

From this we can read off the Feynman rules for Yang–Mills theories, including matter fields. They are summarized in Fig. 2.3.

2.3.1 Gauge fixing in theories with broken gauge symmetry

Gauge fixing in theories with broken gauge symmetries raises some new issues. We consider first a $U(1)$ gauge theory with a single charged scalar field ϕ. We suppose that the potential is such that $\langle\phi\rangle = v/\sqrt{2}$. We call e the gauge coupling and take the conventional scaling for the gauge kinetic terms. We can, again, parameterize the field ϕ as

$$\phi = \frac{1}{\sqrt{2}}[v + \sigma(x)]e^{i\pi/v}. \tag{2.57}$$

Then we can again choose a gauge in which $\pi(x) = 0$. This gauge is known as the unitary gauge since, as we have seen, in this gauge we have exactly the degrees of freedom we expect physically: a massive gauge boson and a single real scalar. But this gauge is not convenient for calculations. The gauge boson propagator in this gauge is

$$\langle A_\mu A_\nu\rangle = -\frac{i}{k^2 - M_V^2}\left(\eta_{\mu\nu} - \frac{k_\mu k_\nu}{M_V^2}\right). \tag{2.58}$$

Because of the momentum factors in the second term, individual Feynman diagrams have a bad high-energy behavior. A more convenient set of gauges, known as R_ξ gauges, avoids this difficulty at the price of keeping the π field (sometimes misleadingly called the

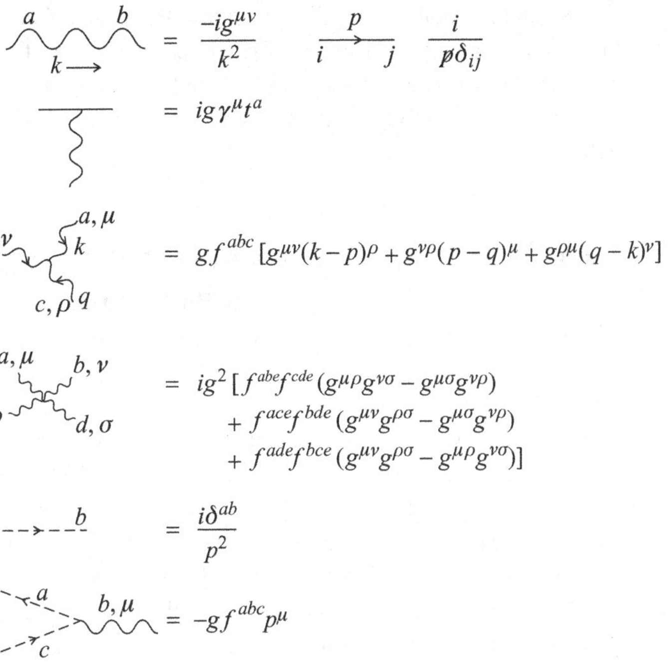

$a \quad b$, $k \longrightarrow$ $= \dfrac{-ig^{\mu\nu}}{k^2}$ $\quad$ $i \xrightarrow{p} j$ $\dfrac{i}{\not p \delta_{ij}}$

$= ig\gamma^\mu t^a$

b,ν; a,μ; c,ρ; p, k, q $= gf^{abc}\,[g^{\mu\nu}(k-p)^\rho + g^{\nu\rho}(p-q)^\mu + g^{\rho\mu}(q-k)^\nu]$

a,μ; b,ν; c,ρ; d,σ

$$= ig^2\,[f^{abe}f^{cde}(g^{\mu\rho}g^{\nu\sigma} - g^{\mu\sigma}g^{\nu\rho}) + f^{ace}f^{bde}(g^{\mu\nu}g^{\rho\sigma} - g^{\mu\sigma}g^{\nu\rho}) + f^{ade}f^{bce}(g^{\mu\nu}g^{\rho\sigma} - g^{\mu\rho}g^{\nu\sigma})]$$

$a \dashrightarrow b$ $= \dfrac{i\delta^{ab}}{p^2}$

p, a, c, b,μ $= -gf^{abc}p^\mu$

Fig. 2.3 Feynman rules for Yang–Mills theory.

Goldstone particle) in the Feynman rules. We take, in the path integral, the gauge-fixing function

$$G = \frac{1}{\sqrt{\xi}}[\partial_\mu A^\mu \xi - ev\pi(x)]. \tag{2.59}$$

The extra term has been judiciously chosen so that when we exponentiate the gauge condition, as in Eq. (2.53), the $A^\mu \partial_\mu \pi$ terms in the action cancel. Explicitly, we have

$$\begin{aligned}\mathcal{L} = &-\frac{1}{2}A_\mu\left[\eta^{\mu\nu}\partial^2 - \left(1-\frac{1}{\xi}\right)\partial^\mu\partial^\nu - (e^2v^2)\eta^{\mu\nu}\right]A_\nu \\ &+ \frac{1}{2}(\partial_\mu\sigma)^2 - \frac{1}{2}m_\sigma^2\sigma^2 + \frac{1}{2}(\partial_\mu\pi)^2 - \frac{\xi}{2}(ev)^2\pi^2 + \mathcal{O}(\phi^3).\end{aligned} \tag{2.60}$$

If we choose $\xi = 1$ (corresponding to the 't Hooft–Feynman gauge), the propagator for the gauge boson is then simply

$$\langle A_\mu A_\nu\rangle = \frac{-i}{k^2 - M_V^2}\eta_{\mu\nu} \tag{2.61}$$

with $M_V^2 = e^2v^2$, but we have also the field π explicitly in the Lagrangian, and it has the propagator

$$\langle \pi\pi\rangle = \frac{i}{k^2 - M_V^2}. \tag{2.62}$$

The mass here is just the mass of the vector boson (for other choices of ξ, this is not true).

This gauge choice is readily extended to non-Abelian theories with similar results: the gauge bosons have simple propagators, like those of massive scalars but multiplied by $\eta_{\mu\nu}$. The Goldstone bosons appear explicitly in perturbation theory, with propagators appropriate to massive fields. The Faddeev–Popov ghosts have couplings to the scalar fields.

2.4 The particles and fields of the Standard Model: gauge bosons and fermions

We are now in a position to write down the Standard Model. It is amazing that, at a microscopic level, almost everything we know about nature is described by such a simple structure. The gauge group is $SU(3)_{\rm c} \times SU(2)_{\rm L} \times U(1)_Y$. The subscript c denotes color, L means left-handed and Y is the hypercharge. Corresponding to these different gauge groups, there are gauge bosons: A^a_μ, $a = 1, \ldots, 8$; W^i_μ, $i = 1, 2, 3$; and B_μ.

One of the most striking features of the weak interactions is the violation of parity. In terms of four-component fields, this means that factors of $1 - \gamma_5$ appear in the couplings of fermions to the gauge bosons. In such a situation it is more natural to work with two-component spinors. For the reader unfamiliar with such spinors, a simple introduction appears in Appendix A. These spinors are the basic building blocks of the four-dimensional spinor representations of the Lorentz group. All spinors can be described as two-component quantities, with various quantum numbers. For example, quantum electrodynamics, which is parity invariant and has a massive fermion, can be described in terms of two left-handed fermions, e and $\bar{e}$, with electric charges $-e$ and $+e$ respectively. The Lagrangian takes the form

$$\mathcal{L} = ie\sigma^\mu D_\mu e^* + i\bar{e}\sigma^\mu D_\mu \bar{e}^* - m\bar{e}e - m\bar{e}^* e^*. \tag{2.63}$$

The covariant derivatives are those appropriate to fields of charge e and $-e$. Parity is symmetry under $\vec{x} \to -\vec{x}$, $e \leftrightarrow \bar{e}^*$ and $\vec{A} \to -\vec{A}$.

We can specify the fermion content of the Standard Model by giving the gauge quantum numbers of the left-handed spinors. So, for example, there are quark doublets which are in the 3 (fundamental) representation of color and doublets of $SU(2)$ and which have hypercharge $1/3$: $Q = (3, 2)_{1/3}$. The appropriate covariant derivative is:

$$D_\mu Q = \left(\partial_\mu - ig_s A^a_\mu T^a - igW^i_\mu T^i - i\frac{g'}{2}\frac{1}{3}B_\mu \right) Q, \tag{2.64}$$

where g_s is the strong coupling constant. Here the T^i's are the generators of $SU(2)$; $T^i = \sigma^i/2$. These are normalized as follows:

$$\mathrm{Tr}(T^i T^j) = \frac{1}{2}\delta^{ij}. \tag{2.65}$$

The T^a are the generators of $SU(3)$; in terms of Gell-Mann's $SU(3)$ matrices, $T^a = \lambda^a/2$. They are normalized in the same way as the $SU(2)$ matrices: Tr $(T^a T^b) = (1/2)\delta^{ab}$.

Table 2.1 Fermions of the Standard Model and their quantum numbers

	$SU(3)$	$SU(2)$	$U(1)_Y$
Q_f	3	2	1/3
$\bar{u}_f$	$\bar{3}$	1	−4/3
$\bar{d}_f$	$\bar{3}$	1	2/3
L_f	1	2	−1
$\bar{e}_f$	1	1	2

We have followed the customary definition in coupling B_μ to half the hypercharge current. We have also scaled the fields so that the couplings appear in the covariant derivative and have labeled the $SU(3)_{\rm c}$, $SU(2)_{\rm L}$, and $U(1)_Y$ coupling constants as g_s, g, and g', respectively. Using matrix-valued fields, defined with the couplings in front of the gauge kinetic terms, this covariant derivative can be written in a very compact manner:

$$D_\mu Q = \left(\partial_\mu - iA_\mu - iW_\mu - \frac{i}{2}\frac{1}{3}B_\mu\right)Q. \tag{2.66}$$

As another example, the Standard Model contains lepton fields L with no $SU(3)$ quantum numbers but which are $SU(2)$ doublets with hypercharge -1. The covariant derivative is

$$D_\mu L = \left(\partial_\mu - igW^i_\mu T^i - \frac{ig'}{2}B_\mu\right)L. \tag{2.67}$$

We have summarized the fermion content in the Standard Model in Table 2.1. Here f labels the quark or lepton flavor, i.e. the generation number: $f = 1, 2, 3$. For example,

$$L_1 = \begin{pmatrix}\nu_e \\ e\end{pmatrix}, \quad L_2 = \begin{pmatrix}\nu_\mu \\ \mu\end{pmatrix}, \quad L_3 = \begin{pmatrix}\nu_\tau \\ \tau\end{pmatrix}. \tag{2.68}$$

The reason why there is this repetitive structure, these three generations, is one of the great puzzles of the Standard Model, to which we will return. In terms of these two-component fields (indicated generically by ψ_i), the gauge-invariant kinetic terms have the form

$$\mathcal{L}_{f,k} = -i\sum_i \psi_i D_\mu \sigma^\mu \psi_i^*, \tag{2.69}$$

where the covariant derivatives are those appropriate to the representation of the gauge group.

Unlike QED (where, in two-component language, parity interchanges e and $\bar{e}^*$), the model does not have a parity symmetry. The fields Q and $\bar{u}$, $\bar{d}$ transform under different representations of the gauge group. There is simply no discrete symmetry that one can find which is the analog of the parity symmetry in QED.

2.5 The particles and fields of the Standard Model: Higgs scalars and the complete Standard Model

In order to account for the masses of the W and Z bosons and those of the quarks and leptons, the simplest approach is to include a scalar, ϕ, which transforms as a $(1,2)_1$ representation of the Standard Model gauge group. This Higgs field possesses both self-couplings and also Yukawa couplings to the fermions. Its kinetic term is simply

$$\mathcal{L}_{\phi,k} = |D_\mu \phi|^2. \tag{2.70}$$

The Higgs potential is similar to that of our toy model (2.24):

$$V(\phi) = \mu^2 |\phi|^2 + \lambda |\phi|^4. \tag{2.71}$$

This is completely gauge invariant. But if μ^2 is negative, the gauge symmetry is broken as before. We will describe this breaking, and the mass matrix of the gauge bosons, shortly.

We could consider a more complicated Higgs sector. For example, we could include multiple Higgs doublets. Or, as we will see in Chapter 8, electroweak symmetry breaking might be the result of some new strong dynamics. But the single Higgs doublet is truly the *simplest* possibility, in the sense that it represents the smallest number of degrees of freedom we can include that will give rise to the observed pattern of gauge boson masses. As of this writing, at the level of precision of the two major LHC experiments, there is evidence for one such doublet and no evidence for additional doublets. Any additional scalars are likely to be heavy compared with the observed Higgs particle and so, if discovered or required by some other theoretical considerations, they can properly be referred to as Beyond the Standard Model physics.

At this point we have written down the most general renormalizable self-couplings of the scalar fields. Renormalizability and gauge invariance permit one other set of couplings in the Standard Model: Yukawa couplings of the scalars to the fermions. The most general such couplings are given by

$$\mathcal{L}_{\text{Yuk}} = y^U_{f,f'} Q_f \bar{u}_{f'} \sigma_2 \phi^* + y^D_{f,f'} Q_f \bar{d}_{f'} \phi + y^L_{f,f'} L_f \bar{e}_{f'} \phi. \tag{2.72}$$

Here y^U, y^D and y^L are general matrices in the space of flavors.

We can simplify the Yukawa coupling matrices significantly by redefining fields. Any 3×3 matrix can be diagonalized by separate left and right $U(3)$ matrices. To see this, suppose that one has some matrix M, not necessarily Hermitian. The matrices

$$A = MM^\dagger, \quad B = M^\dagger M \tag{2.73}$$

will be Hermitian; A can be diagonalized by a unitary transformation U_{L}, say, and B by a unitary transformation U_{R}. In other words

$$U_{\text{L}} M U_{\text{R}}^\dagger, \quad U_{\text{R}} M^\dagger U_{\text{L}}^\dagger \tag{2.74}$$

are diagonal. By redefining fields, we can take y_U as diagonal and $M_d = V_{\text{CKM}} M'_d$ as diagonal; V_{CKM} is the Cabibbo–Kobayashi–Maskawa (CKM) matrix. This matrix is not unique, and we will present various conventional forms in Section 3.3.

To summarize, the entire Lagrangian of the Standard Model consists of the following:

1. gauge-invariant kinetic terms for the gauge fields,

$$\mathcal{L}_a = -\frac{1}{4g_s^2} G_{\mu\nu}^2 - \frac{1}{4g^2} W_{\mu\nu}^2 - \frac{1}{4g'^2} F_{\mu\nu}^2 \tag{2.75}$$

(here we have returned to our scaling with the couplings in front and $G_{\mu\nu}$, $W_{\mu,\nu}$ and $F_{\mu\nu}$ are the $SU(3), SU(2)$ and $U(1)$ field strengths);

2. gauge-invariant kinetic terms for the fermion and Higgs fields, $\mathcal{L}_{f,k}$, $\mathcal{L}_{\phi,k}$;
3. Yukawa couplings of the fermions to the Higgs field, $\mathcal{L}_{\text{Yuk}}$;
4. the potential for the Higgs field, $V(\phi)$.

If we require renormalizability, i.e. that all the terms in the Lagrangian be of dimension four or less, then this is all that we can write down. It is extraordinary that this simple structure incorporates over a century of investigation of elementary particles.

2.6 The gauge boson masses

The field ϕ has an expectation value, which we can take to be as follows:

$$\langle\phi\rangle = \frac{1}{\sqrt{2}}\begin{pmatrix}0\\ v\end{pmatrix}, \tag{2.76}$$

where $v = \mu/\sqrt{\lambda}$. Expanding around this expectation value, the Higgs field can be written as

$$\phi = e^{i\vec{\pi}(x)\cdot\vec{\sigma}/2v}\frac{1}{\sqrt{2}}\begin{pmatrix}0\\ v+\sigma(x)\end{pmatrix}. \tag{2.77}$$

By a gauge transformation we can set $\vec{\pi} = 0$. Not all the gauge symmetry is broken by $\langle\phi\rangle$. It is invariant under the $U(1)$ symmetry generated by

$$Q = T_3 + \frac{Y}{2}. \tag{2.78}$$

This is the electric charge. If we write:

$$L = \begin{pmatrix}\nu\\ e\end{pmatrix}, \quad Q = \begin{pmatrix}u\\ d\end{pmatrix} \tag{2.79}$$

then ν has charge 0 and e has charge -1; u has charge $2/3$ and d has charge $-1/3$. The charges of the singlets also work out correctly.

With this gauge choice we will examine the scalar kinetic terms in order to determine the gauge boson masses. Keeping only terms quadratic in the fluctuating fields (σ and the gauge fields), these now have the form

$$|D_\mu\phi|^2 = \frac{1}{2}(\partial_\mu\sigma^2) + \frac{1}{2}(0\ v)\left(igW_\mu^i\frac{\sigma^i}{2} + \frac{ig'}{2}B_\mu\right)\left(-igW^{\mu j}\frac{\sigma^j}{2} - \frac{ig'}{2}B^\mu\right)\begin{pmatrix}0\\ v\end{pmatrix}. \tag{2.80}$$

It is convenient to define the complex fields

$$W^{\pm}_{\mu} = \frac{1}{\sqrt{2}}(W^1_{\mu} \pm iW^2_{\mu}) \tag{2.81}$$

These are states of definite charge, since they carry zero hypercharge and $T_3 = \pm 1$. In terms of these fields, the gauge boson mass and kinetic terms take the form

$$\begin{aligned} &\partial_{\mu} W^{+}_{\nu} \partial^{\mu} W^{\nu -} + \frac{1}{2}\partial_{\mu} W^{3}_{\nu} \partial^{\mu} W^{\nu 3} + \frac{1}{2}\partial_{\mu} B_{\nu} \partial^{\mu} B^{\nu} \\ &+ \frac{1}{4} g^2 v^2 W^{+}_{\mu} W^{\mu -} + \frac{1}{8} v^2 (g W^3_{\mu} - g' B_{\mu})^2 . \end{aligned} \tag{2.82}$$

Examining the terms involving the neutral fields, B_{μ} and W^3_{μ}, it is natural to redefine

$$A_{\mu} = \cos\theta_{\rm W}\, B_{\mu} + \sin\theta_{\rm W}\, W^3_{\mu}, \quad Z_{\mu} = \sin\theta_{\rm W}\, B_{\mu} + \cos\theta_{\rm W}\, W^3_{\mu} \tag{2.83}$$

where

$$\sin\theta_{\rm W} = \frac{g'}{\sqrt{g^2 + g'^2}} \tag{2.84}$$

is known as the Weinberg angle. The field A_{μ} is massless, while the Ws and Zs have the following masses:

$$M^2_W = \frac{1}{4} g^2 v^2, \quad M^2_Z = \frac{1}{4}(g^2 + g'^2) v^2 = \frac{M^2_W}{\cos^2\theta_{\rm W}}. \tag{2.85}$$

We can immediately see that A_{μ} couples to the current

$$\begin{aligned} j^{\mu}_{\rm em} &= g' \cos\theta_{\rm W} \frac{1}{2} j^Y_{\mu} + g \sin\theta_{\rm W} j^3_{\mu} \\ &= e\left(\frac{1}{2} j^Y_{\mu} + j^3_{\mu}\right), \end{aligned} \tag{2.86}$$

where

$$e = \frac{g g'}{\sqrt{g^2 + g'^2}} \tag{2.87}$$

is the electric charge. So A_{μ} couples precisely as we expect the photon to couple and $W^{\mu\pm}$ couple to the charged currents of the four-fermion theory. The Z boson couples to:

$$j^Z_{\mu} = -g' \sin\theta_{\rm W} \frac{1}{2} j^Y_{\mu} + g \cos\theta_{\rm W} j^3_{\mu}. \tag{2.88}$$

2.7 Quark and lepton masses

On substituting the expectation value for the Higgs field into the expression for the quark and lepton Yukawa couplings, Eq. (2.72) leads directly to masses for the quarks and leptons. The lepton masses and the masses for the u quarks follow immediately:

$$m_{ef} = y_{ef}\frac{v}{\sqrt{2}}, \quad m_{uf} = y_{uf}\frac{v}{\sqrt{2}}. \tag{2.89}$$

So, for example, the Yukawa coupling of the electron is $m_e\sqrt{2}/v$.

The masses for the d quarks are somewhat more complicated. Because y_D is not diagonal, we have a matrix in flavor space for the d quark masses:

$$(m_d)_{ff'} = (y_d)_{ff'}\frac{v}{\sqrt{2}}. \tag{2.90}$$

As we have seen, any matrix can be diagonalized by separate unitary transformations acting on from left or the right. So we can diagonalize this matrix by separate rotations of the d quarks (within the quark doublets) and of the $\bar{d}$ quarks. The rotation of the $\bar{d}$ quarks corresponds to a simple redefinition of these fields. But the rotation of the d quarks is more significant, since it does not commute with $SU(2)_L$. In other words the quark masses are not diagonal in a basis in which the W boson couplings are diagonal. The basis in which the mass matrix is diagonal is known as the *mass basis* (the corresponding fields are often called mass eigenstates).

The unitary matrix V acting on the d quarks is known as the Cabibbo–Kobayashi–Maskawa, or CKM, matrix. In terms of this matrix the coupling of the quarks to the $W^{\pm}$ fields can be written as

$$W^-_\mu u_f \sigma^\mu d^*_{f'} V_{ff'} + W^+_\mu d_f \sigma^\mu u^*_{f'} V^*_{f'f}. \tag{2.91}$$

There is a variety of parameterizations of V, which we will discuss shortly. One interesting feature of the model is the Z couplings. Because V is unitary, these are diagonal in flavor. This explains why Z bosons do not mediate processes which change flavor, such as $K_L \to \mu^+\mu^-$. The suppression of these *flavor-changing neutral currents* was one of the early, and critical, successes of the Standard Model.

2.8 The Higgs field and its couplings

In the simplest Higgs theory, the couplings of the Higgs are fixed. This includes the couplings to gauge bosons, to fermions and to the Higgs field itself. At tree, or classical, level these can be read off the Lagrangian, as follows.

1. There is a Higgs–ZZ coupling and a Higgs–W^+W^- coupling arising from the replacement of ϕ by $\frac{1}{\sqrt{2}}(v+\sigma)$ in the Higgs kinetic term.
2. There is a Yukawa coupling to all fermions, which is proportional to their masses.
3. There are cubic and quartic self-couplings of the Higgs.

We will discuss these couplings in the context of the Higgs search in the next chapter.

Suggested reading

There are a number of textbooks with good discussions of the Standard Model, including those of Peskin and Schroeder (1995), Weinberg (1995), Cottingham and Greenwood (1998), Donoghue *et al.* (1992) and Seiden (2005). We cannot give a full bibliography of the Standard Model here, but the reader may want to examine some original papers, including the discovery of non-Abelian gauge theory by Yang and Mills (1954); the Higgs mechanism by Englert and Brout (1964), Guralnik *et al.* (1964) and Higgs (1964); Salam and Ward (1964), Weinberg (1967) and Glashow *et al.* (1970) on weak interaction theory; 't Hooft (1971), Gross and Wilczek (1973) and Politzer (1973) on asymptotic freedom of the strong interactions. For discussion of the various phases found in gauge theories, see 't Hooft (1980) and Fradkin and Shenker (1979).

Exercises

(1) *The Georgi–Glashow model* Consider a gauge theory based on $SU(2)$, with the Higgs field $\vec{\phi}$ in the adjoint representation. Assuming that ϕ attains an expectation value, determine the gauge boson masses. Identify the photon and the $W^{\pm}$ bosons. Is there a candidate for the Z boson?

(2) Consider the Standard Model with two generations. Show that there is no CP violation and that the CKM matrices can be described in terms of a single angle, known as the Cabibbo angle.

3 Phenomenology of the Standard Model

With the discovery of the Higgs boson in 2012, the Standard Model may well be complete. More precisely, it may be that we know all nature's degrees of freedom up to energy scales of order one TeV and fully understand their interactions (there might be other degrees of freedom with couplings to quarks, leptons and gauge bosons which are significantly suppressed). The predictions of the Standard Model have been subjected to experimental tests in a broad range of processes. In experiments involving leptons alone, or hadrons at high-momentum transfers, detailed and precise predictions are possible. In processes involving hadrons at low momentum, it is often possible to make progress using symmetry arguments. In still other cases one can at least formulate a qualitative picture. In recent years, developments in lattice gauge theory have yielded reliable and precise predictions for at least some features of the large-distance behavior of hadrons. Since 2012 the Higgs boson itself has begun to provide a testing ground for many elements of the Standard Model. There exist excellent texts and reviews treating all these topics. Here we will give only a brief survey, attempting to introduce ideas and techniques which are important in understanding what may lie beyond the Standard Model.

3.1 The weak interactions

We are now in a position to describe weak interactions within the Standard Model. Summarizing our results for the W and Z masses, we have at tree level

$$M_W^2 = \frac{\pi \alpha}{\sqrt{2} G_F \sin^2 \theta_W}, \quad M_Z^2 = \frac{\pi \alpha}{\sqrt{2} G_F \sin^2 \theta_W \cos^2 \theta_W}, \tag{3.1}$$

where θ_W is given by Eq. (2.84) and α is the fine-structure constant. Note in particular that, in the leading approximation,

$$\frac{M_W^2}{M_Z^2} = \cos^2 \theta_W. \tag{3.2}$$

In these expressions the Fermi constant is related to the W mass and the gauge coupling, through

$$G_F = \sqrt{2} \frac{g^2}{8 M_W^2}; \quad G_F = 1.166 \times 10^{-5}\ \mathrm{GeV}^{-2}. \tag{3.3}$$

The Weinberg angle θ_W is given by

$$\sin^2\theta_W = 0.231\,20(15). \tag{3.4}$$

The measured values of the W and Z masses are

$$M_W = 80.425(38)\text{ GeV}, \quad M_Z = 91.1876(21)\text{ GeV}. \tag{3.5}$$

One can see that the experimental quantities satisfy the theoretical relations to good accuracy. They are all in agreement at the one part in 10^2–10^3 level when radiative corrections are included.

The effective Lagrangian for the quarks and leptons obtained by integrating out the W and Z particles is

$$\mathcal{L}_W + \mathcal{L}_Z = \frac{8G_F}{\sqrt{2}}\left[(J^1_\mu)^2 + (J^2_\mu)^2 + (J^3_\mu - \sin^2\theta_W J_{\mu EM})^2\right]. \tag{3.6}$$

The first two terms correspond to the exchange of the charged $W^\pm$ fields. The last term represents the effect of Z boson exchange. This structure has been tested extensively.

The most precise tests of the weak interaction theory involve the Z bosons. Experiments at the LEP accelerator at CERN and the SLD accelerator at SLAC produced millions of Z bosons. These large samples permitted high-precision studies of the line shape and of the branching ratios to various final states. Care is needed in calculating the radiative corrections; it is important to make consistent definitions of the various quantities. Detailed comparisons of theory and experiment can be found on the website of the Particle Data Group (http://pdg.lbl.gov). As inputs, one generally takes the value of G_F measured in μ decays, the measured mass of the Z and the fine structure constant. Outputs include the Z boson total width:

$$\text{experiment, } \Gamma_Z = 2.4952 \pm 0.0023; \quad \text{theory, } \Gamma_Z = 2.4955 \pm 0.0009. \tag{3.7}$$

The decay width of the Z to hadrons and leptons is also in close agreement (see Fig. 3.1). The W mass can also be computed with the above inputs and has been measured quite precisely, particularly at the Tevatron and LEP2 (below we quote first the LEP result and then the Tevatron result):

$$\begin{aligned} \text{experiment, } M_W &= 80.376 \pm 0.033,\ 80.387 \pm 0.016 \pm 0.0023; \\ \text{theory, } M_W &= 80.363 \pm 0.06. \end{aligned} \tag{3.8}$$

The W width, similarly, is:

$$\begin{aligned} \text{experiment, } \Gamma_W &= 2.196 \pm 0.083,\ 2.046 \pm 0.0049; \\ \text{theory } \Gamma_W &= 2.090 \pm 0.001. \end{aligned} \tag{3.9}$$

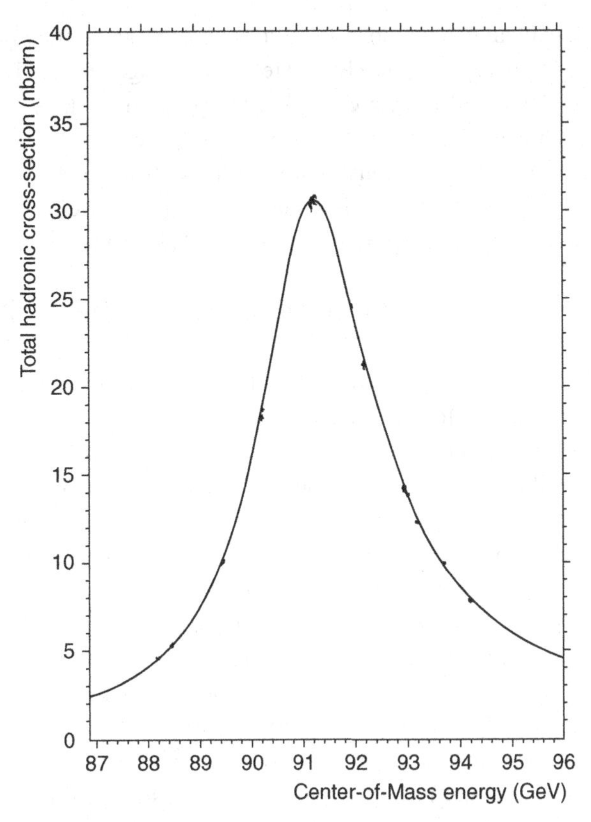

Fig. 3.1 OPAL results for the Z line shape. The solid line corresponds to the theory; the dots give the data.

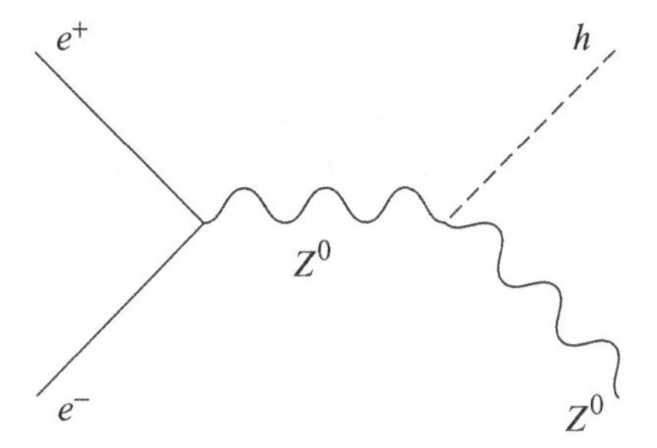

Fig. 3.2 The Higgs can be produced in e^+e^- annihilation, in association with a Z^0 particle.

3.2 Discovery of the Higgs

The simplest possible realization of the Higgs mechanism within the Standard Model is through a single Higgs doublet. In 2012 the two large detectors at the Large Hadron Collider, ATLAS and CMS, reported the discovery of a scalar particle behaving like the Higgs field of this minimal model. The mass of this particle is 125.6 ± 0.4 GeV. As we will explain in a bit more detail shortly, as of this writing both the production cross section and the decays of the Higgs are in rough agreement (10%–20% for several channels) with Standard Model predictions. The precision of these measurements and the quality of Standard Model tests will improve over the next few years. Any model for physics beyond the Standard Model must reproduce these features. It is likely (as we will discuss in the next chapter) that there is a range of energies where the Standard Model is completely described by the Lagrangian of the previous chapter.

3.2.1 Testing the Standard Model with the Higgs

The discovery of the Higgs boson, exciting in itself, brought together many aspects of the Standard Model. The Higgs was discovered in high-energy proton–proton collisions, and understanding the signal requires the full machinery of perturbative QCD (which we will review shortly) including parton distribution functions and higher-order radiative corrections. Higgs production arises through processes including *gluon fusion* (Fig. 3.3), the collision of a gluon from each of the two protons to produce a virtual top quark pair, which then couples to the Higgs, as well as a smaller contribution from quark collisions. There is an equally rich story with the decay channels. Large numbers of Higgs particles are produced at the LHC. The Higgs decays predominantly to $b\bar{b}$ pairs, however, and it is difficult to isolate these decays from the many other sources of such pairs in proton collisions. The original discovery was made in the two-photon channel, whose branching ratio is far smaller but where it is easier (but still challenging) to separate the signal from the background. Indeed, a simple-minded estimate suggests the branching ratio should be of an order given by

$$\frac{\Gamma(H \to \gamma\gamma)}{\Gamma(H \to b\bar{b})} \sim \left(\frac{\alpha}{4\pi}\right)^2 \frac{m_H^3 v^2}{m_t^2 m_b^2} \approx 10^{-4}. \tag{3.10}$$

Comparisons of theory and experiment in the two-photon channel are indicated in Figs. 3.4 and 3.5. Other channels in which comparisons can be made, as of the time of writing,

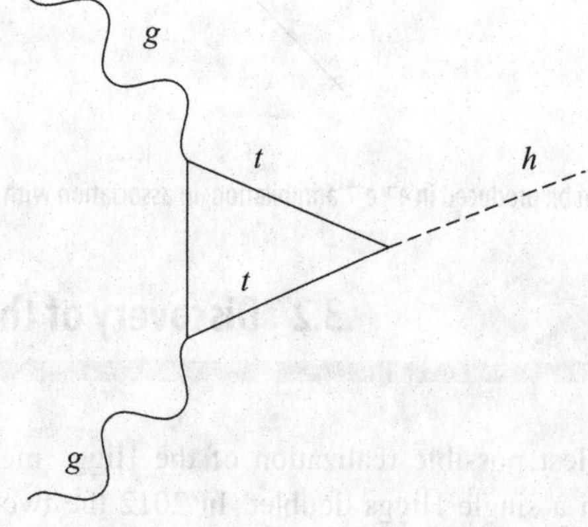

Fig. 3.3 In hadron colliders Higgs particles can be produced by several mechanisms. The diagram above illustrates production by gluons.

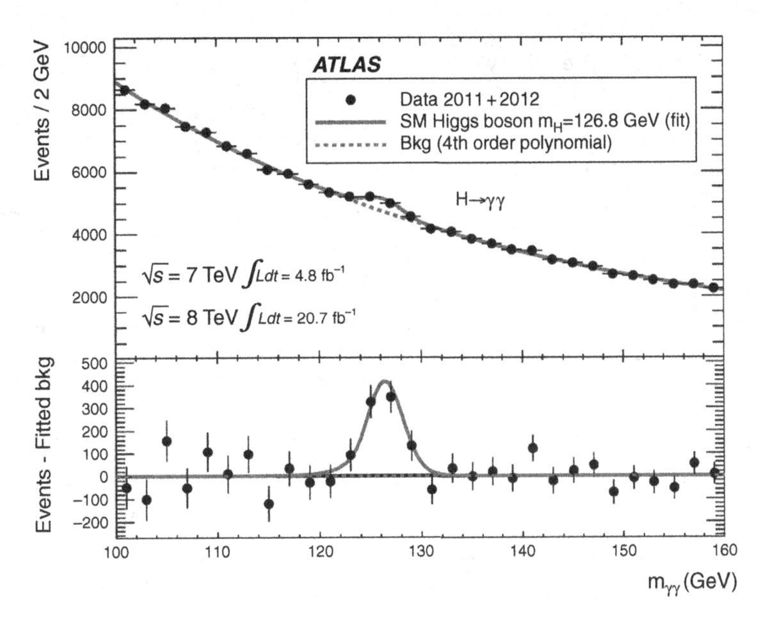

Fig. 3.4 ATLAS data on two-photon production.

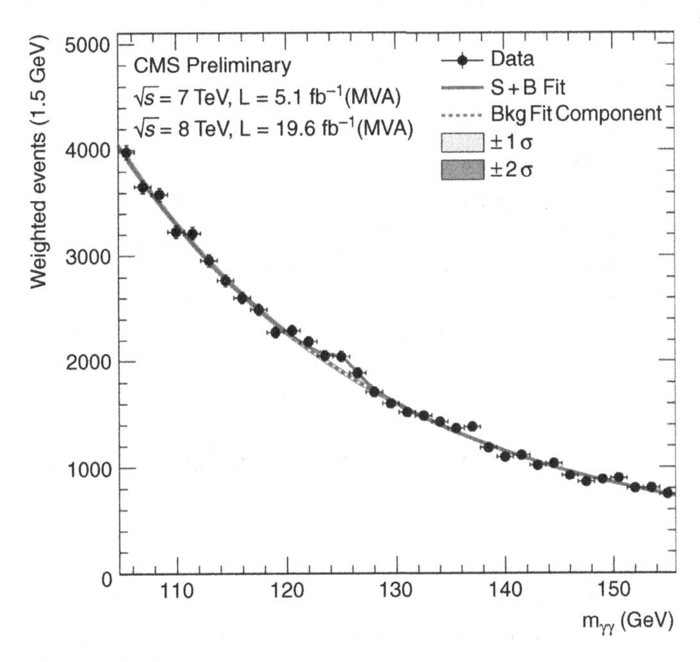

Fig. 3.5 CMS data on two-photon production.

are the ZZ channel (with four observed leptons from the Z decays) and the WW channel. Comparisons, again, with Standard Model expectations appear in Fig. 3.6.

Future runs of the LHC, with higher energies and higher luminosities, will increase the precision of these studies, in many cases at the 5%–10% level. An electron–positron linear

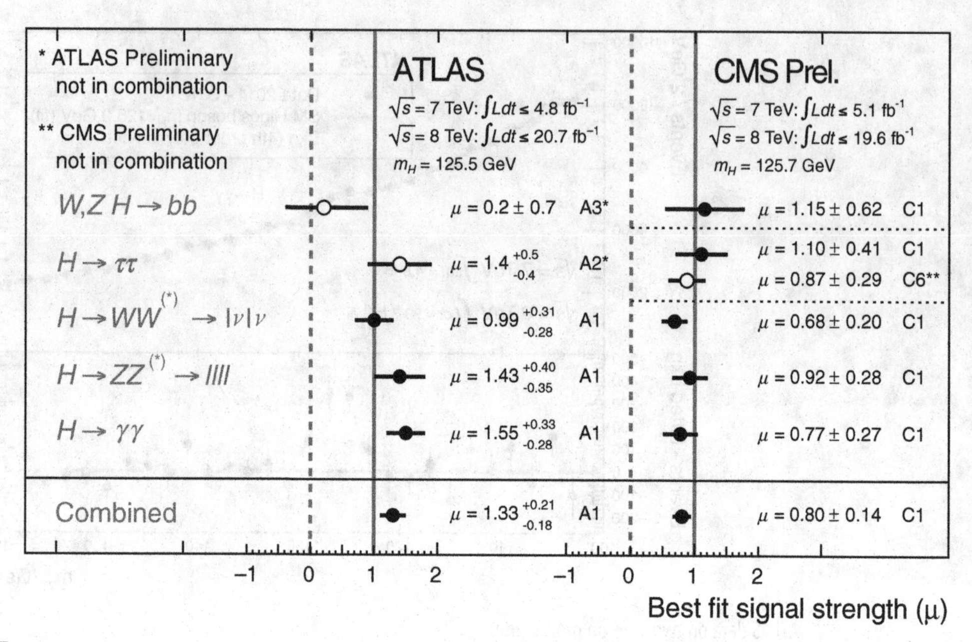

Fig. 3.6 For comparison, ATLAS and CMS measurements of Higgs, in several channels. Used by permission of the Particle Data Group.

collider, or other contemplated high-energy lepton machines, could improve the precision to the few percent level.

In any case, at present it appears that the Standard Model may be complete; any degrees of freedom in nature beyond those of the theory may well be significantly heavier than the Higgs. This clearly has implications for the possible physics we might hope to see beyond the Standard Model. We will discuss this further when we consider supersymmetric models, which predict multiple Higgs doublets.

3.3 The quark and lepton mass matrices

Before considering the small neutrino masses, we note that the lepton–Yukawa couplings can simply be taken as diagonal: there is no mixing. Their extraction from the experimental data is reasonably straightforward. The lepton masses are

$$m_e = 0.511 \text{ MeV}, \quad m_\mu = 113 \text{ MeV}, \quad m_\tau = 1.777 \text{ GeV}. \tag{3.11}$$

The quark masses and mixings pose more severe challenges. First, there is the question of mixing. We have seen that we can take the Yukawa coupling y_u for the u quarks to be diagonal, but we cannot simultaneously diagonalize the couplings y_d for the d quarks. As a result, when the Higgs field acquires an expectation value v, the u quark masses are given by

$$m_{uf} = \frac{(y_u)_f}{\sqrt{2}} v. \tag{3.12}$$

These are automatically diagonal. But the d quark masses are described by a 3×3 mass matrix,

$$m_{dff'} = \frac{(y_d)_{ff'}}{\sqrt{2}} v. \tag{3.13}$$

We can diagonalize this matrix by separate unitary transformations of the $\bar{d}$ and d fields. Because the $\bar{d}$ quarks are singlets of $SU(2)$, the transformation of the $\bar{d}$ field leaves the kinetic terms and gauge interactions for these quarks unchanged. But the transformation for the d quarks does not commute with $SU(2)$, so the couplings of the gauge bosons to these quarks are more complicated. The unitary transformation between the mass or flavor eigenstates and the weak interaction eigenstates is known as the CKM matrix. Denoting the mass eigenstates as u', d', etc., the transformation has the form

$$\begin{pmatrix} d' \\ s' \\ b' \end{pmatrix} = \begin{pmatrix} V_{ud} & V_{us} & V_{ub} \\ V_{cd} & V_{cs} & V_{cb} \\ V_{td} & V_{ts} & V_{tb} \end{pmatrix} \begin{pmatrix} d \\ s \\ b \end{pmatrix}. \tag{3.14}$$

There are various ways of parameterizing the CKM matrix. One standard form, which makes its unitarity manifest, is given as follows:

$$V = \begin{pmatrix} c_{12}c_{13} & s_{12}c_{13} & s_{13}e^{-i\delta} \\ -s_{12}c_{23} - c_{12}s_{23}s_{13}e^{i\delta} & c_{12}c_{23} - s_{12}c_{23}s_{13}e^{i\delta} & s_{23}c_{13} \\ s_{12}s_{23} - c_{12}c_{23}s_{13}e^{i\delta} & -c_{12}s_{23} - s_{12}c_{23}s_{13}e^{i\delta} & c_{23}c_{13} \end{pmatrix}. \tag{3.15}$$

The matrix V is real unless δ is non-zero. Thus δ provides a measure of CP violation.

Experimentally, all the off-diagonal matrix elements are small and in fact are hierarchically so. Wolfenstein developed a convenient parameterization:

$$V = \begin{pmatrix} 1 - \lambda^2/2 & \lambda & A\lambda^3(\rho - i\eta) \\ -\lambda & 1 - \lambda^2/2 & A\lambda^2 \\ A\lambda^3(1 - \rho - i\eta) & -A\lambda^2 & 1 \end{pmatrix} + \mathcal{O}(\lambda^4). \tag{3.16}$$

The Babar and Belle experiments improved significantly our knowledge of these quantities, and in particular of the CP-violating parameter. They demonstrated that, indeed, V is nearly unitary, which constrains possible new physics. The magnitudes of the matrix elements of V are as follows:

$$V = \begin{pmatrix} 0.974\,27 \pm 0.0014 & 0.225\,36 \pm 0.000\,61 & 0.003\,55 \pm 0.000\,15 \\ 0.225\,22 \pm 0.000\,61 & 0.973\,43 \pm 0.000\,15 & 0.0414 \pm 0.0012 \\ 0.008\,86^{+0.000\,33}_{-0.000\,32} & 0.0405^{+0.000\,11}_{-0.000\,12} & 0.999\,14 \pm 0.0005 \end{pmatrix}. \tag{3.17}$$

The Wolfenstein parameters are

$$\lambda = 0.225\,37 \pm 0.000\,61, \quad A = 0.814^{+0.023}_{-0.024},$$
$$\bar{\rho} = 0.117 \pm 0.021, \quad \bar{\eta} = 0.353 \pm 0.013. \tag{3.18}$$

(Here we are following the conventions of the Particle Data Group; $\bar{\rho} = \rho(1 - \lambda^2/2)$.) Note, in particular, that the CP-violating parameter $\bar{\eta}$ is not small (corresponding to δ of order one).

From unitarity follow a number of relations among the elements of the matrix. For example,

$$V_{ud}V_{ub}^* + V_{cd}V_{cb}^* + V_{td}V_{tb}^* = 0. \tag{3.19}$$

From $V_{ud} \approx V_{cb} \approx V_{tb} \approx 1$, this becomes a relation between three complex numbers which says that they form a triangle the *unitarity triangle*. Determining from experiment that these quantities do indeed form a triangle is an important test of this model for the quark masses.

We should also discuss the values of the quark masses themselves. This is somewhat subtle, since we do not observe free quarks; the masses are Lagrangian parameters, related to experimental quantities in a way which depends on a scheme (i.e. a definition) and an energy scale, much as one must specify the scheme and energy scale of the gauge coupling in QCD. For the lighter quarks (u, d and s) these masses can be obtained, at present, only from lattice QCD. As we will discuss further in Section 3.8 on lattice gauge theory, this is a subtle and complex process. However, over the past decade, reliable computations have become possible, with errors at the level of 10% or smaller. With a scale of order 2 GeV, in the $\overline{MS}$ scheme the Particle Data Group, combining results from different lattice collaborations, quotes the following quark masses:

$$\begin{aligned} m_u &= 2.15(15)\text{ MeV}, \quad m_d = 4.7(20)\text{ MeV}, \quad m_s = 93.5(2.5)\text{ MeV}, \\ m_c &\approx 1.15\text{–}1.35\text{ GeV}, \quad m_b \approx 4.1\text{–}4.4\text{ GeV}, \quad m_t \approx 174.3 \pm 5\text{ GeV}. \end{aligned} \tag{3.20}$$

Overall, the picture of the quark and lepton masses is quite puzzling. They vary over nearly five orders of magnitude. Correspondingly, the dimensionless Yukawa couplings have widely disparate values. At the same time the mixing among the quarks is small and hierarchical. Understanding these features might well be a clue to what lies beyond the Standard Model.

We will discuss the question of neutrino masses in Chapter 4, when we discuss the Standard Model as an effective field theory, and in particular the non-renormalizable operators which might arise from integrating out the Beyond the Standard Model physics. We will see that the pattern of neutrino masses does not resemble that of the quarks and charged leptons; they appear anarchical, rather than hierarchical.

3.4 The strong interactions

The strong interactions, as their name implies, are characterized by strong coupling. As a result, perturbative methods are not suitable for most questions. In comparing theory and experiment it is necessary to focus on a few phenomena which are accessible to theoretical analysis. By itself this is not particularly disturbing. A parallel with the quantum mechanics of electrons interacting with nuclei is perhaps helpful. We can understand simple atoms in detail; atoms with very large Z can be treated by Hartree–Fock or other methods. Atoms with intermediate Z, however, can be dealt with only by, at best, detailed numerical analysis accompanied by educated guesswork. Molecules are even more problematic, not to mention solids. But we are able to make detailed tests of the theory (and its extension in quantum electrodynamics) from the simpler systems, and develop a qualitative understanding of the more complicated systems. In many cases we can do a quantitative analysis of small fluctuations about the ground states of the complicated system.

In the theory of strong interactions, as we will see, many problems are hopelessly complicated. Low-lying spectra are hard to deal with; detailed exclusive cross sections in high-energy scattering are essentially impossible. There are many questions we can answer, though. Rates for inclusive questions at very high energy and momentum transfer can be calculated with high precision. Qualitative features of the low-lying spectra of hadron systems and their interactions at low energies can be understood in a qualitative (and sometimes quantitative) fashion by symmetry arguments. Such systems include those in which heavy quarks are bound to light quarks. Recently, progress in lattice gauge theory has made it possible to perform calculations which previously seemed impossible, for features of spectra and even for interaction rates that are important for understanding weak interactions.

3.4.1 Asymptotic freedom

The coupling of a gauge theory (and more generally of a field theory) is a function of energy or length scale. If a typical momentum transfer in a process is q, and if M denotes the cutoff scale, then

$$\frac{8\pi^2}{g^2(q^2)} = \frac{8\pi^2}{g^2(M)} + b_0 \ln \frac{q^2}{M^2}. \tag{3.21}$$

Here

$$b_0 = \frac{11}{3}C_{\rm A} - \frac{2}{3}c_i n_f^{(i)} - \frac{1}{3}c_i n_\phi^{(i)}. \tag{3.22}$$

In this expression $n_f^{(i)}$ is the number of left-handed fermions in the ith representation, while $n_\phi^{(i)}$ is the number of scalars; $C_{\rm A}$ is the quadratic Casimir operator of the adjoint representation and c_i is the quadratic Casimir operator of the ith representation. Thus

$$f^{acd}f^{bcd} = C_{\rm A}\delta^{ab}, \quad \mathrm{Tr}(T^aT^b) = c_i\delta^{ab}. \tag{3.23}$$

These formulas are valid if the masses of the fermions and scalars are negligible at scale q^2. For example, in QCD, at scales of order the Z boson mass, the masses of all but the top quark can be neglected. All the quarks are in the fundamental representation, and there are no scalars. So $b_0 = 22/3$. As a result, g^2 gets smaller as q^2 gets larger and, conversely, g^2 gets larger as q^2 gets smaller. Since momentum transfer is inversely proportional to a typical distance scale, one can say that the strong force gets weaker at short distances, and stronger at large distances. We will calculate b_0 in Section 3.5.

This is quite striking. In the case of QCD it means that hadrons, when probed at very large momentum transfer, behave as collections of free quarks and gluons. Perturbation theory can be used to make precise predictions. However, viewed at large distances hadrons are strongly interacting entities. Perturbation theory is not a useful tool, and other methods must be employed. The most striking phenomena in this regime are confinement – the fact that one cannot observe free quarks – and, closely related, the existence of a mass gap. Neither of these phenomena can be observed in perturbation theory.

3.5 The renormalization group

In thinking about physics beyond the Standard Model, by definition we are considering phenomena involving degrees of freedom to which we have, as yet, no direct experimental access. The question of degrees of freedom which are as yet unknown is the heart of the problem of renormalization. In the early days of quantum field theory it was often argued that one should be able to take a formal limit of infinite cutoff, $\Lambda \rightarrow \infty$. Ken Wilson promulgated a more reasonable view: real quantum field theories describe physics below some characteristic scale Λ. In a condensed matter system this might be the scale of the underlying lattice, below which the system may often be described by a continuum quantum field theory. In the Standard Model, a natural scale is the scale of the W and Z bosons. Below this scale the system can be described by a renormalizable field theory, QED plus QCD, along with certain non-renormalizable interactions – the four-fermion couplings of the weak interactions. In defining this theory, one can take the cutoff to be, say, M_W, or aM_W for some $a < 1$. Depending on the choice of a, the values of the couplings will vary. The parameters of the low-energy effective Lagrangian must depend on a in such a way that physical quantities are independent of this choice. The process of determining the values of couplings in an effective theory which reproduce the effects of some more microscopic theory is often referred to as *matching*.

Knowing how physical couplings depend on the cutoff, one can determine how physical quantities behave in the long-wavelength, infrared, regime by simple dimensional analysis. Quantities associated with operators of dimension less than four will grow in the infrared. They are said to be *relevant*. Those with dimension four will vary as powers of logarithms; they are said to be *marginal*. Quantities with dimension greater than four, those conventionally referred to as non-renormalizable operators, will become less and less important as the energy is lowered. They are said to be *irrelevant*. In strongly interacting theories, the dimensions of operators can be significantly different than those expected from naive classical considerations. The classification of operators as relevant, marginal, or irrelevant applies to their quantum behavior.

At sufficiently low energies we can ignore the irrelevant, non-renormalizable, couplings. Alternatively, by choosing the matching scale M to be low enough, only the marginal and relevant couplings will be important. In a theory with only dimensionless couplings, the variation of the coupling with q^2 is closely related to its variation with the cutoff, M. Physical quantities are independent of the cutoff, so any explicit dependence on the cutoff must be compensated by the dependence of the couplings on M. On dimensional grounds M^2 must appear with q^2, so a knowledge of the dependence of couplings on M permits a derivation of their dependence on q^2. More precisely, in studying, say, a cross section, any explicit dependence on the cutoff must be compensated by a dependence of the coupling on the cutoff. Calling the cross section or other physical quantity σ, we can express this dependence as a differential equation, the *renormalization group equation*:

$$\left(M\frac{\partial}{\partial M} + \beta(g)\frac{\partial}{\partial g}\right)\sigma = 0. \tag{3.24}$$

Here the beta function (or β-function) is given by

$$\beta(g) = M\frac{\partial}{\partial M}g. \tag{3.25}$$

We can evaluate the beta function from our explicit expression, Eq. (3.21), for g^2:

$$\beta(g) = -b_0\frac{g^2}{16\pi^2}g. \tag{3.26}$$

We will compute b_0 in the next section. This equation has corrections in each order of perturbation theory and non-perturbative corrections as well.

So far we have expressed the coupling in terms of a cutoff and a physical scale. In old-fashioned language, the coupling $g^2(M)$ is the "bare" coupling. We can define a "renormalized coupling" $g^2(\mu)$ at a scale μ^2:

$$\frac{8\pi^2}{g^2(\mu^2)} = \frac{8\pi^2}{g^2(M)} + b_0 \ln\frac{\mu}{M}. \tag{3.27}$$

In practice it is necessary to give a more precise definition. We will discuss this when we compute the beta function in the next section. Because of this need to give a precise definition of the renormalized coupling, care is required in comparing theory and experiment. As we will review shortly, there is a variety of definitions in common use and it is important to be consistent.

Quantities like Green's functions are not physical, and obey an inhomogeneous equation. One can obtain this equation in a variety of ways. For simplicity, consider first a Green's function with n scalar fields, such as

$$G(x_1, \ldots, x_n) = \langle\phi(x_1)\cdots\phi(x_n)\rangle. \tag{3.28}$$

This Green's function is related to the *renormalized* Green's function as follows. If the theory is defined at a scale μ, the effective Lagrangian takes the form

$$\mathcal{L}_\mu = Z^{-1}(\mu)(\partial_\mu\phi^2) + \cdots. \tag{3.29}$$

Here the factor Z^{-1} arises from integrating out the physics above the scale μ. It will typically include ultraviolet-divergent loop effects. Rescaling ϕ in such a way that the kinetic term is canonical, $\phi = Z^{1/2}\phi_r$, we have that

$$G(x_1, \ldots, x_n) = Z(\mu)^{n/2}G_r(x_1, \ldots, x_n). \tag{3.30}$$

The left-hand side is independent of μ, so we can write an equation for G_r,

$$\left(\mu\frac{\partial}{\partial\mu} + \beta(g)\frac{\partial}{\partial g} + n\gamma\right) G_r = 0, \tag{3.31}$$

where γ, known as the *anomalous dimension*, is given by

$$\gamma = \frac{1}{2}\mu\frac{\partial}{\partial\mu}\ln Z. \tag{3.32}$$

If these are several different fields, e.g. gauge fields, fermions and scalars, this equation is readily generalized. There is an anomalous dimension for each field, and the $n\gamma$ term is replaced by the appropriate number of fields of each type and their anomalous dimensions.

The effective action obeys a similar equation. Starting with

$$\Gamma(x_1,\ldots,x_n) = Z(\mu)^{-n/2}\Gamma_r(x_1,\ldots,x_n), \tag{3.33}$$

we have

$$\left(\mu\frac{\partial}{\partial\mu} + \beta(g)\frac{\partial}{\partial g} - n\gamma\right)\Gamma_r = 0, \tag{3.34}$$

These equations are readily solved. We could write down the solution immediately, but an analogy with the motion of a fluid is helpful. A typical equation, for example, for the density of a component of a fluid (e.g. the density of bacteria in the fluid) would take the form

$$\left[\frac{\partial}{\partial t} + v(x)\frac{\partial}{\partial x} - \rho(x)\right] D(t,x) = 0, \tag{3.35}$$

where $D(t,x)$ is the density as a function of position and time and $v(x)$ is the velocity of the fluid at x; ρ represents a source term (e.g. the growth due to the presence of yeast or a variable temperature). To solve this equation one first solves for the motion of an element of fluid initially at x, i.e. one solves:

$$\frac{d}{dt}\bar{x}(t;x) = v(\bar{x}(t;x)), \quad \bar{x}(0;x) = x. \tag{3.36}$$

In terms of $\bar{x}$ we can immediately write down a solution for D:

$$\begin{aligned} D(t,x) &= D_0(\bar{x}(t;x)) \exp\left[\int_0^t dt'\,\rho(\bar{x}(t';x))\right] \\ &= D_0(\bar{x}(t;x)) \exp\left[\int_{\bar{x}(t)}^x dx'\frac{\rho(x')}{v(x')}\right]. \end{aligned} \tag{3.37}$$

Here D_0 is the initial density. One can check this solution by plugging it into Eq. (3.35) directly, but each piece has a clear physical interpretation. For example, if there were no source ($\rho = 0$), the solution would become $D_0(\bar{x}(t;x))$. With no velocity, the source would lead to just the expected growth in the density.

Let us apply this to Green's functions. Consider, for example, a two-point function, $G(p) = p^{-2}ih(p^2/\mu^2)$. In our fluid dynamics analogy the coupling g is the analog of the velocity; the log of the scale, $t = \ln(p/\mu)$, plays the role of the time. The equation for g is then

$$\left[\frac{\partial}{\partial t} - \beta(g)\frac{\partial}{\partial g} - 2\gamma(g)\right] h(t) = 0. \tag{3.38}$$

Define $\bar{g}(\mu)$ as the solution of

$$\mu \frac{\partial}{\partial \mu}\bar{g}(\mu) = \beta(\bar{g}). \tag{3.39}$$

At lowest order, this is solved by Eq. (3.27). Then

$$h(p, g) = h(\bar{g}(t)) \exp\left[2 \int_{t_0}^{t} dt' \frac{\gamma(\bar{g}(t', g))}{\beta(\bar{g}(t', g))}\right]. \tag{3.40}$$

One can write the solution in the form

$$G(p, \lambda) = \frac{i}{p^2} G(\bar{g}(t; g)) \exp\left[2 \int_{g}^{\bar{g}} dg' \frac{\gamma(g')}{\beta(g')}\right]. \tag{3.41}$$

3.6 Calculating the beta function

In the previous section we presented the one-loop result for the beta function and used it in various applications. In this section we actually compute this result. There are a number of ways to determine the variation of the gauge coupling with energy scale. One way is to calculate the potential for a very heavy quark–antiquark pair as a function of their separation R (we use the term quark here loosely for a field in the N-dimensional, or *fundamental*, representation of $SU(N)$). The potential is a renormalization-group-invariant quantity. At lowest order it is given by

$$V(R) = -\frac{g^2 C_F}{R} \tag{3.42}$$

where

$$C_F = \sum_{a=1}^{N^2-1} T^a T^a; \tag{3.43}$$

here C_F refers to the fundamental representation and T refers to the adjoint representation. The potential is a physical quantity; as a result it is renormalization-group invariant. In perturbation theory it has corrections behaving as $g^2(\mu)\ln(R\Lambda)$. This follows simply from dimensional analysis. So, if we choose $R = \mu^{-1}$ then the logarithmic terms disappear and we have

$$V(R) = -g^2(R)\frac{C_F}{R}\left[1 + \mathcal{O}(g^2(R))\right]. \tag{3.44}$$

In an asymptotically free theory such as QCD, where the coupling gets smaller with distance, Eq. (3.41) becomes more and more reliable as R gets smaller. This result has physical applications. In the case of a bound state of a top quark and antiquark, one might hope that this would be a reasonable approximation and would describe the binding of the system. Taking $\alpha_s(R) \sim 0.1$, for example, would give a typical radius of order $(17\ \text{GeV})^{-1}$, a length scale where one might expect perturbation theory to be reliable (and for which $\alpha_s(R) \sim 0.1$). By analogy with the hydrogen atom, one would expect the binding energy to be of order 2 GeV. In practice, however, this is not directly relevant, since the width of the top quark is of the same order: the top quark decays before it has time to form a bound state. Still, it should be possible to see evidence for such QCD effects in the production of $t\bar{t}$ pairs near the threshold in e^+e^- annihilation.

A second approach is to study Green's functions in momentum space. The calculation is straightforward, if slightly more tedious than the analogous calculation in a $U(1)$ gauge theory (QED). The main complication is the three-gauge-boson vertex, which has many terms (at one loop, one can use symmetries to simplify greatly the algebra). It is necessary to have a suitable regulator for the integrals. By far the most efficient is the dimensional regularization technique of 't Hooft and Veltman. Here one initially allows the space–time dimensionality d to be arbitrary and takes $d \to 4 - \epsilon$. For convenience, we include the two most frequently needed integration formulas below; their derivation can be found in many textbooks.

$$\int \frac{d^d k}{(k^2 + M^2)^n} = \frac{\pi^{d/2}\Gamma(n - d/2)}{\Gamma(n)}(M^2)^{d/2-n}, \tag{3.45}$$

$$\int \frac{d^d k\, k^2}{(k^2 + M^2)^n} = \frac{\pi^{d/2}\Gamma(n - d/2 - 1)}{\Gamma(n)}(M^2)^{d/2-n+1}. \tag{3.46}$$

Ultraviolet divergences, such as would occur for $n = 2$ in the first integral, give rise to poles in the limit $\epsilon \to 0$. If we were simply to cut off the integral at $k^2 = \Lambda^2$, we would find

$$\int \frac{d^4 k}{(2\pi)^4} \frac{1}{(k^2 + M^2)^2} \approx \frac{1}{16\pi^2} \ln \frac{\Lambda}{M}. \tag{3.47}$$

In dimensional regularization this behaves as follows:

$$\int \frac{d^4 k}{(2\pi)^4} \frac{1}{(k^2 + M^2)^2} = \frac{1}{16\pi^2}\Gamma\left(\frac{\epsilon}{2}\right) \approx \frac{1}{8\pi^2 \epsilon}. \tag{3.48}$$

So ϵ should be thought of as $\ln \Lambda^2$. The computation of the Yang–Mills beta function by studying momentum-space Feynman diagrams can be found in many textbooks and is outlined in the exercises at the end of the chapter.

Here we follow a different approach, known as the background field method. This technique is closely tied to the path integral, which will play an important role in this book. It is also closely tied to the Wilsonian view of renormalization. We break up a field A into a long-wavelength part $\mathcal{A}$ and a shorter-wavelength, fluctuating, quantum part a:

$$A^\mu = \mathcal{A}^\mu + a^\mu. \tag{3.49}$$

We can think of $\mathcal{A}^\mu$ as corresponding to modes of the field with momenta below the scale q and a^μ as corresponding to higher momenta. We wish to compute an effective action for $\mathcal{A}^\mu$, integrating out the high-momentum modes:

$$\int [d\mathcal{A}] \int [da]\, e^{iS(\mathcal{A},a)} = \int [d\mathcal{A}]\, e^{iS_{\rm eff}(\mathcal{A})}; \tag{3.50}$$

(see Appendix C for an explanation of the terminology). In calculating the effective action we are treating $\mathcal{A}^\mu$ as a fixed, classical, background. In this approach one can work entirely in Euclidean space, which greatly simplifies the calculation.

Our first task is to write down $e^{iS(\mathcal{A},a)}$. For this purpose, it is convenient to suppose that $\mathcal{A}$ satisfies its equation of motion. (Otherwise, it is necessary to introduce a source for a.) A convenient choice of gauge is known as the background field gauge,

$$D_\mu a^\mu = 0, \tag{3.51}$$

where D_μ is the covariant derivative defined with respect to the background field $\mathcal{A}$. At one loop we only need to work out the action to second order in the fluctuating fields a^μ, ψ, ϕ. Consider, first, the fermion action. To quadratic order we can set $a^\mu = 0$ in the Dirac Lagrangian. The same holds for scalars. So from the fermions and scalars we obtain

$$\det(\not{D})^{n_f} \det(D^2)^{-n_\phi/2}. \tag{3.52}$$

The fermion functional determinant can be greatly simplified; it is convenient, for this computation, to work with four-component Dirac fermions. Then

$$\begin{aligned} \det(\not{D}) &= \det(\not{D}\not{D})^{1/2} \\ &= \det\left(D^2 + \frac{1}{2} D_\mu D_\nu [\gamma^\mu, \gamma^\nu]\right) \\ &= \det(D^2 + \mathcal{F}^{\mu\nu} \mathcal{J}_{\mu\nu}). \end{aligned} \tag{3.53}$$

Here $\mathcal{F}^{\mu\nu}$ is the field strength associated with $\mathcal{A}$ (we have used the connection between the field strength and the commutator of covariant derivatives, Eq. (2.14)) and $\mathcal{J}_{\mu\nu}$ is the generator of Lorentz transformations in the fermion representation.

What is interesting is that we can write the gauge boson determinant, in the background field gauge, in a similar fashion.[1] With a little algebra, the gauge part of the action can be shown to be

$$\mathcal{L}_{\rm gauge} = -\frac{1}{4g^2}\left(\mathrm{Tr}\, \mathcal{F}^{a\,2}_{\mu\nu} - 2g^2 a^a_\mu D^2 a^{\mu\, a} - 2 a^a_\mu f^{abc} \mathcal{F}^{b\mu\nu} a^c_\nu\right). \tag{3.54}$$

Here we have used the A^a_μ notation in order to be completely explicit about the gauge indices. Recalling the form of the Lorentz generators for the vector representation,

$$(\mathcal{J}^{\rho\sigma})_{\alpha\beta} = i(\delta^\rho_\alpha \delta^\sigma_\beta - \delta^\sigma_\alpha \delta^\rho_\beta), \tag{3.55}$$

we see that this object has the same formal structure as the fermion action,

$$\mathcal{L}_{\rm gauge} = -\frac{1}{2g^2}\left\{ a^a_\mu \left[(-D^2)^{ac} g^{\mu\nu} + 2\left(\frac{1}{2}\mathcal{F}^b_{\rho\sigma} \mathcal{J}^{\rho\sigma}\right)^{\mu\nu} (t^b_G)^{ac} \right] a^c_\nu \right\}. \tag{3.56}$$

[1] The details of these computations are outlined in the exercises. Here we are following closely the presentation in the text by Peskin and Schroeder (1995).

Finally, the Faddeev–Popov Lagrangian is just

$$\mathcal{L}_c = \bar{c}^a[-(D^2)^{ab}]c^b. \tag{3.57}$$

Since the ghost fields are Lorentz scalars, this Lagrangian has the same form as the others. We need, then, to evaluate a product of determinants of the form

$$\det\left[-D^2 + 2\left(\frac{1}{2}\mathcal{F}^b_{\rho\sigma}\mathcal{J}^{\rho\sigma}\right)t^b\right] \tag{3.58}$$

with t and $\mathcal{J}$ the generators appropriate to the representation.

The term in parentheses can be written as

$$\Delta_{r,j} = -\partial^2 + \Delta^{(1)} + \Delta^{(2)} + \Delta^{(\mathcal{J})} \tag{3.59}$$

with

$$\begin{aligned}\Delta^{(1)} &= i(\partial^\mu A^a_\mu t^a + \mathcal{A}^a_\mu t^a \partial^\mu)\\ \Delta^{(2)} &= \mathcal{A}^{a\mu}t^a\mathcal{A}^b_\mu t^b.\\ \Delta^{(\mathcal{J})} &= 2\left(\frac{1}{2}\mathcal{F}^b_{\rho\sigma}\mathcal{J}^{\rho\sigma}\right)t^b.\end{aligned} \tag{3.60}$$

The action we are seeking is the log of the determinant. We are interested in this action expanded to second order in $\mathcal{A}$ and second order in ∂^2:

$$\begin{aligned}\ln\det(\Delta_{r,j}) = \ln\det(-\partial^2) + \mathrm{tr}\Big[&(-\partial^2)^{-1}(\Delta^{(1)} + \Delta^{(2)} + \Delta^{(\mathcal{J})})\\ &- \frac{1}{2}((-\partial^2)^{-1}\Delta^{(1)}(-\partial^2)^{-1}\Delta^{(1)})\Big],\end{aligned} \tag{3.61}$$

where $1/(-\partial^2)$ is the propagator for a scalar field. So this has the structure of a set of one-loop diagrams in a scalar field theory. Since we are working to quadratic order, we can take the $\mathcal{A}$ field to carry momentum k. The term involving two factors $\Delta^{(1)}$ is in some ways the most complicated to evaluate. Note that the trace is a trace in coordinate space and over the gauge and Lorentz indices. In momentum space the space–time trace is just an integral over momenta. We take all the momenta to be Euclidean. So the result is given, in momentum space, by

$$\frac{1}{2}\int\frac{d^dk}{(2\pi)^d}\mathcal{A}^a_\mu(k)\mathcal{A}^b_\nu(-k)\int\frac{d^dp}{(2\pi)^d}\,\mathrm{Tr}\left[\frac{1}{p^2}(2p+k)^\mu t^a\frac{1}{(p+k)^2}(2p+k)^\nu t^b\right]. \tag{3.62}$$

This has precisely the structure of one of the vacuum polarization diagrams of scalar electrodynamics (see Fig. 3.7). The other contribution arises from the factor $\Delta^{(2)}$. Combining the two contributions, and performing the integral by dimensional regularization gives

$$\frac{1}{2}\int\frac{d^dk}{(2\pi)^d}\mathcal{A}^a_\mu(-k)\mathcal{A}^b_\nu(k)(k^2g^{\mu\nu} - k^\mu k^\nu)\left[\frac{C(r)d(j)}{3(4\pi)^2}\Gamma\left(2-\frac{d}{2}\right)(k^2)^{2-d/2}\right], \tag{3.63}$$

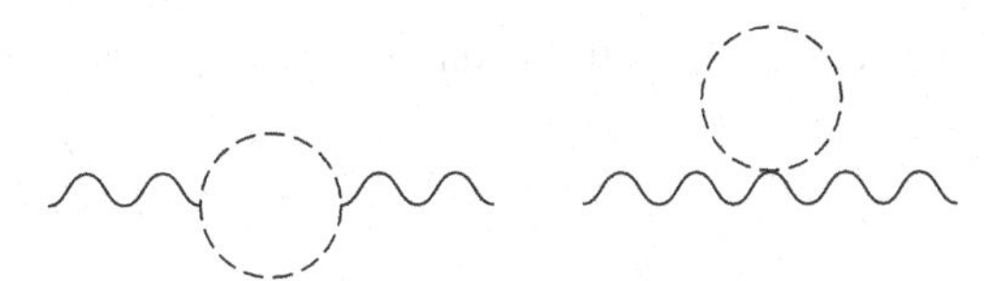

Fig. 3.7 The background-field calculation has the structure of scalar electrodynamics.

where $C(r)$ is a Casimir operator, encountered previously. The quantities $C(j)$ are similar quantities for the Lorentz group: $C(j) = 0$ for scalars, 1 for Dirac spinors and 2 for four-vectors. To quadratic order in the external fields, the transverse terms above give $(\mathcal{F}^{\mu\nu})^2$.

The contribution involving $\Delta^{(\mathcal{J})}$ in Eq. (3.61) is even simpler to evaluate, since the needed factors of momentum (which are derivatives) are already included in $\mathcal{F}$. The rest is bookkeeping; the required action has the form

$$\mathcal{L}_{\text{eff}} = -\frac{1}{4}\left[\frac{1}{g^2} + \frac{1}{2}\left(C_G - C_c - \frac{n_f}{2}C_{n_f}\right)\right]F^2_{\mu\nu} \tag{3.64}$$

where

$$C_i = c_i\frac{1}{16\pi^2}\left(\frac{2}{\epsilon} - \ln k^2\right), \quad c_G = -\frac{20}{3}, \quad c_c = 3, \quad c_{n_f} = -\frac{1}{3}. \tag{3.65}$$

This gives precisely Eq. (3.21).

3.7 The strong interactions and dimensional transmutation

In QCD the only parameters at the classical level with the dimensions of mass are the quark masses. In a world with just two light quarks, u and d, we would not expect the properties of hadrons to be very different from the observed properties of the non-strange hadrons. However, the masses of the up and down quarks are quite small; in fact, as we will see, too small to account for the masses of the non-strange hadrons such as the proton and neutron. In other words, in the limit of zero quark mass these hadrons would not become massless. How can a mass arise in a theory with no classical mass parameters?

While classically QCD is scale invariant, this is not true quantum mechanically. We have seen that we must specify the value of the gauge coupling at a particular energy scale; in the language we have used up to now, the theory is specified by giving the Lagrangian associated with a particular cutoff scale. If we change this scale, we have to change the values of the parameters, and physical quantities such as the proton mass $m_p = u$, should be unaffected. Using our experience with the renormalization group we can write down a differential equation which expresses how such a mass depends on g and μ, so that the mass is independent of which scale we choose to define our theory:

$$\left[\mu\frac{\partial}{\partial\mu} + \beta(g)\frac{\partial}{\partial g}\right]m_p = 0. \tag{3.66}$$

We know the solution of this equation:

$$m_p = C\mu \exp\left[-\int \frac{dg'}{\beta(g')}\right]. \tag{3.67}$$

To lowest order in the coupling,

$$m_p = C\mu \exp\left[-\frac{8\pi^2}{g^2(\mu)}\right]. \tag{3.68}$$

This phenomenon, that a physical mass scale can appear as a result of the need to introduce a cutoff in the quantum theory, is called *dimensional transmutation*. In the next section we will discuss this phenomenon as it occurs in lattice gauge theory. Later we will describe a two-dimensional model with which we can do a simple computation that exhibits the dynamical appearance of a mass scale.

3.8 Confinement and lattice gauge theory

The fact that QCD becomes weakly coupled at high momentum transfers has allowed rigorous comparison with experiment. Despite the fact that the variation of the coupling is only logarithmic, experiments are sufficiently sensitive, and have covered a sufficiently broad range of q^2, that such comparisons are possible. Still, many of the most interesting questions of hadronic physics – and some of the most interesting challenges of quantum field theory – are problems of low momentum transfer. Here one encounters the flip side of asymptotic freedom: at large distances, the theory is necessarily strongly coupled and perturbative methods are not useful. It is, perhaps, frustrating that we cannot compute the masses of the low-lying hadrons in a fashion analogous to the calculation of the properties of simple atoms. Perhaps even more disturbing is that we cannot give a simple argument that quarks are confined or that QCD exhibits a mass gap. To deal with these questions, we will first ask a somewhat naive question: what can we say about the path integral, or for that matter the Hamiltonian, in the limit in which the coupling constant becomes very large? This question is naive in that the coupling constant is not really a parameter of this theory. It is a function of the scale, and the important scale for binding hadrons is that where the coupling becomes of order one. Let us consider the problem anyway. We will start with a pure gauge theory, i.e. a theory without fermions or scalars. Consider, first, the path integral. To extract the spectrum, it should be adequate to consider the Euclidean version:

$$Z = \int [dA_\mu] \exp\left(-\frac{1}{4g^2} F_{\mu\nu}^2\right). \tag{3.69}$$

Let us contrast the weak- and strong-coupling limits of this expression. At weak coupling $1/g^2$ is large, so fluctuations are highly damped; we might expect the action to be controlled by the stationary points. The simplest such stationary point occurs where $F_{\mu\nu} = 0$, and this is the basis of perturbation theory. Later we will see that there are other interesting stationary points – classical solutions of the Euclidean equations.

Now consider strong coupling. As $g \to \infty$ the action vanishes – there is no damping of the quantum fluctuations. It is not obvious how one can develop any sort of approximation scheme. We can consider this problem, alternatively, from a Hamiltonian point of view. A convenient gauge for this purpose is the gauge $A_0 = 0$. In this gauge Gauss's law is a constraint that must be imposed on states. As we will discuss shortly, Gauss's law is (almost) equivalent to the condition that the quantum states must be invariant under time-independent gauge transformations. In the $A^0 = 0$ gauge, the canonical momenta are very simple:

$$\Pi^i = \frac{\partial \mathcal{L}}{\partial \dot{A}^i} = -\frac{1}{g^2} E^i. \tag{3.70}$$

So, the Hamiltonian is

$$\mathcal{H} = \frac{g^2}{2} \vec{\Pi}^2 + \frac{1}{2} \vec{B}^2. \tag{3.71}$$

In the limit $g^2 \to \infty$, the magnetic terms are unimportant and the Π^2 terms dominate. So we should somehow work, in lowest order, with states which are eigenstates of $\vec{E}$. In any approach which respects even rotational covariance, it is unclear how to proceed.

The solution to both dilemmas is to replace the space–time continuum with a discrete lattice of points. In the Lagrangian approach one introduces a space–time lattice. In the Hamiltonian approach one keeps the time continuous but makes space discrete. Clearly there is a large price for such a move: one gives up Lorentz invariance, even rotational invariance. At best, Lorentz invariance is something which one can hope to recover in the limit where the lattice spacing is small compared with the relevant physical distances. There are several rewards, however.

1. One has a complete definition of the theory which does not rely on perturbation theory.
2. The lattice, at strong coupling, gives a simple model of confinement.
3. One obtains a precise procedure in which to calculate the properties of hadrons. With large enough computing power one can in principle calculate the properties of low-lying hadrons with arbitrary precision.

There are other difficulties which must be overcome. Not only is rotational symmetry lost, but other approximate symmetries – particularly chiral symmetries – are complicated. But, over time, combining ingenuity and growing computer power there has been enormous progress in numerical lattice computations. Lattice gauge theory has developed into a highly specialized field of its own, and we will not do justice to it here. However, given the importance of field theories – often strongly coupled field theories – not only for our understanding of QCD but for any understanding of physics beyond the Standard Model, it is worthwhile to briefly introduce the subject here.

3.8.1 Wilson's formulation of lattice gauge theory

In introducing a lattice the hope is that, as one allows the lattice spacing a to become small, one will recover Lorentz invariance. A little thought is required to understand what is meant by *small*. The only scale in the problem is the lattice spacing. But there is another important parameter: the gauge coupling. The value of this coupling, we might expect, should be thought of as the QCD coupling at scale a. So, taking small lattice spacing means physically taking the gauge coupling to be weak. At small lattice spacing, the short-distance Green's functions will be well approximated by their perturbative expansions. On the other hand, the smaller the lattice, the more numerical power required to compute the physically interesting, long-distance, quantities.

There is one symmetry which one might hope to preserve as one introduces a space–time lattice: gauge invariance. Without it, there are many sorts of operators which could appear in the continuum limit and recovering the theory of interest would be likely to be very complicated. Wilson pointed out that there is a natural set of variables to work with; there are known as *Wilson lines*. Consider, first, a $U(1)$ gauge theory. Under a gauge transformation $A_\mu(x) \to A_\mu + ig(x)\partial_\mu g^\dagger(x)$, where $g(x) = e^{i\alpha(x)}$, the object

$$U(x_1, x_2) = \exp\left(i \int_{x_1}^{x_2} dx_\mu A^\mu\right),$$

transforms as follows:

$$U(x_1, x_2) \to g(x_1) U(x_1, x_2) g^\dagger(x_2). \tag{3.72}$$

So, for example, for a charged fermion field $\psi(x)$ transforming as $\psi(x) \to g(x)\psi(x)$, a gauge-invariant operator is

$$\psi^\dagger(x_1) U(x_1, x_2) \psi(x_2). \tag{3.73}$$

From gauge fields alone one can construct an even simpler gauge-invariant object, a Wilson line beginning and ending at some point x:

$$U(x, x) = \exp\left(i \oint_C dx^\mu A_\mu\right), \tag{3.74}$$

where U is called a Wilson loop.

These objects have a simple generalization in non-Abelian gauge theories. Using the matrix form for A_μ the main issue is one of ordering. The required ordering prescription is a path ordering, P:

$$U(x_1, x_2) = P \exp\left(i \int_{x_1}^{x_2} dx^\mu A_\mu\right). \tag{3.75}$$

It is not hard to show that the transformation law for the Abelian case generalizes to the non-Abelian case:

$$U(x_1, x_2) \to g(x_1) U(x_1, x_2) g^{-1}(x_2). \tag{3.76}$$

To see this, note first that path ordering is like time ordering so, if s is the parameter of the path, U satisfies

$$\frac{d}{ds} U(x_1(s), x_2) = \left(ig \frac{dx^\mu}{ds} A_\mu(x_1(s))\right) U(x_1(s), x_2) \tag{3.77}$$

or, more elegantly,

$$\frac{dx_1^\mu}{ds} D_\mu U(x_1, x_2) = 0. \tag{3.78}$$

Now suppose that $U(x_1, x_2)$ satisfies the transformation law (Eq. (3.72)). Then it is straightforward to check, from Eq. (3.77), that $U(x_1 + dx_1, x_2)$ satisfies the correct equation. Since U satisfies a first-order differential equation, this is enough.

Again, the integral around a closed loop, C, is gauge invariant, provided that now one takes the trace:

$$U(x_1, x_1) = \mathrm{Tr}\,\exp\left(i \oint_C dx^\mu A_\mu\right). \tag{3.79}$$

Wilson used these objects to construct a discretized version of the usual path integral. Take the lattice to be a simple hypercube, with points $x^\mu = an^\mu$, where n^μ is a vector of integers and a is called the lattice spacing. At any point x one can construct a simple Wilson line $U(x)_{\mu\nu}$, known as a *plaquette*. This is just the product of Wilson lines around a unit square. Letting n_μ denote a unit vector in the μ direction, we denote the Wilson line $U(x, x + an^\mu)$ by $U(x)_\mu$. These are the basic variables; as they are associated with the lines linking two lattice points they are called *link variables*. Then the Wilson loops about each plaquette are denoted as follows:

$$U(x)_{\mu\nu} = U(x)_\mu U(x + an^\mu)_\nu U(x + an^\mu + an^\nu)_{-\mu} U(x + an^\nu)_{-\nu}. \tag{3.80}$$

In the non-Abelian case, a trace is understood to be taken. For small a, in the Abelian case it is easy to expand $U_{\mu\nu}$ in powers of a and to show that

$$U(x)_{\mu\nu} \approx \exp\left[ia^2 F_{\mu\nu}(x)\right]. \tag{3.81}$$

So, we can write down an action which in the limit of small lattice spacing goes over to the Yang–Mills action:

$$S_{\text{Wilson}} = \frac{1}{4g^2} \sum_{x,\mu,\nu} U(x)_{\mu,\nu}. \tag{3.82}$$

In the non-Abelian case this same expression holds, except with the factor 4 replaced by 2 and a trace over the U matrices.

How might we investigate the question of confinement with this action? Here, Wilson also made a proposal. Consider the amplitude for a process in which a very heavy (infinitely heavy) quark–antiquark pair, separated by a distance R, was produced in the far past and allowed to propagate for a long time T after which the pair annihilates. In Minkowski space the amplitude for this would be given by

$$\langle f|e^{-iHT}|i\rangle, \tag{3.83}$$

where H is the Hamiltonian for the process. If we transform to Euclidean space and insert a complete set of states, for each state we have a factor $\exp(-E_n T)$. As $T \to \infty$ this becomes $e^{-E_0 T}$, where E_0 is the ground state of the system with two infinitely massive quarks separated by a distance R, and is what we would naturally identify with the potential of the quark–antiquark system.

In the path integral this expectation value is precisely the Wilson loop U_P, where P is the path from the point of production to the point of annihilation and back. If the quarks only experience a Coulomb force, one expects the Wilson loop to behave as

$$\langle U_P \rangle \propto e^{-\alpha T/R} \tag{3.84}$$

for a constant α. In other words, the exponential behaves as the *perimeter* of the loop. If the quarks are confined, with a linear confining force, the exponential behaves as e^{-bTR}, i.e. as the *area* of the loop. So Wilson proposed to measure the expectation value of the Wilson loop and determine whether it obeyed a perimeter or area law.

In strong coupling it is a simple matter to do the computation in the lattice gauge theory. We are interested in

$$\int \prod dU(x)_\mu \, \exp\left(-S_{\text{lattice}} + i \prod_{\text{P}} U\right). \tag{3.85}$$

We can evaluate this by expanding the exponent in powers of $1/g^2$. Because

$$\int dU_\mu \, U_\mu = 0, \qquad \int dU_\mu \, U_\mu U^\dagger_\mu = \text{const} \tag{3.86}$$

(you can check this easily in the Abelian case), in order to obtain a non-vanishing result we need to tile the path with plaquettes, as indicated in Fig. 3.8. So the result is exponential in the area,

$$\langle U_P \rangle = \left(\frac{\text{const}}{g^2}\right)^A, \tag{3.87}$$

and the force law is

$$V(R) = \text{const} \times \frac{g^2}{a^2} R. \tag{3.88}$$

This is not a proof of confinement in QCD. First note that this result holds in the strong coupling limit of either an Abelian or a non-Abelian gauge theory. This is possible because

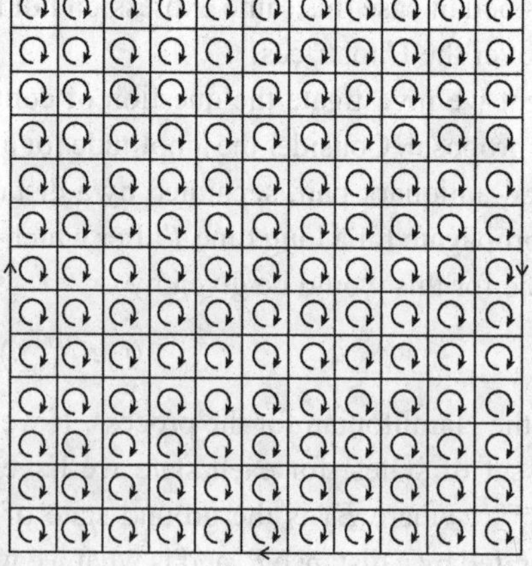

Fig. 3.8 Leading non-vanishing contribution to the Wilson loop in strong coupling lattice gauge theory.

even the pure gauge Abelian lattice theory is an interacting theory. From this we learn that the strong coupling behavior of a lattice theory can be very different than the weak coupling behavior. For QCD we would like to choose the lattice spacing to correspond to a small physical scale, say $a = (4\text{ GeV})^{-1}$, where the gauge coupling is small, and then study the behavior of the correlation functions, Wilson loops and other quantities on much larger scales. At present this requires numerical techniques.

3.8.2 Hamiltonian lattice gauge theory

Before discussing Hamiltonian lattice gauge theories, it is interesting to see how the strong-coupling result arises from a Hamiltonian viewpoint. To simplify the computation we consider a $U(1)$ gauge theory. In the Hamiltonian approach the basic dynamical variables are the matrices U_i associated with the spatial directions. There is also the gauge field A_0. As in continuum field theory, we can choose $A_0 = 0$. In this gauge, in the continuum the dynamical variables are A_i and their conjugate momenta are E_i; on the lattice, the momenta conjugate to the U_i are the E_i. The Hamiltonian has the form

$$H = \sum \frac{g^2 \vec{\Pi}(x)^2}{a} + \frac{1}{4g^2} \sum U_{ij}(\vec{x}) \frac{1}{a}. \tag{3.89}$$

The U_is are compact variables, so the $\Pi(x)$s at each point are like angular momenta. At strong coupling this is a system of decoupled rotors. The ground state of the system has a vanishing value of these angular momenta.

Now introduce a heavy quark–antiquark pair to the system, separated by a distance R in the z direction. In the $A_0 = 0$ gauge, states must be gauge invariant (we will discuss this further when we consider instantons, in the next chapter). So, a candidate state has the form

$$|\Psi\rangle = q^\dagger(0) U_z(0, R) \bar{q}^\dagger(R) |0\rangle. \tag{3.90}$$

Here

$$U_z(0, R) = U_z(0, 1) U_z(1, 2) \cdots U_z(N-1, N), \tag{3.91}$$

where $R = Na$. Now we can evaluate the expectation value of the Hamiltonian in this state. At strong coupling we can ignore the magnetic terms. The effect of the U_z operators is to raise the "angular momentum" associated with each link by one unit (in the $U(1)$ case, $U_z(n, n+1) = e^{i\theta_{n+1}}$). So the energy of the state is just

$$a^{-1} g^2 N, \tag{3.92}$$

and the potential grows linearly with separation.

3.8.3 Numerical methods in lattice gauge theory; introduction of fermions

We have seen that the strong coupling analysis, while providing a model for confinement, is hardly satisfactory. It predicts confinement in lattice QED as well as QCD. It turns out that in QED there is a phase transition (a discontinuous change of behavior) between the strong- and weak-coupling phases. To be sure that the same does not occur in QCD, we need to evaluate the Wilson line on a very fine lattice, at large separation. This means we need to work with an action having a *small* coupling. To put it another way, to reliably describe, say, a proton we need to use a lattice on which the spacing a is much smaller than the QCD scale. At present such studies can only be undertaken by evaluating the lattice path integral numerically. In principle, since the lattice theory reduces space–time to a finite number of points, the required path integral is just an ordinary integral, albeit with a huge number of dimensions. For example, if we have a $10 \times 10 \times 10 \times 10$ lattice, with of order 10^4 links (each a 3×3 matrix), and quarks at each site, it is clear that a straightforward numerical evaluation involves an exponentially large number of operations. In practice it is necessary to use Monte Carlo (statistical sampling) methods to evaluate the integrals. These techniques are now sufficiently powerful to demonstrate convincingly an area law at weak coupling. The constant in the area law, the coefficient of the linear term in the quark–antiquark potential, is a dimensionful parameter. It must be renormalization-group invariant. As a result, it must take the form

$$T = ca^{-2} \exp\left[-\int \frac{dg'}{\beta(g')}\right]. \tag{3.93}$$

At weak coupling we know the form of the beta function, so we know how T should behave as we vary the lattice spacing and coupling. The results of numerical studies are in good agreement with these expressions.

However, we would like to study real QCD, with fermions. Fermions introduce additional challenges. These are of two types. First, one needs a strategy to deal with Grassman integrations in the functional integral. The usual strategy is to hold the bosonic variables fixed while first performing the integral over the fermions. This yields a determinant (in general multiplied by some Green's functions), which must be evaluated for every value of the bosonic integrand. These are determinants of enormous matrices and must themselves be evaluated by statistical techniques. In the early years of lattice gauge theory, such computations were out of reach and so numerical work generally simply dropped the determinant (such calculations were said to be *quenched*). But, by the early years of the new millennium, both algorithms for these computations and computer power had developed to the point that such computations were feasible.

As we will see further in Chapter 5, it is crucial to our understanding of the strong interactions that the u, d and s quarks are light compared with the characteristic scales of the strong interactions and in particular compared with quantities such as the pion decay constant and the ρ meson mass. However, massless or light fermions, on the lattice, are problematic. The difficulties are associated with the fact that their kinetic terms are first order in derivatives. Writing the derivative as a naive difference leads to the problem of fermion "doubling".

To see the difficulty, consider first the kinetic terms for a free boson. Label the lattice points (in a Euclidean lattice) by four vectors $n_\mu a$, where a is the lattice spacing, i.e. $x_\nu = n_\nu a$. Then

$$\partial_\mu \phi(x) \to [\phi(x + n_\mu a) - \phi(x - n_\nu a)]/(2a). \tag{3.94}$$

Now we write a Fourier expansion in terms of

$$\phi_k e^{ik_\mu x_\mu}, \tag{3.95}$$

where $-\pi/a \leq k^\mu \leq \pi/a$. This is the analog of the familiar problem of a particle in a box of size L with periodic boundary conditions. There $k = 2\pi n/L$. Now the roles of x and k are reversed: $x = na$, so k lies in an interval of size $2\pi/a$, as above. Then, for scalars, the second derivative term, defined as above, is proportional to

$$|f_k|^2(1 - \cos k_\mu a) \tag{3.96}$$

which is consistent with the size of the k interval.

However, for fermions, a term such as $\partial_\mu \gamma_\mu$ is proportional to

$$f_k \gamma_\mu \sin k_\mu a \tag{3.97}$$

which has zeros (corresponding to poles in the propagator) not only at $k_\mu = 0$ but also at points where the components $k_\mu = \pi/a$. The appearance of these extra light degrees of freedom is called the *fermion doubling problem*. General theorems show that it is unavoidable. In practice this problem is dealt with in either of two ways. One can attempt to treat the extra fermions as additional light flavors, or one can add a term to the action which gives mass to the extra fermions, typically a term proportional to a parameter and $1 - \cos ka$, known as the Wilson term. The price of the first method is that one must extract results for the actual number of flavors (three) from a theory with more flavors. This has been the approach of the MILC collaboration, one of the large lattice simulation efforts. In the second method one has the difficulty that the parameters must be tuned, as one approaches the continuum limit, in such a way that one obtains the expected symmetry structure of actual QCD. This method has been used by the BMW collaboration and others. Considerable success has been achieved with both, and there is remarkable agreement. A third method is known as the *domain wall fermion* method. Here one introduces a fifth dimension, with fields of opposite chirality living on two walls. This method shows promise but imposes additional computational challenges and to date has been numerically less extensively studied.

3.9 Strong interaction processes at high momentum transfer

Quantum chromodynamics has been tested with high precision in a variety of processes at high momentum transfer (short distances). It is by now an important tool in probing for new physics in particle colliders. Indeed, our understanding of perturbative QCD was crucial to the discovery of the Higgs boson. It is these processes to which one can apply ordinary perturbation theory. If Q^2 is the typical momentum transfer of a process, cross sections are given by a power series in $\alpha_s(Q^2)$. The application of perturbation theory, however, is subtle. In accelerators we observe hadrons; using perturbation theory we compute the production rate for quarks and gluons. We will briefly survey some applications in this section. The simplest process to analyze is e^+e^- annihilation, and we discuss it first.

Then we turn to processes involving the deep inelastic scattering of leptons by hadrons and follow this by considering by processes involving hadrons only. Finally, we describe recent progress in QCD computations for processes involving complicated final states (many gluons) and/or higher orders in perturbation theory.

3.9.1 e^+e^- annihilation

At the level of quarks and gluons, the first few diagrams contributing to the production cross section are exhibited in Fig. 3.9. There is, in perturbation theory, a variety of final states, $q\bar{q}$, $q\bar{q}g$, $q\bar{q}gg$, $q\bar{q}q\bar{q}$ and so on. We do not understand, in any detail, how these quarks and gluons materialize as the observed hadrons. But we might imagine that this occurs as in Fig. 3.10. The initial quarks radiate gluons which can in turn radiate quark–antiquark pairs. As the cascade develops, quarks and antiquarks can pair to form mesons, qqq combinations can form baryons and so on. In these complex processes (called *hadronization*) we can construct many relativistic invariants and many of these will be small, so that perturbation theory cannot be trusted. In a sense this is good; otherwise, we would be able to show that free quarks and gluons were produced in the final states. But if we only ask about the *total* cross section, each term in the series is a function only of the center of mass energy s. As

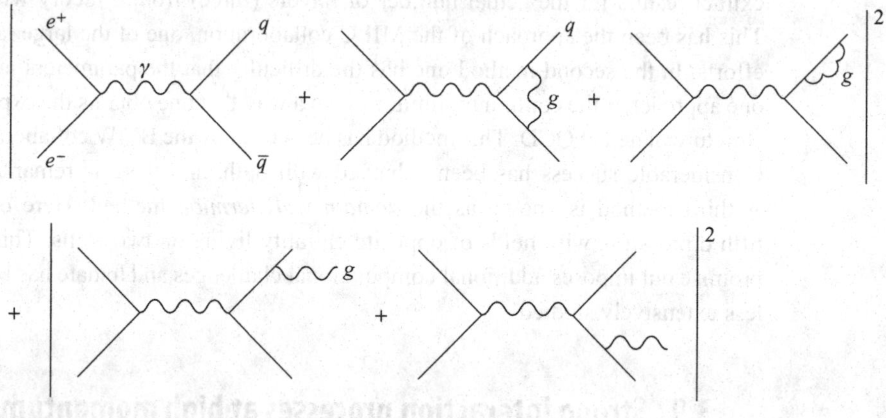

Fig. 3.9 Low-order contributions to e^+e^- annihilation.

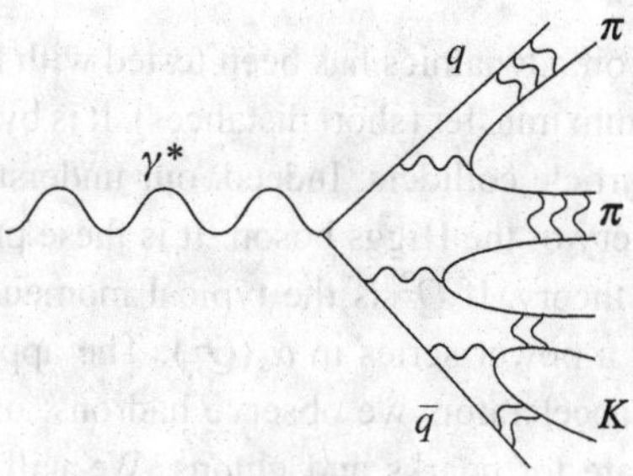

Fig. 3.10 Emission of gluons and quarks leads to the formation of hadrons.

a result, if we simply choose s for the renormalization scale, the cross section is given by a power series in $\alpha_s(s)$. One way to see this is to note that the cross section is proportional to the imaginary part of the photon vacuum polarization tensor, $\sigma(s) \propto \text{Im}\,\Pi$. One can calculate Π in Euclidean space and then analytically continue. In the Euclidean calculation there are no infrared divergences, so the only scales are s and the cutoff (or renormalization scale). It is convenient to consider the ratio

$$R(e^+e^- \to \text{hadrons}) = \frac{\sigma(e^+e^- \to \text{hadrons})}{\sigma(e^+e^- \to \mu^+\mu^-)}. \tag{3.98}$$

The lowest-order (α_s^0) contribution can be written down without any work:

$$R(e^+e^- \to \text{hadrons}) = 3\sum Q_f^2, \tag{3.99}$$

where we have explicitly pulled out a factor 3 for color and the sum is over those quark flavors light enough to be produced at energy $\sqrt{s}$. So, for example, above the charm quark threshold and below the bottom quark threshold this would give

$$R(e^+e^- \to \text{hadrons}) = \frac{10}{3}. \tag{3.100}$$

Before comparing with the data we should consider corrections. The cross section has been calculated through order α_s^3, where $\alpha_s = g_s^2/(4\pi)$; g_s, the strong coupling constant was introduced in Eq. (2.64). Here we quote just the first two orders:

$$R(e^+e^- \to \text{hadrons}) = 3\sum Q_f^2\left(1 + \frac{\alpha_s}{\pi}\right). \tag{3.101}$$

This may be compared with the data in Fig. 3.11.

This calculation has other applications. Among these are applications to the widths of the Z and of the τ lepton. The decays of Z_s to hadrons involve essentially the same Feynman diagrams as before (Fig. 3.12), except for the different Z couplings to the quarks. This may be compared with experiment using Table 3.1.

3.9.2 Jets in e^+e^- annihilation

Much more is measured in e^+e^- annihilation than the total cross section, and clearly we would like to extract further predictions from QCD. If we are to use perturbation theory then it is important that we limit our questions to processes for which all momentum transfers are large. It is also important that perturbation theory should *fail* for some questions. After all, we know that the final states observed in accelerators contain hadrons, not quarks and gluons. If perturbation theory were good for sufficiently precise descriptions of the final state, the theory would simply be wrong.

To understand the issues, let us briefly recall some features of QED for a process like $e^+e^- \to \mu^+\mu^-$. At lowest order one just has the production of a $\mu^+\mu^-$ pair. At order α, however, one has final states with an additional photon and loop corrections to the muon lines (also to the electron or positron lines), as indicated in Fig. 3.13. Both the loop corrections and the total cross section for final states with a photon are infrared divergent. In QED the answer to this problem is *resolution*. In an experiment one cannot detect a photon

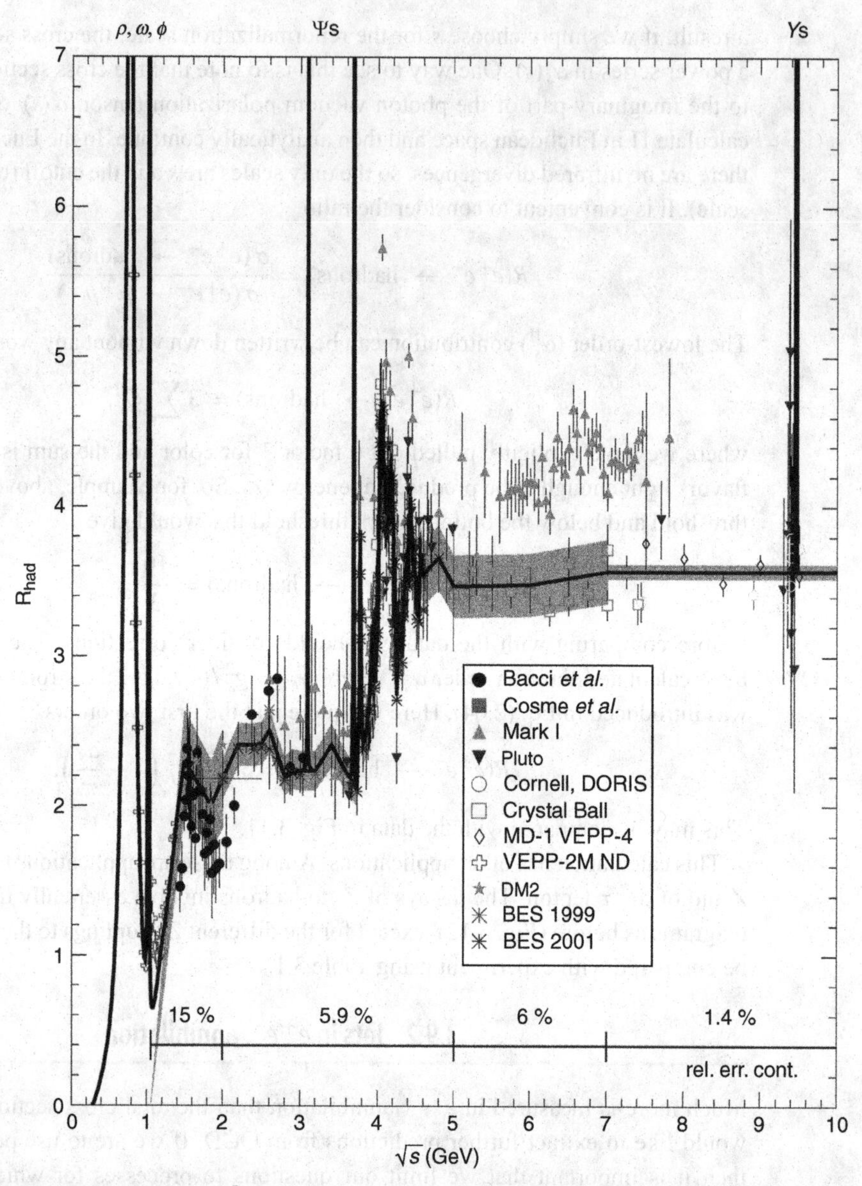

Fig. 3.11 Experimental data for the ratio R in e^+e^- annihilation, together with the theoretical prediction from Eq. (3.101). Reproduced from H. Burkhardt and B. Pietrzyk, *Phys. Lett. B* **513**, 46 (2001). Copyright 2001, with permission from Elsevier.

of arbitrarily low energy. So, in comparing the theory with the observed cross section for $\mu^+\mu^-$ (with no photon), one must allow for the possibility that a very-low-momentum photon is emitted and not detected. By including some energy resolution ΔE the cross sections for each possible final state are made finite. If the energy is very large one also has to keep in mind that experimental detectors cannot resolve photons that are nearly

Table 3.1 Experimental and theoretical values of properties of the Z boson. Note the close agreement at the one part in 10^2—10^3 level. Reprinted from *Electroweak Model and Constraints on New Physics*, Particle Data Group (2005), and S. Eidelman *et al.*, *Phys. Lett. B*, **592**, 1 (2004) (used with permission of the Particle Data Group and Elsevier)

Quantity	Value	Standard Model	Pull
m_t (GeV)	176.1 ± 7.4	176.96 ± 4.0	-0.1
	180.1 ± 5.4		0.6
M_W (GeV)	80.454 ± 0.059	80.390 ± 0.018	1.1
	80.412 ± 0.042		0.5
M_Z (GeV)	91.1876 ± 0.0021	91.1874 ± 0.0021	0.1
Γ_Z (GeV)	2.4952 ± 0.0023	2.4972 ± 0.0012	-0.9
Γ(had) (GeV)	1.7444 ± 0.0020	1.7435 ± 0.0011	—
Γ (inv) (MeV)	499.0 ± 1.5	501.81 ± 0.13	—
$\Gamma(\ell^+\ell^-)$ (MeV)	83.984 ± 0.086	84.024 ± 0.025	—
σ_{had} (nb)	41.341 ± 0.037	41.472 ± 0.000	1.9

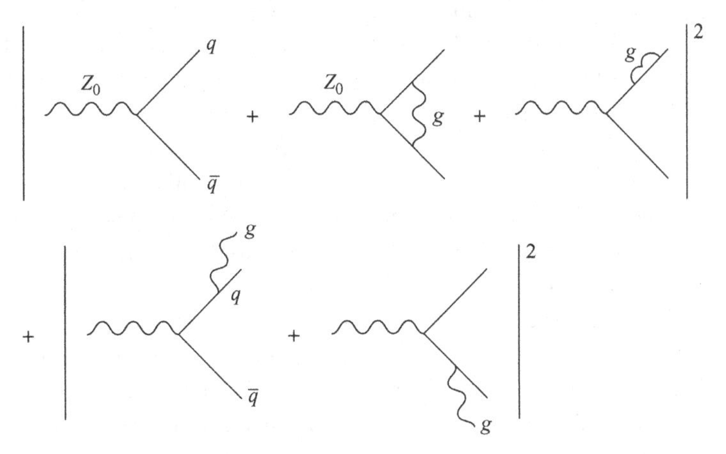

Fig. 3.12 Feynman diagrams contributing to Z decay are similar to those in e^+e^- annihilation.

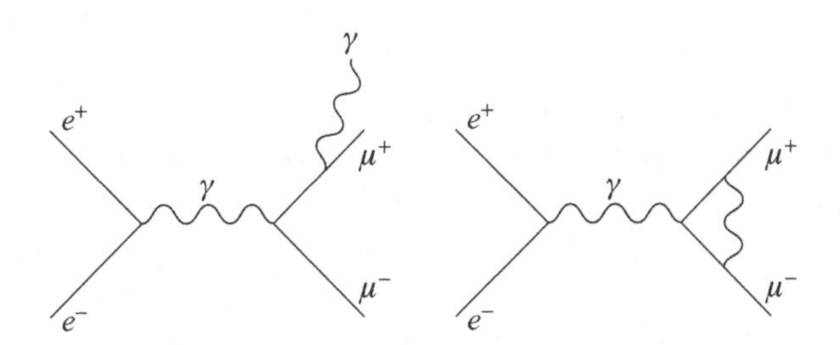

Fig. 3.13 The infrared problem.

parallel to one or other of the outgoing muons. The cross section, again, for each type of final state has large logarithms, $\ln(E/m_\mu)$. These are often called *collinear singularities* or *mass singularities*. So one must allow for the finite angular resolution of real experiments. Roughly speaking, then, the radiative corrections for these processes involve

$$\delta\sigma \propto \frac{\alpha}{4\pi} \ln \frac{E}{\Delta E} \ln \Delta\theta. \tag{3.102}$$

As one makes the energy resolution, or the angular resolution smaller, perturbation theory becomes poorer. In QED it is possible to sum these large double-logarithmic terms.

In QCD these same issues arise. Partial cross sections are infrared divergent. One obtains finite results if one includes an energy and angular resolution. But now the coupling is not as small as in QED, and it grows with energy. In other words, if one takes an energy resolution much smaller than the typical energy in the process, or an angular resolution which is very small, the logarithms which appear in the perturbation expansion signal that the expansion parameter is not $\alpha_s(s)$ but something more like $\alpha_s(\Delta E)$ or $\alpha_s(\Delta\theta s)$. So perturbation theory eventually breaks down.

However, if one does not make ΔE or $\Delta\theta$ too small then perturbation theory should be valid. Consider, again, e^+e^- annihilation to hadrons. One might imagine that on the one hand the processes which lead to the observed final states would involve the emission of many gluon and quark–antiquark pairs from the initial outgoing $q\bar{q}$ pair, as in Fig. 3.10. The final emissions will involve energies and momentum transfers of order the masses of pions and other light hadrons, and perturbation theory will not be useful. On the other hand, we can restrict our attention to the kinematic regime where the gluon is emitted at a large angle relative to the quark and has a substantial energy. There are no large logarithms in this computation, nor in the computation of the $q\bar{q}$ final state. We can give a similar definition for the $q\bar{q}g$ final state. From an experimental point of view, this means that we expect to see jets of particles (or of energy–momentum) that are reasonably collimated, and that we should be able to calculate the cross sections for the emission of such jets. These calculations are similar to those of QED. Such jets are observed in e^+e^- annihilation, and their angular distribution agrees well with theoretical prediction. When first observed, these three-jet events were described, appropriately, as the discovery of the gluon.

3.9.3 Deep inelastic scattering

Deep inelastic scattering was one of the first processes to be studied theoretically in QCD. These are experiments in which a lepton is scattered at high momentum transfer from a nucleus. The lepton can be an electron, a muon or a neutrino; the exchanged particle can be a γ, $W^\pm$ or Z (Fig. 3.14). One does not ask about the details of the final hadronic state but simply how many leptons are scattered at a given angle. Conceptually, these experiments are like Rutherford's experiment which discovered the atomic nucleus. In much the same way, they showed that nucleons contain quarks, having just the charges predicted by the quark model.

In the early days of QCD this process was attractive to study theoretically, because one can analyze it without worrying about issues about defining jets and the like. The inclusive

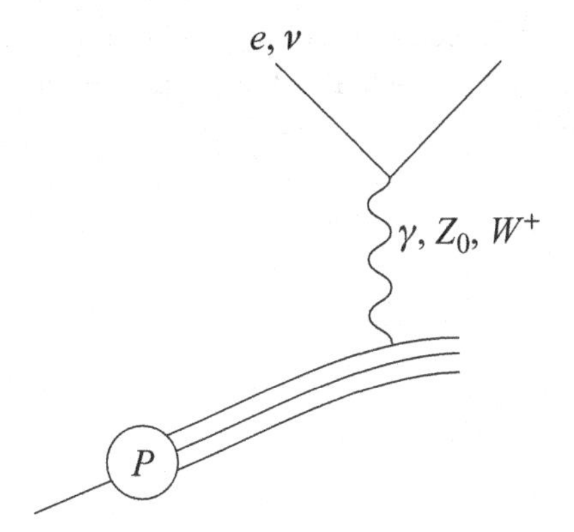

Fig. 3.14 The deep inelastic scattering of leptons from a nucleon.

cross section can be related, by unitarity, to a correlation function of two currents: the electromagnetic current, in the case of the photon and the weak currents in the case of the weak gauge bosons. The currents are space-like separated, and this separation becomes small as the momentum transfer Q^2 becomes large. This analysis is described in many textbooks. Here we will adopt a different viewpoint, which allows a description of the process that generalizes to other processes involving hadrons at high momentum transfer.

Feynman and Bjorken suggested that we could view the incoming proton as a collection of quarks and gluons, which they collectively referred to as *partons*. They argued that one could define the probability $f_i(x)$ of finding a parton of type i carrying a fraction x of the proton momentum (and similarly for neutrons). At high momentum transfer, they argued that the scattering of the virtual photon (or other particle) off the nucleon would actually involve the scattering of this object off one of the partons, the others being "spectators" (Fig. 3.14). In other words, the cross section for deep inelastic scattering would be given by

$$\begin{aligned}&\sigma(e^-(k) + p(P) \to e^-(k') + X) \\ &= \int dx \sum_f f_f(x) \sigma(e^-(k) + q(xP) \to e^-(k') + q_f(p')).\end{aligned} \tag{3.103}$$

This assumption may – should – seem surprising. After all, the scattering process is described by the rules of quantum mechanics and so there should be all sorts of complicated interference effects. We will discuss this question below, but, for now, suffice it to say that the above picture does become correct in QCD for large momentum transfers.

For the case of a virtual photon, the cross section for the parton process can be calculated just as in QED:

$$\frac{d\sigma}{d\hat{t}}(e^- q \to e^- q) = \frac{2\pi\alpha^2 Q_f^2}{\hat{s}^2}\left(\frac{\hat{s}^2 + \hat{u}^2}{\hat{t}^2}\right). \tag{3.104}$$

Here $\hat{s}, \hat{t}, \hat{u}$ are the kinematic invariants of the elementary parton process. For example, if we neglect the mass of the lepton and the incoming nucleon:

$$\hat{s} = 2p \cdot k = 2\zeta P \cdot k = \zeta s. \tag{3.105}$$

If the scattered electron momentum is measured then q is known and we can relate the proton momentum fraction x for the process to measured quantities. From momentum conservation,

$$(\zeta P + q)^2 = 0 \tag{3.106}$$

or

$$q^2 + 2\zeta P \cdot q = 0. \tag{3.107}$$

Solving for ζ:

$$\zeta = x = -\frac{q^2}{2P \cdot q}. \tag{3.108}$$

It is convenient to introduce another kinematic variable,

$$y = \frac{2P \cdot q}{s} = \frac{2P \cdot q}{2P \cdot k}. \tag{3.109}$$

Then $Q^2 = q^2 = xys$, and we can write down the differential cross section:

$$\frac{d^2\sigma}{dxdy}(e^- P \to e^- X) = \left(\sum_f x f_f(x) Q_f^2 \right) \frac{2\pi \alpha^2 s}{Q^4} [1 + (1-y)^2]. \tag{3.110}$$

This and related predictions were observed to hold in the first deep inelastic scattering experiments at SLAC, which provided the first persuasive experimental evidence for the reality of quarks. Note, in particular, the scaling implied by these relations. For fixed y the cross section is a function only of x.

In QCD these notions need a crucial refinement. The distribution functions are no longer independent of Q^2:

$$f_f(x) \to f_f(x, Q^2). \tag{3.111}$$

To understand this, we return to the question: why should a probabilistic model of partons work at all in these very quantum processes? Consider, for example, the Feynman diagrams of Fig. 3.15. Clearly there are complicated interference terms when one squares the amplitude. But it turns out that, in certain gauges, the interference diagrams are suppressed and the cross section is just given by the squares of terms, as in Fig. 3.16. So one finds a probabilistic description of the process, just as Feynman and Bjorken suggested, the distribution function being the result of the sequence of interactions in the figure. These diagrams depend on Q^2. One can write integro-differential equations for these functions, the *Altarelli–Parisi equations*. To explain the data, one determines these distribution functions at one value of Q^2 from experiment and then evolves them to other values. By now, the distribution functions have been studied over a broad range of Q^2. The structure functions must be measured at some Q^2; they can then be evolved to higher Q^2. This program has been very successful, as indicated in Fig. 3.17.

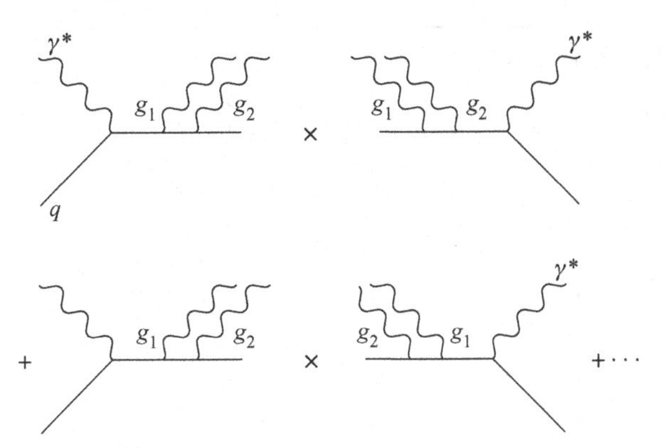

Fig. 3.15 Diagrams contributing to the total rate. The diagrams on the right are complex conjugates of the corresponding amplitudes on the left. The second term represents a complicated interference.

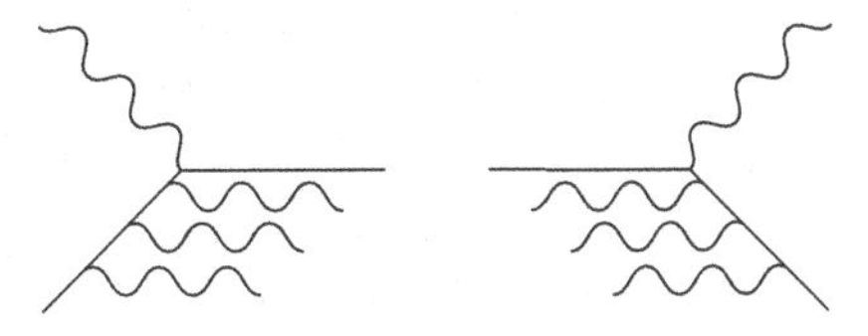

Fig. 3.16 In suitable gauges, deep inelastic scattering is dominated by the absolute squares of amplitudes (interference is unimportant).

3.9.4 Other high-momentum processes

These ideas have been applied to other processes. The analysis which provides a diagrammatic understanding of deep inelastic scattering shows that the same structure functions are relevant to other high-momentum-transfer processes, though care is required in their definitions. Examples include lepton pair production in hadronic collisions (Fig. 3.18) and jet production in hadron collisions, for which a comparison of theory and experiment can be made using Fig. 3.17. But, beyond testing QCD, such processes are crucial to the search for new physics. They have played a critical role in the discovery and study of the Higgs boson and in the exclusion of many possible types of new physics.

3.9.5 QCD beyond the leading order

For many questions it is crucial to compute QCD corrections beyond the leading order. This has been particularly important at the Tevatron and, more recently, the LHC. Such computations present serious challenges, and conventional Feynman diagram analyses are often inadequate. For example, we may be interested in initial states involving two gluons and final states involving two, three, four or more gluons. Already in the computation of the beta function, as discussed in Section 3.5, the three-gluon vertex adds significantly to the algebraic tedium (we avoided some of this by using the background field method).

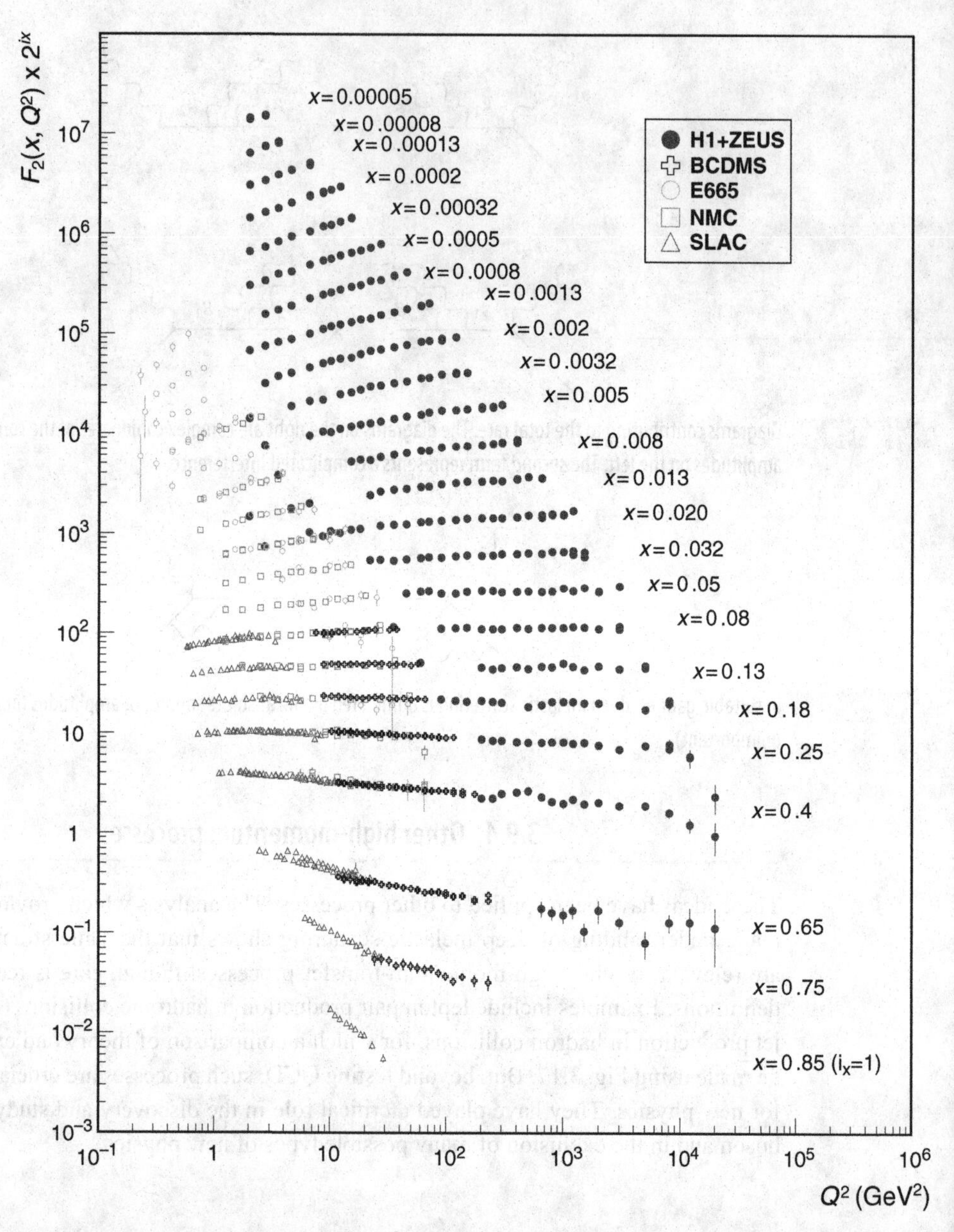

Fig. 3.17 The proton structure function F_2 as a function of Q^2 at fixed x, as determined by several experiments (reproduced by permission of the Particle Data Group).

For cross sections, however, the increase in labor is dramatic, particularly if we follow the standard method of squaring the amplitude and doing polarization and color sums (perhaps with projections). The labor grows essentially exponentially as we add more gluons. Without some cleverness, one quickly exhausts the capabilities of even powerful computers. With the Tevatron, and especially the LHC, programs, the need for such

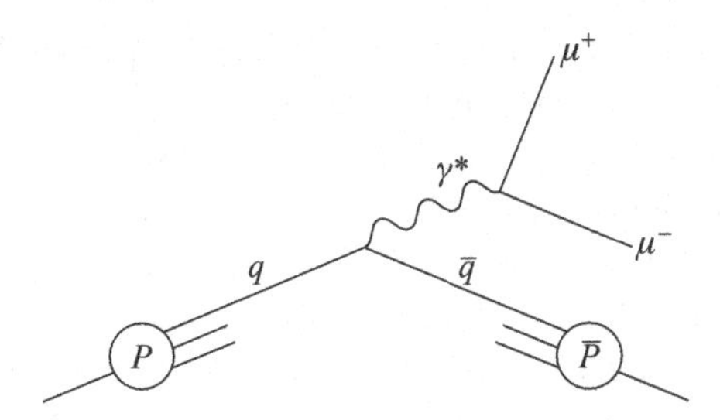

Fig. 3.18 Diagram showing $P\bar{P}$ annihilation with μ-pair production (the Drell–Yan process).

computations, in order to understand the background to possible non-Standard-Model physics has grown dramatically.

Fortunately, there has been significant progress in this arena. A critical aspect of the simplification has been a focus on *amplitudes*, i.e. obtaining the full scattering amplitude before squaring. A simplification of this sort is suggested by string theory, where, as we will see, one computes the scattering amplitude directly and, for example for closed string theories, there is just one diagram at each order. Initially investigators extracted QCD amplitudes from the low-energy limits of such processes, but it soon became clear how to obtain such simplifications directly in field theory. Elements contributing to this progress include the spinor helicity formalism. Here one trades four-vectors for products of spinors. For massless particles these spinors are themselves massless; working with them leads to vast simplifications. Progress in radiative corrections has relied heavily on unitarity, allowing one to compute higher-order diagrams by combining lower-order diagrams. Other important elements include trace-based color descriptions (much as we will see for large N in Chapter 5) and the use of on-shell recursion relations.

Processes involving the collisions of two particles that produce n particles have been calculated at leading order (LO) Amplitudes including one-loop corrections (next to leading order, or NLO) are known for $e^+e^- \rightarrow$ seven jets, $pp \rightarrow W +$ five jets, $pp \rightarrow$ five jets, $W + H$, $H + H$ and $\gamma\gamma$. These computations are now automated and public codes are available, such as GoSam, OpenLoops, Black Hat, Recola and Rocket. Amplitudes including two-loop corrections (NNLO) are known for three-jets production in e^+e^- annihilation and, in pp and $p\bar{p}$ collisions, for the production of Higgs bosons H, $W + H$, $H + H$ and photon pairs.

Suggested reading

There are a number of excellent texts on the Standard Model. *An Introduction to Quantum Field Theory* by Peskin and Schroeder (1995) provides a good introduction both to weak interactions and also to strong interactions, including deep inelastic scattering, parton distributions and the like. Other excellent texts include the books by Cheng and Li (1984), Donoghue *et al.* (1992), Pokorski (2000), Weinberg (1995), Bailin and Love (1993) and Cottingham and Greenwood (1998). More recently Srednicki (2007) and Schwartz (2013) introduced many of the more modern techniques for calculating QCD amplitudes, and the latter provides a more up-to-date survey of Standard Model computations generally. More detail about QCD amplitudes is presented in the lectures by Dixon (2013), who provides many additional references. An elegant calculation of the beta function in QCD, which uses the Wilson loop to determine the potential perturbatively, appears in the lectures of Susskind (1977). These lectures, as well as Wilson's original paper (1974) and the text of Creutz (1983), provide a good introduction to lattice gauge theory. An important subject which we have not discussed in this chapter is that of heavy-quark physics. This is experimentally important and theoretically accessible. A good introduction is provided in the book by Manohar and Wise (2000). The Particle Data Group website provides excellent reviews about a range of Standard Model (as well as Beyond the Standard Model) topics.

Exercises

(1) Add to the Lagrangian of Eq. (2.41) a term

$$\delta\mathcal{L} = \epsilon \, \mathrm{Tr}\, M \tag{3.112}$$

for small ϵ. Show that, in the presence of ϵ, the expectation values of the $\vec{\pi}$ fields are fixed and have a simple physical explanation. Compute the masses of the $\vec{\pi}$ fields directly from the Lagrangian.

(2) Verify Eqs. (3.48)–(3.56).

(3) Compute the mass of the Higgs field as a function of μ and λ (see Eqs. (2.70), (2.71)). Discuss the production of Higgs particles (you do not need to do detailed calculations, but should indicate the relevant Feynman graphs and make crude estimates at least of the cross sections) in e^+e^-, $\mu^+\mu^-$ and $P\bar{P}$ annihilation. Keep in mind that, because some of the Yukawa couplings are extremely small, there may be processes generated by loop effects that are bigger than processes that arise at tree level.

(4) Using the formula for the e^+e^- cross section, determine the branching ratio for decay of the Z into hadrons:

$$B(Z \to \text{ hadrons}) = \frac{\Gamma(Z \to \text{ hadrons})}{\Gamma(Z \to \text{ all})}. \tag{3.113}$$

4 The Standard Model as an effective field theory

The Standard Model has some remarkable properties. Among these, the renormalizable terms respect a variety of symmetries, all of which are observed to hold to a high degree in nature:

- baryon number symmetry;

$$Q \to e^{i\alpha/3}Q, \quad \bar{u} \to e^{-i\alpha/3}\bar{u}, \quad \bar{d} \to e^{-i\alpha/3}\bar{d}; \tag{4.1}$$

- three separate lepton number symmetries,

$$L_f \to e^{-i\alpha_f}L_f, \quad \bar{e}_f \to e^{i\alpha_f}\bar{e}_f. \tag{4.2}$$

It is not necessary to *impose* these symmetries. They are simply consequences of gauge invariance and the fact that there are only so many renormalizable terms that one can write down. These symmetries are said to be "accidental", since they do not seem to result from any deep underlying principle.

This is already a triumph. As we will see when we consider possible extensions of the Standard Model, this did not have to be the case. But this success raises the question: why should we impose the requirement of renormalizability?

4.1 Integrating out massive fields

In the early days of quantum field theory, renormalizability was sometimes presented as a sacred principle. There was a view that field theories were fundamental and should make sense in and of themselves. Much effort was devoted to understanding whether the theories still existed in the limit where the cutoff was taken to infinity.

But there was an alternative paradigm for understanding field theories, provided by Fermi's original theory of weak interactions. In this theory, weak interactions are described by a Lagrangian of the form

$$\mathcal{L}_{\text{weak}} = \frac{G_f}{\sqrt{2}} J^\mu J_\mu. \tag{4.3}$$

Here the currents J^μ are bilinear in the fermions; they include terms like $Q\sigma^\mu T^a Q^*$. This theory, like the Standard Model, was very successful. It took some time to actually determine the form of the currents but, for more than 40 years, all experiments in weak interactions could be summarized in a Lagrangian of this form. Only as the energies of bosons in e^+e^- experiments approached the Z boson mass were deviations observed.

The four-fermion theory is non-renormalizable. Taken seriously as a fundamental theory, it predicts violations of unitarity at TeV energy scales. But, from the beginning, the theory was viewed as an *effective* field theory, valid only at low energies. When Fermi first proposed the theory he assumed that the weak forces were caused by the exchange of particles – what we now know as the W and Z bosons.

4.1.1 Integrating out the *W* and *Z* bosons

Within the Standard Model we can derive the Fermi theory and also understand the deviations. A traditional approach is to examine the Feynman diagram of Fig. 4.1. This can be understood as a contribution to a scattering amplitude, but it is best understood here as a contribution to the effective action of the quarks and leptons. The currents of the Fermi theory are just the gauge currents which describe the coupling at each vertex. The propagator, in the limit of very small momentum transfer, is just a constant. In coordinate space this corresponds to a space–time δ-function; the interaction is local. The effect is just to give the four-fermion Lagrangian. One can consider the effects of small finite momentum by expanding the propagator in powers of q^2. This will give four-fermion operators with derivatives. These are suppressed by powers of M_W and their effects are very tiny at low energies. Still, in principle, they are there and in fact the measurement of such terms at energies that are a significant fraction of M_Z provided the first hints of the existence of the Z boson.

This effective action can also be derived in the path integral approach. Here we literally integrate out the heavy fields, the W and Z. In other words, for fixed values of the light fields, which we denote by ϕ, we perform a path integral over W and Z, expressing the result as an effective action for the ϕ fields (see Appendix C):

$$\int [d\phi] e^{iS_{\text{eff}}} = \int [d\phi] \int [dW_\mu][dW^*_\mu] \Delta_{FP} \times \exp\left[i \int d^4x (W^\dagger_\mu (\partial^2 + M^2_W) W^\mu + J^\mu W^\dagger_\mu + J^{\mu\dagger} W_\mu) \right]. \tag{4.4}$$

Here, for simplicity, we have omitted the Z particle. We have chosen the Feynman–'t Hooft gauge. The currents J^μ and $J^{\mu\dagger}$ are the usual weak currents. They are constructed out of

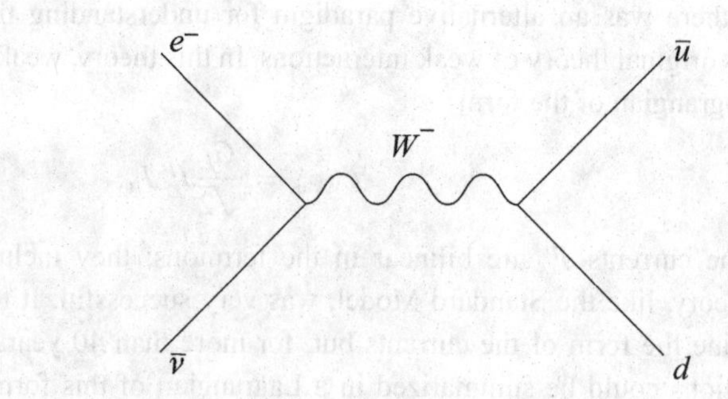

Fig. 4.1 Exchange of the massive W boson gives rise to the four-fermion interaction.

the various light fields, the quarks and leptons, which we have grouped, generically, into the set of fields ϕ. Written in this way, this is the most basic field theory path integral, and we are familiar with the result:

$$e^{iS_{\text{eff}}} = \exp\left[\int d^4x d^4y J^{\mu}(x)\Delta(x,y)J_{\mu}(y)\right]. \tag{4.5}$$

Here $\Delta(x,y)$ denotes the propagator for a scalar of mass M_W. In the limit $M \to \infty$ this is just a δ-function (one can compute this or see it directly from the path integral; if we neglect the derivative terms in the action, the propagator is just a constant in momentum space):

$$\Delta(x,y) = \frac{i}{M_W^2}\delta(x-y). \tag{4.6}$$

So

$$S_{\text{eff}} = \frac{1}{M_W^2}J^{\mu\dagger}J_{\mu}. \tag{4.7}$$

The lesson is that, up to the late 1970s, one could view QED + QCD + the Fermi theory as a perfectly acceptable theory of particle interactions. The theory had to be understood, however, as an effective theory, valid only up to an energy scale of order 100 GeV or so. Sufficiently precise experiments would require the inclusion of operators of dimension higher than four. The natural scale for these operators would be the weak scale. The Fermi theory is ultraviolet divergent. These divergences would be cut off at scales of order the W boson mass.

4.1.2 The simplest Higgs boson, obtained from integrating out other physics at higher energies

It is possible that the Higgs boson is precisely the doublet of the minimal Standard Model, and that upcoming experiments will simply verify that its couplings to quarks, leptons, gauge bosons and itself are exactly those expected. But they might show deviations and, in any case, at least at the LHC these measurements will probably be good only to the 5%–10% level, leaving some room for possible deviations.

If there is new physics at scales of order a few TeV or less, these might affect the properties of the Higgs. One simple possibility is that there is a second Higgs doublet. In other words, there might be two Higgs doublets, ϕ_1 and ϕ_2, with a potential $V(\phi_1,\phi_2)$ and Yukawa couplings to the quarks and leptons. There are strong restrictions on these couplings from low-energy physics (and especially from phenomena like K–$\bar{K}$ mixing). These are satisfied, for example, if one Higgs doublet couples only to up quarks and the other only to down quarks. We will see, for example, in Chapter 11 that in supersymmetric theories these conditions are automatically satisfied, at least at tree level. But there are now further restrictions from the success of the Standard Model in accounting for the properties of the observed Higgs.

To see how these constraints might be satisfied and to see the connection with notions from effective field theory we will focus on the mass matrix for the Higgs fields. Take the quadratic terms to have the form

$$V_m = \mu_1^2|\phi_1|^2 + \mu_2^2|\phi_2|^2 + m^2\phi_1\phi_2. \tag{4.8}$$

Suppose that the mass-squared matrix has one positive and one negative eigenvalue. Take ϕ to correspond to the negative eigenvalue and H to correspond to the positive eigenvalue:

$$V = -\mu^2|\phi|^2 + m^2|H|^2 + \text{quartic}. \tag{4.9}$$

If $m^2 \gg \mu^2$ then we can integrate out H to obtain a potential for ϕ. This limit is referred to as the *decoupling limit* of the two-Higgs-doublet model; if there is a second Higgs doublet, either this, or so-called "alignment", must hold for consistency with the present experimental constraints.

At tree level the potential for $|\phi|^2$ includes a negative quadratic term and a positive quartic. There are also sixth- and higher-order terms, suppressed by powers of m^2. Loop corrections involving the heavy field provide further modifications. The Yukawa couplings are also of Standard Model type. Again, at tree level, if

$$\phi_1 = \cos\alpha\phi - \sin_\alpha \phi_2 H, \quad \phi_2 = \sin\alpha\phi + \cos\alpha H \tag{4.10}$$

then ϕ_1 and ϕ_2 have the Yukawa couplings

$$\mathcal{L}_y = y_1\phi_1 Q\bar{u} + y_2\phi_2 Q\bar{d}, \tag{4.11}$$

where y_1 and y_2 are matrices in the space of generations. It follows that the Yukawa couplings to the up quarks are $y_u \cos\alpha$ and those to the down quarks are $y_d \sin\alpha$.

4.1.3 What might the Standard Model come from?

As successful as the Standard Model is, and despite the fact that it is renormalizable, it is likely that, like the four-fermion theory, it is the low-energy limit of some underlying, more fundamental, theory. In the second half of this book our model for this theory will be string theory. Consistent theories of strings, for reasons which are somewhat mysterious, are theories which describe general relativity and gauge interactions. Unlike field theory, string theory is finite. It does not require a cutoff for its definition. In principle, all physical questions have well-defined answers within the theory. If this is the correct picture for the origin of the laws of nature at extremely short distances, then the Standard Model is just its low-energy limit. When we study string theory we will understand in some detail how such a structure can emerge. For now, the main lesson we should take concerns the requirement of renormalizability: the Standard Model should be viewed as an effective theory, valid up to some energy scale Λ. Renormalizability is not a constraint we impose upon the theory; rather, we *should* include operators of dimension five or higher, with coefficients scaled by inverse powers of Λ. The value of Λ is an experimental question. From the success of the Standard Model, as we will see, we know that the cutoff is large. From string theory we might imagine that $\Lambda \approx M_p = 1.2 \times 10^{18}$ GeV. But, as we will now describe, we have experimental evidence that there is new physics which we must include at scales well below M_p. We will also see that there are theoretical reasons to believe that there should be new physics at TeV energy scales.

4.2 Lepton and baryon number violation; neutrino mass

We have remarked that, at the level of renormalizable operators, baryon number and lepton number are conserved in the Standard Model. Viewed as an effective theory, however, we should include higher-dimension operators with dimensionful couplings. We would expect such operators to arise, as in the case of the four-fermion theory, as a result of new phenomena and interactions at very high energy scales. The coefficients of these operators would be determined by this dynamics.

There would seem, at first, to be a vast array of possibilities for operators which might be included in the Standard Model Lagrangian. But we can organize the possible terms in two ways. First, if $M_{\rm bsm}$ is the scale of some new physics, operators of progressively higher dimension will be suppressed by progressively larger powers of $M_{\rm bsm}$. Second, the most interesting and readily detectable operators are those which violate the symmetries of the renormalizable Lagrangian. This is already familiar in the weak interaction theory. In the Standard Model the symmetries are precisely baryon number and lepton number.

The existence of the neutrino mass is now well established, and several parameters governing these masses are known. As we will see, if the only degrees of freedom involved are the three known two-component neutrinos, the structure of the leading lepton-number-violating operators is known. Several combinations of parameters are determined by the current data, and measuring the remaining ones is a central component of the international (and especially the US) high-energy physics program for the next few decades. Determining whether there are additional degrees of freedom is another major component.

4.2.1 Dimension five: lepton number violation and neutrino mass

To proceed systematically, we should write down operators of dimension five, six and so on. At the level of dimension five, we can write several terms which violate lepton number:

$$\mathcal{L} = \frac{1}{M_{\rm bsm}} \gamma_{f,f'} \phi\phi L_f L'_f + \text{c.c.} \tag{4.12}$$

Here ϕ again denotes the Higgs doublet and the indices are contracted suitably. With non-zero ϕ these terms give rise to neutrino masses. This type of mass term is usually called a *Majorana mass*. In nature these masses are quite small. For example, if $M_{\rm bsm} = 10^{16}$ GeV, which we will see is a plausible scale, then the neutrino masses would be of order 10^{-3} eV. In typical astrophysical and experimental situations, neutrinos are produced with energies of order MeV or larger, so it is difficult to measure these masses by studying the energy–momentum dispersion relation (very sensitive measurements of the end-point spectra beta decay are sensitive to electronvolt-scale neutrino masses). More promising are oscillation experiments, in which these operators give rise to transitions between one type of neutrino and another, which are similar to the phenomenon of K meson oscillations. Roughly speaking, in the β-decay of a d quark, say, one produces the neutrino partner of the electron. However, the mass (energy) eigenstate is a linear combination of the three types of neutrino (as we will see, typically it is principally a combination of two). So, experiments or observations downstream from the production point will measure processes in which neutrinos produce muons or taus. The oscillation periods are of order $E/\Delta m^2$. For MeV neutrinos and $\Delta m \sim 10^{-3}$ eV, this corresponds to distances of order kilometers, which is of interest for neutrinos in the atmosphere or those observed near nuclear reactors; for lighter neutrinos, effects at solar system scales become of interest.

Evidence that neutrinos do have non-zero masses and mixings comes from the study of neutrinos coming from the Sun (the solar neutrinos) and neutrinos produced in the upper atmosphere by cosmic rays (which produce pions that subsequently decay to muons and ν_μs, whose decays in turn produce electrons, ν_μs, and ν_es). Accelerator and reactor experiments have provided dramatic and beautiful evidence in support of this picture. It developed as a result of heroic experimental and theoretical work over more than four decades. The pioneering experiments were those of Ray Davis who, along with John Bahcall, conceived of neutrinos as a tool for the study of the interior of the Sun. His observation of neutrinos at rates lower than those expected in the standard solar model prompted the study of the mixing hypothesis and a range of other experiments. Later, studies of neutrinos from cosmic rays failed to yield the predicted fractions of ν_μs and ν_es. Dedicated studies of neutrinos from nuclear reactors and accelerators have provided further support for the mixing hypothesis and precise measurements of several parameters.

The masses and mixings of the neutrinos can be characterized by a unitary matrix, similar to the CKM matrix for the quarks, known as the Pontecorvo–Maki–Nakagawa–Sakata (PMNS) matrix. It can be parameterized as follows:

$$V = \begin{pmatrix} c_{12}c_{13} & s_{12}c_{13} & s_{13}e^{-i\delta} \\ -s_{12}c_{23} - c_{12}s_{23}s_{13}e^{i\delta} & c_{12}c_{23} - s_{12}c_{23}s_{13}e^{i\delta} & s_{23}c_{13} \\ s_{12}s_{23} - c_{12}c_{23}s_{13}e^{i\delta} & -c_{12}s_{23} - s_{12}c_{23}s_{13}e^{i\delta} & c_{23}c_{13} \end{pmatrix} \times \mathrm{diag}(1, e^{i\alpha_{21}/2}, e^{i\alpha_{31}/2}). \tag{4.13}$$

From the range of experiments described above, we know that

$$\begin{aligned} (\delta m^2)_{21} &= 7.54^{+0.26}_{-0.22} \times 10^{-5}\ \mathrm{eV}^2, \\ \delta m^2 = (\Delta m^2)_{31} - \Delta m^2_{12} &= 2.43 \pm 0.06 \times 10^{-3}\ \mathrm{eV}^2, \end{aligned} \tag{4.14}$$

where the second line holds if $m_1 < m_2$. With the same hierarchy, i.e. ordering of the masses, one has:

$$\begin{aligned} \sin^2\theta_{12} = 0.308 \pm 0.017, \quad \sin^2\theta_{23} &= 0.437^{.033}_{-0.023}, \quad \sin^2\theta_{13} = 0.0234^{+0.0020}_{-0.0019}, \\ \frac{\delta}{\pi} &= 1.39^{+0.38}_{-0.27}. \end{aligned} \tag{4.15}$$

More detail can be found in the references cited at the end of this chapter.

It is conceivable that these masses are not described by the Lagrangian of Eq. (4.12). Instead, the masses might be Dirac, by which one means that there might be additional degrees of freedom; by analogy to the $\bar{e}$ fields we could label these by $\bar{\nu}$, and they would have very tiny Yukawa couplings to the normal neutrinos. This would truly represent a breakdown of the Standard Model: even at low energies, we would be missing basic degrees of freedom. But this does not seem likely. If there are singlet neutrinos N, nothing would prevent them from gaining a Majorana mass m_N, so that

$$\mathcal{L}_{\rm Maj} = m_N NN. \tag{4.16}$$

As for the leptons and quarks, there would also be a coupling of ν to the field N. There would now be a *mass matrix* for the neutrinos, involving both N and ν. For simplicity, consider the case of just one generation. Then this matrix would have the form

$$m_\nu = \begin{pmatrix} m_N & yv \\ yv & 0 \end{pmatrix}. \tag{4.17}$$

Such a matrix has one large eigenvalue, of order m_n, and one small eigen value, of order y^2v^2/M_N. This provides a natural way to understand the smallness of the neutrino mass; it is referred to as the *seesaw mechanism*. Alternatively, we could consider of integrating out the right-handed neutrino and generating the operator of Eq. (4.12).

It seems more plausible that the observed neutrino mass is Majorana than Dirac, but this is a question that hopefully will be settled in time by experiments searching for *neutrinoless double beta decay*, $n+n \to p+p+e^- + e^-$. If it is Majorana, this suggests that there is another scale in physics that is well below the Planck scale. For, even if the new Yukawa couplings are of order one, the neutrino mass is of order

$$m_\nu = 10^{-5}\ \text{eV}(M_{\rm p}/\Lambda), \tag{4.18}$$

where Λ is another scale that is well below the Planck scale and M_P is the Planck mass. If the Yukawas are small, as are many of the quark Yukawa couplings, the scale can be much smaller.

4.2.2 Other symmetry-breaking dimension-five operators

There is another class of symmetry-violating dimension-five operators which can appear in the effective Lagrangian. These are electric and magnetic dipole moment operators. For example, the operator

$$\mathcal{L}_{\mu e} = \frac{e}{M_{\rm bsm}} F_{\mu\nu} \bar{\mu} \sigma^{\mu\nu} e \tag{4.19}$$

(we are using a four-component notation) would lead to the decay of the muon to an electron and a photon. Here $M_{\rm bsm}$ denotes the scale relating to Beyond the standard model physics. There are stringent experimental limits on such muon-number-violating processes, for example:

$$\text{branching ratio}(\mu \to e\gamma) < 1.2 \times 10^{-11}. \tag{4.20}$$

Other operators of this type include those which would generate lepton-number-violating τ decays, on which the limits are far less stringent.

In the Standard Model, CP is an approximate symmetry. We have explained that three generations of quarks are required to violate CP within the Standard Model. So, amplitudes which violate CP must involve all three generations and are typically highly suppressed. From an effective-Lagrangian viewpoint, if we integrate out the W and Z bosons then the operators which violate CP are of dimension six and typically have coefficients suppressed by quark masses and mixing angles, as well as loop factors. As a result, new physics at relatively modest scales has the potential for dramatic effects. Electric dipole moment operators for quarks or leptons would arise from operators of the form

$$\mathcal{L}_{\text{d}} = \frac{em_q}{M_{\text{bsm}}^2}\tilde{F}_{\mu\nu}\bar{q}\sigma^{\mu\nu}q + \text{ c.c.}, \tag{4.21}$$

where

$$\tilde{F}_{\mu\nu} = \frac{1}{2}\epsilon_{\mu\nu\rho\sigma}F^{\rho\sigma}. \tag{4.22}$$

Here $\epsilon_{\mu\nu\rho\sigma}$ is the completely antisymmetric tensor with four indices; $\epsilon_{0123} = 1$. The presence of the ϵ symbol is the signal of CP violation, as the reader can check. In the non-relativistic limit, this is $\vec{\sigma} \cdot \vec{E}$. These would lead, for example, to a neutron electric dipole moment of order

$$d_{\text{n}} = \frac{e}{M_{\text{bsm}}}. \tag{4.23}$$

Searches for such dipole moments set a limits $d_{\text{n}} < 10^{-25}e$ cm. So, unless there is some source of suppression, M_{bsm} in CP-violating processes is larger than about 10^2 TeV.

4.2.3 Irrelevant operators and high-precision experiments

There are a number of dimension-five operators on which it is possible to set somewhat less stringent limits, and in one case there is a possible discrepancy. Corrections to the muon magnetic moment could arise from

$$\mathcal{L}_{g-2} = \frac{e}{M_{\text{bsm}}}F_{\mu\nu}\bar{\mu}\sigma^{\mu\nu}\mu + \text{c.c.}, \tag{4.24}$$

where $F_{\mu\nu}$ is the electromagnetic field (in terms of the fundamental $SU(2)$ and $U(1)$ fields, one can write similar gauge-invariant combinations which reduce to this at low energies). The muon magnetic moment has been measured to extremely high precision, and its Standard Model contribution is calculated with comparable precision; as of the time of writing there is a 2.6σ discrepancy between the two. Whether this reflects new physics is uncertain. We will encounter one candidate for this physics when we discuss supersymmetry.

There are other operators on which we can set TeV-scale limits. The success of QCD in describing jet physics allows one to constrain four-quark operators which would give rise to a hard component in the scattering amplitude. Such operators might arise, for example, if quarks were composite. Constraints on flavor-changing processes provide tight constraints on a variety of operators. Operators such as

$$\mathcal{L}_{\text{fc}} = \frac{1}{M_{\text{bsm}}^2}s\sigma^{\mu}d^* s\sigma_{\mu}d^* \tag{4.25}$$

(where we have switched to a two-component notation) would contribute to $K\bar{K}$ mixing and other processes. This would constrain M_{bsm} to be larger than 100 TeV or so. Any new physics at the TeV scale must explain why such an operator is so severely suppressed.

4.2.4 Dimension-six operators: proton decay

Proceeding to dimension six we can write down numerous terms which violate baryon number, as well as additional lepton-number-violating interactions:

$$\mathcal{L}_{\text{bv}} = \frac{1}{M_{\text{bsm}}^2} Q\sigma^{\mu}\bar{u}^* L\sigma_{\mu}\bar{d}^* + \cdots . \tag{4.26}$$

This can lead to processes such as $p \to \pi e$. Experiments deep underground set limits of order 10^{33} years on this process. Correspondingly, the scale M_{bsm} must be larger than 10^{15} GeV.

So, viewing the Standard Model as an effective-field theory, we see that there are many possible non-renormalizable operators which might appear but most have scales which are tightly constrained by experiment. One might hope – or despair – that the Standard Model will provide a complete description of nature up to scales many orders of magnitude larger than we can hope to probe in experiment.

However, there are a number of reasons to think that the Standard Model is incomplete, and at least one which suggests that it will be significantly modified at scales not far above the weak scale.

4.3 Challenges for the Standard Model

On the one hand, the Standard Model is tremendously successful. With the discovery of the Higgs particle, it can be said to describe the physics of strong, weak and electromagnetic interactions with great precision to energies of order 100 GeV or distances as small as 10^{-17} cm. It explains why baryon number and the separate lepton numbers are conserved, with only one assumption: there is no interesting new physics up to some high-energy scale. As of the end of the 8 TeV run at the LHC, there are almost no discrepancies between theory and experiment.

On the other hand, the Standard Model cannot be a complete theory. The existence of neutrino mass requires at least additional states (if these masses are Dirac), and more likely some new physics at a high-energy scale which accounts for the Majorana neutrino masses. This scale is probably not larger than 10^{16} GeV, well below the Planck scale. The existence of gravity means that there is certainly something missing from the theory. The plethora of parameters – there are 19, counting those of the minimal Higgs sector and the θ parameter (see the next subsection) – suggests that there is a deeper structure. More directly, features of the big bang cosmology which are now well established cannot be accommodated within the Standard Model.

4.3.1 The strong CP problem

In the Standard Model there is a puzzle even at the level of dimension-four operators. Consider

$$\mathcal{L}_{\theta} = \theta F\tilde{F}, \tag{4.27}$$

where θ is a dimensionless parameter and

$$\tilde{F}_{\mu\nu} = \frac{1}{2}\epsilon_{\mu\nu\rho\sigma}F^{\rho\sigma}. \tag{4.28}$$

We usually ignore such operators because classically they are inconsequential; they are total derivatives and do not modify the equations of motion. In a $U(1)$ theory, for example,

$$F\tilde{F} = 2\epsilon^{\mu\nu\rho\sigma}\,\partial_\mu A_\nu \partial_\rho A_\sigma = 2\partial_\mu(\epsilon^{\mu\nu\rho\sigma} A_\nu \partial_\rho A_\sigma). \tag{4.29}$$

In the next chapter we will see that this has a non-Abelian generalization, but that, despite constituting a total divergence, these terms have real effects at the quantum level. In QCD they turn out to be highly constrained. From the limits on the neutron electric dipole moment, we will show in Chapter 5 that $\theta < 10^{-9}$. This is the first real puzzle we have encountered. Why is it such a small dimensionless number? Answering this question, as we will see in Chapter 5, may point to new physics, likely at some very high energy scale.

4.3.2 The hierarchy problem and the question of naturalness

The second very puzzling feature in the Standard Model is the Higgs field. The fact that the model seems to be described by a single Higgs scalar is itself puzzling. We could have included several doublets or perhaps tried to explain the breaking of the gauge symmetry through some more complicated dynamics, as we will discuss in Chapter 8. But there is a more serious question associated with fundamental scalar fields, raised long ago by Ken Wilson. This problem is often referred to as the *hierarchy problem* or the *naturalness problem*.

Consider, first, the one-loop corrections to the electron mass in QED. These are logarithmically divergent. In other words,

$$\delta m = a m_0 \frac{\alpha}{4\pi} \ln \Lambda. \tag{4.30}$$

We can understand this result in simple terms. In the limit $m_0 \to 0$ the theory has an additional symmetry, a chiral symmetry, under which e and $\bar{e}$ transform by independent phases. This symmetry forbids a mass term, so the result must be linear in the (bare) mass. So, on dimensional grounds, any divergence is at most logarithmic. This actually resolves a puzzle of *classical* electrodynamics. Lorentz modeled the electron as a uniformly charged sphere of radius a. As $a \to 0$ the electrostatic energy diverges. In modern terms, we would say that we know a is smaller than 10^{-17} cm, corresponding to a self-energy far larger than the electron mass itself. But we see that in the quantum theory the cutoff occurs at a scale of order the electron mass, and there is no large self-energy correction.

For scalars, however, there is no such symmetry and corrections to masses are quadratically divergent. One can see this easily for the Higgs self-coupling, which gives rise to a mass correction of the form

$$\delta m^2 = \lambda \int \frac{d^4k}{(2\pi)^4(k^2+m^2)}, \tag{4.31}$$

with similar corrections from the top quark loop correction, gauge loops, and others. If we view the Standard Model as an effective-field theory, these integrals should be cut off at a scale where new physics enters. We have argued that this might occur at, say, 10^{14} GeV. But in this case the correction to the Higgs mass would be gigantic compared with the Higgs mass itself. Given that $y_t^2 > \lambda$, we would expect even larger effects from top quark loops.

It is hard to see how this puzzle can be resolved without introducing new physics at a scale not much larger than 1 TeV. Exploring candidates for this new physics will be one of the major subjects of this book. After discussing another fine tuning problem in our current understanding of the laws of nature, we will elevate these concerns to a principle that we might wish to impose on our theories: the principle of naturalness.

4.3.3 The universe: the baryon density, dark matter and dark energy

As we will discuss in Chapter 18, we have good evidence that the energy density of the universe occurs largely in unfamiliar forms: about 27% in non-baryonic pressureless matter (dark matter) and about 68% in some form having with negative pressure (dark energy), with only the remaining 5% comprising ordinary baryons. The dark energy is likely to be a cosmological constant (of which more later).

As we will discuss, particularly in Chapter 19, we might hope to understand the dark matter in terms of some type(s) of new particle. A particle with mass of order 1 TeV (give or take factors of 10) and roughly weak-interaction cross sections would be produced in suitable quantities in the early universe. Beyond the hierarchy problem, this might be another pointer to new physics in the TeV energy range. Alternatively the axion, a much lighter and more weakly interacting particle proposed to solve the strong CP problem, might play this role and would lead to different types of experimental signals.

The baryon density, as we will also see, cannot arise from the Standard Model itself. We will consider a number of possible new physics mechanisms by which it might arise. Without strong assumptions about the history of the universe, it is difficult to pin down the relevant energy scale.

The dark energy raises puzzles which do not point in any obvious way to a particular energy scale. If the dark energy is a cosmological constant then this represents, from the perspective of our effective Lagrangian, a term of dimension zero, whose coefficient has dimensions of $(\text{mass})^4$. Dimensional analysis would suggest that it should be of order the largest possible scale to the fourth power. If this is the Planck scale then dimensional analysis fails by 120 orders of magnitude. In a sense our analysis of the effective action seems back to front. We began with a discussion of dimension-five and dimension-six operators, operators which are irrelevant, and then turned our attention to the Higgs mass, a dimension-two, relevant, operator. We still have not considered the most relevant operator of all, the unit operator.

In quantum field theory, consistently with dimensional analysis, this energy is *quartically* divergent; it is the first divergence one encounters in any quantum field theory textbook. At one loop it is given by an expression of the form

$$\Lambda = \sum_i (-1)^F \int \frac{d^3k}{(2\pi)^3} \frac{1}{2}\sqrt{k^2 + m_i^2}, \tag{4.32}$$

where the sum is over all particle species (including spins). This is just the sum of the zero-point energies of the oscillators of each momentum. If one cuts this off, again at 10^{14} GeV, one gets a result of order

$$\Lambda = 10^{54}\ \text{GeV}^4. \tag{4.33}$$

The measured value of the dark-energy density is by contrast,

$$\Lambda = 10^{-47}\ \text{GeV}^4. \tag{4.34}$$

This wide discrepancy is probably one of the most troubling problems facing fundamental physics today.

4.4 The naturalness principle

Both the Higgs mass and the cosmological constant appear to be finely tuned; they are much smaller than the values we would have guessed from dimensional analysis, and we have seen that quantum corrections are likely to be much larger than the observed parameters themselves. In contrast, we have noted that the electron mass (and the masses of the leptons and quarks more generally), while surprisingly small, does not receive large quantum corrections.

While many physicists were uncomfortable with these tunings, it was 't Hooft who framed this question in terms of a principle, which he dubbed the *naturalness condition*. He argued that a parameter in nature should be small only if the underlying theory becomes more symmetric as the parameter tends to zero. The electron mass in QED provides an illustration of this principle: as it tends to zero, the theory, as we have described, develops a new symmetry, a $U(1)$ chiral symmetry. All the small Yukawa couplings of the Standard Model are similarly natural. We will see that the small masses (relative to the Planck scale) of the hadrons are also compatible with the principle.

Our two puzzling quantities do not satisfy this criterion. The Standard Model does not become more symmetric if one sets the Higgs mass to zero. Similarly, general relativity (as we will see) does not become more symmetric as the cosmological constant tends to zero. The small value of the θ parameter, which violates CP conservation in strong interactions, also poses puzzles. Because the Standard Model violates CP even in the absence of θ, this would seem another violation of naturalness.

These issues each suggest that there should be some new degrees of freedom, or symmetries, or both, beyond those of the Standard Model. This has motivated a broad range of proposals for new physics. These will be the subject of much of this book. But, in recent years, at least one alternative picture for how the parameters of the Standard Model might arise has gained traction. We will consider this idea, known as the *landscape*, in Chapter 30.

4.5 Summary: successes and limitations of the Standard Model

Overall, we face a tension between the striking successes of the Standard Model and its limitations. On the one hand, the model successfully accounts for almost all the phenomena observed in accelerators. On the other hand, it fails to account for some of the most basic phenomena of the universe: dark matter, dark energy and the existence of gravity itself. As a theoretical structure, it also explains successfully what might be viewed as mysterious conservation laws: baryon number and lepton number. But it has 17 parameters – 16 of which are pure numbers, with values which range "all over the map". The rest of this book explores possible solutions of these puzzles, and their implications for particle physics, astrophysics and cosmology.

Suggested reading

The texts by by Peskin and Schroeder (1995) and Schwartz (2014) provide a good introduction both to weak interactions and also to the strong interactions; it includes deep inelastic scattering, parton distributions and the like. Other excellent texts include the books by Cheng and Li (1984), Donoghue *et al.* (1992), Pokorski (2000) and Bailin and Love (1993) among many others. For summaries of data on neutrino oscillations, the Particle Data Group website provides up-to-date reviews; the text by Barger *et al.* (2012) provides a first-rate pedagogical introduction.

5 Anomalies, instantons and the strong CP problem

While perturbation theory is a powerful and useful tool in understanding field theories, for our exploration of physics beyond the Standard Model an understanding of non-perturbative physics will be crucial. There are many reasons for this.

1. One of the great mysteries of the Standard Model is non-perturbative in nature: the smallness of the θ parameter.
2. Strongly interacting field theories will figure in many proposals to understand other mysteries of the Standard Model.
3. The interesting dynamical properties of supersymmetric theories, both those directly related to possible models of nature and those which provide insights into broad physics issues, are non-perturbative in nature.
4. If string theory describes nature, non-perturbative effects are necessarily of critical importance.

We have introduced lattice gauge theory, which is perhaps our only tool for doing systematic calculations in strongly coupled theories. But, as a tool, its value is quite limited. Only a small number of calculations are tractable in practice, and the difficult numerical challenges sometimes obscure the underlying physics. Fortunately, there is a surprising amount that one can learn from symmetry considerations, from semiclassical arguments and from our experimental knowledge of one strongly coupled theory, QCD. In each of these, an important role is played by the phenomena known as anomalies and, related to these, a set of semiclassical field configurations known as *instantons*.

Usually, the term "anomaly" is used to refer to the quantum mechanical violation of a symmetry which is valid classically. Instantons are finite-action solutions of the Euclidean equations of motion, typically associated with tunneling phenomena. Anomalies are crucial to understanding the decay of the π^0 in QCD. Anomalies and instantons account for the absence of a ninth light pseudoscalar meson in the hadron spectrum. Within the weak-interaction theory, anomalies and instantons lead to violations of baryon and lepton number; these effects are unimaginably tiny at the current time but were important in the early universe. The absence of anomalies in gauge currents is important to the consistency of theoretical structures, including both field theories and string theories. The cancelation of anomalies within the Standard Model itself is quite non-trivial. Similar constraints on possible extensions of the Standard Model will be very important. The θ parameter of QCD was mentioned in the previous chapter. The θ term seems innocuous, but, owing to anomalies and instantons, its potential effects are real. Because the θ term violates CP, they are also dramatic. The problem of the smallness of the θ parameter – the *strong CP problem* – forcibly suggests new phenomena beyond the Standard Model, and this will be a recurring theme in this book. In the present chapter we explain how anomalies arise and some of the roles which they play. The discussion is meant to provide the reader with a good working knowledge of these subjects, but it is not encyclopedic. A guide to texts and reviews on the subject appears at the end of the chapter.

5.1 The chiral anomaly

Before discussing real QCD, let us consider a non-Abelian gauge theory theory, with only a single flavor of quark. Before making any field redefinitions, the Lagrangian takes the form:

$$\mathcal{L} = -\frac{1}{4g^2}F_{\mu\nu}^2 + i\bar{q}D^\mu\sigma_\mu\bar{q}^* + iqD^\mu\sigma_\mu q^* + m\bar{q}q + m^*\bar{q}^*q^*. \tag{5.1}$$

The Lagrangian is written here in terms of two-component fermions (see Appendix A). The fermion mass need not be real:

$$m = |m|e^{i\theta}. \tag{5.2}$$

In this chapter it will sometimes be convenient to work with four-component fermions, and it is valuable to make contact with this language in any case. In terms of these, the mass contribution is

$$\mathcal{L}_m = (\mathrm{Re}\, m)\, \bar{q}q + (\mathrm{Im}\, m)\, \bar{q}\gamma_5 q. \tag{5.3}$$

In order to bring this mass contribution to the conventional form, with no γ_5s, one could try to redefine the fermions; switching back to the two-component notation we have

$$q \to e^{-i\theta/2}q, \quad \bar{q} \to e^{-i\theta/2}\bar{q}. \tag{5.4}$$

However, in field theory transformations of this kind are potentially fraught with difficulties because of the infinite number of degrees of freedom.

A simple calculation uncovers one of the simplest manifestations of an anomaly. Suppose, first, that m is very large, $m \to M$. In that case we need to integrate out the quarks and obtain a low-energy effective theory. To do this, we study the path integral (see Appendix C)

$$Z = \int [dA_\mu] \int [dq][d\bar{q}]e^{iS}. \tag{5.5}$$

Suppose that $M = e^{i\theta}|M|$. In order to make M real, we can again make the transformations $q \to qe^{-i\theta/2}, \bar{q} \to \bar{q}e^{-i\theta/2}$ (in four-component language, this is $q \to e^{-i\theta/2\gamma_5}q$)). The result of integrating out the quark, i.e. of performing the path integral over q and $\bar{q}$, can be written in the form

$$Z = \int [dA_\mu] \int e^{iS_{\mathrm{eff}}}. \tag{5.6}$$

Here S_{eff} is the effective action which describes the interactions of gluons at scales well below M.

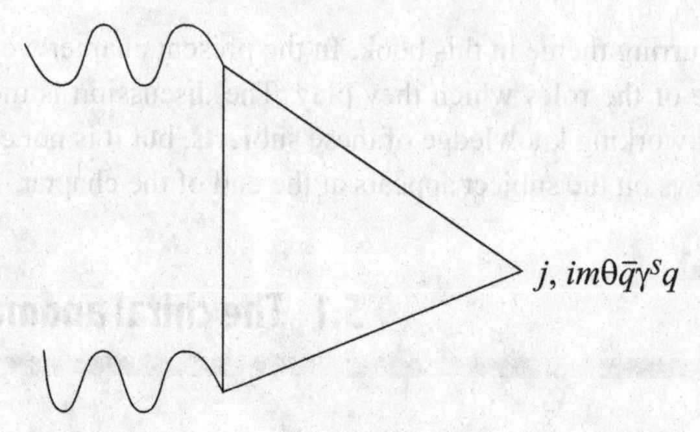

Fig. 5.1 The triangle diagram associated with the four-dimensional anomaly. At the right-hand vertex, one has insertions of the axial current and the chiral density.

Because the field redefinition which eliminates θ amounts to just a change of variables in the path integral, one might expect that there can be no θ-dependence in the effective action. But this is not the case. To see this, suppose that θ is small and, instead of redefining the field treat the θ term as a small perturbation by expanding the exponential. Now consider a term in the effective action with two external gauge bosons. This is obtained from the Feynman diagram in Fig. 5.1. The corresponding term in the action is given by (see Eq. (2.17))

$$\delta\mathcal{L}_{\text{eff}} = -i\frac{\theta}{2}M\,\mathrm{Tr}(T^aT^b)\int\frac{d^4p}{(2\pi)^4}\,\mathrm{Tr}\left(\gamma_5\frac{1}{\not p+\not k_1-M}\not\epsilon_1\frac{1}{\not p-M}\not\epsilon_2\frac{1}{\not p-\not k_2-M}\right). \tag{5.7}$$

Here, the k_is are the momenta of the two gluons, while the ϵs are their polarizations and a and b are their color indices. Introducing Feynman parameters and shifting the p integral gives

$$\begin{aligned}\delta\mathcal{L}_{\text{eff}} = &-i\theta g^2M\,\mathrm{Tr}(T^aT^b)\int d\alpha_1 d\alpha_2\int\frac{d^4p}{(2\pi)^4}\mathrm{Tr}\Big(\gamma_5(\not p-\alpha_1\not k_1+\alpha_2\not k_2+\not k_1+M)\\ &\times\frac{\not\epsilon_1(\not p-\alpha_1\not k_1+\alpha_2\not k_2+M)\not\epsilon_2(\not p-\alpha_1\not k_1+\alpha_2\not k_2-\not k_2+M)}{[p^2-M^2+O(k_i^2)]^3}\Big).\end{aligned} \tag{5.8}$$

For small k_i we can neglect the k-dependence of the denominator. The trace in the numerator is easy to evaluate, since we can drop terms linear in p. This gives, after performing the integrals over the αs,

$$\delta\mathcal{L}_{\text{eff}} = g^2M^2\theta\,\mathrm{Tr}(T^aT^b)\,\epsilon_{\mu\nu\rho\sigma}k_1^\mu k_2^\nu\epsilon_1^\rho\epsilon_2^\sigma\int\frac{d^4p}{(2\pi)^4}\frac{1}{(p^2-M^2)^3}. \tag{5.9}$$

This corresponds to a term in the effective action, which, after performing the integral over p and including a combinatoric factor two from the different ways to contract the gauge bosons, is given by

$$\delta\mathcal{L}_{\text{eff}} = \frac{1}{32\pi^2}\theta\,\mathrm{Tr}(F\tilde{F}). \tag{5.10}$$

Now why does this happen? On the one hand, at the level of the path integral the transformation would seem to amount to a simple change of variables, and it is hard to see why this should have any effect. On the other hand, if one examines the diagram of Fig. 5.1 then one sees that it contains terms which are linearly divergent and thus it should be regulated. A simple way to regulate this diagram is to introduce a Pauli–Villars regulator, which means that one subtracts off a corresponding amplitude with some very large mass Λ. However, our expression above is independent of Λ. So the θ-dependence from the regulator fields cancels that of Eq. (5.10). This sort of behavior is characteristic of an anomaly.

Consider now the case where $m \ll \Lambda_{\rm QCD}$. In this case we should not integrate out the quarks, but we still need to take into account the regulator diagrams. So, if we redefine the fields so, that the quark mass is real (γ_5-free, in the four-component description), the low-energy theory contains light quarks and the θ term of Eq. (5.10).

We can describe this in a fashion which indicates why this is referred to as an *anomaly*. For small m the classical theory has an approximate symmetry under which

$$q \to e^{i\alpha} q, \quad \bar{q} \to e^{i\alpha} \bar{q} \tag{5.11}$$

(in four-component language, $q \to e^{i\alpha\gamma_5} q$). In particular we can define a current

$$j_5^\mu = \bar{q}\gamma_5\gamma_\mu q \tag{5.12}$$

and, classically,

$$\partial_\mu j_5^\mu = m\bar{q}\gamma_5 q. \tag{5.13}$$

Under a transformation by an infinitesimal angle α one would expect that

$$\delta L = \alpha\partial_\mu j_5^\mu = m\alpha\bar{q}\gamma_5 q. \tag{5.14}$$

But the divergence of the current contains another, m-independent, term:

$$\partial_\mu j_5^\mu = m\bar{q}\gamma_5 q + \frac{1}{32\pi^2} F\tilde{F}. \tag{5.15}$$

The first term follows from the equations of motion. To see why the second term is present, we will study a three-point function involving the current and two gauge bosons A_μ and will ignore the quark mass:

$$\Gamma^{AAj} = T\langle \partial_\mu j^{5\mu} A_\rho A_\sigma \rangle. \tag{5.16}$$

This is essentially the calculation we encountered above. Again the diagram is linearly divergent and requires regularization. Let us first consider the graph without the regulator mass. The graph of Fig. 5.1 actually implies two graphs, because we must include the interchange of the two external gluons. The combination is easily seen to vanish, by the sorts of manipulations one usually uses to prove Ward identities:

$$\frac{g^2}{(2\pi)^4} \int d^4p \, \mathrm{Tr}\left(\not{q}\gamma_5 \frac{1}{\not{p} + \not{k}_1} \not{\epsilon}_1 \frac{1}{\not{p}} \not{\epsilon}_2 \frac{1}{\not{p} - \not{k}_2} + (1 \leftrightarrow 2) \right). \tag{5.17}$$

Writing

$$\not{q}\gamma_5 = -\gamma_5(\not{k}_1 + \not{p}) - (\not{p} - \not{k}_2)\gamma_5 \tag{5.18}$$

and using the cyclic property of the trace, one can cancel a propagator in each term. This leaves

$$\int d^4 p \, \mathrm{Tr}\left(-\gamma_5 \, \not{\epsilon}_1 \frac{1}{\not{p}} \, \not{\epsilon}_2 \frac{1}{\not{p} - \not{k}_2} - \gamma_5 \frac{1}{\not{p} + \not{k}_1} \, \not{\epsilon}_1 \frac{1}{\not{p}} \, \not{\epsilon}_2 + (1 \leftrightarrow 2) \right). \tag{5.19}$$

Now making the shift $p \to p + k_2$ in the first term and $p \to p + k_1$ in the second, one finds a pairwise cancelation.

These manipulations, however, are not reliable. In particular, in a highly divergent expression the shifts do not necessarily leave the result unchanged. With a Pauli–Villars regulator the integrals are convergent and the shifts are reliable, but the regulator diagram is non-vanishing and gives the anomaly equation above. One can see this by a direct computation or relate it to our previous calculation, including the masses for the quark and noting that $\not{q}\gamma_5$, in the diagrams with massive quarks, can be replaced by $M\gamma_5$.

This anomaly can be derived in a number of other ways. One can define, for example, the current by *point splitting*, i.e. separating the two fields in the current by an amount ϵ and inserting a Wilson line to ensure gauge invariance.

$$j_5^\mu = \bar{q}(x+\epsilon) \exp\left(i \int_x^{x+\epsilon} dx^\mu A_\mu \right) q(x). \tag{5.20}$$

Because the operators in quantum field theory are singular at short distances, the Wilson line makes a finite contribution. Expanding the exponential carefully, one recovers the same expression for the current. We will do this shortly in two dimensions, leaving the four-dimensional case for the end-of-chapter exercises. A beautiful derivation, closely related to that performed above, is due to Fujikawa. Here one considers the anomaly as arising from a lack of invariance of the path integral measure. One carefully evaluates the Jacobian associated with the change of variables $q \to q(1 + i\gamma_5\alpha)$ and shows that it yields the same result. We will do a calculation along these lines in a two-dimensional model shortly, leaving the four-dimensional case for the exercises.

5.1.1 Applications of the anomaly in four dimensions

The anomaly has a number of important consequences for real physics.

- π^0 *decay* The divergence of the axial isospin current

$$(j_5^3)^\mu = \bar{u}\gamma_5\gamma^\mu \bar{u} - \bar{d}\gamma_5\gamma^\mu d \tag{5.21}$$

has an anomaly due to electromagnetism. This gives rise to a coupling of the π^0 to two photons, and the correct prediction of the lifetime was one of the early triumphs of the color theory of quarks. The computation of the π^0 decay rate appears in the exercises.

- *Anomalies in gauge currents signal an inconsistency in a theory* They mean that gauge invariance, which is crucial to the whole structure of gauge theories (e.g. to the fact that they are simultaneously unitary and Lorentz invariant) is lost. The absence of gauge anomalies is one of the striking ingredients of the Standard Model, and it is also crucial in extensions such as string theory.

- *The anomaly considered here*, as we have indicated above, accounts for the absence of a ninth axial Goldstone boson in the QCD spectrum.

5.1.2 Return to QCD

What we have just learned is that if in our simple model above we require that the quark masses are real then we must allow for the possible appearance, in the Lagrangian of the Standard Model, of the θ term in Eq. (5.10). In weak interactions this term does not have physical consequences. At the level of the renormalizable terms, we have seen that the theory respects separate B and L symmetries; B, for example, is anomalous. So, if we simply redefine the quark fields by a B transformation, we can remove θ from the Lagrangian.

For the θ angles of QCD and QED we have no such symmetry. In the case of QED we do not really have a non-perturbative definition of the theory, and the effects of θ are hard to assess, but one might expect that, when embedded in any consistent structure (such as a grand unified theory (GUT) or string theory) they will be very small, possibly zero. As we saw, $F\tilde{F}$ gives a total divergence. The right-hand side of Eq. (4.24) is not gauge invariant, however, so one might imagine that it could be important. But, as long as A falls off at least as fast as $1/r$ (i.e. F falls off faster than $1/r^2$), the surface term behaves as $1/r^4$ and so vanishes.

In the case of non-Abelian gauge theories, the situation is more subtle. It is again true that $F\tilde{F}$ can be written as a total divergence:

$$F\tilde{F} = \partial^\mu K_\mu, \quad K_\mu = \epsilon_{\mu\nu\rho\sigma}\left(A^a_\nu F^a_{\rho\sigma} - \frac{2}{3} f^{abc} A^a_\nu A^b_\rho A^c_\sigma\right). \tag{5.22}$$

However, the statement that F falls off faster than $1/r^2$ does not permit an equally strong statement about A. We will see shortly that there are finite-action classical solutions for which $F \sim 1/r^4$ but $A \to 1/r$, so that the surface term cannot be neglected. These solutions are instantons. This is the reason that θ can have real physical effects.

5.2 A two-dimensional detour

There are many questions in four dimensions which we cannot answer except by using numerical lattice calculation. These include the problem of dimensional transmutation and the effects of the anomaly on the hadron spectrum. There is a class of models in two dimensions which are asymptotically free and in which one can study these questions in a controlled approximation. Two dimensions often form a poor analog for four but, for some of the issues we are facing here, the parallels are extremely close. In these two-dimensional examples the physics is more manageable, but still rich. In four dimensions, the calculations are qualitatively similar; they are only more difficult because the Dirac algebra and the various integrals are more involved.

5.2.1 The anomaly in two dimensions

First we investigate the anomaly in the quantum electrodynamics of a massless fermion in two dimensions; this will be an important ingredient in the full analysis. The point-splitting method is particularly convenient here. Just as in four dimensions, we write

$$j_5^\mu = \bar{\psi}(x+\epsilon)\exp\left(i\int_x^{x+\epsilon} A_\rho dx^\rho\right)\gamma^\mu\gamma^5\psi(x). \tag{5.23}$$

Naively, one can set $\epsilon = 0$ and then the divergence vanishes by the equations of motion. In quantum field theory, however, products of operators become singular as the operators come close together. For very small ϵ we can pick up the leading singularity in the product of $\psi(x+\epsilon)\psi(x)$ by using the operator product expansion (OPE). The OPE states that the product of two operators at short distances can be written as a series of local operators of progressively higher dimension, with coefficients that are less and less singular. For our case this means that

$$\bar{\psi}(x+\epsilon)\gamma^\mu\gamma^5\psi(x) = \sum \frac{c_n}{\epsilon^{1-n}}\mathcal{O}_n(x), \tag{5.24}$$

where $\mathcal{O}_n$ is an operator of dimension n. The leading term comes from the unit operator. To evaluate its coefficient we can take the vacuum expectation value of both sides of this equation. On the left-hand side, this is just the propagator.

It is not hard to work out the fermion propagator in coordinate space in two dimensions. For simplicity we work with space-like separations, so that we can Wick-rotate to Euclidean space. Start with the scalar propagator

$$\begin{aligned}\langle\phi(x)\phi(0)\rangle &= \int \frac{d^2p}{(2\pi)^2}\frac{1}{k^2}e^{-ip\cdot x}\\ &= \frac{1}{2\pi}\ln(|x|\mu),\end{aligned} \tag{5.25}$$

where μ is an infrared cutoff. (When we come to string theory this propagator, with its infrared sensitivity, will play a crucial role.) Correspondingly, the fermion propagator is

$$\langle\bar{\psi}(x+\epsilon)\psi(x)\rangle = \partial\!\!\!/\langle\phi(x)\phi(0)\rangle = \frac{1}{2\pi}\frac{\epsilon\!\!/}{\epsilon^2}. \tag{5.26}$$

Expanding the factor in the exponential to order ϵ gives

$$\partial_\mu j_5^\mu = \text{classical term} + \frac{i}{2\pi}\partial_\mu\epsilon_\rho A^\rho\ \mathrm{Tr}\left(\frac{\epsilon\!\!/}{\epsilon^2}\gamma^\mu\gamma^5\right). \tag{5.27}$$

Evaluating the trace gives $\epsilon_{\mu\nu}\epsilon^\nu$; averaging ϵ over angles ($\langle\epsilon_\mu\epsilon_\nu\rangle = \frac{1}{2}\eta_{\mu\nu}\epsilon^2$) yields

$$\partial_\mu j_5^\mu = \frac{1}{2\pi}\epsilon_{\mu\nu}F^{\mu\nu}. \tag{5.28}$$

This is parallel to the situation in four dimensions. The divergence of the current is itself a total derivative:

$$\partial_\mu j_5^\mu = \frac{1}{2\pi}\epsilon_{\mu\nu}\partial^\mu A^\nu. \tag{5.29}$$

So, it is possible to define a new current which is conserved:

$$J^\mu = j_5^\mu - \frac{1}{2\pi}\epsilon^\mu_\nu A^\nu. \tag{5.30}$$

However, just as in the four-dimensional case, this current is not gauge invariant. There is a familiar field configuration for which A does not fall off at infinity: the field of a point charge. If one has charges $\pm\theta$ at infinity, they give rise to a constant electric field, $F_{0i} = \pm e\theta$. So θ has a very simple interpretation in this theory.

It is easy to see that the physics is periodic in θ. For $\theta > q$ it is energetically favorable to produce a pair of charges from the vacuum which shield the charge at ∞.

5.2.2 Path integral computation of the anomaly

One can also do this calculation using the path integral, following Fujikawa. The redefinition of the fields which eliminates the phase in the fermion mass matrix is, from this point of view, just a change of variables. The question is: what is the Jacobian? The Euclidean path integral is defined by expanding the fields:

$$\psi(x) = \sum a_n \psi_n(x), \tag{5.31}$$

where

$$\not{D}\psi_n(x) = \lambda_n \psi_n(x) \tag{5.32}$$

and the measure is

$$\int \Pi\, da_n da_n^*. \tag{5.33}$$

Here, for normalized functions ψ_n,

$$a_n = \int d^2x\, \psi_n^*(x)\psi(x). \tag{5.34}$$

So, under an infinitesimal γ_5 transformation, we have

$$\delta\psi = i\theta\gamma_5\psi, \tag{5.35}$$

$$\delta a_n = i\theta \int d^2x\, \psi_n(x)\gamma_5\psi_m(x) a_m. \tag{5.36}$$

The required Jacobian is then

$$\det\left(\delta_{nn'} + i\theta \int d^2x\, \bar{\psi}_{n'}\gamma_5\psi_n\right). \tag{5.37}$$

To evaluate this determinant we write $\det(M) = e^{\text{Tr}\log M}$. To linear order in θ, we need to evaluate

$$\text{Tr}\,(i\theta\gamma_5). \tag{5.38}$$

This trace must be regularized. A simple procedure is to replace the determinant by

$$\text{Tr}\left(i\theta\gamma_5 e^{-\lambda_n^2/M^2}\right). \tag{5.39}$$

At the end of the calculation we take $M \to \infty$. We can replace λ_n^2 by

$$\not{D}\not{D} = D^2 + \frac{1}{2}\sigma_{\mu\nu}F^{\mu\nu}. \tag{5.40}$$

Expanding in powers of $F^{\mu\nu}$, it is only necessary to work to first order (in the analogous calculation in four dimensions, it is necessary to work to second order). In other words, we expand the exponent to first order in $F^{\mu\nu}$ and make the replacement $D^2 \to p^2$. The required trace is given by

$$i\theta\int\frac{d^2p}{p^2}\,\text{Tr}(\gamma_5\sigma_{\mu\nu})\,\frac{F^{\mu\nu}}{M}e^{-p^2/M^2}. \tag{5.41}$$

The trace in this expression now just refers to a trace over the Dirac indices. The momentum integral is elementary, and we obtain

$$\int \Pi\, da_n da_n^* \to \int \Pi\, da_n da_n^* \exp\left(i\frac{\theta}{2\pi}\int d^2x\,\epsilon_{\mu\nu}F^{\mu\nu}\right). \tag{5.42}$$

Interpreting the divergence of the current as the variation of the effective Lagrangian, we see that we have recovered the anomaly equation (5.15). The anomaly in four and other dimensions can also be calculated in this way. The exercises at the end of the chapter provide more details of these computations.

5.2.3 The CP^N model: an asymptotically free theory

The model we have considered so far is not quite like QCD in at least two ways. First, there are no instantons; second, the coupling e is dimensionful. We can obtain a theory closer to QCD by considering a class of theories with dimensionless couplings, the *non-linear sigma models*. These are models whose fields are the coordinates of some smooth manifold. They can be, for example, the coordinates of an n-dimensional sphere. An interesting case is the CP^N model; here the CP stands for "complex projective" space. This space is described by a set of coordinates z_i, $i = 1,\ldots,N+1$, where z_i is identified with αz_i and α is any complex constant. Alternatively, we can define the space through the constraint

$$\sum_i |z_i|^2 = 1, \tag{5.43}$$

where the point z_i is equivalent to $e^{i\alpha}z_i$. In the field theory, the z_is become two-dimensional fields $z_i(x)$. To implement the first constraint, we can add to the action a Lagrange multiplier field $\lambda(x)$. For the second, we observe that the identification of points in the "target space"

CP^N must hold at every point in *ordinary space–time*, so this is a $U(1)$ gauge symmetry. Introducing a gauge field A_μ and the corresponding covariant derivative, we want to study the Lagrangian

$$\mathcal{L} = \frac{1}{g^2}\left[|D_\mu z_i|^2 - \lambda(x)(|z_i|^2 - 1)\right]. \tag{5.44}$$

Note that there is no kinetic term for A_μ, so we can simply eliminate it from the action using its equations of motion. This yields

$$\mathcal{L} = \frac{1}{g^2}\left(|\partial_\mu z_j|^2 + |z_j^* \partial_\mu z_j|^2\right). \tag{5.45}$$

It is easier, however, to proceed keeping A_μ in the action. In this case the action is quadratic in z, and we can integrate out the z fields:

$$\begin{aligned} Z &= \int [dA][d\lambda][dz_j] \exp(-S) = \int [dA][d\lambda] \exp\left(-\int d^2x\, \Gamma_{\text{eff}}[A,\lambda]\right) \\ &= \int [dA][d\lambda] \exp\left(-N \operatorname{Tr}\log(-D^2 - \lambda) - \frac{1}{g^2}\int d^2x\, \lambda\right). \end{aligned} \tag{5.46}$$

5.2.4 The large-N limit

By itself, the result in Eq. (5.46) is still rather complicated. The fields A_μ and λ have non-linear and non-local interactions. Things become much simpler if one takes the large-N limit, $N \to \infty$ with g^2N fixed. In this case the interactions of λ and A_μ are suppressed by powers of N. For large N the path integral is dominated by a single field configuration, which solves

$$\frac{\delta \Gamma_{\text{eff}}}{\delta \lambda} = 0, \tag{5.47}$$

or, setting the gauge field to zero,

$$N \int \frac{d^2k}{(2\pi)^2} \frac{1}{k^2 + \lambda} = \frac{1}{g^2}. \tag{5.48}$$

The integral on the left-hand side is ultraviolet divergent. We will simply cut it off at scale M. This gives

$$\lambda = m^2 = M \exp\left(-\frac{2\pi}{g^2 N}\right). \tag{5.49}$$

Here, a theory which is classically scale invariant exhibits a mass gap. This is the phenomenon of *dimensional transmutation*. These masses are related in a renormalization-group-invariant fashion to the cutoff. So the theory is quite analogous to QCD. We can read off the leading term in the beta function from the familiar formula

$$m = M \exp\left(-\int \frac{dg}{\beta(g)}\right). \tag{5.50}$$

So, with

$$\beta(g) = -\frac{1}{2\pi} g^3 b_0, \tag{5.51}$$

we have $b_0 = 1$.

Most important for our purposes is the question of θ-dependence. Just as in $(1+1)$-dimensional electrodynamics we can introduce a θ term,

$$S_\theta = \frac{\theta}{2\pi} \int d^2x\, \epsilon_{\mu\nu} F^{\mu\nu}. \tag{5.52}$$

Here $F_{\mu\nu}$ can be expressed in terms of the fundamental fields z_j. As usual, this is the integral of a total divergence. But, precisely as in the case of $(1+1)$-dimensional electrodynamics discussed above, this term is physically important. In a perturbation theory approach to the model, this is not entirely obvious; however, using our reorganization of the theory at large N, it is. The lowest-order action for A_μ is trivial, but at one loop (order $1/N$) a kinetic term for A is generated through the vacuum polarization loop:

$$\mathcal{L}_{\text{kin}} = \frac{N}{2\pi m^2} F^2_{\mu\nu}. \tag{5.53}$$

At this order, then, the effective theory consists of the gauge field with coupling $e^2 = 2\pi m^2/N$ and some coupling to a set of charged massive fields z. As we have already argued, θ corresponds to a non-zero background electric field due to charges at infinity, and the theory clearly has a non-trivial θ-dependence.

To this model one can add massless fermions. In this case one has an anomalous $U(1)$ symmetry, as in QCD. There is then no θ-dependence; by redefining the fermions according to $\psi \to e^{ia\theta}\psi$ one can eliminate θ. In this model the absence of a θ-dependence can be understood more physically: θ represents a charge at ∞, and it is possible to shield any such charge with massless fermions. But there is a non-trivial breaking of the $U(1)$ symmetry. At low energies, one has now a theory with a fermion coupled to a dynamical $U(1)$ gauge field. The breaking of the associated $U(1)$ symmetry in such a theory is a well-studied phenomenon, which we will not pursue here.

5.2.5 The role of instantons

There is another way to think about the breaking of the $U(1)$ symmetry and the θ-dependence in this theory. If one considers the Euclidean functional integral, it is natural to look for stationary points of the integration, i.e. for classical solutions of the Euclidean equations of motion. Since they are potentially important it is necessary that these solutions have a finite action, which means that they must be localized in Euclidean space and time. For this reason, such solutions were dubbed "instantons" by 't Hooft. Instantons are not difficult to find in the CP^N model; we will describe them below. These solutions carry non-zero values of the *topological charge*,

$$\frac{1}{2\pi} \int d^2x\, \epsilon_{\mu\nu} F_{\mu\nu} = n, \tag{5.54}$$

and have an action $2\pi n$. If we write $z_i = z_{i\,\mathrm{cl}} + \delta z_i$ then the functional integral, in the presence of a θ term, has the form

$$Z_{\mathrm{inst}} = e^{\frac{-2\pi n}{g^2}} e^{in\theta} \int [d\delta z_j] \exp\left(-\delta z_i \frac{\delta^2 S}{\delta z_i \delta z_j} \delta z_j + \cdots\right). \tag{5.55}$$

It is easy to construct the instanton solution in the case of CP^1. Rather than write the theory in terms of a gauge field, as we have done above, it is convenient to parameterize it in terms of a single complex field Z. One can, for example, define Z as z_1/z_2 and let $\bar{Z}$ denote its complex conjugate. Then, with a bit of algebra, one can show that the action for Z which follows from Eq. (5.45) takes the following form (it is easiest to work backwards, starting with the equation below and deriving Eq. (5.45)):

$$\mathcal{L} = \frac{\partial_\mu Z \partial_\mu \bar{Z}}{(1 + \bar{Z}Z)^2}. \tag{5.56}$$

The function

$$g_{Z\bar{Z}} = \frac{1}{(1 + \bar{Z}Z)^2} \tag{5.57}$$

has an interesting significance. There is a well-known mapping of the unit sphere $x_1^2 + x_2^2 + x_3^2 = 1$ onto the complex plane:

$$z = \frac{x_1 + ix_2}{1 - x_3}. \tag{5.58}$$

The inverse is

$$x_1 = \frac{z + \bar{z}}{1 + |z|^2}, \quad x_2 = \frac{z - \bar{z}}{i(1 + |z|^2)}, \quad x_3 = \frac{|z|^2 - 1}{|z|^2 + 1}. \tag{5.59}$$

The line element on the sphere is mapped in a non-trivial way onto the plane:

$$ds^2 = dx_1^2 + dx_2^2 + dx_3^3 = g_{z\bar{z}} dz d\bar{z} = \frac{1}{(1 + \bar{z}z)^z} dz d\bar{z}. \tag{5.60}$$

So, the model describes a field that is constrained to move on a sphere; g is the metric of the sphere. In general, such a model is called a non-linear sigma model. This is an example of a Kahler geometry, a type of geometry which will figure significantly in our discussion of string compactification.

It is straightforward to write down the equations of motion:

$$\partial^2 Z g_{Z\bar{Z}} + \partial_\mu Z \left(\partial_\mu \bar{Z} \frac{\partial g}{\partial \bar{Z}} + \partial_\mu \phi \frac{\partial g}{\partial Z}\right) = 0, \tag{5.61}$$

or

$$\partial_z \partial_{\bar{z}} Z - \frac{2 \partial_z Z \partial_{\bar{z}} \bar{Z}}{1 + \bar{Z}Z} = 0. \tag{5.62}$$

Now using space–time coordinates $z = x_1 + ix_2$, $\bar{z} = x_1 - ix_2$, we see that if Z is anti-analytic then the equations of motion are satisfied! So a simple solution, which, as you can check, has finite action, is

$$Z(\bar{z}) = \rho \bar{z}. \tag{5.63}$$

In addition to evaluating the action you can evaluate the topological charge,

$$\frac{1}{2\pi}\int d^2x\,\epsilon_{\mu\nu}F^{\mu\nu} = 1, \tag{5.64}$$

for this solution. More generally, the topological charge measures the number of times that Z maps the complex plane into the complex plane; $Z = z^n$ has charge n.

We can generalize these solutions. The solution of Eq. (5.63) breaks several symmetries of the action: translation invariance, two-dimensional rotational invariance and the scale invariance of the classical equations. So we should be able to generate new solutions by translating, rotating and dilating the solution. You can check that

$$Z(z) = \frac{az+b}{cz+d} \tag{5.65}$$

is a solution with action 2π. The parameters $a, \ldots, d$ are called collective coordinates. They correspond to the symmetries of translations, dilations and rotations and special conformal transformations (forming the group $SL(2, C)$). In other words, any given finite-action solution breaks the symmetries. In the path integral the symmetry of Green's functions is recovered when one integrates over the collective coordinates. For translations this is particularly simple. Integrating over x_0, the instanton position,

$$\langle Z(x)Z(y)\rangle \approx \int d^2x_0\,\phi_{\text{cl}}(x-x_0)\phi_{\text{cl}}(y-x_0)e^{-S_0}. \tag{5.66}$$

(The precise measure is obtained by the Faddeev–Popov method.) Similarly, integration over the parameter ρ yields a factor

$$\int d\rho\,\rho^{-1}\exp\left(-\frac{2\pi}{g^2(\rho)}\right). \tag{5.67}$$

Here the first factor follows on dimensional grounds. The second follows from renormalization-group considerations. It can be found by explicit evaluation of the functional determinant. Note that, because of asymptotic freedom, this means that typical Green's functions will be divergent in the infrared.

There are many other features of this instanton that one can consider. For example, one can add massless fermions to the model; the resulting theory has a chiral $U(1)$ symmetry, which is anomalous. The instanton gives rise to non-zero Green's functions, which violate the $U(1)$ symmetry. We will leave investigation of fermions in this model to the exercises and turn to the theory of interest, which exhibits phenomena parallel to this simple theory.

5.3 Real QCD

The model of the previous section mimics many features of real QCD. Indeed, we will see that much of our discussion can be carried over, almost word for word, to the observed strong interactions. This analogy is helpful, given that in QCD we have no approximation which gives us control over the theory comparable with that which we found in the large-N limit of the CP^N model. As in that theory, we have the following.

- There is a θ parameter, which appears as an integral over the divergence of a non-gauge invariant current.
- There are instantons, which indicate that physical quantities should be θ-dependent. However, instanton effects cannot be considered in a controlled approximation, and there is no clear sense in which θ-dependence can be understood as arising from instantons.
- In QCD there is also a large-N expansion but, while it produces significant simplification, one cannot solve the theory even in the leading large-N approximation. Instead, an understanding of the underlying symmetries, and experimental information about chiral symmetry breaking, provides critical information about the behavior of the strongly coupled theory and allows computations of the physical effects of θ.

5.3.1 The theory and its symmetries

In order to understand the effects of θ it is sufficient to focus on the light quark sector of QCD. For simplicity in writing down some of the formulas, we will consider a simplified theory with two light quarks; it is not difficult to generalize the resulting analysis to the case of three. It is believed that the masses of the u and d quarks are of order 5 MeV and 10 MeV, respectively, much smaller than the scale of QCD. So we first consider an idealization of the theory in which these masses are set to zero. In this limit, the theory has a symmetry $SU(2)_{\rm L} \times SU(2)_{\rm R}$. Calling

$$q = \begin{pmatrix} u \\ d \end{pmatrix}, \quad \bar{q} = \begin{pmatrix} \bar{u} \\ \bar{d} \end{pmatrix}, \tag{5.68}$$

the two $SU(2)$ symmetries act separately on q and $\bar{q}$ (thought of as left-handed fermions),

$$q^T \to q^T U_{\rm L}, \quad \bar{q} \to U_{\rm R}\bar{q}. \tag{5.69}$$

This symmetry is spontaneously broken. The order parameter for the symmetry breaking is believed to be an expectation value for the quark bilinear product:

$$\mathcal{M} = \bar{q}q. \tag{5.70}$$

Under the original symmetry,

$$\mathcal{M} \to U_{\rm R}\mathcal{M}U_{\rm L}. \tag{5.71}$$

The expectation value (condensate) of $\mathcal{M}$ is

$$\langle \mathcal{M} \rangle = c\Lambda_{\rm QCD}^3 \begin{pmatrix} 1 & 0 \\ 0 & 1 \end{pmatrix}. \tag{5.72}$$

This breaks some of the original symmetry but preserves the symmetry $U_{\rm L} = U_{\rm R}$. This symmetry is just the $SU(2)$ isospin symmetry. The Goldstone bosons associated with the three broken symmetry generators must transform in a representation of the unbroken symmetry: these are the pions, which an form isospin vector. One can think of the

Goldstone bosons as being associated with a slow variation of the expectation value in space, so we can introduce

$$\mathcal{M} = \bar{q}q = M_0 \exp\left[i\frac{\pi_a(x)\tau_a}{f_\pi}\right]\begin{pmatrix} 1 & 0 \\ 0 & 1 \end{pmatrix} \tag{5.73}$$

The quark mass term in the Lagrangian is then (for simplicity taking $m_u = m_d = m_q$)

$$m_q \mathcal{M}. \tag{5.74}$$

Replacing $\mathcal{M}$ by the expression (5.73) gives a potential for the pion fields. Expanding $\mathcal{M}$ in powers of π/f_π, the minimum of the potential occurs for $\pi_a = 0$. Expanding to second order, one has

$$m_\pi^2 f_\pi^2 = m_q M_0. \tag{5.75}$$

We have been a bit cavalier about the symmetries. The theory also has two $U(1)$ symmetries:

$$q \to e^{i\alpha} q, \quad \bar{q} \to e^{i\alpha}\bar{q}, \tag{5.76}$$

$$q \to e^{i\alpha} q, \quad \bar{q} \to e^{-i\alpha}\bar{q}. \tag{5.77}$$

The first of these is baryon number symmetry and it is not chiral (and is not broken by the condensate). The second is the axial $U(1)_5$ symmetry; it is broken by the condensate. So, in addition to the pions there should be another approximate Goldstone boson. But there is no good candidate among the known hadrons. The η has the right quantum numbers but, as we will see below, it is too heavy to be interpreted in this way. The absence of this fourth (or, in the case of three light quarks, ninth) Goldstone boson is called the $U(1)$ problem.

The $U(1)_5$ symmetry suffers from an anomaly, however, and we might hope that this has something to do with the absence of a corresponding Goldstone boson. The anomaly is given by

$$\partial_\mu j_5^\mu = \frac{1}{16\pi^2} F\tilde{F}. \tag{5.78}$$

Again, we can write the right-hand side as a total divergence

$$F\tilde{F} = \partial_\mu K^\mu, \tag{5.79}$$

where

$$K_\mu = \epsilon_{\mu\nu\rho\sigma}\left(A_\nu^a F_{\rho\sigma}^a - \frac{2}{3} f^{abc} A_\nu^a A_\rho^b A_\sigma^c\right). \tag{5.80}$$

This accounts for the fact that in perturbation theory the axial $U(1)$ symmetry is conserved. Non-perturbatively, as we will now show, there are important configurations in the functional integral for which the right-hand side does not vanish rapidly at infinity.

5.3.2 Instantons in QCD

In the Euclidean functional integral

$$Z = \int [dA][dq][d\bar{q}]e^{-S} \tag{5.81}$$

it is natural to look for stationary points of the effective action, i.e. finite-action classical solutions of the theory in imaginary time. The Yang–Mills equations are complicated non-linear equations, but it turns out that, much as in the CP^N model, the instanton solutions can be found rather easily. The following tricks simplify the construction and turn out to yield the general solution. First, note that the Yang–Mills action satisfies an inequality, the Bogomol'nyi bound:

$$\int (F \pm \tilde{F})^2 = \int (F^2 + \tilde{F}^2 \pm 2F\tilde{F}) = \int (2F^2 \pm 2F\tilde{F}) \geq 0. \tag{5.82}$$

So, the action is bounded by $| \int F\tilde{F}|$, the bound being saturated when

$$F = \pm\tilde{F}, \tag{5.83}$$

i.e. if the gauge field is (anti-)self-dual.[1] This equation is a first-order equation, and it is easy to solve if one first restricts to an $SU(2)$ subgroup of the full gauge group. One makes the ansatz that the solution should be invariant under a combination of ordinary rotations and global $SU(2)$ gauge transformations. Take

$$g(x) = \frac{x_4 + i\vec{x} \cdot \vec{\tau}}{r} \tag{5.84}$$

and

$$A_\mu = f(r^2)g\partial_\mu g^{-1}. \tag{5.85}$$

Then, substituting in to the Yang–Mills equations yields

$$f = \frac{-ir^2}{r^2 + \rho^2}, \tag{5.86}$$

where ρ is an arbitrary quantity with dimensions of length. The choice of origin here is also arbitrary; this can be remedied by simply replacing x by $x - x_0$ everywhere in these expressions, where x_0 represents the location of the instanton.

From this solution, it is clear why $\int \partial_\mu K^\mu$ does not vanish for the solution: while A is a pure gauge at infinity, it falls only as $1/r$. Indeed, since $F = \tilde{F}$, for this solution we have

$$\int F^2 = \int d^4x4\tilde{F}^2 = 32\pi^2. \tag{5.87}$$

[1] This is not an accident, nor was the analyticity condition in the CP^N case. In both cases we can add fermions so that the model becomes supersymmetric. Then one can show that if some supersymmetry generators Q_α annihilate a field configuration then the configuration is a solution. This is a first-order condition; in the Yang–Mills case it implies self-duality and in the CP^N case it requires analyticity.

This result can also be understood topologically. Note that g defines a mapping from the "sphere at infinity" into the gauge group. It is straightforward to show that

$$\frac{1}{32\pi^2}\int d^4x F\tilde{F} \tag{5.88}$$

counts the number of times that g maps the sphere at infinity into the group (once for this specific example; n times more generally). In the exercises and suggested reading, features of the instanton are explored in more detail.

The expression in Eq. (5.85) is, by its nature, gauge-dependent and other presentations of the solution are sometimes convenient. For example, if one formally transforms by g^{-1}, one obtains a solution which falls more rapidly to zero but which is singular at the origin.

The instanton was presented by 't Hooft in a fashion which is often more useful for actual computations. Defining the symbol η as follows,

$$\eta_{aij} = \epsilon_{aij},\ \eta_{a4i} = -\eta_{ai4} = -\delta_{ai},\quad \bar{\eta}_{a\mu\nu} = (-1)^{\delta_{a\mu}+\delta_{a\nu}}\eta_{a\mu\nu}, \tag{5.89}$$

the instanton takes the simple form

$$A^a_\mu = \frac{2\eta_{a\mu\nu}x^\nu}{x^2+\rho^2} \tag{5.90}$$

while the field strength is given by

$$F^a_{\mu\nu} = \frac{4\eta_{a\mu\nu}\rho^2}{(x^2+\rho^2)^2}. \tag{5.91}$$

That this configuration solves the equations of motion follows from

$$\eta_{a\mu\nu} = \frac{1}{2}\epsilon_{\mu\nu\alpha\beta}\eta_{a\alpha\beta}. \tag{5.92}$$

The alert reader will note that the η symbols are connected to the embedding of $SU(2)$ of the gauge group into an $SU(2)$ subgroup of $O(4) = SU(2)\times SU(2)$. This can be understood by noting that

$$\eta_{a\mu\nu} = \frac{1}{2}\,\mathrm{Tr}(\sigma^a\sigma_{\mu\nu}),\quad \bar{\eta} = \mathrm{Tr}(\sigma^a\bar{\sigma}_{\mu\nu}). \tag{5.93}$$

In this form it is easy to check that $F = \tilde{F}$, so the equations are satisfied. Note the $1/r$ falloff of A^μ, as opposed to the $1/r^4$ falloff of $F_{\mu\nu}$.

So, we have exhibited potentially important contributions to the path integral which violate the $U(1)$ symmetry. How does this symmetry violation show up? Let us consider the path integral more carefully. Having found a classical solution, we want to integrate over small fluctuations over it. Including the θ term these have the form

$$\langle \bar{u}u\bar{d}d\rangle = e^{-8\pi^2/g^2}e^{i\theta}\int [d\delta A][dq][d\bar{q}]\exp\left(-\frac{\delta^2 S}{\delta A^2}\delta A^2 - S_{q,\bar{q}}\right)\bar{u}u\bar{d}d. \tag{5.94}$$

Now S contains an explicit factor $1/g^2$. As a result the fluctuations are formally suppressed by g^2 relative to the leading contribution. The one-loop functional integral yields a product of determinants for the fermions and a product of inverse square root determinants for the bosons.

Consider the integral over the fermions. It is straightforward, if challenging, to evaluate the determinants. However, if the quark masses are zero then the fermion functional integrals are also zero, because there is a zero mode for each of the fermions, i.e. for both q and $\bar{q}$ there is a normalizable solution of the equations

$$\not{D}u = 0, \quad \not{D}\bar{u} = 0 \tag{5.95}$$

and similarly for d and $\bar{d}$. It is straightforward to construct the solutions

$$u = \frac{\rho}{[\rho^2 + (x - x_0)^2]^{3/2}}\zeta, \tag{5.96}$$

where ζ is a constant spinor, and similarly for $\bar{u}$, etc.

Let's understand this a bit more precisely. Euclidean path integrals are conceptually simple. Consider some classical solution, $\Phi_{\rm cl}(x)$ (here Φ denotes collectively the various bosonic fields; we will treat, for now, the fermions as vanishing in the classical solutions). In the path integral, at small coupling we are interested in small fluctuations about the classical solution,

$$\Phi = \Phi_{\rm cl} + \delta\Phi. \tag{5.97}$$

Because the action is stationary at the classical solution,

$$S = S_{\rm cl} + \int d^4x\, \delta\Phi \frac{\partial^2 \mathcal{L}}{\partial \Phi^2} \delta\Phi + \cdots . \tag{5.98}$$

The second derivative here is a shorthand for a second-order differential operator, which we will simply denote by S'' and refer to as the *quadratic fluctuation operator*. We can expand $\delta\Phi$ in (normalizable) eigenfunctions of this operator Φ_n with eigenvalues λ_n, $\Phi = c_n\Phi_n$. The result of the functional integral is then $\prod \lambda_n^{-1/2}$. This is the leading correction to the classical limit. Higher-order corrections are suppressed by powers of g^2. This is most easily seen by working in the scaling where the action has a factor $1/g^2$. Then one can derive the perturbation theory from the path integral in the usual way; the main difference from the usual treatment with zero background fields is that the propagators are more complicated. The propagators for various fields in the instanton background are in fact known in closed form.

The form of the differential operator is familiar from our calculation of the beta function in the background field method (using the background field gauge). For the gauge bosons, in a suitable (background field) gauge it is

$$S'' = \mathcal{D}^2 + \mathcal{J}_{\mu\nu}F^{\mu\nu}. \tag{5.99}$$

Here $\mathcal{D}$ is just the covariant derivative, the vector potential corresponds to the classical solution (an instanton) and similarly for the field strength; $\mathcal{J}_{\mu\nu}$ is the generator of Lorentz transformations in the vector representation. The eigenvalue problem was completely solved by 't Hooft.

Both the bosonic and fermionic quadratic fluctuation operators have zero eigenvalues. For the bosons, these potentially give infinite contributions to the functional integral and they must be treated separately. The difficulty is that among the variations of the fields are symmetry transformations, which comprise changes in the location of the instanton (translations), rotations of the instanton and scale transformations. Consider translations. For every solution there corresponds an infinite set of other solutions obtained by shifting the origin (varying x_0). Thus, instead of integrating over a coefficient c_0, we integrate over the *collective coordinate* x_0 (one must also include a suitable Jacobian factor). The effect of this is to restore translational invariance in the Green's functions. We will see this explicitly shortly. Similarly, the instanton breaks the rotational invariance of the theory; correspondingly, we can find a three-parameter set of solutions and zero modes. Integrating over these rotational collective coordinates restores rotational invariance. (The instanton also breaks a global gauge symmetry, but a combination of rotations and gauge transformations is preserved.)

Finally, the classical theory is scale invariant; this is the origin of the parameter ρ in the solution. Again, one must treat ρ as a collective coordinate and integrate over ρ. There is a power of ρ arising from the Jacobian, which can be determined on dimensional grounds. For the Green's function Eq. (5.90), for example, which has dimension six, we have (if all the fields are evaluated at the same point),

$$\int d\rho\, \rho^{-7}. \tag{5.100}$$

However, there is additional ρ-dependence because the quantum theory violates scale symmetry. This can be understood by replacing g^2 by $g^2(\rho)$ in the functional integral and using

$$e^{-8\pi^2 g^2(\rho)} \approx (\rho M)^{b_0} \tag{5.101}$$

for small ρ. For three-flavor QCD, for example, $b_0 = 9$ and the ρ integral diverges for large ρ. This relation simply states that the integral is dominated by the infrared, where the QCD coupling becomes strong.

Fermion functional integrals introduce a new feature. In four-component language, it is necessary to treat q and $\bar{q}$ as independent fields. This rule gives the functional integral as a determinant rather than as, say, the square root of a determinant. (In two-component language, this corresponds to treating q and q^* as independent fields.) So, at the one-loop order, we need to study

$$\not{D} q_n = \lambda_n q_n, \quad \not{D} \bar{q}_n = \lambda_n \bar{q}_n. \tag{5.102}$$

For non-zero λ_n there is a pairing of solutions with opposite eigenvalues of γ_5. In four-component notation one can see this from

$$\not{D} q_n = \lambda_n q_n \ \rightarrow\ \not{D} \gamma_5 q_n = -\lambda_n \gamma_5 q_n. \tag{5.103}$$

Zero eigenvalues, however, are special. There is no corresponding pairing. This has implications for the fermion functional integral. Writing

$$q(x) = \sum a_n q_n(x), \tag{5.104}$$

$$S = \sum \lambda_n a_n^* a_n \tag{5.105}$$

we have

$$\int [dq][d\bar{q}] e^{-S} = \prod_{n=0}^{\infty} da_n da_n^* \exp\left(-\sum_{n\neq 0} \lambda_n a_n^* a_n\right). \tag{5.106}$$

Because the zero modes do not contribute to the action, many Green's functions vanish. For example, $\langle 1 \rangle = 0$. In order to obtain a non-vanishing result, we need enough insertions of q to "soak up" all the zero modes.

We have seen that, in the instanton background, there are normalizable fermion zero modes, one for each left-handed field. This means that, in order for the path integral to be non-vanishing, we need to include insertions of enough qs and $\bar{q}$s to soak up all the zero modes. In other words, in two-flavor QCD, non-vanishing Green's functions have the form

$$\langle \bar{u}u\bar{d}d \rangle \tag{5.107}$$

and violate the symmetry. Note that the symmetry violation is just as predicted from the anomaly equation,

$$\Delta Q_5 = \frac{2}{16\pi^2} \int d^4x F\tilde{F} = 4. \tag{5.108}$$

This is a particular example of an important mathematical theorem known as the Atiyah–Singer index theorem.

We can put all this together to evaluate a Green's function which violates the classical $U(1)$ symmetry of the massless theory, $\langle \bar{u}(x)u(x)\bar{d}(x)d(x) \rangle$. Taking the gauge group to be $SU(2)$ there is one zero mode for each of u, $\bar{u}$, d and $\bar{d}$. The fields in this expectation value can soak up all these zero modes. The effect of the integration over x_0 is to give a result that is independent of x, since the zero modes are functions only of $x - x_0$. The integration over the rotational zero modes gives a non-zero result only if the Lorentz indices are contracted in a rotationally invariant manner (the same applies to the gauge indices). The integration over the instanton scale size – the conformal collective coordinate – is more problematic, exhibiting precisely the infrared divergence of Eq. (5.100).

So, we have provided some evidence that the $U(1)$ problem is solved in QCD, but no reliable calculation. What about the θ-dependence? Let us ask first about the θ-dependence of the vacuum energy. In order to get a non-zero result, we need to allow that the quarks are massive. Treating the mass as a perturbation, we obtain a result of the form

$$E(\theta) = C\Lambda_{\text{QCD}}^9 m_u m_d \cos\theta \int d\rho\, \rho^{-3} \rho^9. \tag{5.109}$$

So, as in the CP^N model, we have evidence for θ-dependence but cannot do a reliable calculation. That we cannot do a calculation should not be a surprise. There is no small parameter in QCD to use as an expansion parameter. Fortunately, we can use other facts which we know about the strong interactions to get a better handle on both the $U(1)$ problem and the θ-dependence question.

Before continuing, however, let us consider the weak interactions. Here there is a small parameter and there are no infrared difficulties, so we might expect instanton effects to be small. The analog of the $U(1)_5$ symmetry in this case is baryon number. Baryon number has an anomaly in the standard model, since all the quark doublets have the same sign of the baryon number. 't Hooft showed that one could actually use instantons, in this case, to compute the violation of baryon number. Technically, there are no finite-action Euclidean solutions in this theory; this follows, as we will see in a moment, from a simple scaling argument. However, 't Hooft realized that one can construct important configurations having non-zero topological charge by starting with the instantons of the pure gauge theory and perturbing them. For the Higgs boson, one solves the equation

$$D^2\phi = V'(\phi). \tag{5.110}$$

For a light boson, one can neglect the right-hand side. Then this equation is solved by

$$\phi(x) = i\bar{\sigma}^\mu x^\mu \left(\frac{1}{x^2+\rho^2}\right)^{1/2} \langle\phi\rangle. \tag{5.111}$$

Note that at large x, this has the form $g(x)\langle\phi\rangle$. As a result, the action of the configuration is finite. One finds the following correction to the action:

$$\delta S = \frac{1}{g^2} v^2 \rho^2. \tag{5.112}$$

Including this in the exponential damps the ρ integral at large ρ, and leads to a convergent result.

Now including the fermions, there is a zero mode for each $SU(2)$ doublet. So, one obtains a non-zero expectation value for correlation functions of the form $\langle QQQLLL\rangle$, where the color and $SU(2)$ indices are contracted in a gauge-invariant way and the flavors for the Qs and Ls are all different. The coefficient is

$$\mathcal{A}_{\text{bv}} = C\, e^{-2\pi/\alpha_{\text{w}}}. \tag{5.113}$$

From this, one can see that baryon number violation occurs in the Standard Model but at an incredibly small rate. One can also calculate a term in the effective action, involving three quarks and three leptons, with a similar coefficient by studying Green's functions in which all the fields are widely separated. We will encounter this sort of computation later, when we discuss instantons in supersymmetric theories.

5.3.3 Physical interpretation of the instanton solution

We have derived dramatic physical effects from the instanton solution by direct calculation, but we have not provided a physical picture of the phenomena that the instanton describes. Already in quantum mechanics imaginary-time solutions of the classical equations of motion are familiar in the Wentzel–Kramers–Brillouin (WKB) analysis of tunneling, and the Yang–Mills instanton (and the CP^N instanton) also describe tunneling phenomena. In this subsection we will confine our attention to pure gauge theories. The generalization to theories with fermions and/or scalars is straightforward and interesting.

To understand the instanton in terms of tunneling, it is helpful to work in a non-covariant gauge, in which there is a Hamiltonian description. The gauge $A_0 = 0$ is particularly useful. In this gauge the canonical coordinates are the A_is and their conjugate momenta are the E_is (with a minus sign). This is too many degrees of freedom if all are treated as independent. The resolution lies in the need to enforce Gauss's law, which is now to be viewed as an operator constraint on states. For example, in a $U(1)$ theory,

$$G(\vec{x})|\Psi\rangle = (\vec{\nabla}\cdot\vec{E} - \rho)|\Psi\rangle = 0. \tag{5.114}$$

The left-hand side is almost the generator of gauge transformations. On the gauge fields, for example,

$$\left[\int d^3x\,\omega(\vec{x})G(\vec{x}), A_i(\vec{y})\right] = -\int d^3x\,\partial_j\omega(\vec{x})[E(\vec{x})_j, A(\vec{y})_i] = \partial_i\omega(\vec{y}). \tag{5.115}$$

In the second step we have integrated by parts and dropped a possible surface term. This requires that $\omega \to 0$ fast enough at infinity. Such gauge transformations are called "small". We have learned that, in the $A_0 = 0$ gauge, states must be invariant under time-independent, small, gauge transformations.

In electrodynamics this is not particularly interesting. But the same manipulations hold in non-Abelian theories, and in this case there are interesting large gauge transformations. An example is

$$g(\vec{x}) = \exp\left(i\pi\frac{\vec{x}\cdot\vec{\sigma}}{\sqrt{\vec{x}^2 + a^2}}\right). \tag{5.116}$$

We can also consider powers g^n of g. We can think of g as mapping three-dimensional space into the group $SU(2)$. The number of times that the mapping wraps around the gauge group is known as the winding number, and it can be written as

$$n = \frac{1}{24\pi^2}\int d^3x\,\epsilon_{ijk}\,\mathrm{Tr}(\partial_i g\partial_j g\partial_k g). \tag{5.117}$$

However, g_n is not unique; we can multiply by any small gauge transformation without changing n. The zero-energy states consist of $A_i = ig^{-n}\partial_i g^n$ averaged over the small gauge transformations in such a way as to make them invariant.

With just a little algebra one can show that $n = \int d^3xK_0$, where K^μ is the topological current encountered in Eq. (5.80). So an instanton, in $A_0 = 0$ gauge, corresponds to a tunneling between states of different n. More precisely, there is a non-zero matrix element of the Hamiltonian between states of different n,

$$\langle n|H|n \pm 1\rangle = \epsilon. \tag{5.118}$$

This is analogous to the situation in crystals, and the energy eigenstates are similar to Bloch waves,

$$|\theta\rangle = \sum_n e^{in\theta}|n\rangle, \tag{5.119}$$

with energy $\epsilon\cos\theta$. This θ is precisely the quantity which entered as a parameter in the Lagrangian.

5.3.4 QCD and the $U(1)$ problem

In real QCD we have seen that, on the one hand, instanton configurations violate the axial $U(1)$ symmetry. In general, there is no small parameter which governs the size of this breaking, so there is no reason to expect a light (pseudo)Goldstone. Consistent with this, explicit calculations are infrared divergent. Again, this is not a surprise; there is no small parameter which would justify the use of a semiclassical approximation, but the instanton analysis we have described makes clear that there is no reason to expect that there is a light Goldstone boson. Actually, while there is no obvious reason why perturbative and semiclassical (instanton) techniques should give reliable results, there are two approximation method techniques available. The first is for large N, where one now allows the N of $SU(N)$ to be large, with g^2N fixed. In contrast with the case of CP^N, this does not give enough simplification to permit explicit computations, but it does allow one to make qualitative statements about the theory. Witten has pointed out a way in which one can relate the mass of the η (or η' if one is thinking in terms of $SU(3) \times SU(3)$ current algebra) to quantities in a theory without quarks. The anomaly is then an effect suppressed by a power of N, in the large-N limit, because the loop diagram contains a factor g^2 but not a factor N. So, for large N it can be treated as a perturbation and the η is almost massless. The quantity $\partial_\mu j_5^\mu$ acts as a creation operator for η (just as $\partial_\mu j_5^{\mu\,3}$ is a creation operator for the π meson), so one can compute the mass if one knows the correlation function at zero momentum,

$$\langle \partial_\mu j_5^\mu(x) \partial_\mu j_5^\mu(y) \rangle \propto \frac{1}{N^2} \langle F(x)\tilde{F}(x)F(y)\tilde{F}(y) \rangle. \tag{5.120}$$

To leading order in the $1/N$ expansion, the $F\tilde{F}$ correlation function can be computed in the theory without quarks. Witten argued that, while it vanishes order by order in perturbation theory, there is no reason that this correlation function need vanish in the full theory. Attempts have been made to compute this quantity both in lattice gauge theory and using the anti-de Sitter–conformal-theory (AdS–CFT) correspondence discovered in string theory and discussed later in this text. Both methods give promising results.

So, the $U(1)$ problem should be viewed as solved, in the sense that in the absence of any argument to the contrary, there is no reason to think that there should be an extra Goldstone boson in QCD.

The second approximation scheme which gives some control of QCD is known as chiral perturbation theory. The masses of the u, d and s quarks are small compared with the QCD scale, and the mass terms for these quarks in the Lagrangian can be treated as perturbations. This will figure in our discussion in the next section.

5.4 The strong CP problem

5.4.1 The θ-dependence of the vacuum energy

The assumption that the anomaly resolves the $U(1)$ problem in QCD raises another issue. Given that $\int d^4x\, F\tilde{F}$ has physical effects, a θ term in the action has physical effects as well. Since this term is CP odd, this means that there is the potential for strong CP-violating effects. These effects should vanish in the limit of zero quark mass since, in this case, by a field redefinition we can remove θ from the Lagrangian. In the presence of quark masses, the θ-dependence of many quantities can be computed. Consider, for example, the vacuum energy. In QCD, the quark mass term in the Lagrangian has the form

$$\mathcal{L}_\mathrm{m} = m_u \bar{u}u + m_d \bar{d}d + \mathrm{h.c.} \tag{5.121}$$

Were it not for the anomaly we could, by redefining the quark fields, take m_u and m_d to be real. Instead, we can define these fields in such a way that there is no $\theta F\tilde{F}$ term in the action but a phase in m_u and m_d. Clearly, we have some freedom in making this choice. In the case where m_u and m_d are equal, it is natural to choose these phases to be the same. We will explain shortly how one proceeds when the masses are different (as they are in nature). So

$$\mathcal{L}_\mathrm{m} = (m_u \bar{u}u + m_d \bar{d}d)e^{i\theta} + \mathrm{h.c.} \tag{5.122}$$

Now we want to treat this term as a perturbation. At first order, it makes a contribution to the ground-state energy proportional to its expectation value. We have already argued that the quark bilinear forms have non-zero vacuum expectation values, so

$$E(\theta) = (m_u + m_d)\cos\theta\langle\bar{q}q\rangle. \tag{5.123}$$

While without a difficult non-perturbative calculation we cannot calculate the separate quantities on the right-hand side of this expression, we can, using current algebra, relate them to measured quantities. It is shown in Appendix B that

$$m_{\pi^2} f_{\pi^2} = \mathrm{Tr}\,(M_q\langle\mathcal{M}\rangle) = (m_u + m_d)\langle\bar{q}q\rangle. \tag{5.124}$$

Replacing the quark mass terms in the Lagrangian by their expectation values, we can immediately read off the energy of the vacuum as a function of θ:

$$E(\theta) = m_\pi^2 f_\pi^2 \cos\theta. \tag{5.125}$$

This expression can readily be generalized to the case of three light quarks, by similar methods. So, we see that there is real physics in θ even if we do not understand how to do an instanton calculation. In the next section we will calculate a more interesting quantity: the neutron electric dipole moment as a function of θ.

5.4.2 The neutron electric dipole moment

The most interesting physical quantities to study in connection with CP violation are electric dipole moments, particularly that of the neutron, d_n. If CP were badly violated in strong interactions, one might expect $d_n \approx e\,\mathrm{fm} \approx 10^{-14}$ cm (here e is the electron charge). But the experimental limit on the dipole moment is striking,

$$d_n < 10^{-25}\ e\ \mathrm{cm}. \tag{5.126}$$

Using current algebra the leading contribution to the neutron electric dipole moment due to θ can be calculated, and one obtains a limit $\theta < 10^{-9}$. Here we outline the main steps in the calculation; I urge you to work out the details following the reference in the suggested reading. We will simplify the analysis by working in an exact $SU(2)$-symmetric limit, i.e. by taking $m_u = m_d = m$. We again treat the Lagrangian of Eq. (5.122) as a perturbation. We can understand how this term depends on the π fields by making an axial $SU(2)$ transformation on the quark fields. In other words, a background π field can be thought

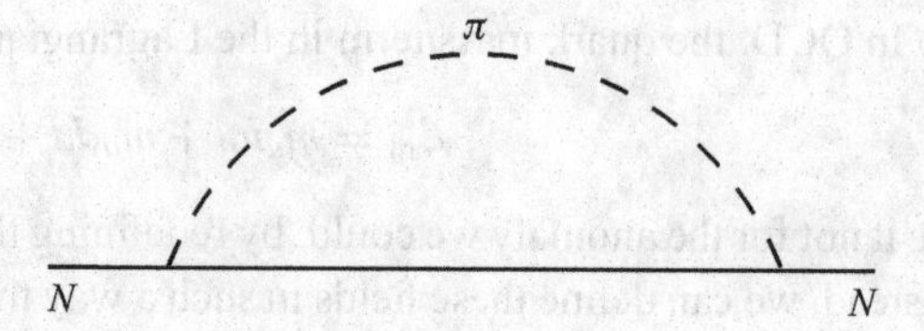

Fig. 5.2 Diagram in which CP-violating coupling of the pion contributes a newtron electric dipole moment d_n.

of as a small chiral transformation on the vacuum. Then, for example, for the τ_3 direction, $q \to (1 + i\pi_3\tau_3)q$ (the π field parameterizes the transformation), so the action becomes

$$\frac{m}{f_\pi}\pi_3(\bar{q}\gamma_5 q + \theta\bar{q}q). \tag{5.127}$$

The second term gives rise to a CP-violating coupling, $\bar{g}_{\pi NN}\pi^a \bar{N}\tau^a N$, of the pions and nucleons N. This is related to the matrix elements of $\bar{q}\tau^a q$ between nucleons. These, in turn, can be estimated by noting that at zero moment they are the matrix elements of an isospin charge operator between nucleons. The latter matrix elements can be estimated using the Gell-Mann and Ne'eman $SU(3)$ symmetry (a similar operator with coefficient m_s is responsible for the splitting between the members of the baryon octet). One obtains, in this way,

$$\bar{g}_{\pi NN} \approx -\theta\frac{(m_\Xi - m_N)m_u m_d}{2f_\pi(m_u + m_d)m_s} \approx 0.38. \tag{5.128}$$

This coupling is difficult to measure directly, but it gives rise, in a calculable fashion, to a neutron electric dipole moment. Consider the graph of Fig. 5.2. This graph generates a neutron electric dipole moment, if we take one coupling to be the standard pion–nucleon coupling and the other the coupling we have computed above. The resulting Feynman graph is infrared divergent; we cut this off at m_π while cutting off the integral in the ultraviolet at the QCD scale. The low-energy calculation is reliable in the limit that m_π is small, so that $\ln(m_\pi/\Lambda_{\rm QCD})$ is large compared to unity. The result is

$$d_{\rm n} = \frac{g_{\pi NN}\bar{g}_{\pi NN}}{4\pi^2 m_N}\ln\frac{M_N}{m_\pi}. \tag{5.129}$$

The matrix element can be estimated using the $SU(3)$ symmetry of Gell-Mann and Ne'eman, as mentioned above, yielding $d_{\rm n} = 5.2 \times 10^{-16}\theta$ cm. The experimental bound gives $\theta < 10^{-9}$–10^{-10}. Understanding why CP violation is so small in strong interactions is known as the strong CP problem.

5.5 Possible solutions of the strong CP problem

What should our attitude towards this problem be? We might argue that, on the one hand, some Yukawa couplings are as small as 10^{-5}, so why is 10^{-9} so bad? On the other hand, we suspect that the smallness of the Yukawa couplings is related to approximate symmetries, and that these Yukawa couplings are telling us something. Perhaps there is some explanation of the smallness of θ, and perhaps this is a clue to new physics. In this section we review some of the solutions which have been proposed to understand the smallness of θ.

5.5.1 Zero *u* quark mass

Suppose that the mass of the u quark were zero. In this case, by a field redefinition of the u quark

$$u \to e^{-i\theta} u, \tag{5.130}$$

one could make the θ term vanish as a consequence of the anomaly. This would be a simple enough explanation, but there are two issues. First, why should we make this redefinition? We might imagine that it is the result of a symmetry, but this symmetry cannot be a real symmetry of the underlying theory since it is violated by QCD (through the anomaly). We will see later in this book that discrete symmetries, with anomalies of the kind required to understand a vanishing u quark mass, do in fact frequently arise in string theory. So, perhaps this sort of explanation is plausible. We would not, then, expect that the u quark mass should be *exactly* zero but, instead, examining our formula for the neutron electric dipole moment, we would require that the ratio m_u/m_d should be less than about 10^{-10}.

As we described in Chapter 3, however, lattice gauge theory computations establish a non-zero value of the u quark mass with large statistical significance. It is worth noting why researchers in the past contemplated this possibility. Examining the mass spectrum of the pseudoscalar mesons, using the methods of current algebra or chiral Lagrangians (we will discuss these further in Chapter 8), one obtains $m_u/m_d \approx 0.5$. The question, however, is which mass values should actually appear in this formula? In particular, in a theory in which $m_u = 0$ at some high scale, instantons will generate a non-zero mass for m_u at lower scales. The resulting expression is infrared divergent, but we take as the main lesson that it is proportional to $m_d m_s$. Because m_s is not so different from the characteristic scales of QCD, one might imagine that an effective mass of the needed size could be found. It is this possibility which has been excluded by modern lattice computations.

5.5.2 Spontaneous CP violation

Suppose that the underlying theory respects CP and that the observed CP violation is spontaneous. Because θ is CP odd, the underlying theory has $\theta = 0$. One might hope that this feature would be preserved when the symmetry is spontaneously broken. Satisfying this condition and simultaneously generating an order-one CP-violating angle in the CKM matrix is a model-building challenge which we will not review here. Suffice it to say that this can be achieved at tree level. However, existing realizations rely on model-building cleverness and do not have a clear conceptual basis. So, one must ask how plausible is this possibility, and does it survive quantum corrections.

There are a number of ways in which θ might be generated in the low-energy theory. First, suppose that CP is broken by the expectation value of a complex field Φ. There might well be direct couplings such as

$$\frac{1}{16\pi^2}\,(\mathrm{Im}\,\Phi)\,F\tilde{F}. \tag{5.131}$$

Note that Φ might also couple to fermions, giving them a large mass through its expectation value. When these fermions are integrated out this would also generate an effective θ. This is likely, simply because of the anomalous field redefinitions which may be required to make the masses of these fields real. There do exist, however, models which, while complicated, meet the requirements of small θ.

5.5.3 The axion

Perhaps the most compelling explanation of the smallness of θ involves a hypothetical particle called the axion. We present here a slightly updated version of the original idea of Peccei and Quinn.

Consider the vacuum energy as a function of θ (Eq. (5.123)). This energy has a minimum at $\theta = 0$, i.e. at the CP-conserving point. As Weinberg noted long ago, this is almost automatic: points of higher symmetry are necessarily stationary points. As it stands this observation is not particularly useful, since θ is a parameter, not a dynamical variable. But, suppose that one has a field a with coupling to QCD:

$$\mathcal{L}_{\text{axion}} = (\partial_\mu a)^2 + \frac{a/f_a + \theta}{32\pi^2} F\tilde{F}, \tag{5.132}$$

where f_a is known as the axion decay constant. Suppose, in addition, that the rest of the theory possesses a symmetry, called the Peccei–Quinn symmetry,

$$a \to a + \alpha \tag{5.133}$$

for constant α. Then, by a shift in a one can eliminate θ. What we have previously called the vacuum energy as a function of θ, $E(\theta)$, is now $V(a/f_a)$, the potential energy of the axion. It has a minimum at $\theta = 0$. The strong CP problem is solved.

One can estimate the axion mass by simply examining $E(\theta)$, (Eq. 5.125):

$$m_a^2 \approx \frac{m_\pi^2 f_\pi{}^2}{f_a^2}. \tag{5.134}$$

If $f_a \sim$ TeV, this yields a mass of order keV. If $f_a \sim 10^{16}$ GeV, this gives a mass of order 10^{-9} eV.

There are several questions one can raise about this proposal.

- Should the axion already have been observed? The couplings of the axion to matter can be worked out in a given model in a straightforward way, using the methods of current algebra (in particular non-linear Lagrangians). All the couplings of the axion are suppressed by powers of f_a. This is characteristic of a Goldstone boson. At zero momentum a change in the field is like a symmetry transformation so, before including the QCD effects which explicitly break the symmetry, axion couplings are suppressed by powers of momentum over f_a; QCD effects are suppressed by Λ_{QCD}/f_a. Thus if f_a is large enough then the axion is difficult to see. The strongest limit turns out to come from red giant stars. The production of axions is "semiweak", i.e. it is suppressed only by one power of f_a rather than two powers of m_W; as a result, axion emission is competitive with neutrino emission until $f_a > 10^{10}$ GeV or so.

- As we will describe in more detail in Chapter 18, the axion could have been copiously produced in the early universe. As a result there is an *upper bound* on the axion decay constant, of about 10^{11} GeV. If this bound is saturated, the axion constitutes the dark matter. We will discuss this bound in detail in Chapter 19.
- Can one search for the axion experimentally? Typically, the axion couples not only to the $F\tilde{F}$ of QCD but also to the same object in QED. This means that in a strong magnetic field an axion can convert to a photon. Precisely this effect is being searched for by the ADMX experiment at the University of Washington. The basic idea is to suppose that the dark matter in the halo of our galaxy consists principally of axions. Using a (superconducting) resonant cavity with a high Q value in a large magnetic field, one searches for the conversion of these axions into excitations of the cavity due to the coupling of the axion to the electromagnetic field, $F\tilde{F} = \vec{E} \cdot \vec{B}$. The experiments have already reached a level where they set interesting limits; the next generation of experiments will cut a significant swath in the presently allowed parameter space.
- The coupling of the axion to $F\tilde{F}$ violates the shift symmetry; this is why the axion can develop a potential. But this seems rather paradoxical: one is postulating a symmetry, preserved to some high degree of approximation but which is not a symmetry: it is at the least broken by tiny QCD effects. Is this reasonable? To understand the nature of the problem, consider one of the ways in which an axion can arise. In some approximation we can suppose that we have a global symmetry under which a scalar field ϕ transforms as $\phi \to e^{i\alpha}\phi$. Suppose, further, that ϕ has an expectation value. This could arise due to a potential, $V(\phi) = -\mu^2|\phi|^2 + \lambda|\phi|^4$. Associated with the symmetry breaking would be a (pseudo)-Goldstone boson, a. We can parameterize ϕ as follows:

$$\phi = f_a e^{ia/f_a}, \quad |\langle\phi\rangle| = f_a. \tag{5.135}$$

If this field couples to fermions, they gain mass from its expectation value. At one loop, the same diagrams as those discussed in our anomaly analysis generate a coupling $aF\tilde{F}$, from integrating out the fermions. This calculation is identical to the corresponding calculation for pions discussed earlier. But we usually assume that global symmetries in nature are accidents. For example, baryon number is conserved in the Standard Model simply because there are no gauge-invariant renormalizable operators which violate the symmetry. We believe it is violated by higher-dimensional terms. The global symmetry we postulate here is presumably an accident of the same sort. But for the axion, the symmetry must be *extremely* good. We can introduce an axion quality Q_a,

$$Q_a = \frac{1}{m_a^2 f_a} \frac{\partial V}{\partial a}, \tag{5.136}$$

which must be less than 10^{-10}. Suppose, for example, one has a symmetry breaking operator $\phi^{n+4}/M_{\rm p}^n$. Such a term gives a linear contribution to the axion potential of order $f_a^{n+3}/M_{\rm p}^n$. If $f_a \sim 10^{11}$, this swamps the would-be QCD contribution $m_\pi^2 f_\pi^2/f_a$ unless $n > 12$!

This last objection finds an answer in string theory. In this theory there are axions with just the right properties, i.e. there are symmetries in the theory which are *exact* in perturbation theory, but which are broken by exponentially small non-perturbative effects. The most natural value for f_a would appear to be of order $M_{\rm GUT}$ or $M_{\rm p}$. Whether this can be made compatible with cosmology, or whether one can obtain a lower scale, is an open question to which we will return.

Suggested reading

There are a number of excellent books and reviews on anomalies, as well as good treatments in quantum field theory textbooks. The texts of Peskin and Schroeder (1995), Pokorski (2000) and Weinberg (1995) have excellent treatments of different aspects of anomalies. The string textbook of Green *et al.* (1987) provides a good introduction to anomalies in higher dimensions. One of the best introductions to the physics of instantons is provided in the article of Coleman (1985). The $U(1)$ problem in two-dimensional electrodynamics, and its role as a model for confinement, was discussed by Casher *et al.* (1974). The serious reader should study 't Hooft's instanton paper from 1976, in which he both uncovers much of the physical significance of the instanton solution and also performs a detailed evaluation of the determinant. The propagators in the instanton background are given in Brown *et al.* (1978). Instantons in CP^N models were studied by Affleck (1980). The dependence of $d_{\rm n}$ on θ was calculated by Crewther *et al.* (1979) in a short and quite readable paper.

Exercises

(1) Derive Eq. (5.15).

(2) Calculate the decay rate of the π^0 to two photons. You will need the matrix element

$$\langle \pi(q)|\partial_\mu j_5^{\mu 3}|0\rangle = f_\pi q^\mu e^{iq\cdot x}, \tag{5.137}$$

where $f_\pi = 93\,\mathrm{MeV}$. You will need also to compute the anomaly in the third component of the axial isospin current.

(3) Fill in the details of the anomaly computation in two dimensions, being careful about signs and factors of 2.

(4) Fill in the details of the Fujikawa computation of the anomaly, in the CP^N model, again being careful about factors of 2. Make sure that you understand why one is calculating

a determinant and why the factors appear in the exponential. Verify that the action of Eq. (5.56) is equal to

$$\mathcal{L} = g_{\phi,\phi^*} \partial_\mu \phi \partial_\mu \phi^*, \tag{5.138}$$

where g is the metric of the sphere in complex coordinates, i.e. it is the line element $dx_1^2 + dx_2^2 + dx_3^2$ expressed as $g_{z,z} dzdz + g_{z,z^*} dzdz^* + g_{z^*z} dz^* dz + g_{dz^* dz^*} dz^* dz^*$. A model with an action of this form is called a non-linear sigma model; the idea is that the fields live on some "target" space, with metric g. Verify Eqs. (5.56) and (5.59).

(5) Check that Eqs. (5.85) and (5.86) solve (5.83).

6 Grand unification

One of the troubling features of the Standard Model is the plethora of coupling constants; overall there are 18, counting θ. It seems puzzling that a theory which purports to be a fundamental theory should have so many parameters. Another is the puzzle of charge quantization: why are the hypercharges all rational multiplets of one another (and, as a result, the electric charges rational multiples of one another)? Finally, the gauge group itself is rather puzzling. Why is it semi-simple rather than simple?

Georgi and Glashow put forward the *grand unification* proposal which answers some of these questions. They suggested that the underlying gauge symmetry of nature is a simple group, broken at some high-energy scale down to the gauge group of the Standard Model. The Standard Model gauge group has rank 4 (there are four commuting generators); $SU(N)$ groups have rank $N-1$. So the simplest group among the $SU(N)$ groups which might incorporate the Standard Model is $SU(5)$. Without any fancy group theory, it is easy to see how to embed $SU(3) \times SU(2) \times U(1)$ in $SU(5)$. Consider the gauge bosons. These are in the adjoint representation of the group. Written as matrices, under infinitesimal space–time independent gauge transformations we have

$$\delta A_\mu = i\omega^a [T^a, A_\mu]. \tag{6.1}$$

The T_as are 5×5 traceless Hermitian matrices; altogether, there are 24 of them. We can then break up the gauge generators in the following way. Writing indices on T^a as $(T^a)_i^j$, the T^as act on the fundamental five-dimensional representation ("the 5") as

$$(T^a)_i^j 5_j. \tag{6.2}$$

So, if we think of the 5 as

$$5 = \begin{pmatrix} q_1 \\ q_2 \\ q_3 \\ L_1 \\ L_2 \end{pmatrix} \tag{6.3}$$

then the T^as can be broken up into a set of $SU(3)$ generators and a set of $SU(2)$ generators:

$$T^a = \begin{pmatrix} \lambda^a/2 & 0 \\ 0 & 0 \end{pmatrix}, \quad T^i = \begin{pmatrix} 0 & 0 \\ 0 & \sigma^i/2 \end{pmatrix}. \tag{6.4}$$

Here the λ^as are Gell-Mann's $SU(3)$ matrices and the σ^is are the Pauli matrices. There are three commuting matrices among these. The remaining, diagonal, matrix can be taken to be

$$\tilde{Y} = \frac{1}{\sqrt{60}} \begin{pmatrix} -2 & 0 & 0 & 0 & 0 \\ 0 & -2 & 0 & 0 & 0 \\ 0 & 0 & -2 & 0 & 0 \\ 0 & 0 & 0 & 3 & 0 \\ 0 & 0 & 0 & 0 & 3 \end{pmatrix}. \tag{6.5}$$

Finally, there are 12 off-diagonal matrices:

$$\left(X^i_a\right)^b_j = \delta^i_j \delta^b_a \tag{6.6}$$

where $a, b = 1, 2, 3$; $i, j = 1, 2$. These are not Hermitian; they are analogous to the raising and lowering operators in $SU(2)$. One can readily form Hermitian linear combinations. The associated vector mesons must be very heavy; they mediate B-violating processes, as in Fig. 6.1. These can lead, for example, to $p \to \pi^0 e^+$.

We want to claim that $\tilde{Y}$ is proportional to the ordinary hypercharge and determine the proportionality constant. To do this, we consider, not the 5 but the $\bar{5}$ and make the identification

$$\bar{5} = \begin{pmatrix} \bar{d}_1 \\ \bar{d}_2 \\ \bar{d}_3 \\ L_1 \\ L_2 \end{pmatrix}. \tag{6.7}$$

Now, the generators of $SU(5)$ acting on the $\bar{5}$ are $-T^{aT}$. So we can read off immediately that $Y = \sqrt{60}\tilde{Y}/3$. Since the gauge groups are unified in a single group, the gauge couplings are all the same, so we can compute the Weinberg angle. Calling g the $SU(5)$ coupling,

$$g\tilde{Y} = \frac{g'}{2} Y, \tag{6.8}$$

where g' is the hypercharge coupling of the Standard Model. From this, $g^2 = (5/3)g'^2$. The Weinberg angle is given by

$$\sin^2 \theta_W = \frac{g'^2}{g^2 + g'^2} = \frac{3}{8}. \tag{6.9}$$

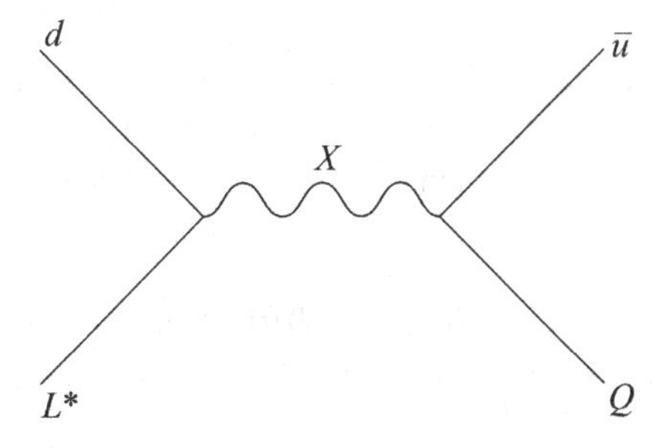

Fig. 6.1 The exchange of heavy vector particles in GUTs violates B and L. It can lead to processes such as $p \to \pi^0 e^+$.

So we have two dramatic predictions, if we assume that the Standard Model is unified in this way:

1. the $SU(3)$ and $SU(2)$ gauge couplings are equal;
2. the Weinberg angle satisfies $\sin^2\theta_W = 3/8$.

Before assessing these predictions, let us first figure out where we would put the rest of the quarks and leptons. In a single generation of the Standard Model, there are 15 fields. The group $SU(5)$ has a ten-dimensional representation, the antisymmetric product of two 5s. It can be written as an antisymmetric matrix, 10_{ij}. If i and j are both $SU(3)$ indices, we obtain a $(\bar{3},1)_{-4/3}$ of $SU(3)$. If one is an $SU(3)$ and one an $SU(2)$ index, we obtain a $(3,2)_{1/3}$. If both are $SU(2)$ indices, we obtain a $(1,1)_2$. Here the subscripts denote the ordinary hypercharge, related to $\tilde{Y}$ as above. These are just the quantum numbers of the quark doublet Q, of $\bar{u}$ and of $\bar{e}$. As a matrix,

$$10 = \begin{pmatrix} 0 & \bar{u}^3 & -\bar{u}^2 & Q_1^1 & Q_1^2 \\ -\bar{u}^3 & 0 & \bar{u}^1 & Q_2^1 & Q_2^2 \\ \bar{u}^2 & -\bar{u}^1 & 0 & Q_3^1 & Q_3^2 \\ -Q_1^1 & -Q_2^1 & -Q_3^1 & 0 & \bar{e} \\ -Q_1^2 & -Q_2^2 & -Q_3^2 & -\bar{e} & 0 \end{pmatrix}. \tag{6.10}$$

So, a single generation of quarks and leptons fits neatly into a $\bar{5}$ and 10 of $SU(5)$.

6.1 Cancelation of anomalies

An anomaly in a gauge symmetry would represent a breakdown of gauge invariance. The consistency of gauge symmetries rests, however, on gauge invariance. For example, to demonstrate that such theories are both unitary and Lorentz invariant we have used different gauges. The cancelation of anomalies is crucial, and the absence of anomalies in the Standard Model is surely no accident.

It is not hard to check that in $SU(5)$ the anomaly of the $\bar{5}$ cancels that of the 10. In general, the anomalies in a gauge theory are proportional to d_{abc}, where

$$\{T_a, T_b\} = d_{abc}T_c. \tag{6.11}$$

One can organize the anticommutator above in terms of the various types of generator, for example $SU(3)$, $SU(2)$, $U(1)$, and the off-diagonal generators, which transform as $(3,2)$ of $SU(3)\times SU(2)$, and then check each class. We leave the details for the exercises.

6.2 Renormalization of couplings

If we are going to describe the Standard Model, $SU(5)$ must break at some high-energy scale to $SU(3)\times SU(2)\times U(1)$. Above this scale, the full $SU(5)$ symmetry holds to a good approximation, and all couplings renormalize in the same way. Below this scale the couplings renormalize differently. We can write down the equations for the renormalization of the three separate couplings:

$$\alpha_i^{-1}(\mu) = \alpha_{\rm gut}^{-1}(M_{\rm gut}) + \frac{b_0^i}{4\pi} \ln \frac{\mu}{M_{\rm gut}}. \tag{6.12}$$

We can calculate the beta functions at one loop starting with the usual formula:

$$b_0 = \frac{11}{3} C_A - \frac{4}{3} c_{\rm f}^{(i)} N_{\rm f}^{(i)} - \frac{1}{3} c_\phi^{(i)} N_\phi^{(i)}, \tag{6.13}$$

where $N_{\rm f}^{(i)}$ is the number of fermions in the ith representation; $N_\phi^{(i)}$ is the number of scalars. For $SU(N)$ $C_{\rm A} = N$ and, for fermions or scalars in the fundamental representation, $c_{\rm f} = c_\phi = 1/2$.

For the $SU(3)$ and $SU(2)$ couplings the beta function coefficients b_0^i are readily computed. For $U(1)$, we need to remember the relative normalization computed above:

$$b_0^2 = \frac{181}{6}, \quad b_0^3 = 7, \quad b_0^1 = \frac{61}{15}. \tag{6.14}$$

We can run these equations backwards. The $SU(2)$ and $U(1)$ couplings are the best measured, so it makes sense to start with these and run them up to the unification scale. This determines $\alpha_{\rm gut}$ and $M_{\rm gut}$. We can then predict the value of the $SU(3)$ coupling at, say, M_Z. One finds that the unification scale, $M_{\rm gut}$, is about 10^{15} GeV and that α_3 is off by about seven standard deviations. In the exercises you will have the opportunity to perform this calculation in detail. We will see later that low-energy supersymmetry greatly improves this.

6.3 Breaking to $SU(3) \times SU(2) \times U(1)$

In $SU(5)$, it is relatively easy to introduce a set of Higgs fields which break the gauge symmetry down to $SU(3) \times SU(2) \times U(1)$. Consider a Hermitian scalar field Φ in the adjoint representation. Writing Φ as a matrix, we have the transformation law

$$\delta\Phi = \omega^a [T^a, \Phi]. \tag{6.15}$$

Suppose that the minimum of the Φ potential lies at a point where

$$\Phi = v\tilde{Y}. \tag{6.16}$$

Then the $SU(3)$, $SU(2)$ and $U(1)$ generators all commute with $\langle\Phi\rangle$, but those for the X bosons do not.

Consider the most general $SU(5)$-invariant potential:

$$V = -m^2 \,{\rm Tr}\,\Phi^2 + \frac{\lambda}{4} {\rm Tr}\,\Phi^4 + \frac{\lambda'}{4} ({\rm Tr}\,\Phi^2)^2. \tag{6.17}$$

One can find the minimum of this potential by first using an $SU(5)$ transformation to diagonalize Φ, obtaining

$$\Phi = \mathrm{diag}(a_1, a_2, a_3, a_4, a_5). \tag{6.18}$$

The potential is a function of the a_is, which one wants to minimize subject to the constraint of vanishing trace. This can be done by using a Lagrange multiplier.

To establish that one has a local minimum of the form Eq. (6.16), one can proceed more simply. Write the potential as a function of v:

$$V = -\frac{1}{2}m^2 v^2 + \frac{a\lambda}{4} + \frac{b\lambda'}{4}v^4, \tag{6.19}$$

where $a = 7/120$, $b = 1/4$. Then the extremum with respect to v occurs for

$$v = \frac{\mu}{\sqrt{a\lambda + b\lambda'}}. \tag{6.20}$$

To establish that this is a local minimum, we need to show that the eigenvalues of the scalar mass-squared matrix are all positive. We can investigate this by considering small fluctuations about the stationary point. This point preserves $SU(3)\times SU(2)\times U(1)$. Writing $\Phi = \langle\Phi\rangle + \delta\Phi$, $\delta\Phi$ can be decomposed under $SU(3) \times SU(2) \times U(1)$ as follows:

$$\delta\Phi = (1,1) + (8,1) + (1,3) + (3,2) + (\bar{3},2). \tag{6.21}$$

The point (6.20) is certainly stationary; because of the symmetry, only the $(1,1)$ term can appear linearly in the potential, and it is this piece whose minimum we have just found. To establish that the point (6.20) is in fact a local minimum, one needs to show that the quadratic terms in the fluctuations are all positive. This is done in the exercises.

6.4 $SU(2) \times U(1)$ breaking

In addition to the adjoint, it is necessary to include a 5 representations of the Higgs H in order to break $SU(2) \times U(1)$ down to the $U(1)$ of electromagnetism and to give mass to the quarks and leptons. The Higgs has the form

$$H = \begin{pmatrix} H_c \\ H_d \end{pmatrix}, \tag{6.22}$$

where H_c is a color triplet of scalars and H_d is the ordinary Higgs doublet. For H one might have been tempted to write a potential of the form

$$V(H) = -\mu^2|H|^2 + \frac{\lambda}{4}|H|^4. \tag{6.23}$$

However, this would lead to a number of difficulties. Perhaps the most important is that, when included in the larger theory with the adjoint field Φ, this potential has too much symmetry; there is an extra $SU(5)$ which would lead to an assortment of unwanted Goldstone bosons. At the same time the scale μ must be of order the scale of electroweak symmetry breaking (as long as λ is not too much larger than unity). So, the Higgs triplets will have masses of order the weak scale. But if the doublet couples to quarks and leptons, the triplet will have *baryon- and lepton-number-violating* couplings to the quarks and leptons. So the triplet must be very massive.

Both problems can be solved if we couple Φ to H. The allowed couplings include:

$$V_{\Phi H} = \Gamma H^* \Phi H + \lambda' H^* H \,\mathrm{Tr}\, \Phi^2 + \lambda'' H^* \Phi^2 H. \tag{6.24}$$

If we carefully adjust the constants Γ, λ', λ'' and μ^2, we can arrange that the doublets are light and the triplets are heavy. For example, if we choose $\lambda = \lambda' = 0$ and $\mu^2 = -3(\Gamma/\sqrt{60})v - \epsilon$ then the Higgs doublets have mass-squared $-\epsilon$ in the Lagrangian, while the triplets have mass of order M_{gut}. This tuning of parameters, which must be performed in each order of perturbation theory, provides an explicit realization of the hierarchy problem.

Turning to the fermion masses, we are led to an interesting realization: not only does grand unification make predictions for the gauge couplings, it can predict relations among fermion masses as well. The gange group $SU(5)$ permits the following couplings:

$$\mathcal{L}_y = y_1 \epsilon_{ijklm} H^i 10^{jk} 10^{lm} + y_2 H_i^* \bar{5}_j 10^{ij}. \tag{6.25}$$

Here the ys are matrices in the space of generations. When H acquires an expectation value, it gives mass to the quarks and leptons. The first coupling gives mass to the up-type quarks. The second coupling gives mass to both the down-type quarks and the leptons. If we consider only the heaviest generation, we then have the tree level prediction

$$m_b = m_\tau. \tag{6.26}$$

This prediction is off by a factor 3 but, like the prediction of the coupling constant, it can be corrected by renormalization to roughly the observed amount. For the lightest quarks and leptons the prediction fails. However, unlike the unification of gauge couplings, such predictions can be modified if there are additional Higgs fields in other representations. In addition, for the lightest fermions, higher-dimensional operators, suppressed by powers of the Planck mass, can make significant contributions to masses. In supersymmetric grand unified theories, the ratio of the GUT scale to the Planck scale is about 10^{-2}, whereas the lightest quarks and leptons have masses four orders of magnitude below the weak scale. We will postpone a numerical study of these corrections since the simplest $SU(5)$ theory does not correctly predict the values of the coupling constants, and will return to this subject when we discuss supersymmetric grand unified theories, which do successfully predict the observed values of the couplings.

6.5 Charge quantization and magnetic monopoles

While we must postpone success with the calculation of the unified couplings to our chapters on supersymmetry, we should pause and note two triumphs. First, we have a possible explanation for one of physics' greatest mysteries: why is electric charge quantized? Here it is automatic; electric charge, an $SU(5)$ generator, is quantized, just as color and isospin are quantized.

However, Dirac long ago offered another explanation of electric charge quantization: magnetic monopoles. He realized that the consistency of quantum mechanics demands that if even a single monopole exists in the universe, electric charges must all be integer multiples of a fundamental charge. So we might suspect that magnetic monopoles are hidden somewhere in this story. Indeed they are; this are discussed in Chapter 7.

6.6 Proton decay

We have discussed the dimension-six operators which can arise in the Standard Model and violate baryon number. Exchanges of the X bosons generate operators such as

$$\frac{g^2}{M_X^2} Q\sigma_\mu \bar{u}^* Q\sigma^\mu \bar{e}^*. \tag{6.27}$$

This leads to the decay $p \to \pi^0 e^+$. In this model, one predicts a proton lifetime of order 10^{28} years if $M_{\rm gut} \approx 10^{15}$ GeV. The current limit on this decay mode is 5×10^{33} years. We will discuss the situation for supersymmetric models later.

The realization that baryon-number violation is likely in any more fundamental theory opens up a vista on a fundamental question about nature: why is there more matter than antimatter in the universe? If, at some very early time, there were equal amounts of matter and antimatter then, if baryon number is violated, one has the possibility of producing an excess. Other conditions must be satisfied as well; we will describe this in the chapter on cosmology.

6.7 Other groups

While $SU(5)$ may in some respects be the simplest group for unification, once one has set off in this direction there are many possibilities. Perhaps the next simplest is unification in the group $O(10)$. As $O(10)$ has rank 5, there is one extra commuting generator; presumably this symmetry must be broken at some scale. More interesting, though, is the fact that a single generation fits neatly into an irreducible representation: the 16. The group $O(10)$ has an $SU(5)$ subgroup, under which the 16 decomposes as a $10 + \bar{5} + 1$. The singlet has precisely the right Standard Model quantum numbers – none – to play the role of the right-handed neutrino in the seesaw mechanism; see below Eq. (4.17).

We will not review the group theory of O groups in detail, but we can describe some of the important features. We will focus specifically on $O(10)$, but much of the discussion here is easily generalized to other groups. The generators of $O(10)$ are 10×10 antisymmetric matrices. There are 45 of these. We are particularly interested in how they transform under the Standard Model group. The embedding of the Standard Model in $SU(5)$, as we have learned, is very simple, so a useful way to proceed to understand $O(10)$ is to find its $SU(5)$ subgroup.

One way to think of $O(10)$ is as the group of rotations of ten-dimensional vectors. Call the components of such a vector x^A, $A = 1, \ldots, 10$. Transformations in $SU(5)$ are "rotations" of complex five-dimensional vectors z^i. So, we define

$$z^1 = x^1 + ix^2, \quad z^2 = x^3 + ix^4, \quad z^3 = x^5 + ix^6 \tag{6.28}$$

and so on. With this correspondence it is easy to see that there is a subgroup of $O(10)$ transformations that preserves the product $z \cdot z'^*$. This is the $SU(5)$ subgroup of $O(10)$.

From our construction, it follows that the 10 of $O(10)$ transforms as a $5 + \bar{5}$ of $SU(5)$. We can determine the decomposition of the adjoint by writing

$$A^{AB} = A^{i\bar{i}} + A^{ij} + A^{\bar{i}\bar{j}}. \tag{6.29}$$

The labeling here is meant to indicate the types of complex index that the matrix A can carry. The first term is just the 24-dimensional representation of $SU(5)$, plus an additional singlet. This singlet is associated with a $U(1)$ subgroup of $O(10)$, which rotates all the objects with i-type indices by one phase and all those with $\bar{i}$ type indices by the opposite phase. Note that A^{ij} is antisymmetric in its indices; in our study of $SU(5)$ we learned that this is the 10 representation. We can take it to carry charge 2 under the $U(1)$ subgroup. Then $A^{\bar{i}\bar{j}}$ corresponds to the $\overline{10}$ representation, with charge -2. This accounts for all 45 fields.

But where is the 16-dimensional representation? We are familiar, from our experience with ordinary rotations in three and (Euclidean four) dimensions as well as from the Lorentz group, with the fact that O groups may have spinor representations. To construct these we need to introduce the equivalent of the Dirac gamma matrices Γ, satisfying

$$\{\Gamma^I, \Gamma^J\} = 2\delta^{IJ}. \tag{6.30}$$

It is not hard to construct explicit matrices which satisfy these anticommutation relations but there is a simpler approach, which also makes the $SU(5)$ embedding clear. The anticommutation relations are similar to the relations for fermion creation and annihilation operators. So, define

$$a^1 = \frac{1}{2}(\Gamma^1 + i\Gamma^2), \quad a^2 = \frac{1}{2}(\Gamma^3 + i\Gamma^4) \tag{6.31}$$

and so on, and similarly for their complex conjugates. Note that the a^is form a 5 of $SU(5)$, with charge $+1$ under the $U(1)$. These operators satisfy the algebra

$$\{a^i, a^{\bar{j}}\} = \delta^{i\bar{j}}. \tag{6.32}$$

These are the anticommutation relations for five pairs of fermion creation–annihilation operators. We know how to construct the corresponding "states", i.e. the representations of the algebra. We define a state $|0\rangle$ annihilated by the a^is. Then there are five states created by the action of $a^{\bar{i}}$ on this state:

$$\bar{5}_{-1} = a^{\bar{i}}|0\rangle. \tag{6.33}$$

The main symbol $\bar{5}$ indicates the $SU(5)$ representation and the subscript indicates the $U(1)$ charge. We could now construct the states obtained with two creation operators, but let us construct the states built using an odd number:

$$10_{-3} = a^{\dagger i} a^{\dagger j} a^{\dagger k}|0\rangle, \quad 1_{-5} = a^{\dagger 1} a^{\dagger 2} a^{\dagger 3} a^{\dagger 4} a^{\dagger 5}|0\rangle. \tag{6.34}$$

We have indicated that the first representation transforms like a 10 of $SU(5)$, while the second transforms like a singlet.

The states which involve even numbers of creation operators transform like a 5, a $\overline{10}$, and a singlet. Why do we distinguish these two sets? Remember, the goal of this construction is to obtain *irreducible* representations of the group $O(10)$. As in the Dirac theory, we can construct the symmetry generators from the Dirac matrices,

$$S^{IJ} = \frac{i}{4}[\Gamma^I, \Gamma^J]. \tag{6.35}$$

These, too, can be decomposed on a complex basis, like A^{IJ}. But, as for the usual Dirac matrices, there is another Γ matrix that we can construct, which is the analog of Γ^5: Γ^{11}. This matrix anticommutes with all the Γs, and so with the a's. Thus the states with even numbers of creation operators are eigenstates with eigenvalue $+1$ under Γ^{11}, while those with odd numbers are eigenstates with eigenvalue -1. Since Γ^{11} commutes with the symmetry generators, these two representations are irreducible.

A similar construction works for other groups. When we come to discuss string theories in ten dimensions, we will be especially interested in the representations of $O(8)$. Here the same construction yields two eight-dimensional representations, denoted 8 and 8′.

The embedding of the states of the Standard Model in $O(10)$ is clear, since we already know how to embed them in a $\bar{5}+10$ of $SU(5)$. But what of the other state in the 16? This is a Standard Model singlet. We do not yet have a candidate in the particle data book for this. However, there are two observations we can make. First, the symmetries of the Standard Model do not forbid a mass for this particle. What does forbid a mass is the extra $U(1)$. So, if this symmetry is broken at very high energies, perhaps with the initial breaking of the gauge symmetry, this particle can gain a large mass. We will not explore the possible Higgs fields in $O(10)$ but, as in $SU(5)$, there are many possibilities and the $U(1)$ can readily be broken. Second, this particle has the right quantum numbers to couple to the left-handed neutrino of the Standard Model. So this particle can naturally lead to a "seesaw" neutrino mass. This mass might be expected to be of order some typical Yukawa coupling squared divided by the unification scale. It is also possible that this extra $U(1)$ is broken at some lower scale, yielding a larger value for the neutrino mass.

Suggested reading

There is any number of good books and reviews on the subject of grand unification. The books by Ross (1984), Mohapatra (2003) and Ramond (1999) all treat the topics introduced in this chapter in great detail. The reader will find his or her interest in this topic increases after studying some aspects of supersymmetry.

Exercises

(1) Verify the cancelation of anomalies between the $\bar{5}$ and 10 representations of $SU(5)$.

(2) Establish the conditions for the solution of Eq. (6.16) to be a local minimum of the potential.

(3) Perform the calculation of coupling unification in the $SU(5)$ model. Verify Eqs. (6.14) for the $SU(3)$, $SU(2)$ and $U(1)$ beta functions. Start with the measured values of the $SU(2)$ and $U(1)$ couplings, being careful about the differing normalizations in the Standard Model and in $SU(5)$. Compute the value of the unification scale (the point where these two couplings are equal); then determine the value of α_3 at M_Z. Compare with the value given by the Particle Data Group. You need only study the equations to one-loop order. In practice, two-loop corrections, as well as threshold effects and higher-order corrections to the beta function, are often included.

(4) Add to the Higgs sector of the $SU(5)$ theory a set of scalars in the 45 representation. Show that in this case all the quark masses are free parameters.

7 Magnetic monopoles and solitons

Anyone who has inspected Maxwell's equations even briefly has probably speculated about the existence of magnetic monopoles. There is no experimental evidence for magnetic monopoles, but the equations would be far more symmetric if they existed. It was Dirac who first considered carefully the implications of monopoles, and he came to a striking conclusion: the existence of monopoles would require that *electric charge* be quantized in terms of a fundamental unit. The problem of describing a monopole lies in writing $\vec{B} = \vec{\nabla} \times \vec{A}$. We could simply give up this identification, but Dirac recognized that $\vec{A}$ is essential in formulating quantum mechanics. To resolve the problem we can follow Wu and Yang and maintain that $\vec{B} = \vec{\nabla} \times \vec{A}$ but not require that the vector potential be single valued. Suppose that we have a monopole located at the origin. In the northern hemisphere we can take

$$\vec{A}_{\rm N} = \frac{g}{4\pi r} \frac{1 - \cos\theta}{\sin\theta} \hat{e}_\phi, \tag{7.1}$$

while in the southern hemisphere we take

$$\vec{A}_{\rm S} = -\frac{g}{4\pi r} \frac{1 + \cos\theta}{\sin\theta} \hat{e}_\phi. \tag{7.2}$$

By looking up the formulae for the curl operator in spherical coordinates, you can check that, in both hemispheres,

$$\vec{B} = \frac{g}{4\pi r^2} \hat{r}, \tag{7.3}$$

so indeed this does describe a magnetic monopole.

Each of expressions (7.1) and (7.2) is singular along a half-line: $A_{\rm N}$ is singular along $\theta = \pi$; $A_{\rm S}$ is singular along $\theta = 0$. These string-like singularities are known as *Dirac strings*. They are suitable vector potentials to describe long thin solenoids which start at the origin and go to infinity along the negative or positive z axis. With discontinuous $\vec{A}$, though, we need to ask whether quantum mechanics is consistent. Consider the equator ($\theta = \pi/2$). We have

$$\vec{A}_{\rm N} - \vec{A}_{\rm S} = \frac{g}{2\pi r} \hat{e}_\phi = -\vec{\nabla}\chi, \quad \chi = -\frac{g}{2\pi}\phi, \tag{7.4}$$

where ϕ is the azimuthal angle and χ is a general function. So, the difference has the form of a gauge transformation. But to be a gauge transformation it must act sensibly on particles of definite charge. In particular, it must be single valued. As such a particle circumnavigates the sphere, its wave function acquires a phase

$$\exp\left(ie \int d\vec{x} \cdot \vec{A} \right). \tag{7.5}$$

Potentially, this phase is different for $\vec{A}_{\rm N}$ and $\vec{A}_{\rm S}$, in which case the string would be a detectable, real, object. But the phases are the same if

$$\exp\left(i\frac{eg}{2\pi}\oint d\vec{x}\cdot\vec{\nabla}\phi\right) = 1 \quad \text{or} \quad eg = 2\pi n. \tag{7.6}$$

This is known as the *Dirac quantization condition*. Dirac argued that, since e can be the charge of any charged particle, if there is even one monopole somewhere in the universe, this result shows that charge must be quantized.

In pure electrodynamics the status of magnetic monopoles is obscure; the $\vec{B}$ field is singular and the energy is infinite. In non-Abelian gauge theories with scalar fields (Higgs fields), however, monopoles often arise as finite-energy non-dissipative solutions of the classical equations. Such solutions cannot arise in linear theories like electrodynamics; all configurations in such a theory spread with time. Non-dissipative solutions can only arise in non-linear theories, and even then, such solutions – known as solitons – can only arise in special circumstances.

The simplest theory which exhibits monopole solutions is $SU(2)$ (more precisely $O(3)$) Yang–Mills theory with a single Higgs particle in the adjoint representation. But, before considering this case, which is somewhat complicated, it is helpful to consider solitons in lower-dimensional situations.

7.1 Solitons in 1 + 1 dimensions

Consider a quantum field theory in 1 + 1 dimensions, with

$$\mathcal{L} = \frac{1}{2}(\partial_\mu\phi)^2 - V(\phi). \tag{7.7}$$

Here

$$V(\phi) = -\frac{1}{2}m^2\phi^2 + \frac{\lambda}{4}\phi^4. \tag{7.8}$$

This potential, which is symmetric under $\phi \to -\phi$, has two degenerate minima, $\pm\phi_0$. Normally, we would choose as our vacuum a state localized about one or the other minimum. These correspond to trivial solutions of the equations of motion. We can consider a more interesting configuration, a localized finite-energy solution known as a *soliton*, for which

$$\phi(x \to \pm\infty) \to \pm\phi_0. \tag{7.9}$$

Such a solution interpolates between the two different vacua. We can construct this solution much as one solves analogous problems in classical mechanics, by quadrature. Finding the solution for this particular model, known as a *kink*, is left for the exercises; the result is

$$\phi_{\rm kink} = \phi_0 \tanh[(x - x_0)m]. \tag{7.10}$$

This solution is shown in Fig. 7.1. The kink has finite energy. As we have indicated, there is a continuous infinity of solutions, corresponding to the fact that this kink can be

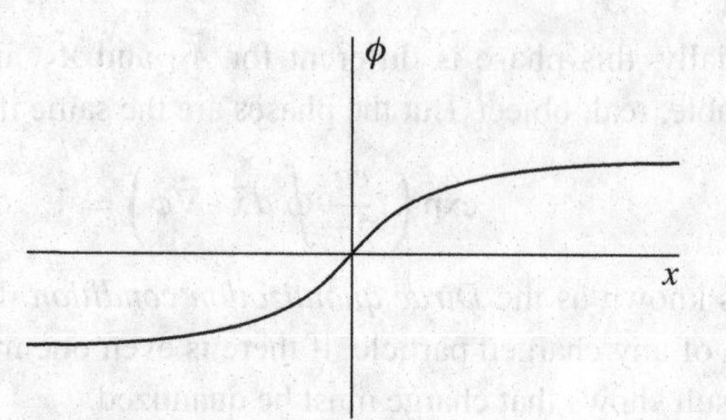

Fig. 7.1 Kink solution of the two-dimensional field theory.

located anywhere; this is a consequence of the underlying translational invariance. We can use this to understand in what sense the kink is a particle. Consider configurations which are not quite solutions of the equations of motion, in which x_0 is allowed to be a slowly varying function $x_0(t)$ of t. We can write down the action for these configurations:

$$S_{\text{kink}} = \int dt \int dx \left[\frac{1}{2}(\partial_\mu \phi_{\text{kink}})^2 - V(\phi_{\text{kink}}) \right]. \tag{7.11}$$

Only the $\dot{\phi}$ term contributes. The result is

$$S_{\text{kink}} = \int dt \frac{M}{2} \dot{x}_0^2. \tag{7.12}$$

Here M is precisely the energy of the kink. So, the kink truly acts as a particle. The quantity x_0 is called a *collective coordinate*. We will see that such collective coordinates arise for each symmetry broken by the soliton. These are similar to the collective coordinates we encountered in the Euclidean problem of the instanton.

7.2 Solitons in 2 + 1 dimensions: strings or vortices

As we go up in dimension, the possible solitons become more interesting. Consider a $U(1)$ gauge theory in $2+1$ dimensions, with a single charged scalar field ϕ. This model is often called the Abelian Higgs model. The Lagrangian is

$$\mathcal{L} = |D_\mu \phi|^2 - V(|\phi|). \tag{7.13}$$

We assume that the potential is such that

$$\langle \phi \rangle = v. \tag{7.14}$$

Now we have a possibility that we have not considered before. Working in plane polar coordinates r, θ, if we consider only the potential then we can imagine obtaining finite-energy configurations for which, at large r,

$$\phi \to e^{in\theta} v. \tag{7.15}$$

Because the potential tends to its minimum at infinity, such a configuration has finite potential energy. However, the kinetic energy diverges, since $\partial_\mu \phi$ includes $(1/r)\partial_\theta \phi$. We can try to cancel this with a non-vanishing gauge field. At ∞, the scalar field is a gauge transformation of the constant configuration, so to achieve finite energy we want to gauge-transform the gauge field as well,

$$A_\theta \to n; \tag{7.16}$$

consequently, at ∞, $D_\mu \phi \to 1/r^2$ or a higher negative power of r. It is not hard to construct such solutions numerically. As for the kinks, these configurations again have collective coordinates, corresponding to the two translational degrees of freedom and a rotational (or charge) degree of freedom.

We can take these configurations as configurations in a (3+1)-dimensional theory, which are constant with respect to z. Viewed in this way, these are vortices, or strings. One has collective coordinates corresponding to transverse motions of the string, $x_0(z,t), y_0(z,t)$. These string configurations could be quite important in cosmology. Such a broken $U(1)$ theory could lead to the appearance of long strings, which could carry enormous amounts of energy. For a time, these were considered a possible origin of inhomogeneities leading to the formation of galaxies, but the data now disfavors this possibility.

7.3 Magnetic monopoles

Dirac's argument shows that, in the presence of a monopole, electric charges are all multiples of a basic charge. This means that the $U(1)$ symmetry is effectively compact. So, a natural place to look for monopoles is in gauge theories where $U(1)$ is a subgroup of a simple group. The $SU(5)$ grand unified theory is an example, in which electric charge is quantized.

We start, though, with the simplest example of this sort, an $SU(2)$ gauge theory with Higgs fields in the adjoint representation, ϕ^a. Such a theory was first considered by Georgi and Glashow as a model for weak interactions without neutral currents and is known as the Georgi–Glashow model. An expectation value for ϕ, $\phi^3 = v$ or

$$\phi = \frac{v}{2}\begin{pmatrix} 1 & 0 \\ 0 & -1 \end{pmatrix} \tag{7.17}$$

leaves an unbroken $U(1)$. The spectrum includes massive charged gauge bosons $W^\pm$ and a massless gauge boson, which we will call the photon, γ. By analogy to the string or vortex solutions, we require finite energy at ∞:

$$\phi \to gv. \tag{7.18}$$

In the $(2+1)$-dimensional case we could think of the gauge transformation as a mapping from the space at infinity (topologically a circle) onto the gauge group (also a circle).

In three dimensions we want gauge transformations which map the two-sphere S_2 into the gauge group $SU(2)$. For example, we can take

$$g(\vec{x}) = i\frac{\hat{x}^i\sigma^i}{2} \tag{7.19}$$

This suggests the following ansatz (guess) for a solution:

$$\phi^a = \hat{r}^a h(r), \quad A_i^a = -\epsilon_{ij}^a \frac{\hat{r}^j}{r} j(r). \tag{7.20}$$

This solution is very symmetric: it is invariant under a combined rotation in spin and isospin (rather similar to the sorts of symmetry of the instanton solution). Note that h and j satisfy coupled non-linear equations, which in general must be solved numerically. We can see from the form of the action that the mass is of order $1/g^2$. In the next section we show that an analytic solution can be obtained in a particular limit.

We can write down an elegant expression for the number of times $g(x)$ maps the sphere into the gauge group:

$$N = \frac{1}{4\pi}\int dS^i \,\epsilon_{ijk}\,\mathrm{Tr}(g\partial_j g\partial_k g). \tag{7.21}$$

In terms of the field, ϕ,

$$N = \frac{1}{8\pi v^3}\int dS^i \,\epsilon^{ijk}\epsilon^{abc}\partial_i\phi^a\partial_j\phi^b\partial_k\phi^c. \tag{7.22}$$

Finally, we need a definition of the magnetic charge. A natural choice is

$$\int d^3x \frac{1}{v}\partial_i(\phi^a B_i^a) = \frac{4\pi N}{e}.$$

Putting these statements together, we see that this solution, the *'t Hooft–Polyakov monopole*, has one Dirac unit of magnetic charge.

7.4 The BPS limit

Prasad and Sommerfield wrote down an exact monopole solution in the limit $V = 0$. This limit seemed originally rather artificial, but we will see later that some supersymmetric field theories automatically have a vanishing potential for a subset of fields. What simplifies the analysis in this limit is that the equations for the monopole, which are ordinarily second-order non-linear differential equations, become first-order equations. We will shortly see how to understand this in terms of supersymmetry. First, though, we will derive this result by looking directly at the potentials for the gauge and scalar fields. We start by deriving a bound, the Bogomol'nyi–Prasad–Sommerfield (BPS) bound, on the mass of a static field configuration. Again we call the gauge coupling e, to avoid confusion with the magnetic charge g:

$$M_{\mathrm{m}} = \int d^3x \frac{1}{2}\left[\frac{1}{e^2}\vec{B}^a\cdot\vec{B}^a + (\vec{D}\Phi)^a\cdot(\vec{D}\Phi)^a\right]. \tag{7.23}$$

We can compare this with

$$A_\pm = \int d^3x \left[\frac{1}{e}\vec{B}^a \pm (\vec{D}\Phi)^a \right]^2$$
$$= \frac{1}{2}\int d^3x \left[\frac{1}{e^2}\vec{B}^{a2} + (\vec{D}\Phi)^{a2} \right] \pm \int d^3x \frac{1}{e}\vec{B}^a(\vec{D}\Phi)^a. \tag{7.24}$$

We can integrate the last term by parts. You can check that this works for both parts of the covariant derivative, i.e. this term becomes:

$$\frac{1}{e}\int d^3x\,(\vec{D}\cdot\vec{B})^a\Phi^a - \frac{1}{e}\int d^2a\Phi^a\hat{n}\cdot\vec{B}^a. \tag{7.25}$$

The first term vanishes by the Bianchi identity (the Yang–Mills generalization of the equation $\vec{\nabla}\cdot\vec{B}=0$). The second term is v times what we have defined to be the monopole charge, g. So we have

$$A_\pm = \int d^3x \left[\frac{1}{e^2}\vec{B}^a \pm (\vec{D}\Phi)^a \right]^2 = M_{\rm m} \pm \frac{vg}{e}. \tag{7.26}$$

The left-hand side of this equation is clearly greater than zero, so we have shown that

$$M_{\rm m} \geq \left|\frac{vg}{e}\right|. \tag{7.27}$$

This bound, known as the Bogomol'nyi or BPS bound, is saturated when

$$\vec{B}^a = \pm\frac{1}{e}(\vec{D}\Phi)^a. \tag{7.28}$$

Note that while so far in this chapter we have worked in terms of $SU(2)$, this result generalizes to any gauge group with Higgs in the adjoint representation. But let us still focus on $SU(2)$ and try to find a solution which satisfies the Bogomol'nyi bound. As in the case of the 't Hooft–Polyakov monopole, it is convenient to write:

$$\Phi^a = \frac{\hat{r}^a}{er}H(evr), \quad A_i^a = -\epsilon_{ij}^a\frac{\hat{r}^j}{er}[1-K(evr)]. \tag{7.29}$$

Here we are using a dimensionless variable, $u=evr$, in terms of which the Hamiltonian scales simply. We are looking for solutions for which $H\to 0$ and $K\to 1$ as $r\to 0$. Otherwise, the solutions would be singular at the origin. At ∞, we want the configuration to look like a gauge transformation of the vacuum solution, so we require

$$K\to 0, \quad H\to evr \qquad \text{as } r\to\infty. \tag{7.30}$$

We will leave the details to the exercises, but it is straightforward to show that these equations are solved by

$$H(y) = y\coth y - 1, \quad K(y) = \frac{y}{\sinh y}. \tag{7.31}$$

The monopole mass is

$$M_{\rm m} = \frac{vg}{e} = \frac{2\pi v}{e^2}, \tag{7.32}$$

as predicted by the BPS formula.

7.5 Collective coordinates for the monopole solution

In lower-dimensional examples we witnessed the emergence of collective coordinates, which described the translations and other collective motions of the solitons. In the case of the monopole we have similar collective coordinates. Again, the solutions violate translational invariance. As a result we can generate new solutions on replacing $\vec{x}$ by $\vec{x}-\vec{x}_0$. Now viewing x_0 as a slowly varying function of t, we obtain as before the action of a non-relativistic particle of mass $M_{\rm m}$. The particle is non-relativistic in the weak coupling limit because its mass scales as $1/g^2$ and it becomes infinitely heavy as $g \to 0$.

There is another collective coordinate of the monopole solution, which has quite remarkable properties. In the monopole solution, charged fields are excited. So the monopole solution is not invariant under the $U(1)$ gauge transformations of electrodynamics. One might think that this is not important; after all, we have stressed that gauge transformations are not real symmetries but instead represent a redundancy of the description of a system. But we need to be more precise. In interpreting Yang–Mills instantons, we worked in the $A_0 = 0$ gauge. In this gauge the important gauge transformations are time-independent gauge transformations, and these fall into two classes: large gauge transformations and small gauge transformations. The small gauge transformations are those which fall rapidly to zero at infinity, and physical states must be invariant under these. For large gauge transformations this is not the case, and they can correspond to physically distinct configurations.

For the monopole configurations, the interesting gauge transformations are those which tend, at infinity, to a transformation in the unbroken $U(1)$ group. For large r, this direction is determined by the direction of the Higgs fields. We must be careful how we fix the gauge; again we will work in the $A_0 = 0$ gauge. For our collective motion, we want to study gauge transformations in this direction which vary slowly in time. It is important, however, that we remain in the $A_0 = 0$ gauge, so the transformations that we will study are not quite gauge transformations. Specifically, we consider

$$\delta A_i = \frac{D_i[\chi(t)\Phi]}{v}, \tag{7.33}$$

where $\chi(t)$ is a general time-dependent function, but we transform A_0 by

$$\delta A_0 = \frac{D_0(\chi\Phi)}{v} - \frac{\dot{\chi}\Phi}{v} \tag{7.34}$$

and, in order that the Gauss law constraint be satisfied, we require that $\delta\Phi = 0$. The action for χ has the form:

$$S = \frac{C}{2e^2}\dot{\chi}^2. \tag{7.35}$$

Note that χ is bounded between 0 and 2π, i.e. it is an angular variable. Its conjugate variable is like an angular momentum; calling this Q we have

$$Q = p_\chi = \frac{C}{e^2}\dot{\chi}, \quad H = \frac{1}{2C}e^2Q^2. \tag{7.36}$$

In the case of a BPS monopole, the constant C is $e^2M_{\rm m}/(2v^2)$. So, each monopole has a tower of charged excitations, with energies of order e^2 above the ground state. These excitations of the monopole about the ground state are known as *dyons*. The mass formula for these states has the form, in the case of a BPS monopole:

$$M = vg + \frac{vQ^2}{g}. \tag{7.37}$$

We will understand this better when we embed this structure in a supersymmetric field theory.

7.6 The Witten effect: the electric charge in the presence of θ

We have argued that in a $U(1)$ gauge theory it is difficult to see the effects of θ. But, in the presence of a monopole, a θ term (see Section 5.3) has a dramatic effect, pointed out by Witten: the monopole acquires an electric charge that is proportional to θ.

We can see this first in an heuristic way. We will work in a gauge with non-zero A_0 and take all fields as static. Then

$$\vec{E} = -\vec{\nabla}A_0, \quad \vec{B} = \frac{g}{4\pi}\frac{\vec{r}}{r^2} + \vec{\nabla}\times\vec{A}. \tag{7.38}$$

For such a configuration, the θ term,

$$\mathcal{L}_\theta = \frac{\theta e^2}{8\pi^2}\vec{E}\cdot\vec{B}, \tag{7.39}$$

takes the form

$$\mathcal{L}_\theta = -\frac{\theta e^2 g}{32\pi^2}\int d^3r A_0 \vec{\nabla}\cdot\frac{\hat{r}}{r^2} = -\frac{\theta e^2 g}{8\pi^2}\int d^3r A_0 \delta(\vec{r}). \tag{7.40}$$

We started with a magnetic monopole at the origin, but we now also have an electric charge at the origin, $\theta e^2 g/(8\pi^2)$.

One might worry that in this analysis one is dealing with a singular field configuration, but in the non-Abelian case the configuration is non-singular. We can give a more precise argument. Let us go back to the $A_0 = 0$ gauge. In this gauge we can sensibly write down the canonical Hamiltonian. In the absence of θ, the conjugate momentum to $\vec{A}$ is $\vec{E}$. But, in the presence of θ, there is an additional contribution,

$$\vec{\Pi} = -\frac{d\vec{A}}{dt} + \frac{\theta e^2}{8\pi^2}\vec{B}. \tag{7.41}$$

Now we will think about the invariance of states under small gauge transformations. For $\theta = 0$ we saw that the small gauge transformations, with gauge parameter ω, are generated by

$$Q_\omega = \int d^3x \vec{\nabla}\omega\cdot\vec{E}. \tag{7.42}$$

An interesting set of large gauge transformations is those with $\omega^a = \lambda\Phi^a/v$. For these, if we integrate by parts then we obtain a term which vanishes by Gauss's law (Gauss's law is enforced by the invariance under small gauge transformations), and a surface term. This surface term gives the total $U(1)$ charge times λ. We can think of this another way. For the low-lying excitations, multiplication by e^{iQ_ω} corresponds to shifting the dynamical variable χ by a constant, λ. In general the wave functions for χ have the form $e^{iq\chi}$, where q is quantized. So the states pick up a phase $e^{iq\lambda}$. This is just the transformation of a state of charge q under a global gauge transformation with phase λ.

In the presence of θ, however, the operator which implements time-independent gauge transformations is modified. The field $\vec{E}$ is replaced by the canonical momentum above. Now acting on states, the extra term gives a factor $g\theta/(2\pi)$ in the exponent. Even states with $q = 0$ pick up a phase, so there is an additional contribution to the charge,

$$Q = n_e e - \frac{\theta n_m}{2\pi}. \tag{7.43}$$

7.7 Electric–magnetic duality

As mentioned earlier, Maxwell's equations suggest a possible duality between electricity and magnetism. If there were magnetic charges, these equations would take the form

$$\vec{\nabla} \cdot \vec{E} = \rho_e, \quad \vec{\nabla} \cdot \vec{B} = \rho_m, \tag{7.44}$$

$$\vec{\nabla} \times \vec{E} = -\frac{\partial \vec{B}}{\partial t} + \vec{j}_m, \quad \vec{\nabla} \times \vec{B} = \frac{\partial \vec{E}}{\partial t} + \vec{j}_e. \tag{7.45}$$

These equations retain their form if we replace $\vec{E}$ by $-\vec{B}$ and $\vec{B}$ by $\vec{E}$ and also let $\rho_e \to \rho_m$ and $\rho_m \to -\rho_e$ (and similarly for the electric and magnetic currents).

Now that we have a framework for discussing magnetic charges, it is natural to ask whether some theories of electrodynamics really obey such a symmetry. In general, however, this is a difficult problem. We have just learned that electric and magnetic charges, when they both exist, obey a reciprocal relation, $g \propto 1/e$. From the point of view of quantum field theory, this means that exchanging electric and magnetic charges also means replacing the fundamental coupling by its inverse. In other words, if there is such a duality symmetry, *it relates a strongly coupled theory to a weakly coupled theory*. We do not know a great deal about strongly coupled gauge theories, so investigating the possibility of such a duality is a difficult problem. That such a symmetry might exist in theories of the type we have been discussing is not entirely crazy. For example the monopole masses behave, at weak coupling, like $1/g^2$. So as the coupling becomes strong, these particles become light, even as the charged states become heavy. They have complicated quantum numbers (some monopole states are fermionic, for example).

Remarkably, there is a circumstance where such dualities can be studied, namely theories with more than one supersymmetry (in four dimensions): $N=4$ supersymmetric Yang–Mills theory turns out to exhibit an electric–magnetic duality. These theories will be discussed in Chapter 15. Crucial to verifying this duality will be a deeper understanding of the Bogomol'nyi–Prasad–Sommerfield (BPS) condition, which will allow us to establish exact formulas for the masses of certain particles that are valid for all values of the coupling. These formulas will exhibit precisely the expected duality between electricity and magnetism.

Suggested reading

There are many excellent reviews and texts on monopoles. These include Coleman (1981) and Harvey (1996), and this chapter borrows ideas from both. You can find an introduction to the subject in Chapter 6 of Jackson's electrodynamics text (1999).

Exercises

(1) Verify that Eqs. (7.1) and (7.2) are those for infinitely long, thin, solenoids ending at the origin.

(2) Find the kink solution of the $(1 + 1)$-dimensional model. Show that the collective coordinate action is

$$S = \int dt \, \frac{1}{2} M_{\text{kink}} \dot{x}_0^2.$$

(3) Verify that Eqs. (7.31) solve the BPS equations.

8 Technicolor: a first attempt to explain hierarchies

In Chapter 5 we learned a great deal about quantum chromodynamics. In Section 4.5 we argued that the hierarchy problem is one of the puzzles of the Standard Model. The grand unified models of Chapter 6 provided a quite stark realization of the hierarchy problem. In an $SU(5)$ grand unified model we saw that it is necessary to adjust carefully the couplings in the Higgs potential in order to obtain light doublet and heavy color triplet Higgs. This is already true at tree level; loop effects will correct these relations, requiring further delicate adjustments.

Attempts to understand the hierarchy problem in a manner consistent with 't Hooft's naturalness principle fall into three broad categories: the dynamical breaking of electroweak symmetry, supersymmetry (in which it is still possible that the breaking of electroweak symmetry is dynamical), geometric approaches (large extra dimensions or warped space–times) and supersymmetry. The present chapter gives a brief introduction to dynamical models; Chapters 9–16 will deal with supersymmetry both as a possible new symmetry of nature and a possible solution to the hierarchy problem. We will discuss geometric solutions in Chapter 29 after we have learned about theories of space–time, i.e. general relativity and string theory.

The first proposal to resolve the hierarchy problem goes by the name *technicolor*. The technicolor hypothesis exploits our understanding of QCD dynamics. It elegantly explains the breaking of the electroweak symmetry. It has more difficulty accounting for the masses of the quarks and leptons, and simple versions seem incompatible with precision studies of the W and Z particles and now the discovery of a Standard-Model-like Higgs boson. In this chapter we will introduce the basic features of the technicolor hypothesis. We will not attempt to review the many models that have been developed to try to address the difficulties of flavor and precision electroweak experiments. It is probably safe to say that, as of this writing, none is totally successful nor particularly plausible. But it should be kept in mind that this may reflect the limitations of theorists; experiment may yet reveal that nature has chosen this path. In any case, the study of these theories will deepen our understanding of the Standard Model and of strongly coupled quantum field theories and will open our eyes to possibilities for new physics.

We will then turn briefly to dynamical alternatives to technicolor. One of the most interesting of these is the possibility that the Higgs particle is itself an approximate Goldstone particle, the result of the breaking of some accidental global symmetry. By itself this approach does not completely solve the hierarchy problem, but it suppresses the problem of quadratic divergences to higher orders and one might imagine that the phenomenon might arise in some more complete dynamical framework. It has the virtue that in it the Higgs is to a good approximation a fundamental field, as appears to be the case experimentally.

8.1 QCD in a world without Higgs fields

Consider a world with only a single generation of quarks and no Higgs fields. In such a world the quarks would be exactly massless. The $SU(2)_L \times SU(2)_R$ symmetry of QCD would be, in part, a gauge symmetry; $SU(2)_L$ would correspond to the $SU(2)$ symmetry of the weak interactions. The hypercharge Y would include a generator of $SU(2)_R$ and baryon number:

$$Y = 2T_{3R} + B. \tag{8.1}$$

The quark condensate,

$$\langle q_f \bar{q}_{f'} \rangle = \Lambda^3 \delta_{ff'}, \tag{8.2}$$

would break some of the gauge symmetry. Electric charge, however, would be conserved, so $SU(2) \times U(1) \to U(1)$.

In Appendix C it is shown that the quark condensate conserves a vector $SU(2)$ symmetry, ordinary isospin. This $SU(2)$ symmetry is generated by the linear sum

$$T_i = T_{iL} + T_{iR}. \tag{8.3}$$

So, the $SU(2)$ gauge bosons transform as a triplet of the conserved isospin. This guarantees that the successful tree level relation

$$M_W = M_Z \cos\theta \tag{8.4}$$

is satisfied. The $SU(2)$ which accounts for this relation is called a *custodial* symmetry (the Higgs potential of the Standard Model possesses, in fact, an approximate $O(4)$ symmetry which has a suitable $SU(2)$ subgroup).

To understand the masses of the gauge bosons remember that, for a broken symmetry with current j^μ, the coupling of the Goldstone boson to the current is

$$\langle 0 | j^\mu | \pi(p) \rangle = i f_\pi p^\mu. \tag{8.5}$$

This means that there is a non-zero amplitude for a gauge boson to turn into a Goldstone, and vice versa. The diagram of Fig. 8.1 is proportional to

$$g^2 f_\pi^2 p^\mu \frac{i}{p^2} p^\nu. \tag{8.6}$$

As the momentum tends to zero, this tends to a constant – the mass of the gauge boson. For the charged gauge bosons the mass is just

$$m_{W^\pm}^2 = g^2 f_\pi^2, \tag{8.7}$$

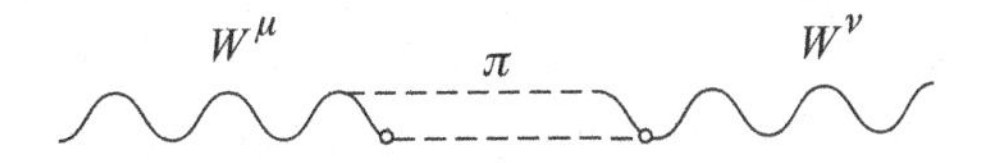

Fig. 8.1 Diagrammatic representation of technicolor.

while for the neutral gauge bosons we have a mass matrix

$$f_{\pi^2}\begin{pmatrix} g^2 & gg' \\ gg' & g'^2 \end{pmatrix}, \tag{8.8}$$

giving one massless gauge boson and one with mass-squared $(g^2 + g'^2)f_\pi^2$.

All this can be nicely described in terms of the non-linear sigma model used to describe pion physics. Recall that the pions could be described in terms of a matrix field,

$$\Sigma = |\langle\overline{\psi}\psi\rangle| e^{i\vec{\pi}\cdot\vec{\tau}/2}, \tag{8.9}$$

which transforms under $SU(2)_{\rm L} \times SU(2)_{\rm R}$ as follows:

$$\Sigma \to U_{\rm L} \Sigma U_{\rm R}^\dagger. \tag{8.10}$$

Changes in the magnitude of the condensate are associated with excitations in QCD that are much more massive than the pion fields (the σ field of our linear sigma model of Section 2.2). So, it is natural to treat this as a constant. The field Σ is then constrained to take values on a manifold. As in our examples in two dimensions, a model based on such a field is called a non-linear sigma model. The Lagrangian is

$$\mathcal{L} = f_\pi^2 \, {\rm Tr}(\partial_\mu \Sigma^\dagger \partial^\mu \Sigma). \tag{8.11}$$

In the context of the physics of light pseudo-Goldstone particles, the virtue of such a model is that it incorporates the effects of broken symmetry in a very simple way. For example, all the results of current algebra can be derived by studying the physics of such a theory and its associated Lagrangian.

In the case of the σ-model we have an identical structure except that we have gauged some of the symmetry, so we need to replace the derivatives by covariant derivatives:

$$\partial_\mu \Sigma \to D_\mu \Sigma = \partial_\mu \Sigma - i\frac{A_\mu^a \sigma_a}{2}\Sigma - i\Sigma\frac{\sigma_3}{2}B_\mu. \tag{8.12}$$

Again, we can choose a unitary gauge; we just set $\Sigma = 1$. The Lagrangian in this gauge is simply

$$\mathcal{L} = {\rm Tr}\left(\frac{A_\mu^a \sigma_a}{2}\Sigma + \Sigma\frac{\sigma_3}{2}B_\mu\right)^2. \tag{8.13}$$

This yields exactly the mass matrix as we wrote down before.

8.2 Fermion masses: extended technicolor

In technicolor models, the Higgs field is replaced by new strong interactions which break $SU(2) \times U(1)$ at a scale $F_\pi = 1$ TeV. However, the Higgs field of the Standard Model gives mass not only to the gauge bosons but to the quarks and leptons as well. In the absence of the Higgs scalar there are chiral symmetries which prohibit masses for any of the quarks and leptons. While our simple model can explain the masses of the Ws and Zs, it has no mechanism to generate mass for the ordinary quarks and leptons.

If we are to avoid introducing fundamental scalars, the only way to break these symmetries is to introduce further gauge interactions. Consider first a single generation of quarks and leptons. Enlarge the gauge group to $SU(3) \times SU(2) \times U(1) \times SU(N+1)$. The technicolor group will be an $SU(N)$ subgroup of the last factor. Take each quark and lepton to be part of an $N+1$ or $\overline{N+1}$ representation of this larger group. To avoid anomalies, we will also include a right-handed neutrino. In other words, our multiplet structure is:

$$\begin{pmatrix} Q \\ q \end{pmatrix}, \quad \begin{pmatrix} \bar{U} \\ \bar{u} \end{pmatrix}, \quad \begin{pmatrix} \bar{D} \\ \bar{d} \end{pmatrix}, \quad \begin{pmatrix} L \\ \ell \end{pmatrix}, \quad \begin{pmatrix} \bar{E} \\ \bar{e} \end{pmatrix} \quad \begin{pmatrix} \bar{N} \\ \bar{\nu} \end{pmatrix}. \tag{8.14}$$

Here $q, \bar{u}, \bar{d}, \ell$, etc., are the usual quarks and leptons; the fields denoted by capital letters are the techniquarks. Now suppose that the $SU(N+1)$ is broken to $SU(N)$ at a scale $\Lambda_{\mathrm{etc}} \gg \Lambda_{\mathrm{tc}}$ by some other gauge interactions, in a manner similar to that of technicolor. Then there is a set of massive gauge bosons with mass of order Λ_{etc}. Exchanges of these bosons give rise to operators such as

$$\mathcal{L}_{4\mathrm{f}} = \frac{1}{\Lambda_{\mathrm{etc}}^2} Q\sigma_\mu q^* \bar{U}\sigma^\mu \bar{u}^* + \mathrm{h.c.} \tag{8.15}$$

Using the following identity for the Pauli matrices,

$$\sum_\mu (\sigma_\mu)_{\alpha\dot{\alpha}} (\sigma^\mu)^{\dot{\beta}\beta} = \delta_\alpha^\beta \delta_{\dot{\alpha}}^{\dot{\beta}}, \tag{8.16}$$

permits us to rewrite the four-fermion interaction as

$$\mathcal{L}_{4\mathrm{f}} = \frac{1}{\Lambda_{\mathrm{ETC}}} Q\bar{U} q^* \bar{u}^* + \mathrm{h.c.} \tag{8.17}$$

We can replace $Q\bar{U}$ by its expectation value, which is of order Λ_{tc}^3. This gives rise to a mass for the u quark. The other quarks and leptons gain mass in a similar fashion.

This particular extended technicolor (ETC) model is clearly unrealistic on many counts: it has only one generation; there is a massive neutrino; there are relations among the masses which are unrealistic; there are approximate global symmetries which lead to unwanted pseudo-Goldstone bosons. Still, it illustrates the basic idea of extended technicolor models: additional gauge interactions break the unwanted chiral symmetries which protect the quark and lepton masses from radiative corrections.

One can try to build realistic models by considering more complicated groups and representations for the extended technicolor (ETC) interactions. Rather than attempt this here, we will consider some issues in a general way. We will imagine that we have a model with three generations. The extended technicolor interactions generate a set of four-fermion interactions which break the chiral symmetries acting on the separate quarks and leptons. In a model of three generations, there are a number of challenges which must be addressed.

1. Perhaps the most serious is the problem of flavor-changing neutral currents. In addition to four-fermion operators which generate mass, there will also be four-fermion operators involving just the ordinary quarks and leptons. These operators will not, in general,

respect flavor symmetries. They are likely to include terms like

$$\mathcal{L}_{\Delta S=2} = \frac{1}{\Lambda_{\rm etc}} \bar{s} d \bar{s}^* d^*, \tag{8.18}$$

which violate strangeness by two units. Unless $\Lambda_{\rm etc}$ is extremely large (of order hundreds of TeV), this will lead to unacceptably large rates for $K^0 \leftrightarrow \bar{K}^0$.

2. Generating the top quark mass is potentially problematic; it is larger than the W and Z masses. If the ETC scale is large, it is hard to see how to achieve this.
3. The problem of pseudo-Goldstone bosons is generic to technicolor models, in just the fashion we saw for the simple model.

The challenge of technicolor model building is to construct models which solve these problems. We will not attempt to review the various approaches which have been put forward here. Models which solve these problems are typically extremely complicated. Instead, we briefly discuss another serious difficulty: the precision measurement of electroweak processes.

8.3 The Higgs discovery and precision electroweak measurements

In Section 4.5 we stressed that the parameters of the electroweak theory have been measured with high precision and compared with detailed theoretical calculations, including radiative corrections. One naturally might wonder whether a strongly interacting Higgs sector could reproduce these results. The answer is that it is difficult. There are two categories of corrections which one needs to consider. The first are, in essence, corrections to the relation

$$M_W = M_Z \cos\theta_W. \tag{8.19}$$

In a general technicolor model these will be large. But we have seen why this relation holds in the minimal Standard Model: there is an approximate global $SU(2)$ symmetry. This is in fact the case of the simplest technicolor model we encountered above. So this problem is likely to have solutions.

There are, however, other corrections as well, resulting from the fact that in these strongly coupled theories the gauge boson propagators are quite different from those in weakly coupled field theories. They have been estimated in many models and are found to be far too large to be consistent with the data. More details about this problem, and speculations on possible solutions, can be found in the suggested reading.

The discovery of a Higgs particle behaving very much as a simple fundamental doublet poses further challenges. In analogy with QCD, in general we would not expect to find scalars much lighter than the TeV scale, and would expect that any such scalars would be quite broad resonances. There is no reason to expect that they should be narrow, with couplings close to those of the Standard Model, never mind couplings as expected in the Minimal Supersymmetric Standard Model.

8.4 The Higgs as a Goldstone particle

An attractive possibility which has received much attention over the years is that the Higgs doublet is a pseudo-Goldstone particle of some approximate global symmetry. If the characteristic scale of the underlying theory is Λ, so that the next lightest excitations have masses of this order while the parameters of the Higgs potential are loop suppressed, we might hope that the doublet will behave like an elementary field up to terms suppressed by powers of Λ.

Necessarily this symmetry is broken by the gauge interactions. This is important, as such symmetry breaking is necessary to obtain a potential for the Higgs field. As an example, we might imagine that the underlying global symmetry is $SU(3)$, and the Goldstone bosons of this $SU(3)$ symmetry can be described by a non-linear sigma model with a field Σ living on the coset $SU(3)/SU(2)$. The components of Σ include the Higgs field. The difficulty with the simplest version is that the scales f (the Goldstone decay constant) and Λ are not appreciably separated. At one loop there are quadratically divergent corrections to the Higgs mass from gauge loops. These are cut off at some scale Λ. From considerations of unitarity – the scale Λ should be such that loop corrections are at most of order one – one expects that $\Lambda^2 < 4\pi f^2$. This is insufficient to explain precision electroweak breaking or the Higgs width.

To avoid this difficulty, models have been constructed with more intricate symmetries. Often, a phenomenon known as *collective symmetry breaking* is invoked. The basic idea is that there are several gauge interactions and only collectively do they break enough symmetry that one can generate a Higgs potential. In the resulting "little Higgs" theories the symmetries prevent a one-loop contribution to the Higgs mass at one-loop order, and the Higgs field appears to be elementary to the required precision.

It is important that the fermions also respect these larger symmetries. This requires, at a minimum, additional vector-like fields. At a more microscopic level one expects that these global symmetries are accidents of the underlying structure. Non-Abelian symmetries acting both on scalars and fermions in the required, rather intricate, ways may be challenging to discover. Some existing models invoke supersymmetry to achieve this.

Suggested reading

An up-to-date set of lectures on technicolor, including the problems of flavor and electroweak precision measurements, are given in the online article of Chivukula (2000). An introduction to the analysis of precision electroweak physics is provided by Peskin (1990); for an application to technicolor theories, see Peskin and Takeuchi (1990). The Particle Data group summary of technicolor theories surveys the status of dynamical models for electroweak symmetry breaking, in light of the Higgs discovery. Little Higgs theories are described in the reviews of Perelstein (2007) and Schmaltz and Tucker-Smith (2005).

Exercises

(1) Determine the relations between the quark and lepton masses in the extended technicolor model above.

(2) What are the symmetries of the extended technicolor model in the limit where we turn off the ordinary $SU(3) \times SU(2) \times U(1)$ gauge interactions? How many of these symmetries are broken by the condensate? Each broken symmetry gives rise to an appropriate Nambu–Goldstone boson. Some of these approximate symmetries are broken explicitly by the ordinary gauge interactions. The corresponding Goldstone bosons will then gain mass, typically of order $\alpha_i \Lambda_{\rm etc}$. Some will not gain mass of this order, however. Which symmetry (or symmetries) will be respected by the ordinary gauge interactions?

PART 2

SUPERSYMMETRY

9 Supersymmetry

In a standard advanced field theory course, one learns about a number of symmetries: Poincaré invariance, global continuous symmetries, discrete symmetries, gauge symmetries, approximate and exact symmetries. These latter symmetries all have the property that they commute with Lorentz transformations and in particular with rotations. So, the multiplets of the symmetries always contain particles of the same spin; in particular, they always consist of either bosons or fermions.

For a long time, it was believed that these were the only allowed types of symmetry; this statement was even embodied in a theorem, known as the Coleman–Mandula theorem. However, physicists studying theories based on strings stumbled on a symmetry which related fields of different spin. Others quickly worked out simple field theories with this new symmetry, called *supersymmetry*.

Supersymmetric field theories can be formulated in dimensions up to eleven. These higher-dimensional theories will be important when we consider string theory. In this chapter we consider theories in four dimensions. The supersymmetry charges, because they change spin, must themselves carry spin – they are spin-$1/2$ operators. They transform as doublets under the Lorentz group, just like the two-component spinors χ and χ^*. (The theory of two-component spinors is reviewed in Appendix A, where our notation, which is essentially that of the text by Wess and Bagger (1992), is explained.) There can be 1, 2, 4 or 8 such spinors; correspondingly, the symmetry is said to be $N = 1, 2, 4$ or 8 supersymmetry. Like the generators of an ordinary group, the supersymmetry generators obey an algebra; unlike an ordinary bosonic group, however, the algebra involves anticommutators as well as commutators (it is said to be "graded").

There are at least four reasons to think that supersymmetry might have something to do with TeV-scale physics. The first is the hierarchy problem: as we will see, supersymmetry can both explain how hierarchies arise, and why there are no large radiative corrections. The second is the unification of couplings. We have seen that while the gauge group of the Standard Model can in a rather natural way be unified in a larger group, the couplings do not unify properly. In the minimal supersymmetric extension of the Standard Model (the minimal supersymmetric Standard Model, or MSSM) the couplings unify nicely if the scale of supersymmetry breaking is about 1 TeV. Third, the assumption of TeV-scale supersymmetry almost automatically yields a suitable candidate for dark matter, with a density in the required range. Finally, low-energy supersymmetry is strongly suggested by string theory, though at present one cannot assert that this is an actual prediction.

9.1 The supersymmetry algebra and its representations

Because the supersymmetry generators are spinors, they do not commute with the Lorentz generators. Perhaps, then, it is not surprising that a supersymmetry algebra involves translation generators Q, $(\bar{Q}_{\dot{\alpha}} = Q^*_{\dot{\alpha}})$[1] with anticommutators

$$\{Q^A_\alpha, \bar{Q}^B_{\dot{\beta}}\} = 2\sigma^\mu_{\alpha\dot{\beta}}\delta^{AB}P_\mu, \tag{9.1}$$

$$\{Q^A_\alpha, \bar{Q}^B_\beta\} = \epsilon_{\alpha\beta}X^{AB}; \tag{9.2}$$

here A, $B = 1, \ldots, N$, where the integer N labels a particular algebra. The X^{AB}s are Lorentz scalars, antisymmetric in A, B, known as *central charges*.

If nature is supersymmetric, it is likely that for the low-energy symmetry $N = 1$, corresponding to only one possible value for the index A above. Only $N = 1$ supersymmetry has chiral representations. Of course, one might imagine that the chiral matter would arise at the point where supersymmetry was broken. As we will see, it is very difficult to break $N > 1$ supersymmetry spontaneously; however, this is not the case for $N = 1$. The smallest irreducible representations of $N = 1$ supersymmetry which can describe massless fields are as follows:

- chiral superfields (ϕ, ψ_α), comprising a complex scalar and a chiral fermion;
- vector superfields (λ, A_μ), comprising a chiral fermion and a vector meson, both, in general, in the adjoint representation of the gauge group;
- the gravity supermultiplet $(\psi_{\mu,\alpha}, g_{\mu\nu})$, compressing a spin-3/2 particle, the *gravitino*, and a spin-2 particle, the *graviton*.

One can work in terms of these fields, writing down supersymmetry transformation laws and constructing invariants. This turns out to be rather complicated; one must use the equations of motion to realize the full algebra. Great simplification is achieved by enlarging space–time to include commuting and anticommuting variables. The result is called *superspace*.

9.2 Superspace

We may conveniently describe $N = 1$ supersymmetric field theories by using superspace. Superspace allows a simple description of the action of the symmetry on fields and provides an efficient algorithm for the construction of invariant Lagrangians. In addition, calculations of Feynman graphs and other quantities are often greatly simplified using superspace, at least in the limit where supersymmetry is unbroken or nearly so.

[1] The notation with the bar over the Qs and θs is helpful here and conforms with that of the classic text of Wess and Bagger. Note that this differs from our notation in earlier chapters, where we used a bar on left-handed *fields* to distinguish particles transforming in, say, the 3 or $\bar{3}$ representation of $SU(3)$.

In superspace, in addition to the ordinary coordinates x^μ one has a set of anticommuting, *Grassmann*, coordinates, θ_α and $\theta^*_{\dot\alpha} = \bar\theta_{\dot\alpha}$. The Grassmann coordinates obey

$$\{\theta_\alpha, \theta_\beta\} = \{\bar\theta_{\dot\alpha}, \bar\theta_{\dot\beta}\} = \{\theta_\alpha, \bar\theta_{\dot\beta}\} = 0. \tag{9.3}$$

Grassmann coordinates provide a representation of the classical configuration space for fermions; they are familiar from the problem of formulating the fermion functional integral. Note that the square of any θ vanishes. The derivatives also anticommute:

$$\left\{\frac{\partial}{\partial\theta_\alpha}, \frac{\partial}{\partial\bar\theta_{\dot\beta}}\right\} = 0, \quad \text{etc.} \tag{9.4}$$

Crucial in the discussion of Grassmann variables is the problem of integration. In discussing the Poincaré invariance of ordinary field-theory Lagrangians, the property of ordinary integrals that

$$\int_{-\infty}^{\infty} dx\, f(x+a) = \int_{-\infty}^{\infty} dx\, f(x) \tag{9.5}$$

is important. We require that the analogous property hold for Grassmann integration (here for one variable):

$$\int d\theta f(\theta + \epsilon) = \int d\theta f(\theta). \tag{9.6}$$

This is satisfied by the integration rule

$$\int d\theta\, (1, \theta) = (0, 1). \tag{9.7}$$

For the case of $\theta_\alpha, \bar\theta_{\dot\alpha}$, one can write a simple integral table:

$$\int d^2\theta\, \theta^2 = 1, \quad \int d^2\bar\theta\, \bar\theta^2 = 1, \tag{9.8}$$

all other such integrals vanish.

One can formulate a superspace description for both local and global supersymmetry. The local case is rather complicated, and we will not deal with it here, referring the interested reader to the suggested reading and confining our attention to the global case.

The goal of the superspace formulation is to provide a classical description of the action of the symmetry on fields, just as one describes the action of the Poincaré generators. Consider a function of the superspace variables, $f(x^\mu, \theta, \bar\theta)$. The supersymmetry generators act on such a function as differential operators:

$$Q_\alpha = \frac{\partial}{\partial\theta_\alpha} - i\sigma^\mu_{\alpha\dot\alpha}\bar\theta^{\dot\alpha}\partial_\mu, \quad \bar Q_{\dot\alpha} = -\frac{\partial}{\partial\bar\theta_{\dot\alpha}} + i\theta^\alpha\sigma^\mu_{\alpha\dot\alpha}\partial_\mu. \tag{9.9}$$

Note that the θs have mass dimension $-1/2$. It is easy to check that the Q_αs obey the algebra. For example,

$$\{Q_\alpha, Q_\beta\} = \left\{\left(\frac{\partial}{\partial\theta_\alpha} - i\sigma^\mu_{\alpha\dot\alpha}\bar\theta^{\dot\alpha}\partial_\mu\right), \left(\frac{\partial}{\partial\theta_\beta} - i\sigma^\nu_{\beta\dot\beta}\bar\theta^{\dot\beta}\partial_\nu\right)\right\} = 0, \tag{9.10}$$

since the θs and their derivatives anticommute. With just slightly more effort one can construct the $\{Q_\alpha, \bar{Q}_{\dot{\alpha}}\}$ anticommutator.

One can think of the Qs as generating infinitesimal transformations in superspace with Grassmann parameter ϵ. One can construct finite transformations as well by exponentiating the Qs; because there are only a finite number of non-vanishing polynomials in the θs, these exponentials contain only a finite number of terms. The result can be expressed elegantly:

$$e^{\epsilon Q+\bar{\epsilon}\bar{Q}}\Phi(x^\mu,\theta,\bar{\theta}) = \Phi(x^\mu - i\epsilon\sigma^\mu\bar{\theta} + i\theta\sigma^\mu\bar{\epsilon}, \theta+\epsilon, \bar{\theta}+\bar{\epsilon}). \tag{9.11}$$

If one expands Φ in powers of θ, there are only a finite number of terms. These can be decomposed into two irreducible representations of the algebra, corresponding to the chiral and vector superfields described above. To understand these, we need to introduce one more set of objects, the covariant derivatives D_α and $\bar{D}_{\dot{\alpha}}$. These are objects which anticommute with the supersymmetry generators and thus are useful for writing down invariant expressions. They are given by

$$D_\alpha = \partial_\alpha + i\sigma^\mu_{\alpha\dot{\alpha}}\bar{\theta}^{\dot{\alpha}}\partial_\mu, \quad \bar{D}_{\dot{\alpha}} = -\partial_{\dot{\alpha}} - i\theta^\alpha\sigma^\mu_{\alpha\dot{\alpha}}\partial_\mu. \tag{9.12}$$

They satisfy the anticommutation relations

$$\{D_\alpha, \bar{D}_{\dot{\alpha}}\} = -2i\sigma^\mu_{\alpha\dot{\alpha}}\partial_\mu, \quad \{D_\alpha, D_\alpha\} = \{\bar{D}_{\dot{\alpha}}, \bar{D}_{\dot{\beta}}\} = 0. \tag{9.13}$$

We can use the Ds to construct irreducible representations of the supersymmetry algebra. Because the Ds anticommute with the Qs, the condition

$$\bar{D}_{\dot{\alpha}}\Phi = 0 \tag{9.14}$$

is invariant under supersymmetry transformations. Fields that satisfy this condition are called chiral fields. To construct such fields, we would like to find combinations of x^μ, θ and $\bar{\theta}$ which are annihilated by $\bar{D}_\alpha$. Writing

$$y^\mu = x^\mu + i\theta\sigma^\mu\bar{\theta}, \tag{9.15}$$

then

$$\Phi = \Phi(y) = \phi(y) + \sqrt{2}\theta\psi(y) + \theta^2 F(y) \tag{9.16}$$

is a chiral (scalar) superfield. Expanding in θ, we see that the expansion terminates:

$$\begin{aligned}\Phi = {} & \phi(x) + i\theta\sigma^\mu\bar{\theta}\partial_\mu\phi + \frac{1}{4}\theta^2\bar{\theta}^2\partial^2\phi \\ & + \sqrt{2}\theta\psi - \frac{i}{\sqrt{2}}\theta\theta\partial_\mu\psi\sigma^\mu\bar{\theta} + \theta^2 F.\end{aligned} \tag{9.17}$$

We can work out the transformation laws. Starting with

$$\delta\Phi = \epsilon^\alpha Q_\alpha\Phi + \epsilon^*_{\dot{\alpha}}\bar{Q}^{\dot{\alpha}}, \tag{9.18}$$

the components transform as follows:

$$\delta\phi = \sqrt{2}\epsilon\psi, \quad \delta\psi = \sqrt{2}\epsilon F + \sqrt{2}i\sigma^\mu\epsilon^*\partial_\mu\phi, \quad \delta F = i\sqrt{2}\epsilon^*\bar{\sigma}^\mu\partial_\mu\psi. \tag{9.19}$$

Vector superfields form another irreducible representation of the algebra; they satisfy the condition

$$V = V^\dagger. \tag{9.20}$$

Again, it is easy to check that this condition is preserved by supersymmetry transformations. A vector superfield V can be expanded in a power series in the θs:

$$V = i\chi - i\chi^\dagger - \theta\sigma^\mu\theta^* A_\mu + i\theta^2\bar{\theta}\bar{\lambda} - i\bar{\theta}^2\theta\lambda + \frac{1}{2}\theta^2\bar{\theta}^2 D. \tag{9.21}$$

Here χ is not quite a chiral field. It is a superfield which is a function of θ only, i.e. it has terms with zero, one or two θs; χ^* is its conjugate.

If V is to describe a massless field, the presence of A_μ indicates that there should be some underlying gauge symmetry, which generalizes the conventional transformation of bosonic theories. In the case of a $U(1)$ theory, gauge transformations act by

$$V \to V + i\Lambda - i\Lambda^\dagger \tag{9.22}$$

where Λ is a chiral field. The $\theta\theta^*$ term in Λ is precisely a conventional gauge transformation of A_μ. In the case of a $U(1)$ theory, one can define a gauge-invariant field strength

$$W_\alpha = -\frac{1}{4}\bar{D}^2 D_\alpha V. \tag{9.23}$$

By a gauge transformation, we can set $\chi = 0$. The resulting gauge is known as the Wess–Zumino gauge. This gauge is analogous to the Coulomb gauge in electrodynamics:

$$W_\alpha = -i\lambda_\alpha + \theta_\alpha D - \sigma_\alpha^{\mu\nu}\beta F_{\mu\nu}\theta_\beta + \theta^2\sigma^\mu_{\alpha\dot{\beta}}\partial_\mu\lambda^{*\dot{\beta}}. \tag{9.24}$$

The gauge transformation of a chiral field of charge q is given by

$$\Phi \to e^{-iq\Lambda}\Phi. \tag{9.25}$$

One can form gauge-invariant combinations using the vector superfield (connection) V:

$$\Phi^\dagger e^{+qV}\Phi. \tag{9.26}$$

We can also define a *gauge-covariant derivative* by

$$\mathcal{D}_\alpha\Phi = D_\alpha\Phi + D_\alpha V\Phi. \tag{9.27}$$

This construction has a non-Abelian generalization. It is most easily motivated by first generalizing the transformation of Φ to

$$\Phi \to e^{-i\Lambda}\Phi, \tag{9.28}$$

where Λ is now a matrix-valued chiral field.

Now we want to combine $\phi^\dagger$ and ϕ in a gauge-invariant way. By analogy with what we did in the Abelian case, we introduce a matrix-valued field V and require that

$$\Phi^\dagger e^V \Phi \tag{9.29}$$

be gauge-invariant. So we require that

$$e^V \to e^{-i\Lambda^*}e^V e^{i\Lambda}. \tag{9.30}$$

From this, we can define a gauge-covariant field strength,

$$W_\alpha = -\frac{1}{4}\bar{D}^2 e^{-V} D_\alpha e^V. \tag{9.31}$$

This transforms under gauge transformations like a chiral field in the adjoint representation:

$$W_\alpha \rightarrow e^{i\Lambda} W_\alpha e^{-i\Lambda}. \tag{9.32}$$

9.3 $N = 1$ Lagrangians

In ordinary field theories we construct Lagrangians that are invariant under translations by integrating densities over all space. The Lagrangian changes by a derivative under translations, so the *action* is invariant. Similarly, if we start with a Lagrangian density in superspace, a supersymmetry transformation acts by differentiation with respect to x or θ. So, integrating the variation over the full superspace gives zero. This is the basic feature of the integration rules that we introduced earlier. In terms of equations we have

$$\delta \int d^4x \int d^4\theta \, h(\Phi, \Phi^\dagger, V) = \int d^4x d^4\theta \, (\epsilon^\alpha Q_\alpha + \epsilon_{\dot{\alpha}} Q^{\dot{\alpha}}) h(\Phi, \Phi^\dagger, V) = 0. \tag{9.33}$$

For chiral fields, integrals over *half* superspace are invariant. If $f(\Phi)$ is a function of chiral fields only, f itself is chiral. As a result,

$$\delta \int d^4x d^2\theta \, f(\Phi) = \int d^4x d^2\theta (\epsilon^\alpha Q_\alpha + \epsilon_{\dot{\alpha}} Q^{\dot{\alpha}}) f(\Phi). \tag{9.34}$$

The integrals over the Q_α terms vanish when integrated over x with respect to $d^2\theta$. The Q^* terms also give zero. To see this, note that $f(\Phi)$ is itself chiral (check), so that

$$\bar{Q}_{\dot{\alpha}} f \propto \theta^\alpha \sigma^\mu{}_{\alpha\dot{\alpha}} \partial_\mu f. \tag{9.35}$$

We can construct a general Lagrangian for a set of chiral fields Φ_i and gauge group $\mathcal{G}$. The chiral fields have dimension one (again, note that the θs have dimension $-1/2$). The vector superfields V are dimensionless, while W_α has dimension $3/2$. With these ingredients, we can write down the most general renormalizable Lagrangian. First, there are terms involving integration over the full superspace:

$$\mathcal{L}_{\text{kin}} = \int d^4\theta \sum_i \Phi_i^\dagger e^V \Phi_i, \tag{9.36}$$

where the factor e^V is in the representation of the gauge group appropriate to the field Φ_i. We can also write down an integral over half of superspace:

$$\mathcal{L}_W = \int d^2\theta \, W(\Phi_i) + \text{c.c.} \tag{9.37}$$

Here $W(\Phi)$ is a holomorphic function of the Φ_is (it is a function of Φ_i, not $\Phi_i^\dagger$), called the superpotential. For a renormalizable theory,

$$W = \frac{1}{2} m_{ij} \Phi_i \Phi_j + \frac{1}{3} \Gamma_{ijk} \Phi_i \Phi_j \Phi_k. \tag{9.38}$$

Finally, for the gauge fields we can write

$$\mathcal{L}_{\text{gauge}} = \frac{1}{g^{(i)2}} \int d^2\theta \, W_\alpha^{(i)2}. \tag{9.39}$$

The full Lagrangian density is

$$\mathcal{L} = \mathcal{L}_{\text{kin}} + \mathcal{L}_W + \mathcal{L}_{\text{gauge}}. \tag{9.40}$$

The superspace formulation has provided us with a remarkably simple way to write the general Lagrangian. In this form, however, the meaning of these various terms is rather opaque. We would like to express them in terms of the component fields. We can do this by using our expressions for the fields in terms of their components, and our simple integration table. We first consider a single chiral field Φ that is neutral under any gauge symmetries. Then

$$\mathcal{L}_{\text{kin}} = |\partial_\mu \Phi|^2 + i\psi_\Phi \, \partial_\mu \sigma^\mu \psi_\Phi^* + F_\Phi^* F_\Phi. \tag{9.41}$$

The field F is referred to as an *auxiliary field*, as it appears without derivatives in the action. Its equation of motion will be algebraic and can be solved easily. It has no dynamics. For several fields, labeled with an index i, the generalization is immediate:

$$\mathcal{L}_{\text{kin}} = |\partial_\mu \phi_i|^2 + i\psi_i \partial_\mu \sigma^\mu \psi_i^* + F_i^* F_i. \tag{9.42}$$

It is also easy to work out the component form of the superpotential terms. We will write this down for several fields:

$$\mathcal{L}_W = \frac{\partial W}{\partial \Phi_i} F_i + \frac{\partial^2 W}{\partial \Phi_i \Phi_j} \psi_i \psi_j. \tag{9.43}$$

For our special choice of superpotential this becomes

$$\mathcal{L}_W = F_i(m_{ij}\Phi_j + \lambda_{ijk}\Phi_j\Phi_k) + (m_{ij} + \lambda_{ijk}\Phi_k)\psi_i\psi_j + \text{c.c.} \tag{9.44}$$

It is a simple matter to solve for the auxiliary fields:

$$F_i^* = -\frac{\partial W}{\partial \Phi_i}. \tag{9.45}$$

Substituting back into the Lagrangian, we obtain

$$V = |F_i|^2 = \left| \frac{\partial W}{\partial \Phi_i} \right|^2. \tag{9.46}$$

To work out the couplings of the gauge fields, it is convenient to choose the Wess–Zumino gauge. Again, this is analogous to the Coulomb gauge, in that it makes manifest the physical degrees of freedom (the gauge bosons and gauginos) but the supersymmetry is not explicit. We will leave performing the integrations over superspace to the exercises, and just quote the full Lagrangian in terms of the component fields:

$$\mathcal{L} = -\frac{1}{4}g_a^{-2}F_{\mu\nu}^{a2} - i\lambda^a\sigma^\mu D_\mu\lambda^{a*} + |D_\mu\phi_i|^2 - i\psi_i\sigma^\mu D_\mu\psi_i^*$$
$$+ \frac{1}{2g^2}(D^a)^2 + D^a\sum_i \phi_i^* T^a\phi_i + F_i^*F_i - F_i\frac{\partial W}{\partial\phi_i} + \text{c.c.}$$
$$+ \sum_{ij}\frac{1}{2}\frac{\partial^2 W}{\partial\phi_i\partial\phi_j}\psi_i\psi_j + i\sqrt{2}\sum\lambda^a\psi_i T^a\phi_i^*. \tag{9.47}$$

The scalar potential is found by solving for the auxiliary D and F fields:

$$V = |F_i|^2 + \frac{1}{2g_a^2}(D^a)^2 \tag{9.48}$$

with

$$F_i = \frac{\partial W}{\partial\phi_i^*}, \quad D^a = \sum_i (g^a\phi_i^* T^a\phi_i). \tag{9.49}$$

In the case where there is a $U(1)$ factor in the gauge group, there is one more term one can include in the Lagrangian, known as the Fayet–Iliopoulos D term. In superspace,

$$\xi\int d^4\theta V \tag{9.50}$$

is supersymmetric and gauge invariant, since the integral $\int d^4\theta\Phi$ vanishes for any chiral field. In components, this is simply a term linear in D, ξD; so, solving for D from its equations of motion, we obtain

$$D = \xi + \sum_i q_i\phi_i^*\phi_i. \tag{9.51}$$

9.4 The supersymmetry currents

We have written down classical expressions for the supersymmetry generators, but for many purposes it is valuable to have expressions for these objects as operators in quantum field theory. We can obtain these by using the Noether procedure. We need to be careful, though, because the Lagrangian is not invariant under supersymmetry transformations but instead transforms by a total derivative. This is similar to the problem of translations in field theory. To see that there is a total derivative in the variation, recall that the Lagrangian has the form, in superspace,

$$\int d^4\theta f(\theta,\bar\theta) + \int d^2\theta W(\theta) + \text{c.c.} \tag{9.52}$$

The supersymmetry generators all involve a $\partial/\partial\theta$ term and a $\theta\partial_\mu$ term. The variation of the Lagrangian is proportional to $\int d^4\theta\epsilon Q f + \cdots$. The term involving $\partial/\partial\theta$ integrates to zero, but the extra term does not; only in the action, obtained by integrating the Lagrangian density over space–time, does the derivative term drop out.

So, in performing the Noether procedure the variation of the Lagrangian will have the form

$$\delta\mathcal{L} = \epsilon\partial_\mu K^\mu + (\partial_\mu\epsilon)T^\mu. \tag{9.53}$$

Integrating by parts, we have that $K^\mu - T^\mu$ is conserved. Taking this into account, for a theory with a single chiral field,

$$j^\mu_\alpha = \sqrt{2}\sigma^\nu_{\alpha\dot\beta}\bar\sigma^{\mu\dot\beta\gamma}\psi_\gamma\partial_\nu\phi^* + i\sqrt{2}F\sigma^{\mu\alpha\dot\alpha}\psi^*_{\dot\alpha} \tag{9.54}$$

and similarly for $j^\mu_{\dot\alpha}$. The generalization for several chiral fields is obvious: one makes the replacements $\psi \to \psi_i$, $\phi \to \phi_i$, etc. and sums over i. One can check that the (anti)commutators of the Qs (which are integrals over j^0) with the various fields gives the correct transformations laws. One can do the same for the gauge fields. Working with the action written in terms of W there are no derivatives, so the variation of the Lagrangian comes entirely from the $\partial_\mu K^\mu$ term in Eq. (9.53). We have already seen that the variation of $\int d^2\theta$ is a total derivative. The current is worked out in the exercises at the end of this chapter.

9.5 The ground state energy in globally supersymmetric theories

One striking feature of the Lagrangian of Eq. (9.47) is that the potential $V \geq 0$. This fact can be traced back to the supersymmetry algebra. Start with the equation

$$\{Q_\alpha, \bar Q_{\dot\beta}\} = 2P_\mu\sigma^\mu_{\alpha\dot\beta}, \tag{9.55}$$

multiply by σ^0 and take the trace:

$$E = \frac{1}{4}Q_\alpha\bar Q_{\dot\alpha} + \bar Q_{\dot\alpha}Q_\alpha. \tag{9.56}$$

Since the left-hand side is positive, the energy is always greater than or equal to zero.

In global supersymmetry, $E = 0$ is very special: the expectation value of the energy is an *order parameter* for supersymmetry breaking. If the supersymmetry is unbroken then $Q_\alpha|0\rangle = 0$, so the ground-state energy vanishes *if and only if* the supersymmetry is unbroken.

Alternatively, consider the supersymmetry transformation laws for λ and ψ. One has, under a supersymmetry transformation with parameter ϵ,

$$\delta\psi = \sqrt{2}\epsilon F + \cdots, \qquad \delta\lambda = i\epsilon D + \cdots. \tag{9.57}$$

In quantum theory the supersymmetry transformation laws become operator equations

$$\delta\psi = i\{Q, \psi\}; \tag{9.58}$$

so, taking the vacuum expectation value of both sides, we see that a non-vanishing field F means broken supersymmetry. Again the vanishing of the energy is an indicator of supersymmetry breaking. So, if either F or D has an expectation value, the supersymmetry is broken.

The signal of ordinary (bosonic) symmetry breakdown is a Goldstone boson. In the case of supersymmetry the signal is the presence of a Goldstone fermion, or *goldstino*. One can prove a goldstino theorem in almost the same way as one proves Goldstone's theorem. We will do this shortly, when we consider simple models of supersymmetry and its breaking.

9.6 Some simple models

In this section we consider some simple models, in order to develop some practice with supersymmetric Lagrangians and to illustrate how supersymmetry is realized in the spectra of these theories.

9.6.1 The Wess–Zumino model

One of the earliest, and simplest, models is the Wess–Zumino model, a theory of a single chiral field (no gauge interactions). For the superpotential we take

$$W = \frac{1}{2}m\phi^2 + \frac{\lambda}{3}\phi^3. \tag{9.59}$$

The scalar potential is (using ϕ for the super-and-scalar field)

$$V = |m\phi + \lambda\phi^2|^2 \tag{9.60}$$

and the ϕ field has mass-squared $|m|^2$. The fermion mass term is

$$\frac{1}{2}m\psi\psi, \tag{9.61}$$

so the fermion also has mass m.

We will now consider the symmetries of the model. First, set $m = 0$. The theory then has a continuous global symmetry. This is perhaps not obvious from the form of the superpotential, $W = (\lambda/3)\phi^3$. But the Lagrangian is an integral over superspace of W, so it is possible for W to transform and for the θs to transform in a compensating fashion. Such a symmetry, which does not commute with supersymmetry, is called an *R symmetry*. If, by convention, we take the θs to carry charge 1 then the $d\theta$s carry charge -1 (think of the integration rules). So the superpotential must carry charge 2. In the present case, this means that ϕ carries charge $2/3$. Note that each component of the superfield transforms differently:

$$\phi \to e^{i(2/3)\alpha}\phi, \quad \psi \to e^{i(2/3-1)\alpha}\psi, \quad F \to e^{i(2/3-2)\alpha}F. \tag{9.62}$$

Now consider the problem of mass renormalization at one loop in this theory. First suppose again that $m = 0$. From our experience with non-supersymmetric theories we might expect a quadratically divergent correction to the scalar mass. But ϕ^2 carries charge $4/3$, and this forbids a mass term in the superpotential. For the fermion the symmetry does not permit us to draw any diagram which corrects the mass. For the boson, however, there are two diagrams, one with intermediate scalars and one with fermions. We will study these in detail later. Consistently with our argument, these two diagrams are found to cancel.

What if, at tree level, $m \neq 0$? We will see shortly that there are still no corrections to the mass term in the superpotential. In fact, perturbatively, there are no corrections to the superpotential at all. There are, however, wave-function renormalizations; rescaling ϕ corrects the masses. In four dimensions, the wave-function corrections are logarithmically divergent, so there are logarithmically divergent corrections to the masses but no quadratic divergences.

9.6.2 A $U(1)$ gauge theory

Consider a $U(1)$ gauge theory, with two charged chiral fields, ϕ^+ and ϕ^-, having charges ± 1, respectively. First suppose that the superpotential vanishes. Our experience with ordinary field theories would suggest that we start developing a perturbation expansion about the point in field space $\phi^\pm = 0$. But, consider the potential in this theory. In the Wess–Zumino gauge we have

$$V(\phi^\pm) = \frac{1}{2}D^2 = \frac{g^2}{2}(|\phi^+|^2 - |\phi^-|^2)^2. \tag{9.63}$$

Zero-energy supersymmetric minima have $D = 0$. By a gauge choice we can set

$$\phi^+ = v, \quad \phi^- = v' e^{i\alpha}, \tag{9.64}$$

with v, v' parameters with dimensions of mass. Then $D = 0$ if $v = v'$. In field theory, as discussed in Section 2.3, when one has such a continuous degeneracy, just as in the case of global symmetry breaking, one must choose a vacuum. Each vacuum is physically distinct – in this case, the spectra are different – and there are no transitions between vacua.

It is instructive to work out the spectrum in a vacuum with a given v. One has, first, the gauge bosons, with masses

$$m_v^2 = 4g^2 v^2. \tag{9.65}$$

This accounts for three degrees of freedom. From the Yukawa couplings of the gaugino λ to the ϕs, one has a term

$$\mathcal{L}_\lambda = \sqrt{2} g v \lambda (\psi_{\phi^+} - \psi_{\phi^-}), \tag{9.66}$$

so we have a Dirac fermion with mass $2gv$. Now we have accounted for three bosonic and two fermionic degrees of freedom. The fourth bosonic degree of freedom is a scalar; one can think of it as the partner of the Higgs, which is eaten in the Higgs phenomenon.

To compute its mass, note that, expanding the scalars as

$$\phi^{\pm} = v + \delta\phi^{\pm}, \tag{9.67}$$

we have

$$D = gv(\delta\phi^{+} + \delta\phi^{+*} - \delta\phi^{-} - \delta\phi^{-*}). \tag{9.68}$$

So D^2 gives a mass to the real part of $\delta\phi^{+} - \delta\phi^{-}$, equal to the mass of the gauge bosons and gauginos. Since the masses differ in states with different v, these states are physically inequivalent.

There is also a massless state: a single chiral field. For the scalars, this follows on physical grounds: the expectation value v is undetermined and one phase is undetermined, so there is a massless complex scalar. For the fermions, the linear combination $\psi_{\phi^+} + \psi_{\phi^-}$ is massless. So we have the correct number of fields to construct a massless chiral multiplet. We can describe this elegantly by introducing the composite chiral superfield or *modulus*

$$\Phi = \phi^{+}\phi^{-} \approx v^2 + v(\delta\phi^{+} + \delta\phi^{-}). \tag{9.69}$$

Its components are precisely the massless complex scalar and the chiral fermion which we identified above.

This is our first encounter with a phenomenon which is nearly ubiquitous in supersymmetric field theories and string theory: there are often continuous sets of vacuum states, at least in some approximation. The set of such physically distinct vacua is known as the *moduli space*. In this example the set of such states is parameterized by the values of the modulus field Φ.

In quantum mechanics, in such a situation we would solve for the wave function of the modulus and the ground state would typically involve a superposition of the different classical ground states. We have seen, though, that in field theory one must choose a value for the modulus field. In the presence of such a degeneracy, for each such value one has, in effect, a different field theory – no physical process leads to transitions between one such state and another. Once the degeneracy is lifted, however, this is no longer the case and transitions, as we will frequently see, are possible.

9.7 Non-renormalization theorems

In ordinary field theories, as we integrate out the physics between one scale and another, we generate every term in the effective action permitted by the symmetries. This is not the case in supersymmetric field theories. This feature gives such theories surprising, and possibly important, properties when we consider questions of naturalness. It also gives us a powerful tool to explore the dynamics of these theories, even at strong coupling. This power comes easily; in this section, we will enumerate these theorems and explain how they arise.

So far, we have restricted our attention to renormalizable field theories. But we have seen that, in considering Beyond the Standard Model physics, we may wish to relax this restriction. It is not hard to write down the most general, globally supersymmetric, theory with at most two derivatives, using the superspace formalism:

$$\mathcal{L} = \int d^4\theta K(\phi_i, \phi_i^\dagger) + \int d^2\theta W(\phi_i) + \text{c.c.} + \int d^2\theta f_a(\phi)(W_\alpha^{(a)})^2 + \text{c.c.} \tag{9.70}$$

The function K is known as the Kahler potential. Its derivatives dictate the form of the kinetic terms for the different fields. The functions W and f_a are holomorphic (what physicists would comfortably call "analytic") functions of the chiral fields. In terms of the component fields (see the exercises) the real part of f couples to $F_{\mu\nu}^2$; the functions W and f_a thus determine the gauge couplings. The imaginary parts couple to the now-familiar operator $F\tilde{F}$. These features of the Lagrangian will be important in much of our discussion of supersymmetric field theories and string theory.

Non-supersymmetric theories have the property that they tend to be generic; any term permitted by symmetries in the theory will appear in the effective action, with an order of magnitude determined by dimensional analysis.[2] Supersymmetric theories are special in that this is not the case. In $N = 1$ theories, there are non-renormalization theorems governing the superpotential and the gauge coupling functions f of Eq. (9.70). These theorems assert that the superpotential is not corrected in perturbation theory beyond its tree level value, while f is at most renormalized at one loop.[3]

Originally, these theorems were proven by the detailed study of Feynman diagrams. Seiberg has pointed out that they can be understood in a much simpler way. Both the superpotential and the functions f are holomorphic functions of the chiral fields, i.e. they are functions of the ϕ_is and not the ϕ_i^*s. This is evident from their construction. Seiberg argued that the coupling constants of a theory may be thought of as *expectation values* of chiral fields and so the superpotential must be a holomorphic function of these as well. For example, consider a theory of a single chiral field Φ with superpotential

$$W = m\Phi^2 + \lambda\Phi^3. \tag{9.71}$$

We can think of λ and m as the expectation values of chiral fields $\lambda(x, \theta)$ and $m(x, \theta)$.

In the Wess–Zumino Lagrangian, if we first set λ to zero then there is an R symmetry under which Φ has R-charge 1 and λ has R-charge -1. Now consider corrections to the effective action in perturbation theory. For example, renormalizations of λ in the superpotential necessarily involve positive powers of λ. But such terms (apart from λ^1) have the wrong R-charge to preserve the symmetry. So there can be no renormalization of this coupling. There can be wave function renormalization, since K is not holomorphic, so $K = K(\lambda^\dagger\lambda)$ is allowed in general.

There are many interesting generalizations of these ideas, and we will not survey them here but will just mention two further examples. First, gauge couplings can be thought of

[2] In some cases, there may be suppression by a few powers of the coupling.

[3] There is an important subtlety connected with these theorems. Both should be interpreted as applying only to a Wilsonian effective action, in which one integrates out the physics above some scale μ. If infrared physics is included, the theorems do not necessarily hold. This is particularly important for the gauge couplings.

in the same way, i.e. we can treat g^{-2} as part of a chiral field. More precisely, we define

$$S = \frac{8\pi^2}{g^2} + ia + \cdots . \tag{9.72}$$

The real part of the scalar field in this multiplet couples to $F_{\mu\nu}^2$ but the imaginary part, a, couples to $F\tilde{F}$. Because $F\tilde{F}$ is a total derivative, in perturbation theory there is a symmetry under constant shifts of a. The effective action should respect this symmetry. Because the gauge coupling function f is holomorphic, this implies that

$$f(g^2) = S + \text{const} = \frac{8\pi^2}{g^2} + \text{const}. \tag{9.73}$$

The first term is just the tree level term. One-loop corrections yield a constant, but there are no higher-order corrections in perturbation theory! This is quite a striking result. It is also paradoxical, since the two-loop beta functions for supersymmetric Yang–Mills theories were computed long ago and are, in general, non-zero. The resolution of this paradox is subtle and interesting. It provides a simple computation of the two-loop beta function. In a particular renormalization scheme, it gives an exact expression for the beta function. This is explained in Appendix D.

Before explaining the resolution of the above paradox, there is one more non-renormalization theorem which we can prove rather trivially here. This is the statement that if there is no Fayet–Iliopoulos D term at tree level, this term can be generated at most at one loop. To prove this, write the D term as

$$\int d^4\theta d(g,\lambda) V. \tag{9.74}$$

Here $d(g,\lambda)$ is some unknown function of the gauge and other couplings in the theory. But, if we think of g and λ as chiral fields then this expression is only gauge invariant if d is a constant, corresponding to a possible one-loop contribution. Such contributions do arise in string theory.

In string theory, all the parameters *are* expectation values of chiral fields. Indeed, non-renormalization theorems in string theory, both for world-sheet and string perturbation theory, were proven by the sort of reasoning we have used above.

9.8 Local supersymmetry: supergravity

If supersymmetry has anything to do with nature, and is not merely an accident, then it must be a local symmetry. There is not space here for a detailed exposition of local supersymmetry. For most purposes, both theoretical and phenomenological, there are fortunately only a few facts we need to know. The field content (in four dimensions) is like that of global supersymmetry, except that now one has a graviton and a gravitino. Note that the number of additional bosonic and fermionic degrees of freedom (a minimal requirement if the theory is to be supersymmetric) is the same. The graviton is described by a traceless symmetric tensor; in $d - 2 = 2$ dimensions, this has two independent components. Similarly, the gravitino ψ_μ has both a vector and a spinor index. It satisfies a constraint similar to tracelessness,

$$\gamma^\mu \psi_\mu = 0. \tag{9.75}$$

In $d - 2$ dimensions, this amounts to two conditions, leaving two physical degrees of freedom.

As in global supersymmetry (without the restriction of renormalizability), the terms in the effective action with at most two derivatives or four fermions are completely specified by three functions:

1. the Kahler potential $K(\phi, \phi^\dagger)$, a function of the chiral fields;
2. the superpotential $W(\phi)$, a holomorphic function of the chiral fields;
3. the gauge coupling functions $f_a(\phi)$, which are also holomorphic functions of the chiral fields.

The Lagrangian which follows from these is quite complicated, as it includes many two- and four-fermion interactions. It can be found in the suggested reading. Our main concern in this text will be the scalar potential. This is given by

$$V = e^K \left[\left(\frac{\partial W}{\partial \phi_i} + \frac{\partial K}{\partial \phi_i} W \right) g^{i\bar{j}} \left(\frac{\partial W^*}{\partial \phi^*_{\bar{j}}} + \frac{\partial K}{\partial \phi^*_{\bar{j}}} W \right) - 3|W|^2 \right], \tag{9.76}$$

where

$$g_{i\bar{j}} = \frac{\partial^2 K}{\partial \phi_i \partial \phi_{\bar{j}}} \tag{9.77}$$

is the (Kahler) metric associated with the Kahler potential. In this equation, we have adopted units in which $M = 1$, where Newton's gravitational constant is given by

$$G_{\mathrm{N}} = \frac{1}{8\pi M^2} \tag{9.78}$$

and $M \approx 2 \times 10^{18}$ GeV is known as the *reduced Planck mass*.

Suggested reading

The text by Wess and Bagger (1992) provides a good introduction to superspace, the fields and Lagrangians of supersymmetric theories in four dimensions and supergravity. Other texts include those by Gates *et al.* (1983) and Mohapatra (2003). Appendix B of Polchinski's (1998) text provides a concise introduction to supersymmetry in higher dimensions. The supergravity Lagrangian is derived and presented in its entirety in Cremmer *et al.* (1979) and Wess and Bagger (1992) and is reviewed in, for example, Nilles (1984). Non-renormalization theorems were first discussed from the viewpoint presented here by Seiberg (1993).

Exercises

(1) Verify the commutators of the Qs and the Ds.

(2) Check that, given the definition Eq. (9.15), Φ is chiral. Show that any function of chiral fields is a chiral field.

(3) Verify that W_α transforms as in Eq. (9.32) and that $\mathrm{Tr}W_\alpha^2$ is gauge invariant.

(4) Derive the expression (9.47) for the component Lagrangian including gauge interactions and the superpotential, by performing the superspace integrals. For an $SU(2)$ theory with a scalar triplet $\vec{\phi}$ and singlet, X, take $W = \lambda(\vec{\phi}^2 - \mu^2)$. Find the ground state and work out the spectrum.

(5) Derive the supersymmetry current for a theory with several chiral fields. For a single field Φ and $W = (1/2)\, m\Phi^2$, verify, using the canonical commutation relations, that the Qs obey the supersymmetry algebra. Work out the supercurrent for a pure supersymmetric gauge theory.

10 A first look at supersymmetry breaking

If supersymmetry has anything to do with the real world, it must be a broken symmetry, as we do not see any degeneracy between bosons and fermions in nature. In the globally supersymmetric framework that we have presented so far, this breaking could be spontaneous or explicit. As we will argue later, once we promote the symmetry to a local symmetry, the breaking of supersymmetry must be spontaneous. The signal of such a breaking is a massless fermion, the *goldstino*, whose interactions are governed by low-energy theorems. However, as we will also see, at low energies the theory can appear to be a globally supersymmetric theory with explicit, "soft", breaking of the symmetry. In this chapter we will discuss some features of both spontaneous and explicit breaking.

10.1 Spontaneous supersymmetry breaking

We have seen that supersymmetry breaking is signaled by a non-zero expectation value of an F component of a chiral superfield or a D component of a vector superfield. Models involving only chiral fields with no supersymmetric ground state are referred to as O'Raifeartaigh models. A simple example has three singlet fields, A, B and X, with superpotential

$$W = \lambda A(X^2 - \mu^2) + mBX. \tag{10.1}$$

With this superpotential, the equations

$$F_A = \frac{\partial W}{\partial A} = 0, \quad F_B = \frac{\partial W}{\partial B} = 0 \tag{10.2}$$

are incompatible. To actually determine the expectation values and the vacuum energy, it is necessary to minimize the potential. There is no problem in satisfying the equation $F_X = 0$. So, we need to minimize

$$V_{\text{eff}} = |F_A|^2 + |F_B|^2 = |\lambda^2||X^2 - \mu^2|^2 + m^2|X|^2. \tag{10.3}$$

Assuming that μ^2 and λ are real, the solutions are given by

$$X = 0, \quad X^2 = \frac{2\lambda^2\mu^2 - m^2}{2\lambda^2}. \tag{10.4}$$

The corresponding vacuum energies are

$$V_0^{(A)} = |\lambda^2\mu^4|, \quad V_0^{(B)} = m^2\mu^2 - \frac{m^4}{4\lambda^2}. \tag{10.5}$$

The vacuum at $X \neq 0$ disappears at a critical value of μ.

Let us consider the spectrum in the first of these (the solution with $X = 0$). We will focus, in particular, on the massless states. First, there is a massless scalar. This arises because at this level not all the fields are fully determined. The equation

$$\frac{\partial W}{\partial X} = 0 \tag{10.6}$$

can be satisfied provided that

$$2\lambda AX + mB = 0. \tag{10.7}$$

This vacuum degeneracy is accidental and, as we will see later, is lifted by quantum corrections.

There is also a massless fermion, ψ_A. This fermion is the goldstino. Replacing the auxiliary fields in the supersymmetry current for this model (Eq. (9.54)) gives

$$j_\mu^\alpha = i\sqrt{2}F_A\sigma^\mu_{\alpha\dot{\alpha}}\psi_A^{*\dot{\alpha}}. \tag{10.8}$$

You should check that the massive states do not form Bose–Fermi degenerate multiplets.

10.1.1 The Fayet–Iliopoulos *D* term

It is also possible to generate an expectation value for a D term. In the case of a $U(1)$ gauge symmetry, we saw that

$$\mu^2 \int d^4\theta\, V = \mu^2 D \tag{10.9}$$

is gauge invariant. Under the transformation $\delta V = \Lambda + \Lambda^\dagger$, the integrals over the chiral and antichiral fields Λ and $\Lambda^\dagger$ are zero. This can be seen either by doing the integrations directly or by noting that differentiation by Grassmann numbers is equivalent to integration (recall our integral table). As a result, for example, $\int d^2\bar{\theta} \propto (\bar{D})^2$. This Fayet–Iliopoulos D term can lead to supersymmetry breaking. For example, if one has two charged fields $\Phi^\pm$ with charges ± 1 and superpotential $m\Phi^+\Phi^-$, one cannot simultaneously make the two auxiliary F fields and the auxiliary D field vanish.

One important feature of both types of model is that at tree level, in the context of global supersymmetry, the spectra are never realistic. They satisfy a sum rule,

$$\sum(-1)^F m^2 = 0. \tag{10.10}$$

Here $(-1)^F = 1$ for bosons and -1 for fermions. This guarantees that there are always light states, and often color and/or electromagnetic symmetry are broken. These statements are not true of radiative corrections or of supergravity, as we will explain later.

It is instructive to prove this sum rule. Consider a theory with chiral fields only (no gauge interactions). The potential is given by

$$V = \sum_i \left| \frac{\partial W}{\partial \phi_i} \right|^2 . \tag{10.11}$$

The boson mass matrix has terms of the form $\phi^*_{\bar{i}} \phi_j$ and $\phi_i \phi_j$ + c.c., where we are using indices $\bar{i}$ and $\bar{j}$ for complex conjugate fields. The latter terms, as we will now see, are connected with supersymmetry breaking. The various terms in the mass matrix can be obtained by differentiating the potential:

$$m^2_{i\bar{j}} = \frac{\partial^2 V}{\partial \phi_i \partial \phi^*_{\bar{j}}} = \frac{\partial^2 W}{\partial \phi_i \partial \phi_k} \frac{\partial^2 W^*}{\partial \phi^*_{\bar{k}} \partial \phi^*_{\bar{j}}}, \tag{10.12}$$

$$m^2_{ij} = \frac{\partial^2 V}{\partial \phi_i \partial \phi_j} = \frac{\partial W}{\partial \phi^*_{\bar{k}}} \frac{\partial^3 W}{\partial \phi_k \partial \phi_i \partial \phi_j}. \tag{10.13}$$

The first term has just the structure of the square of the fermion mass matrix,

$$\mathcal{M}_{\mathrm{F}ij} = \frac{\partial^2 W}{\partial \phi_i \partial \phi_j}. \tag{10.14}$$

So, writing the boson mass matrix $\mathcal{M}^2_{\mathrm{B}}$ in the basis $(\phi_i \; \phi^*_{\bar{j}})$ we see that Eq. (10.10) holds.

The theorem is true whenever a theory can be described by a renormalizable effective action. Various non-renormalizable terms in the effective action can give additional contributions to the mass. For example, in our O'Raifeartaigh model, $\int d^4\theta A^\dagger A Z^\dagger Z$ will violate the tree level sum rule. Such terms arise in renormalizable theories when one integrates out heavy fields to obtain an effective action at some scale. In the context of supergravity, such terms are present already at tree level. This is perhaps not surprising, given that these theories are non-renormalizable and must be viewed as effective theories from the very beginning (perhaps as the effective low-energy description of string theory). We will discuss the construction of realistic models shortly. First, however, we turn to the issues of the *goldstino theorem* (the fermionic analog of Goldstone's theorem) and explicit, soft, supersymmetry breaking.

10.2 The goldstino theorem

In each of the examples of supersymmetry breaking there is a massless fermion in the spectrum. We might expect this, by analogy with Goldstone's theorem. The essence of the usual Goldstone theorem is the statement that, for a spontaneously broken global symmetry, there is a massless scalar. There is a coupling of this scalar to the symmetry current j^μ. From Lorentz invariance (see Appendix B),

$$\langle 0 | j^\mu | \pi(p) \rangle = f p^\mu . \tag{10.15}$$

Correspondingly, in the low-energy effective field theory (valid below the scale of symmetry breaking) the current takes the form

$$j^\mu = f\partial^\mu \pi(x). \tag{10.16}$$

Analogous statements for the spontaneous breaking of global supersymmetry are easy to prove. Suppose that the symmetry is broken by the F component of a chiral field (this can be a composite field). Then we can study

$$\int d^4x\, \partial_\mu \left(e^{iq\cdot x} T\langle j^\mu_\alpha(x)\psi_\Phi(0)\rangle\right) = 0, \tag{10.17}$$

where T is the time-ordering operator and j^μ_α is the supersymmetry current; the integral of j^0_α over space is the supersymmetry charge. This expression vanishes because it is an integral of a total derivative. Now evaluating the derivatives, there are two non-vanishing contributions: one from the exponential and one from the action on the time-ordering symbol. Obtaining these derivatives and then taking the limit $q \to 0$ gives

$$\langle\{Q_\alpha, \psi_\Phi(0)\}\rangle = iq_\mu T\langle j^\mu_\alpha(x)\psi_\Phi(0)\rangle_{\mathrm{FT}}, \tag{10.18}$$

where FT indicates the Fourier transform. The left-hand side is constant, so the Green's function on the right-hand side must be singular as $q \to 0$. By the usual spectral representation analysis, this shows that there is a massless fermion coupled to the supersymmetry current. In weakly coupled theories we can understand this more simply. Recalling the form of the supersymmetry current, if one of the Fs has an expectation value then

$$j^\mu_\alpha = i\sqrt{2}(\sigma^\mu)_{\alpha\dot\alpha}\psi^{*\dot\alpha}F. \tag{10.19}$$

To leading order in the fields, current conservation amounts to just the massless Dirac equation; F, here, is the goldstino decay constant. We can understand the massless fermion which appeared in the O'Raifeartaigh model in terms of this theorem. It is easy to check that

$$\psi_G \propto F_A\psi_A + F_B\psi_B, \tag{10.20}$$

as in Eq. (10.8) for the case $F_B = 0$.

10.3 Loop corrections and the vacuum degeneracy

We saw that in the O'Raifeartaigh model, at the classical level there is a large vacuum degeneracy. To understand the model fully, we need to investigate the fate of this degeneracy in the quantum theory. Consider the vacuum with $X = 0$. In this case, A is undetermined at the classical level. But A is only an approximate modulus. At one loop, quantum corrections generate a potential for A. Our goal is to integrate out the various massive fields to obtain the effective action for A. At one loop, this is particularly easy. The tree level mass spectrum depends on A. The one-loop vacuum energy is

$$\sum_i (-1)^F \int \frac{d^3k}{(2\pi)^3} \frac{1}{2} \sqrt{\vec{k}^2 + m_i^2}. \tag{10.21}$$

Here the sum is over all possible helicity states; again the factor $(-1)^F$ weights bosons with 1 and fermions with -1. In field theory this expression is usually very divergent in the ultraviolet, but in the supersymmetric case it is far less so. If supersymmetry is unbroken, the boson and fermion contributions cancel and the correction simply vanishes. If supersymmetry is broken, the divergence is only logarithmic. To see this we can simply study the integrand at large k, expanding the square root in powers of m^2/k^2. The leading, quartically divergent, term is independent of m^2 and so vanishes. The next term is quadratically divergent, but it vanishes because of the sum rule: $\sum (-1)^F m_i^2 = 0$.

So, at one loop the potential behaves as

$$V(A) = -\sum (-1)^F m_i^4 \int \frac{d^3k}{16(2\pi)^3 k^3} \approx \sum (-1)^F m_i^4 \frac{1}{64\pi^2} \ln \frac{m_i^2}{\Lambda^2}. \tag{10.22}$$

To compute the potential precisely, we need to work out the spectrum as a function of A. We will content ourselves with the limit of large A. Then the spectrum consists of a massive fermion ψ_X, with mass $2\lambda A$, and the real and imaginary parts of the scalar components of X, with masses

$$m_s^2 = 4|\lambda^2 A^2| \pm 2\mu^2\lambda^2 x^2. \tag{10.23}$$

So

$$V(A) = |\lambda^2|\mu^4 \left(1 + \frac{\lambda^2}{8\pi^2} \ln \frac{|\lambda A|^2}{\Lambda^2}\right). \tag{10.24}$$

This result has a simple interpretation. The leading term is the classical energy; the correction corresponds to replacing λ^2 by $\lambda^2(A)$, the running coupling at scale A. In this theory, a more careful study shows that the minimum of the potential is precisely at $A = 0$.

10.4 Explicit soft supersymmetry breaking

Ultimately, if nature is supersymmetric, it is likely that we will want to understand supersymmetry breaking through some dynamical mechanism. But we can be more pragmatic, accept that supersymmetry is broken and parameterize the breaking using the mass differences between the ordinary fields and their superpartners. It turns out that this procedure does not spoil the good ultraviolet properties of the theory. Such mass terms are said to be "soft" for precisely this reason.

We will consider soft breakings in more detail in the next chapter when we discuss the Minimal Supersymmetric Standard Model (MSSM), but we can illustrate the main point simply. Take as a model the Wess–Zumino model, with $m = 0$ in the superpotential. Add

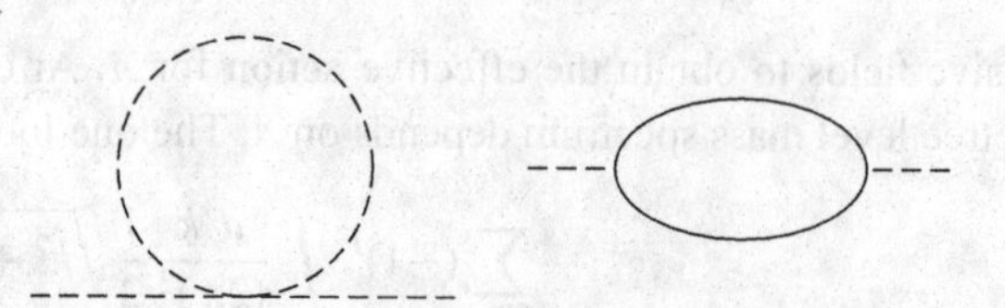

Fig. 10.1 One-loop corrections to scalar masses arising from Yukawa couplings.

to the Lagrangian an explicit mass term $m^2_{\rm soft}|\phi|^2$. Then we can calculate the one-loop correction to the scalar mass from the two graphs of Fig. 10.1. In the supersymmetric case these two graphs cancel. With the soft breaking term, the cancelation is not exact; instead one obtains

$$\delta m^2 = -\frac{|\lambda|^2}{16\pi^2} m^2_{\rm soft} \ln \frac{\Lambda^2}{m^2_{\rm soft}}. \tag{10.25}$$

We can understand this simply on dimensional grounds. We know that for $m^2_{\rm soft} = 0$ there is no correction. Treating the soft term as a perturbation, the result is necessarily proportional to $m^2_{\rm soft}$; at most, then, any divergence must be logarithmic.

In addition to soft masses for scalars, one can add soft masses for gauginos; one can also include trilinear scalar couplings. We can understand how these might arise at a more fundamental level, which also makes clear the sense in which these terms are soft. Suppose that we have a field Z with non-zero F component, as in the O'Raifeartaigh model (but in a more general form). Suppose, further, that at tree level there are no renormalizable couplings between Z and the other fields of the model, which we will denote generically as ϕ. Non-renormalizable couplings, such as

$$\mathcal{L}_Z = \frac{1}{M^2} \int d^4\theta \, Z^\dagger Z \phi^\dagger \phi, \tag{10.26}$$

can be expected to arise as we integrate high-energy processes to obtain the effective Lagrangian; they are not forbidden by any symmetry. Replacing Z by its expectation value, $\langle Z \rangle = \cdots + \theta^2 \langle F_Z \rangle$, gives a mass term for the scalar component of ϕ:

$$\mathcal{L}_Z = \frac{|\langle F \rangle|^2}{M^2} |\phi|^2 + \cdots . \tag{10.27}$$

This is precisely the soft scalar mass we described above; it is soft because it is associated with a high-dimensional operator Similarly, the operator:

$$\int d^2\theta \frac{Z}{M} W_\alpha^2 = \frac{F_Z}{M} \lambda\lambda + \cdots \tag{10.28}$$

gives rise to a mass for gauginos. The term

$$\int d^2\theta \frac{Z}{M} \phi\phi\phi \tag{10.29}$$

leads to a trilinear coupling of the scalars. Simple power counting shows that loop corrections to these couplings due to renormalizable interactions are at most logarithmically divergent.

To summarize, there are three types of soft-breaking term which can appear in a low-energy effective action:

- soft scalar masses, $m_\phi^2|\phi|^2$ and $\tilde{m}\phi^2\phi\phi + c.c.$;
- gaugino masses, $m_\lambda\lambda\lambda$;
- trilinear scalar couplings, $\Gamma\phi\phi\phi$.

All three types of coupling will play an important role when we think about possible supersymmetry phenomenologies.

10.5 Supersymmetry breaking in supergravity models

We stressed in the last chapter that, since nature includes gravity, if supersymmetry is not simply an accident it must be a local symmetry. If the underlying scale of supersymmetry breaking is high enough, supergravity effects will be important. The potential of a supergravity model will be sufficiently important to us that it is worth writing it down again:

$$V = e^K \left[\left(\frac{\partial W}{\partial \phi_i} + \frac{\partial K}{\partial \phi_i} W \right) g^{i\bar{j}} \left(\frac{\partial W}{\partial \phi_{\bar{j}}^*} + \frac{\partial K}{\partial \phi_{\bar{j}}^*} W^* \right) - 3|W|^2 \right]. \tag{10.30}$$

In supergravity the condition for unbroken supersymmetry is that the *Kahler derivative* of the superpotential should vanish:

$$D_i W = \frac{\partial W}{\partial \phi_i} + \frac{\partial K}{\partial \phi_i} W = 0. \tag{10.31}$$

When this is not the case, supersymmetry is broken. If we require the vanishing of the cosmological constant then we have

$$3|W|^2 = \sum_{i,\bar{j}} D_i W D_{\bar{j}} W^* g^{i\bar{j}}. \tag{10.32}$$

In this case the gravitino mass turns out to be

$$m_{3/2} = \langle e^{K/2} W \rangle. \tag{10.33}$$

There is a standard strategy for building supergravity models. One introduces two sets of fields, the hidden-sector fields, denoted by Z_i, and the visible-sector fields, denoted by y_a. The Z_is are assumed to be connected with supersymmetry breaking and to have only very small couplings to the ordinary fields y_a. In other words, one assumes that the superpotential W has the form

$$W = W(Z) + W_y(y), \tag{10.34}$$

at least up to terms suppressed by $1/M$. The y fields should be thought of as the ordinary matter fields and their superpartners.

One also usually assumes that the Kahler potential has a "minimal" form,

$$K = \sum Z_i^\dagger Z_i + \sum y_a^\dagger y_a. \tag{10.35}$$

One chooses (i.e. tunes) the parameters of W_Z in such a way that

$$\langle F_Z \rangle \approx M_W M \tag{10.36}$$

and

$$\langle V \rangle = 0. \tag{10.37}$$

Note that this means that

$$\langle W \rangle \approx M_W M^2. \tag{10.38}$$

The simplest model of the hidden sector is known as the *Polonyi model*. In this model

$$W = m^2(Z + \beta), \tag{10.39}$$

$$\beta = (2 + \sqrt{3})M. \tag{10.40}$$

In global supersymmetry, with only renormalizable terms, this would be a rather trivial superpotential, but this is not so in supergravity. The minimum of the potential for Z lies at

$$Z = (\sqrt{3} - 1)M, \tag{10.41}$$

and

$$m_{3/2} = (m^2/M)e^{(\sqrt{3}-1)^2/2}. \tag{10.42}$$

This symmetry breaking also leads to soft-breaking mass terms for the fields y, i.e. terms of the form

$$m_0^2 |y_i|^2. \tag{10.43}$$

These arise from the $|(\partial_i K)\, W|^2 = |y_i|^2 |W|^2$ terms in the potential. For the simple Kahler potential,

$$m_0^2 = 2\sqrt{3} m_{3/2}^2, \quad A = (3 - \sqrt{3}) m_{3/2}. \tag{10.44}$$

If we now allow for a non-trivial W_y, we also find supersymmetry-violating quadratic and cubic terms in the potential. These are known as the B and A terms and have the form

$$B_{ij} m_{3/2} \phi_i \phi_j + A_{ijk} m_{3/2} \phi_i \phi_j \phi_k. \tag{10.45}$$

For example, if W is homogeneous and of degree three, there are terms in the supergravity potential of the form

$$e^K \frac{\partial W}{\partial y_a} \frac{\partial K}{\partial y_a^*} \langle W \rangle + \text{c.c.} = 3 m_{3/2} W(y). \tag{10.46}$$

Additional contributions arise from

$$e^K \left\langle \frac{\partial W}{\partial z_i} \right\rangle \langle z_i^* \rangle W^* + \text{c.c.} \tag{10.47}$$

There are analogous contributions to the B terms. In the exercises, these are worked out for specific models.

Gaugino masses m_λ (both in local and global supersymmetry) can arise from a non-trivial gauge coupling function

$$f^a = c\frac{Z}{M}, \tag{10.48}$$

which gives

$$m_\lambda = \frac{cF_z}{M}. \tag{10.49}$$

These models have just the correct structure to build a theory of TeV-scale supersymmetry, provided that $m_{3/2} \sim$ TeV. They have soft breakings of the correct order of magnitude. We will discuss their phenomenology further when we discuss the Minimal Supersymmetric Standard Model (MSSM) in the next chapter.

Even without a deep understanding of local supersymmetry, there are a number of interesting observations we can make. Most important, our arguments for the non-renormalization of the superpotential in global supersymmetry remain valid here. This will be particularly important when we come to string theory, which is a locally supersymmetric theory.

Suggested reading

It was Witten (1981) who most clearly laid out the issues of supersymmetry breaking. His paper remains extremely useful and readable today. The notion that one should consider adding soft-breaking parameters to the MSSM was developed by Dimopoulos and Georgi (1981). Good introductions to models with supersymmetry breaking in supergravity are provided by a number of review articles and textbooks, for example those of Mohapatra (2003) and Nilles (1984).

Exercises

(1) Work out out the spectrum of the O'Raifeartaigh model. Show that the spectrum is not supersymmetric, but verify the sum rule $\sum(-1)^F m^2 = 0$.
(2) Work out the spectrum of a model with a Fayet–Iliopoulos D term and supersymmetry breaking. Again verify the sum rule.
(3) Check Eqs. (10.40)–(10.44) for the Polonyi model.

11 The Minimal Supersymmetric Standard Model

We can now very easily construct a supersymmetric version of the Standard Model. For each gauge field of the usual Standard Model we introduce a vector superfield. For each fermion (quark or lepton) we introduce a chiral superfield with the same gauge quantum numbers. Finally, we need at least two Higgs doublet chiral fields; if we introduce only one, as in the simplest version of the Standard Model, the resulting theory possesses gauge anomalies and is inconsistent. So, the theory is specified by the gauge group $SU(3) \times SU(2) \times U(1)$ and enumeration the chiral fields,

$$Q_f, \quad \bar{u}_f, \quad \bar{d}_f, \quad L_f, \quad \bar{e}_f, \quad f = 1, 2, 3; \quad H_U, \quad H_D. \tag{11.1}$$

The gauge-invariant kinetic terms, auxiliary D terms and gaugino–matter Yukawa couplings are completely specified by the gauge symmetries. The superpotential can be taken to be

$$W = H_U(\Gamma_U)_{f,f'} Q_f \bar{U}_{f'} + H_D(\Gamma_D)_{f,f'} Q_f \bar{D}_{f'} + H_D(\Gamma_E)_{f,f'} L_f \bar{e}_{f'}. \tag{11.2}$$

If the Higgs fields obtain suitable expectation values then $SU(2) \times U(1)$ is broken and quarks and leptons acquire mass, just as in the Standard Model.

There are other terms which can also be present in the superpotential. These include the μ term, $\mu H_U H_D$. This is a supersymmetric mass term for the Higgs fields; see Section 11.1.1. We will see later that we need $\mu \gtrsim M_Z$ to have a viable phenomenology. A set of dimension-four terms permitted by the gauge symmetries raise serious issues. For example, one can have the terms

$$\bar{u}_f \bar{d}_g \bar{d}_h \Gamma^{fgh} + Q_f L_g \bar{d}_h \lambda^{fgh}. \tag{11.3}$$

These couplings violate B and L! This is our first serious setback. In the Standard Model, there is no such problem. The leading B- and L-violating operators permitted by gauge invariance possess dimension six, and they will be highly suppressed if the scale of interactions which violates these symmetries is high, as in grand unified theories.

If we are not going to simply give up, we need to suppress B and L violation at the level of dimension-four terms. The simplest approach is to postulate additional symmetries. There are various possibilities one can imagine.

1. *Global continuous symmetries* It is hard to see how such symmetries could be preserved in any quantum theory of gravity, and in string theory there is a theorem which asserts that there are no global continuous symmetries. We will prove this statement, at least for a large subset of known string theories, later.

2. *Discrete symmetries* As we will see later, discrete symmetries can be gauge symmetries. As such they will not be broken in a consistent quantum theory. They are common in string theory. These symmetries are often *R symmetries*, symmetries which do not commute with supersymmetry.

A simple (though not unique) solution to the problem of B- and L-violation by dimension-four operators is to postulate a discrete symmetry known as R-parity. Under this symmetry, all ordinary particles are even while their superpartners are odd. Imposing this symmetry immediately eliminates all the dangerous operators. For example,

$$\int d^2\theta\, \bar{u}\bar{d}\bar{d} \sim \psi_{\bar{u}}\psi_{\bar{d}}\tilde{\bar{d}} \tag{11.4}$$

(we have changed notation again: the tilde here indicates the superpartner of the ordinary field, i.e. the *squark*) is odd under the symmetry.

More formally, we can define this symmetry as the following set of transformations on superfields:

$$\theta_\alpha \to -\theta_\alpha, \tag{11.5}$$

$$(Q_f, \bar{u}_f, \bar{d}_f, L_f, \bar{e}_f) \to -(Q_f, \bar{u}_f, \bar{d}_f, L_f, \bar{e}_f), \tag{11.6}$$

$$(H_U, H_D) \to (H_U, H_D). \tag{11.7}$$

Alternatively, we can describe it as multiplication of the quark and lepton superfields by -1, multiplication of the Higgs fields by 1 and a 2π rotation in space (which rotates all fermions by -1). Because invariance under 2π rotations is automatic in Lorentz-invariant theories, we need only the overall multiplication of the superfields. With this symmetry the full, renormalizable, superpotential is just that in Eq. (11.2).

In addition to solving the problem of very fast proton decay, R-parity has another striking consequence: the lightest of the new particles predicted by supersymmetry, the Lightest supersymmetric particle (LSP), is *stable*. This particle can easily be neutral under the gauge groups. It is then, inevitably, very weakly interacting. This in turn means the following.

- The generic signature of R-parity-conserving supersymmetric theories is the occurrence of events with missing energy.
- Supersymmetry is likely to produce an interesting dark-matter candidate.

This second point is one of the principal reasons that many physicists have found the possibility of low-energy supersymmetry so compelling. If one calculates the dark-matter density then, as we will see in the chapter on cosmology, one automatically finds a density in the right range if the scale of supersymmetry breaking is about 1 TeV. Later, we will see an additional piece of circumstantial evidence for low-energy supersymmetry: the unification of the gauge couplings within the MSSM.

We can imagine more complicated symmetries which would have similar effects, and we will have occasion to discuss these later. We can also relax the assumption of exact R-parity conservation. If, for example, the lepton-number-violating couplings are forbidden then the restrictions on the baryon-number-violating couplings are not so severe and the phenomenological consequences are interesting. In most of what follows we will assume a conserved Z_2 R-parity.

11.1 Soft supersymmetry breaking in the MSSM

If supersymmetry is a feature of the underlying laws of nature then it is certainly broken. The simplest approach to model building with supersymmetry is to add soft-breaking terms to the effective Lagrangian in such a way that the squarks, sleptons and gauginos have sufficiently large masses that they have not yet been observed (or, in the event that they are discovered, to account for their values).

Without a microscopic theory of supersymmetry breaking, all the soft terms are independent. It is of interest to ask how many soft-breaking parameters there are in the MSSM. More precisely, we will count the parameters of the model beyond those of the minimal Standard Model with a single Higgs doublet. Having imposed R-parity, the number of Yukawa couplings is the same in both theories, as are the numbers of gauge couplings and θ parameters. The quartic couplings of the Higgs fields are completely determined by the gauge couplings. So the "new" terms arise from the soft-breaking terms as well as the μ term for the Higgs fields. We will speak loosely of all of this as the soft-breaking Lagrangian. Suppressing flavor indices, we have

$$\begin{aligned}\mathcal{L}_{\rm sb} = {} & \widetilde{Q}^* m_Q^2 \widetilde{Q} + \tilde{\bar{u}}^* m_{\bar{u}}^2 \tilde{\bar{u}} + \tilde{\bar{d}}^* m_{\bar{d}}^2 \tilde{\bar{d}} + \tilde{L}^* m_L^2 \tilde{L} + \tilde{\bar{e}}^* m_{\bar{e}}^2 \tilde{\bar{e}} + H_U \widetilde{Q} A_u \tilde{\bar{u}} + H_D \widetilde{Q} A_d \tilde{\bar{d}} \\ & + H_D \widetilde{L} \widetilde{A}_l \tilde{\bar{e}} + \text{c.c.} + M_i \lambda\lambda + \text{c.c.} + m_{H_U}^2 |H_U|^2 + m_{H_D}^2 |H_D|^2 + \mu B H_U H_D \\ & + \mu \psi_H \psi_H. \end{aligned} \tag{11.8}$$

The matrices m_Q^2, $m_{\bar{u}}^2$ etc. are 3×3 Hermitian matrices, so they have nine independent entries. The matrices A_u, A_d etc. are general 3×3 complex matrices, so they each possess 18 independent entries. Each of the gaugino masses is a complex number, so these introduce six additional parameters. The quantities μ and B are also complex; they add four more. In total, then, there are 111 new parameters. As in the Standard Model, not all these parameters are meaningful; we are free to make field redefinitions. The counting is significantly simplified if we just ask how many parameters there are beyond the usual 18 of the minimal theory.

To understand what redefinitions are possible beyond the transformations on the quarks and leptons which go into defining the CKM parameters, we need to ask what are the symmetries of the MSSM before the introduction of the soft-breaking terms and the μ term (the μ term is more or less on the same footing as the soft-breaking terms, since it is of the same order of magnitude; as we will discuss later, it might well arise from the physics of supersymmetry breaking). Apart from the usual baryon and lepton numbers, there are two more. The first is a Peccei–Quinn symmetry, under which the two Higgs superfields rotate by the same phase while the right-handed quarks and leptons rotate by the opposite phase. The second is an R symmetry, a generalization of the symmetry we found in the Wess–Zumino model (see Section 9.6.1). It is worth describing this in some detail. By definition, an R symmetry is a symmetry of the Hamiltonian which does not commute with the supersymmetry generators. Such symmetries can be continuous or discrete. In the case of continuous R symmetries, by convention we can take the θs to transform by a phase $e^{i\alpha}$. Then the general transformation law takes the form

$$\lambda_i \to e^{i\alpha}\lambda_i \tag{11.9}$$

for the gauginos, while, for the elements of a chiral multiplet, we have

$$\Phi_i(x,\theta) \to e^{ir_i\alpha}\Phi(x,\theta e^{i\alpha}), \tag{11.10}$$

or, in terms of the component fields,

$$\phi_i \to e^{ir_i\alpha}\phi_i, \quad \psi_i \to e^{i(r_i-1)\alpha}\psi_i, \quad F_i \to e^{i(r_i-2)\alpha}F_i. \tag{11.11}$$

In order that the Lagrangian exhibit a continuous R symmetry, the total R charge of all terms in the superpotential must be 2. In the MSSM, we can take $r_i = 2/3$ for all the chiral fields.

The soft-breaking terms, in general, break two of the three lepton-number symmetries, the R-symmetry and the Peccei–Quinn symmetry. So there are four non-trivial field redefinitions which we can perform. In addition, the minimal Standard Model has two Higgs parameters. So from our 111 parameters, we can subtract a total of six, leaving 105 as the number of *new* parameters in the MSSM.

Clearly we would like to have a theory which predicts these parameters. Later, we will study some candidates. To get started, however, it is helpful to make an ansatz. The simplest thing to do is suppose that all the scalar masses are the same, all the gaugino masses the same and so on. It is necessary to specify also a scale at which this ansatz holds, since, even if true at one scale, it will not continue to hold at lower energies. Almost all investigations of supersymmetry phenomenology assume such a degeneracy at a large energy scale, typically the reduced Planck mass M_p. It is often said that degeneracy is automatic in supergravity models, so this is frequently called the supergravity (SUGRA) model but, as we will see, supergravity by itself makes *no* prediction of degeneracy. Some authors, similarly, include this assumption as part of the definition of the MSSM, but in this text we will use the term MSSM to refer to the particle content and the renormalizable interactions. In any case, the ansatz consists of the following statements at the high-energy scale.

1. All the scalar masses are the same, $\widetilde{m}^2 = m_0^2$. This assumption is called the *universality* of the scalar masses.
2. The gaugino masses are the same, $M_i = M_0$. This is referred to as the *GUT relation*, since it holds in simple grand unified models.
3. The soft-breaking cubic terms are assumed to be given by

$$\mathcal{L}_{\text{tri}} = A(H_U Q y_u \bar{u} + H_D Q y_d \bar{d} + H_D L y_l \bar{e}). \tag{11.12}$$

The matrices y_u, y_d etc. are the same as those which appear in the Yukawa couplings. This is the assumption of *proportionality*.

Note that with this ansatz, if we ignore the various phases possibilities, five parameters are required to specify the model $(m_0^2, M_0, A, B_\mu, \mu)$. One of these can be traded for M_Z, so this is quite an improvement in predictive power. In addition, this ansatz automatically satisfies all constraints coming from rare processes. As we will explain, rare decays and flavor violation are suppressed ($b \rightarrow s + \gamma$ is not as strong a constraint, but it requires other relations among soft masses). However, we need to ask: just how plausible are these assumptions? We will try to address this question later.

11.1.1 The μ term

One puzzle in the MSSM is the μ term, the supersymmetric mass term for the Higgs fields. This term is not forbidden by the gauge symmetries, so the first question is: why is it small, of order a few TeV rather than of order M_p or M_{gut}? One possibility is that there is a symmetry which accounts for this. There might, for example, be a discrete symmetry forbidding $H_U H_D$ in the superpotential, spontaneously broken by the fields which also break supersymmetry. Another possibility is related to the non-renormalization theorems. If for some reason, there is no mass term at lowest order for the Higgs fields, one will not be generated perturbatively. The μ term, then, might be the result of the same non-perturbative dynamics, for example, those responsible for supersymmetry breaking. In string theories, as we will see later, it is quite common to find massless particles at tree level, simply "by accident". Such a phenomenon can also be arranged in grand unified theories.

In the absence of a large, tree level, μ term, supersymmetry breaking can quite easily generate a μ term of order $m_{3/2}$. Consider, for example, the Polonyi model. The operator

$$\int d^4\theta \frac{1}{M_p} Z^\dagger H_U H_D \tag{11.13}$$

would generate a μ term of just the correct size. In simple grand unified theories, such a term is often generated.

When we discuss other models for supersymmetry breaking, such as gauge mediation, we will see that the μ term sometimes poses additional challenges.

11.1.2 Cancelation of quadratic divergences in gauge theories

We have already seen that soft supersymmetry-violating mass terms receive only logarithmic divergences. While not essential to our present discussion, it is perhaps helpful to see how the cancelation of quadratic divergences for scalar masses arises in gauge theories like the MSSM.

Take, first, a $U(1)$ theory, with (massless) chiral fields ϕ^+ and ϕ^-. Without doing any computation it is easy to see that, provided we work in a way which preserves supersymmetry, there can be no quadratic divergence. In the limit where the mass term vanishes, the theory has a chiral symmetry under which ϕ^+ and ϕ^- rotate by the same phase,

$$\phi^\pm \rightarrow e^{i\alpha}\phi^\pm. \tag{11.14}$$

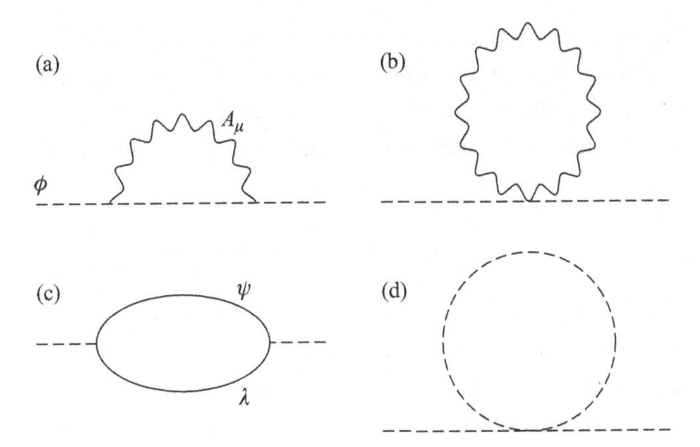

Fig. 11.1 One-loop diagrams contributing to scalar masses in a supersymmetric gauge theory.

This symmetry forbids a mass term $\Lambda\phi^+\phi^-$ in the superpotential the only from in which a supersymmetric mass term could appear. The actual diagrams we need to compute are shown in Fig. 11.1. Since we are interested only in the mass, we can take the external momentum to be zero. It is convenient to choose the Landau gauge for the gauge boson. In this gauge the gauge boson propagator is

$$D_{\mu\nu} = -i\left(g_{\mu\nu} - \frac{q_\mu q_\nu}{q^2}\right)\frac{1}{q^2}, \tag{11.15}$$

so the first diagram vanishes. The second, third and fourth are straightforward to work out from the basic Lagrangian. One finds:

$$I_\mathrm{b} = g^2 i \times i\frac{3}{(2\pi)^4}\int\frac{d^4k}{k^2}, \tag{11.16}$$

$$I_\mathrm{c} = g^2 i \times i\frac{(\sqrt{2})^2}{(2\pi)^4}\int\frac{d^4k}{k^4}\,\mathrm{Tr}(k_\mu\sigma^\mu k_\nu\bar{\sigma}^\nu) \tag{11.17}$$

$$= -\frac{4g^2}{(2\pi)^4}\int\frac{d^4k}{k^2}, \tag{11.18}$$

$$I_\mathrm{d} = g^2(i)(i)\frac{1}{(2\pi)^4}\int\frac{d^4k}{k^2}. \tag{11.19}$$

It is easy to see that the sum $I_\mathrm{a} + I_\mathrm{b} + I_\mathrm{c} + I_\mathrm{d} = 0$.

Including a soft-breaking mass for the scalars only changes I_d:

$$\begin{aligned} I_\mathrm{d} &\to \frac{g^2}{(2\pi)^4}\int\frac{d^4k}{k^2 - \widetilde{m}^2} \\ &= -i\frac{g^2}{(2\pi)^4}\int\frac{d^4k_E}{k_E^2 + \tilde{m}^2} \\ &= \widetilde{m}^2_{\mathrm{independent}} + \frac{ig^2}{16\pi^2}\widetilde{m}^2 \ln\frac{\Lambda^2}{\widetilde{m}^2}. \end{aligned} \tag{11.20}$$

We have worked here in Minkowski space and have indicated the factors i to assist the reader in obtaining the correct signs for the diagrams. In the second line of Eq. (11.20) we have performed a Wick rotation. In the third line we have separated off the mass-independent part, since we know that this is canceled by the other diagrams.

Summarizing, the one-loop mass shift is

$$\delta\widetilde{m}^2 = -\frac{g^2}{16\pi^2}\widetilde{m}^2 \ln\frac{\Lambda^2}{\widetilde{m}^2}. \tag{11.21}$$

Note that the mass shift is proportional to $\widetilde{m}^2$, the supersymmetry-breaking mass, which we expect since supersymmetry is restored as $\widetilde{m}^2 \to 0$. In the context of the Standard Model we see that the scale of supersymmetry breaking cannot be much larger than the Higgs mass scale itself without fine tuning. Roughly speaking, it cannot be much larger than this scale than by a factor of order $1/\sqrt{\alpha_W}$, i.e. of order six. We also see that the correction has a logarithmic sensitivity to the cutoff. So, just as for the gauge and Yukawa couplings, the soft masses run with the energy.

11.2 $SU(2) \times U(1)$ breaking

In the MSSM there are a number of general statements which can be made about the breaking of $SU(2) \times U(1)$. The only quartic couplings of the Higgs fields arise from the $SU(2)$ and $U(1)$ D^2 terms. The general form of the soft-breaking mass terms has been described above. So, before we worry about any detailed ansatz for the soft breakings, we note that the Higgs potential is given quite generally by

$$V_{\text{Higgs}} = m^2_{H_U}|H_U|^2 + m^2_{H_D}|H_D|^2 - m_3^2(H_U H_D + \text{h.c.}) + \frac{1}{8}(g^2 + g'^2)(|H_U|^2 - |H_D|^2)^2 + \frac{1}{2}g^2|H_U H_D|^2. \tag{11.22}$$

This potential by itself conserves CP; a simple field redefinition removes any phase in m_3^2. (As we will discuss shortly, there are many other possible sources of CP violation in the MSSM.) The physical states in the Higgs sector are usually described by assuming that CP is a good symmetry. In that case there are two CP-even scalars, H^0 and h^0, where, by convention, h^0 is the lighter of the two. There are a CP-odd neutral scalar A and charged scalars $H^\pm$. At tree level, one also defines a parameter which is the ratio of the vevs of H_U and H_D or v_1 and v_2:

$$\tan\beta = \frac{|\langle H_U\rangle|}{|\langle H_D\rangle|} \equiv \frac{v_1}{v_2}. \tag{11.23}$$

Note that, with this definition, as $\tan\beta$ grows so does the Yukawa coupling of the b quark.

To obtain a suitable vacuum, there are two constraints which the soft breakings must satisfy.

1. Without the soft-breaking terms, $H_U = H_D$ ($v_1 = v_2 = v$) makes the $SU(2)$ and $U(1)$ D terms vanish, i.e. there is no quartic coupling in this direction. So the energy

is unbounded below, unless

$$m_{H_U}^2 + m_{H_D}^2 - 2|m_3|^2 > 0. \tag{11.24}$$

2. In order to obtain symmetry breaking, the Higgs mass matrix must have a negative eigenvalue. This gives the requirement

$$\left|m_3^2\right|^2 > m_{H_U}^2 m_{H_D}^2. \tag{11.25}$$

When these conditions are satisfied, it is straightforward to minimize the potential and determine the spectrum. One finds that

$$m_A^2 = \frac{m_3^2}{\sin\beta\cos\beta}. \tag{11.26}$$

It is conventional to take m_A^2 as one parameter. Then one finds that the charged Higgs masses are given by

$$m_{H^\pm}^2 = m_W^2 + m_A^2, \tag{11.27}$$

while the neutral Higgs masses are

$$m_{H^0,h^0}^2 = \frac{1}{2}\left[m_A^2 + m_Z^2 \pm \sqrt{\left(m_A^2 + m_Z^2\right)^2 - 4m_Z^2 m_A^2 \cos 2\beta}\,\right]. \tag{11.28}$$

Note the inequalities

$$m_{h^0} \le m_A, \qquad m_{h^0} \le m_Z, \qquad m_{H^\pm} \ge m_W. \tag{11.29}$$

With the discovery of the Higgs at 125 GeV, it would appear that the MSSM is ruled out. However, these are tree level relations. We will shortly turn to the issue of radiative corrections and will see that these can be quite substantial. We will also see, however, that accounting for a Higgs mass of 125 GeV appears to require a significant fine tuning of the parameters.

11.3 Embedding the MSSM in supergravity

In the previous chapter we introduced $N = 1$ supergravity theories. These theories are not renormalizable and must be viewed as effective theories, valid below some energy scale which might be the Planck scale or unification scale (or something else).

The approach we have introduced to model building is quite useful when we are considering models for the origin of supersymmetry breaking in the MSSM. The basic assumptions of this approach were as follows.

- The theory consists of two sets of fields the *visible sector fields* y_a, which in the context of the MSSM would be the quark and lepton superfields, and the *hidden sector fields* z_i, responsible for supersymmetry breaking.

- The superpotential was taken to have the form

$$W(z,y) = W_z(z) + W_y(y). \tag{11.30}$$

- For the Kahler potential we took the simple ansatz

$$K = \sum_a y_a^\dagger y_a + \sum_i z_i^\dagger z_i. \tag{11.31}$$

In this case, we saw that if the supersymmetry-breaking scale was of order

$$M_{\text{int}} = m_{3/2} M_{\text{p}} \tag{11.32}$$

then there was an array of soft-breaking terms of order $m_{3/2}$. In particular, there were universal masses and A terms,

$$am_{3/2}^2 |y_a|^2 + bm_{3/2} W_{ab} y_a y_b + cm_{3/2} W_{abc} y_a y_b y_c. \tag{11.33}$$

Here $W_{ab} = \partial_a \partial_b W$ and $W_{abc} = \partial_a \partial_b \partial_c W$.

Given that the MSSM is at best an effective-low-energy theory, one can ask how natural are our assumptions, and what would be the consequences of relaxing them? The assumption that there is some sort of hidden sector, and that the superpotential breaks up as we have hypothesized, is, as we will see, a reasonable one. It can be enforced by symmetries. The assumption that the Kahler potential takes this simple (often called "minimal") form is a strong one, not justified by symmetry considerations. It turns out not to hold in any general sense in string theory, the only context in which presently we can compute it. If we relax this assumption, we lose the universality of scalar masses and the proportionality of the A terms to the superpotential. As we will see later in this chapter, without these or something close the MSSM is not compatible with experiment.

11.4 Radiative corrections to the Higgs mass limit

We have seen that, in the MSSM, the Higgs mass at tree level is less than the Z mass. This bound is clearly violated in nature. In this section and the next, we will see that a 125 GeV Higgs particle can be accommodated within the MSSM, though it requires either a large scale of supersymmetry breaking or the introduction of new degrees of freedom.

In the MSSM, at tree level, the form of the Higgs potential is highly constrained because the quartic couplings are completely determined by the gauge interactions. Once supersymmetry (susy) is broken, however, there can be corrections to the quartic terms from radiative corrections.These corrections are soft, in that the susy-violating four-point functions vanish rapidly at momenta above the susy-breaking scale. Still, they are important in determining the low-energy properties of the theory, such as the Higgs vacuum expectation values (vevs) and the spectrum.

The largest effect of this kind comes from loops involving top quarks or their scalar partners, the *stops*. It is not hard to get a rough estimate of the effect. In the limit $\tilde{m}_t \gg m_t$, the effective Lagrangian is not supersymmetric below $\tilde{m}_t$. As a result, there can be

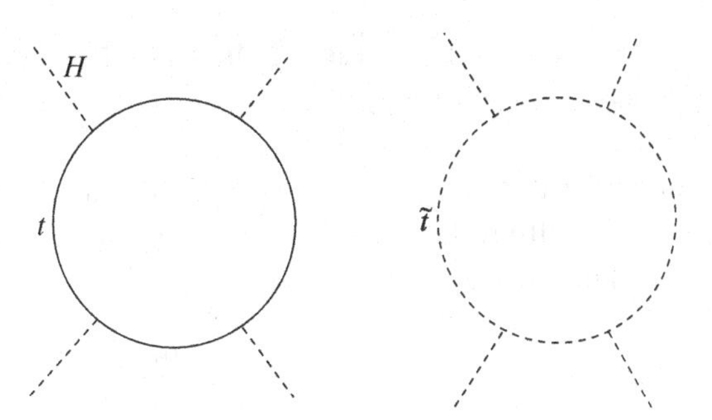

Fig. 11.2 Corrections to quartic Higgs couplings from top loops.

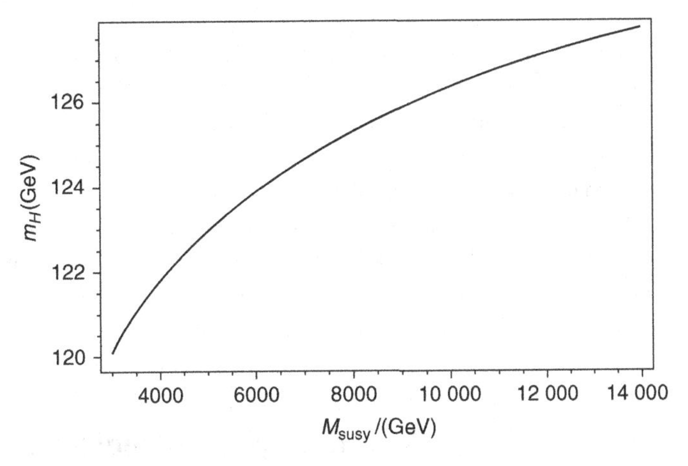

Fig. 11.3 Higgs mass as a function of susy-breaking parameters.

corrections to the Higgs quartic couplings. Consider the diagrams of Fig. 11.2. In this limit we can get a reasonable estimate by just keeping the top quark loop. The result will be logarithmically divergent, and we can take the cutoff to be $\widetilde{m}_t$. So we have

$$\delta\lambda = (-1)y_t^4 \times 3 \int \frac{d^4k}{(2\pi)^4} \mathrm{Tr} \frac{1}{(\not{k} - m_t)^4} \tag{11.34}$$

$$= -\frac{12iy_t^4}{16\pi^2} \ln \frac{\widetilde{m}_t^2}{m_t^2}. \tag{11.35}$$

One can get a better estimate by keeping finite terms and higher-order corrections. There exist online tools to perform these calculations (mentioned in the references at the end of this chapter). For large $\tan\beta$ these corrections are most effective; this corresponds precisely to the decoupling limit discussed in Chapter 3, where the Higgs is principally H_U. A typical plot of m_H as a function of $\tilde{m}_t$, for small values of the A parameter for the stops and for large $\tan\beta$, is that of Fig. 11.3. We see that, for moderate values of the A parameter, a Higgs of 125 GeV corresponds to a stop mass of order 10 TeV. As we will see in the next section, this, in turn, implies a significant tuning of the Higgs mass.

11.5 Fine tuning of the Higgs mass

We saw earlier that in the Wess–Zumino model at one loop there is a negative renormalization of the soft-breaking scalar masses. This calculation can be translated to the MSSM, with a modification for the color and $SU(2)$ factors. One obtains

$$m_{H_U}^2 = (m_{H_U})_0^2 - \frac{6y_t^2}{16\pi^2} \ln \frac{\Lambda^2}{m^2} (\widetilde{m}_t^2 + m_{\bar{t}}^2), \tag{11.36}$$

$$\widetilde{m}_t^2 = (\widetilde{m}_t)_0^2 - \frac{4y_t^2}{16\pi^2} \ln \frac{\Lambda^2}{m^2} \widetilde{m}_H^2. \tag{11.37}$$

So, we see that loop corrections involving the top quark Yukawa coupling reduce both the Higgs and the stop masses. If $\widetilde{m}_t^2 = 10$ TeV, and if $\Lambda \sim M_{\mathrm{p}}$, the correction to the stop mass is of order one but the correction to the Higgs mass is of order $8000 m_H^2$! This suggests a tuning of the parameter $(m_{H_U})_0^2$ at nearly the one part in 10 000 level, and a more refined renormalization group analysis supports this.

Such a tuning of parameters is troubling, given that we introduced supersymmetry in order to avoid such problems with naturalness. It is, at least, not as extreme as the situation without supersymmetry. It is also consistent with the data. In the next section, we will mention a few ideas to ameliorate this tuning.

11.6 Reducing the tuning: the NMSSM

We have seen that in the MSSM the effective Higgs quartic coupling is small because it is determined by the gauge couplings; this is what accounts for the tree level Higgs mass bound. The requirement of a large stop mass was driven by the need to enhance the quartic coupling. One might also hope to enhance the quartic coupling by introducing additional fields with superpotential couplings to the Higgs. The simplest approach yields the Next to Minimal Supersymmetric Standard Model, or NMSSM. In its simplest version the field content of the model is that of the MSSM plus an additional singlet, S. The superpotential includes a term

$$W_{\mathrm{NMSSM}} = \lambda S H_U H_d \tag{11.38}$$

in addition to the Yukawa couplings of the Higgs. This superpotential leads to a quartic coupling

$$\delta V = |\lambda H_U H_d|^2, \tag{11.39}$$

which can increase the Higgs mass. However, λ cannot be arbitrarily large otherwise perturbation theory would break down. Requiring that there be no Landau pole for λ typically implies that $\lambda < 0.7$.

One difficulty with this proposal is that the maximum effect occurs when $\tan\beta \sim 1$, so that H_U and H_D are more or less aligned. In this limit the top quark corrections to the quartic coupling are less effective. Adding other terms to the superpotential, such as $\frac{1}{2}m_S S^2$ and S^3 as well as the various possible soft breakings, yields a large parameter space to explore. One typically finds that fine tuning can be significantly improved over the MSSM, but because of the constraints on λ it is still significantly worse than 10%.

There are other proposals to reduce the tuning of the MSSM by introducing additional degrees of freedom. Additional gauge interactions, for example, can help. Perhaps a compelling model may yet emerge. As we will see in the following sections, however, direct searches for supersymmetric particles, especially with the LHC, have placed stringent lower limits on the masses of supersymmetric partners of ordinary particles.

11.7 Constraints on low-energy supersymmetry: direct searches and rare processes

Naturalness points to supersymmetry at a scale below the TeV scale – arguably of order M_Z. We have already discussed how the Higgs mass points towards a significantly higher scale, somewhere around 10 TeV. Direct searches for supersymmetric particles, as we will briefly review here, also point to a high scale. Current limits on squarks and gluinos are, over much of the parameter space, larger than a TeV and they will become stronger (or evidence for supersymmetry will emerge) during future LHC runs. The limits on leptons, charginos and neutralinos (see below) are significant, though not quite as strong.

There are also strong constraints on the supersymmetry parameters (the 101 parameters we counted in the MSSM, for example) from rare processes.

11.7.1 Direct searches for supersymmetric particles

As mentioned above, direct searches for supersymmetric particles at LEP, the Tevatron and the LHC have placed significant limits on their masses. Among the states in the MSSM which are possible discovery channels for supersymmetry, are the *charginos*, linear combinations of the partners of the $W^\pm$ and $H^\pm$, and the *neutralinos*, linear combinations of the partners of the Z and γ (B and W^3) and the neutral Higgs. The mass matrix for the charginos, $w^\pm$ and $\tilde{h}^\pm$ is given by

$$\begin{aligned}\mathcal{L}_{\text{sb}} = {} & \widetilde{Q}^* m_Q^2 \widetilde{Q} + \tilde{\bar{u}}^* m_{\bar{u}}^2 \tilde{\bar{u}} + \tilde{\bar{d}}^* m_{\bar{d}}^2 \tilde{\bar{d}} + \tilde{L}^* m_L^2 \tilde{L} + \tilde{\bar{e}}^* m_{\bar{e}}^2 \tilde{\bar{e}} + H_U \widetilde{Q} A_u \tilde{\bar{u}} + H_D \widetilde{Q} A_d \tilde{\bar{d}} \\ & + \langle H_U \rangle^2 + m_{H_D}^2 \langle H_D \rangle^2 + \mu B H_U H_D \\ & + \mu \psi_H \psi_H. \end{aligned} \tag{11.40}$$

The matrices m_Q^2, $m_{\bar{U}}^2$ and so on that give mass to the scalar partners of quarks and leptons (*squarks* and *sleptons*) are 3×3 Hermitian matrices, so they have nine independent entries. The matrices A_u, A_d etc. are general 3×3 complex matrices, so they each possess

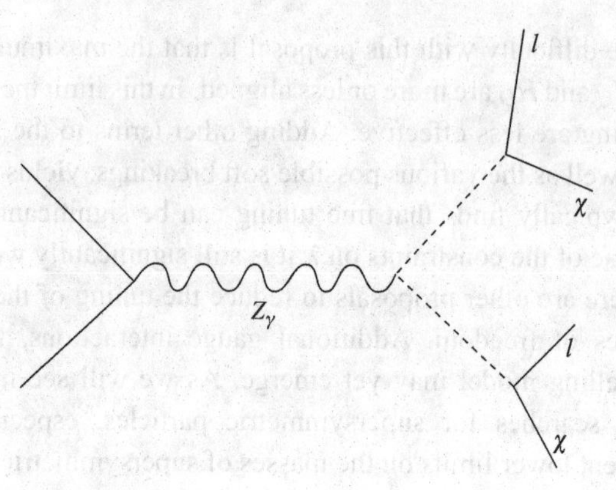

Fig. 11.4 Slepton production in e^+e^- annihilation.

18 independent entries. Each gaugino mass is a complex number, so these introduce six additional parameters; M_1, M_2 and M_3 are Majorana mass terms for the $U(1)$, $SU(2)$ and $SU(3)$ gauginos. The quantities μ and B are also complex and so introduce four more parameters. In total, then, there are 111 new parameters. As in the Standard Model, they are not all meaningful since we are free to make field redefinitions. The counting is significantly simplified if we just ask how many parameters there are beyond the usual 18 of the minimal theory.

For the neutralinos, w^0, b, $\tilde{h}_U^0$, $\tilde{h}_D^0$, there is a 4×4 mass matrix. We will leave the study of these for the exercises. Conventionally, the charginos are denoted $\tilde{\chi}_1^+$, $\tilde{\chi}_1^-$, $\tilde{\chi}_2^+$, $\tilde{\chi}_2^-$, where the label 2 indicates a chargino having greater mass. The neutralinos are denoted $\tilde{\chi}_1^0$, $\tilde{\chi}_2^0$, $\tilde{\chi}_3^0$, $\tilde{\chi}_4^0$, again ordered by increasing mass. The lightest of these states is stable if R-parity is conserved and is a natural dark-matter candidate.

The direct searches are easy to describe, and production and decay rates can be computed given a knowledge of the spectrum since the couplings of the fields are known. If R-parity is conserved then the LSP is stable and weakly interacting, so the characteristic signal for supersymmetry is *missing energy*. For example, in e^+e^- colliders one can produce slepton pairs, if they are light enough, through the diagram of Fig. 11.4. These then decay, typically, to a lepton and a neutralino, as indicated. So the final state contains a pair of acoplanar leptons and missing energy. The LEP ran at center of mass energies as high as $\sqrt{s} = 209$ GeV, setting limits of order 90 GeV on sleptons and 103.5 GeV on charginos. The LHC has strengthened these limits in some regions of the parameter space.

In hadron colliders at high energies, one has the potential to produce colored hadrons – squarks and gluinos – at high rates. As a result the most dramatic limits on supersymmetric particles have been set by the LHC (following earlier searches at the Tevatron). The LHC has run at 7 and 8 GeV, collecting 20 (femtobarns)$^{-1}$ of data per detector at the higher energy, Setting limits, however, on gaugino and squark masses (and those of other states) is a model-dependent process. For example, if gauginos are heavier than squarks, they will first decay to a gluon and a squark; the squark may decay to a quark and a neutralino or to a quark and a chargino, with the chargino in turn decaying by a variety of possible channels. If the squarks are heavier than gluinos, there are alternative decay chains.

Many analyses employ the ansatz we called SUGRA (see Section 11.1), with five parameters. Quite stringent limits can then be set on these different parameters, and correspondingly on the masses of the various superparticles. In recent years this model has been refined somewhat and rebranded as the *Constrained Minimal Supersymmetric Standard Model*, or CMSSM. A more phenomenological variant with assumptions which are not quite as restrictive is the PMSSM. The strategy, in this framework, is to allow the maximum (or close to the maximum) number of parameters consistent with the various facts of low-energy physics. An alternative approach, adopted by many theorists and employed in many experimental analyses, is referred to as the "simplified model" method. Here one focuses on signals, i.e. particular production and decay possibilities, rather than on fitting to models. From all these types of analysis one finds lower limits on gluinos of order 1.2–1.7 TeV and similar limits for squarks.

11.7.2 Constraints from rare processes

Rare processes provide another set of strong constraints on the soft-breaking parameters. In the simple ansatz, all the scalar masses are the same at some very high energy scale. However, even if this is assumed to be true at one scale, it is not true at all scales, i.e. these relations are *renormalized*. Indeed, all 105 parameters are truly parameters and it is not obvious that the assumptions of universality and proportionality are *natural*. However, there are strong experimental constraints which suggest some degree of degeneracy.

As one example, there is no reason, a priori, why the mass matrix for the $\tilde{L}$s (the partners of the lepton doublets) should be diagonal in the same basis as the charged leptons. If it is not then there is no conservation of separate lepton numbers, and the decay $\mu \to e\gamma$ will occur (Fig. 11.5). To see that we are potentially in serious trouble, we can make a crude estimate. The muon lifetime is proportional to $G_F^2 m_\mu^5$. The decay $\mu \to e\gamma$ occurs owing to the operator

$$\mathcal{L}_{\mu e\gamma} = eCF_{\mu\nu}\bar{u}\sigma^{\mu\nu}e. \tag{11.41}$$

If there is no particular suppression, we might expect that

$$C = \frac{\alpha_W}{\pi}\frac{m_\mu}{m_{\text{susy}}^2}. \tag{11.42}$$

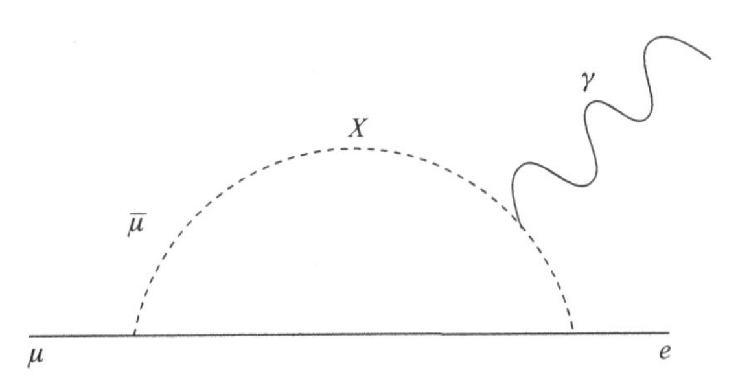

Fig. 11.5 Contribution to $\mu \to e\gamma$.

Therefore the branching ratio, i.e. the ratio of the rate of decay to e_γ and the rate for all decays, would be of order

$$BR = \frac{\Gamma(\mu \to e + \gamma)}{\Gamma(\mu \to \text{all})} = \left(\frac{\alpha_w}{\pi}\right)^2 \left(\frac{M_W}{M_{\text{susy}}}\right)^4. \tag{11.43}$$

This ratio might become as small as 10^{-8}–10^{-9} if the supersymmetry-breaking scale is large, 1 TeV or so. But the current experimental limit is 1.2×10^{-11}. So even in this case it is necessary to suppress the off-diagonal terms. More detailed descriptions of the limits are found in the suggested reading at the end of the chapter.

Another troublesome constraint arises from the neutron and electron electric dipole moments, d_n and d_e. Any non-zero value of these quantities signifies CP violation. Currently, one has $d_n < 2.9 \times 10^{-26} e$ cm and $d_e < 18.7 \times 10^{-29} e$ cm. The soft-breaking terms in the MSSM contain many new sources of CP violation. Even with the assumptions of universality and proportionality, the gaugino mass and the A, μ and B parameters are all complex and can violate CP. At the quark level, the issue is that one-loop diagrams can generate a quark dipole moment, as in Fig. 11.6. Note that this particular diagram is proportional to the phases of the gluino and the A parameter. It is easy to see that, even if $m_{\text{susy}} \sim 500$ GeV, these phases must be smaller than about 10^{-2}. More detailed estimates can be found in the suggested reading at the end of the chapter.

In the real world CP is violated, so it is puzzling that all the soft-supersymmetry-violating terms should preserve CP to such a high degree. This is in contrast with the minimal Standard Model, with a single Higgs field, which can reproduce the observed CP violation with phases of order 1. It is thus a serious challenge to understand why CP should be such a good symmetry if nature is supersymmetric. Various explanations have been offered. We will discuss some of these later, but it should be kept in mind that the smallness of CP violation suggests that either the low-energy supersymmetry hypothesis is wrong or there is some interesting physics which explains the surprisingly small values of the dipole moments.

So far, we have discussed constraints on slepton degeneracy and CP-violating phases. There are also constraints on the squark masses, arising from various flavor-violating processes. In the Standard Model the most famous of these are strangeness-changing processes such as $K\bar{K}$ mixing. One of the early triumphs of the Standard Model was that it successfully explained why this mixing is so small. Indeed, the Standard Model gives

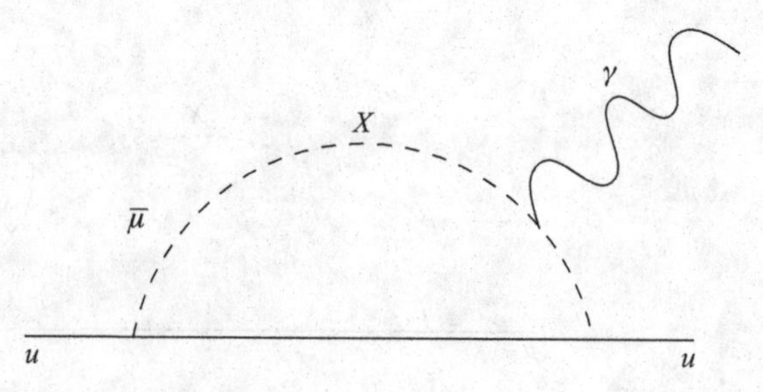

Fig. 11.6 Contribution to d_n in supersymmetric theories.

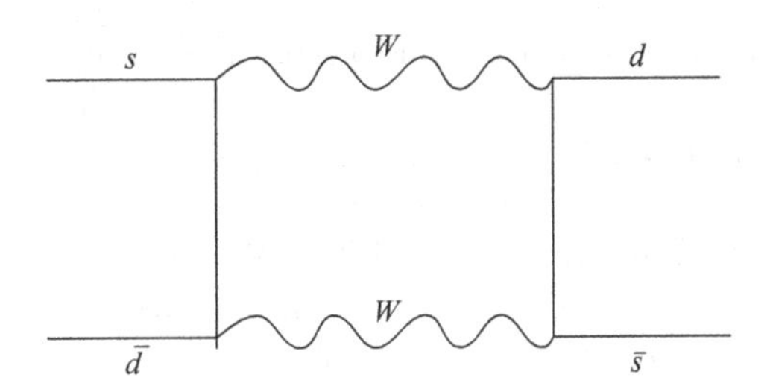

Fig. 11.7 Contribution to $K \leftrightarrow \bar{K}$ in the Standard Model.

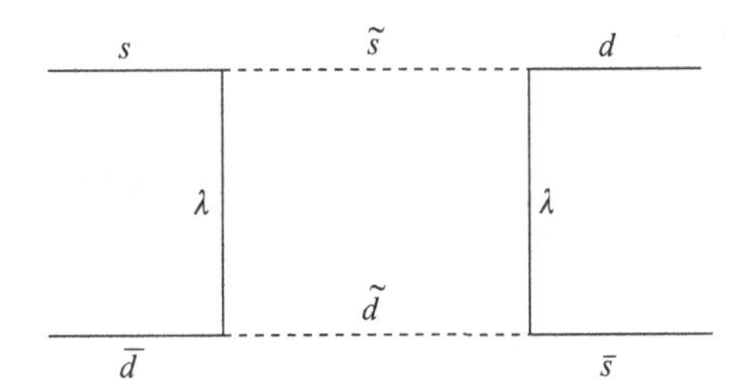

Fig. 11.8 Gluino exchange contribution to $K\bar{K}$ mixing in the MSSM.

a quite good estimate for the mixing. This was originally used to predict – amazingly accurately – the charm quark mass. The mixing receives contributions from box diagrams such as that shown in Fig. 11.7. If we consider only the first two generations and ignore the quark masses (compared with M_W), we have that

$$\mathcal{M}(K^0 \to \bar{K}^0) \propto (V_{di} V^\dagger_{is})(V^\dagger_{sj} V_{jd}) = 0. \tag{11.44}$$

Including fermion masses leads to terms in the low-energy effective action $\mathcal{L}_{\rm eff}$ of order

$$\frac{\alpha_W}{4\pi} \frac{m_c^2}{M_W^2} G_{\rm F} \ln \frac{m_c^2}{m_u^2} (\bar{s}\gamma^\mu \gamma_5 d)(\bar{d}\gamma^\mu \gamma^5 s) + \cdots. \tag{11.45}$$

The matrix element of the operator appearing here can be estimated in various ways, and one finds that this expression roughly saturates the observed value (this was the origin of the prediction by Gaillard and Lee of the value of the charm quark mass). Similarly, the CP-violating parameter in the kaon system (the "ϵ" parameter) is in rough accord with observation for reasonable values of the CKM parameter δ.

In supersymmetric theories, if squarks are degenerate then there are similar cancelations. However, if they are not then there are new, very dangerous, contributions. The most serious is that indicated in Fig. 11.8, arising from the exchange of gluinos and squarks. This is nominally larger than the Standard Model contribution by a factor $(\alpha_s/\alpha_W)^2 \approx 10$. Also, the Standard Model contribution vanishes in the chiral limit whereas the gluino exchange does not, and this leads to an additional enhancement of nearly an order of magnitude. However, the diagram is highly suppressed in the limit of exact universality and proportionality. Proportionality means that the A terms in Eq. (11.8) are suppressed by factors of order the light quark masses, while universality means that the squark propagator $\langle \tilde{q}^* \tilde{q} \rangle$ is proportional to the unit matrix in flavor space. So, on the one hand, there are no appreciable off-diagonal terms which can contribute to the diagram. On the other hand, there is surely some degree of non-degeneracy. One finds that, even if the characteristic susy scale is 1 TeV, one needs degeneracy in the down squark sector at the one part in 30 level.

So the CP-preserving part of the $K\bar{K}$ mass matrix already tightly constrains the down squark mass matrix and the CP-violations part provides even more severe constraints. There are also strong limits on $D\bar{D}$ mixing, which significantly restrict the mass matrix in the up squark sector. Other important constraints on soft breakings come from other rare processes such as $b \to s\gamma$. Again, more details can be found in the references given in the suggested reading.

Suggested reading

The minimal supersymmetric Standard Model is described in most reviews of supersymmetry. Probably the best place to look for up-to-date reviews of the model and the experimental constraints is the Particle Data Group website. A useful collection of renormalization group formulas for supersymmetric theories is provided in the review by Martin and Vaughn (1994). Limits on rare processes are discussed in a number of articles, such as that by Masiero and Silvestrini (1997). The status of the NMSSM, including questions of tuning, is discussed in Hall *et al.* (2012).

Exercises

(1) Derive Eqs. (11.24)–(11.27).

(2) Verify the formula for the top quark corrections to the Higgs mass. Evaluate y_t in terms of m_t and $\sin\beta$. Show that, to this level of accuracy,

$$m_h^2 < m_Z^2 \cos 2\beta + \frac{12g^2}{16\pi^2} \frac{m_t^4}{m_W^2} \ln\left(\widetilde{m}^2 m_t^2\right).$$

(3) Estimate the sizes of the supersymmetric contributions to the quark electric dipole moment, assuming that all the superpartner masses are of order m_{susy} and that δ is a typical phase. Assuming, as well, that the neutron electric dipole moment is of order the quark electric dipole moment, how small do the phases have to be if $m_{\text{susy}} = 500$ GeV?

12 Supersymmetric grand unification

In this brief chapter we discuss one of the most compelling pieces of circumstantial evidence in favor of supersymmetry: the unification of coupling constants. Earlier, we introduced grand unification without supersymmetry. In this chapter we consider how supersymmetry modifies that story.

12.1 A supersymmetric grand unified model

Just as in theories without supersymmetry, the simplest group into which one can unify the gauge group of the Standard Model is $SU(5)$. The quark and lepton superfields of a single generation again fit naturally into a $\bar{5}$ and a 10.

To break $SU(5)$ down to $SU(3) \times SU(2) \times U(1)$, we can again consider a 24-dimensional representation of the Higgs field Σ. If we wish supersymmetry to be unbroken at high energies, the superpotential for this field should not lead to supersymmetry breaking. The simplest renormalizable superpotential is

$$W(\Sigma) = m \operatorname{Tr} \Sigma^2 + \frac{\lambda}{3} \operatorname{Tr} \Sigma^3. \tag{12.1}$$

Treating this as a globally supersymmetric theory (i.e. ignoring supergravity corrections), the equations

$$\frac{\partial W}{\partial \Sigma} = 0 \tag{12.2}$$

are conveniently studied by introducing a Lagrange multiplier to enforce $\operatorname{Tr} \Sigma = 0$. The resulting equations have three solutions:

$$\Sigma = 0, \quad \Sigma = \frac{m}{\lambda} \operatorname{diag}(1, 1, 1, -4), \quad \Sigma = \frac{m}{\lambda} \operatorname{diag}(2, 2, 2, -3, -3). \tag{12.3}$$

These solutions either leave $SU(5)$ unbroken or break $SU(5)$ down to $SU(4) \times U(1)$ or the Standard Model group. Each solution is isolated; you can check that there are no massless fields from Σ in any of these states. At the classical level they are degenerate.

If we include supergravity corrections, however, these states are split in energy. Provided that the unification scale m is substantially below the Planck scale, these corrections can be treated perturbatively. In order to make the cosmological constant vanish in the

$SU(3) \times SU(2) \times U(1)$ (in brief, $(3, 2, 1)$) vacuum, it is necessary to include a constant in the superpotential such that, in this vacuum, the expectation value of the superpotential is zero. As a result the other two states have negative energy (as we will see in the chapter on gravitation, they correspond to solutions in which space–time is not Minkowski but *anti-de Sitter*).

We will leave working out the details of these computations to the exercises and turn to other features of this model. It is necessary to include Higgs fields to break $SU(2) \times U(1)$ down to $U(1)$. The simplest choice for the Higgs is the 5-dimensional representation. As in the MSSM, it is actually necessary to introduce two sets of fields so as to avoid anomalies: a 5 and $\bar{5}$ are the minimal choice. We denote these fields by H and $\bar{H}$.

Once again it is important that the color triplet Higgs fields in these multiplets be massive in the $(3, 2, 1)$ vacuum. The most general renormalizable superpotential that couples the Higgs to the adjoint is

$$m_H H\bar{H} + y\bar{H}\Sigma H. \tag{12.4}$$

By carefully adjusting y (or m) we can arrange that the Higgs doublet is massless. As a result the triplet is automatically massive, with a mass of order m_H. Of course, this represents an extreme fine tuning. We will see that the unification scale is about 10^{16} GeV, so this is a tuning of one part in 10^{13} or so. But it is curious that this tuning only needs be done classically. Because the superpotential is not renormalized, radiative corrections do not lead to large masses for the doublets.

12.2 Coupling constant unification

The calculation of coupling constant unification in supersymmetric theories is quite similar to that in non-supersymmetric theories. We assume that the threshold for the supersymmetric particles is somewhere around 1 TeV. So, up to that scale, we run the renormalization group equations just as in the Standard Model. Above that scale there are new contributions from the superpartners of ordinary particles. The leading terms in the beta functions are as follows:

$$SU(3), b_0 = 3; \quad SU(2), b_0 = -1; \quad U(1), b_0 = -\frac{33}{5}. \tag{12.5}$$

One can be more thorough, including two-loop corrections and threshold effects. The result of such an analysis are shown in Fig. 12.1. One has:

$$M_{\rm gut} = 1.2 \times 10^{16}\text{GeV}, \quad \alpha_{\rm gut} \approx \frac{1}{25}. \tag{12.6}$$

The agreement in the figure is striking. One can view this as a successful prediction of α_s (see below Eq. (3.100)), given the values of the $SU(2)$ and $U(1)$ couplings.

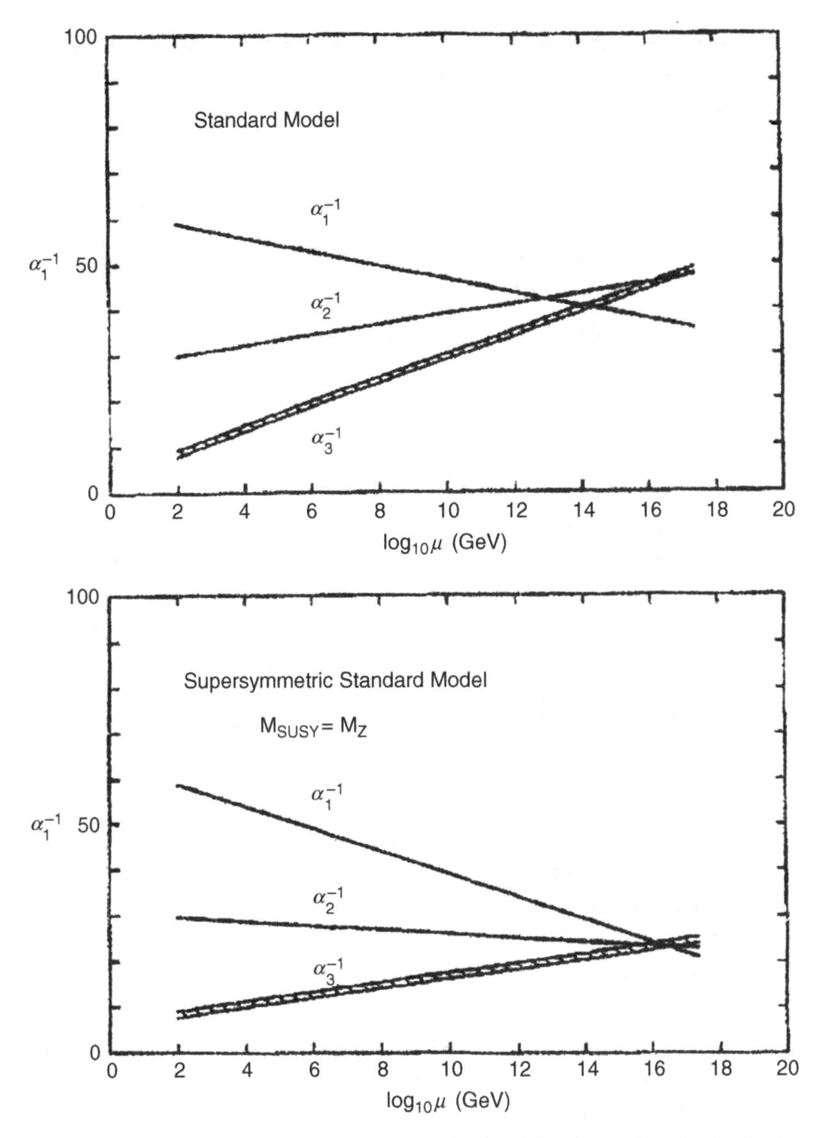

Fig. 12.1 In the Standard Model the couplings do not unify at a point. In the MSSM they do, provided that the threshold for new particle production is at about 1 TeV. Reprinted with permission from P. Langacker and N. Polonsky, Uncertainties in coupling constant unification, *Phys. Rev. D*, **47**, 4028, 1993. Copyright (1993) by the American Physical Society.

12.3 Dimension-five operators and proton decay

We have seen that, in supersymmetric theories, there are dangerous dimension-four operators. These can be forbidden by a simple Z_2 symmetry, i.e. R-parity. But there are also operators of dimension five which can potentially lead to proton decay rates far larger than the experimental limits. The MSSM possesses B- and L-violating dimension-five operators which are permitted by all symmetries. For example, R-parity does not forbid such operators as

$$\mathcal{O}_5^a = \frac{1}{M}\int d^2\theta\, \bar{u}\bar{u}\bar{d}e^+, \quad \mathcal{O}_5^b = \frac{1}{M}\int d^2\theta\, QQQL. \tag{12.7}$$

These are still potentially very dangerous. When one integrates out the squarks and gauginos they will lead to dimension-six B- and L-violating operators in the Standard Model with coefficients (optimistically) of order

$$\frac{\alpha}{4\pi}\frac{1}{Mm_{\text{susy}}}. \tag{12.8}$$

Comparing with the usual minimal $SU(5)$ prediction, and supposing that $M \sim 10^{16}$ GeV, one sees that a suppression of order 10^9 or so is needed.

Fortunately, such a suppression is quite plausible, at least in the framework of supersymmetric GUTs. In a simple $SU(5)$ model, for example, the operators in Eq. (12.7) will be generated by exchange of the color triplet partners of ordinary Higgs fields, and thus one obtains two factors of Yukawa couplings. Also, in order that the operators be $SU(3)$ invariant the color indices must be completely antisymmetrized, so more than one generation must be involved. This suggests that suppression by factors of order the CKM angles is plausible. So we can readily imagine a suppression by factors 10^{-9}–10^{-11}. Proton decay can be used to restrict – and does severely restrict – the parameter space of particular models. The simplest $SU(5)$ model, with TeV-scale squarks and gauginos and the simplest Higgs structure, can be ruled out, for example. But what is quite striking is that we are automatically in the right range to be compatible with experimental constraints, and perhaps even to see something. It is not obvious that things would work out like this.

So far we have phrased this discussion in terms of baryon-violating physics at M_{gut}. But, whatever the underlying theory at M_{p} may be, there is no reason to think that it should preserve baryon number. So one expects that already at scales just below M_{p} these dimension-five terms are present. If their coefficients were simply of order $1/M_{\text{p}}$, the proton decay rate would be enormous, five orders of magnitude or more faster than the current bounds. In any such theory one must also explain the smallness of the Yukawa couplings. One popular approach is to postulate approximate symmetries. Such symmetries could well suppress the dangerous operators at the Planck scale. One might expect that there would be further suppression in any successful underlying theory. After all, the rate from Higgs exchange in GUTs is very small because the Yukawa couplings are small. We do not really know why the Yukawa couplings are small, but it is natural to suspect that this is a consequence of (approximate) symmetries. These same symmetries, if present, would also suppress dimension-five operators from Planck-scale sources, presumably by a comparable amount.

Finally, we mentioned earlier that one can contemplate symmetries that would suppress dimension-four operators beyond a Z_2 R-parity. Such symmetries, as we will see, are common in string theory. One can write down R-symmetries which forbid not only all the dangerous dimension-four operators but some or all the dimension-five operators as well. In this case, proton decay could be unobservable in feasible experiments.

Suggested reading

A good introduction to supersymmetric GUTs is provided in Witten (1981). The reviews and texts which we have mentioned on supersymmetry and grand unification all provide good coverage of the topic. The Particle Data Group website has an excellent survey, including up-to-date unification calculations and constraints on dimension-five operators. Murayama and Pierce (2002) discussed the constraints on minimal $SU(5)$ unification from dimension-five operators.

Exercises

(1) Work through the details of the simplest $SU(5)$ supersymmetric grand unified model. Solve the equations

$$\frac{\partial W}{\partial \Sigma} = 0.$$

Couple the system to supergravity, and determine the value of the constant in the superpotential required to cancel the cosmological constant in the $(3, 2, 1)$ minimum. Determine the resulting value of the vacuum energy in the $SU(5)$ symmetric minimum.

(2) In the simplest $SU(5)$ model, include a 5 and a $\bar{5}$ representation of Higgs fields. Write down the most general renormalizable superpotential for these fields and the 24-dimensional representation, Σ. Find the condition on the parameters of the superpotential such that there is a single light doublet. Using the fact that only the Kahler potential is renormalized, show that this tuning of parameters at tree level ensures that the doublet remains massless to all orders of perturbation theory. Now consider the couplings of quarks and leptons required to generate masses for the fermions. Show that exchanges of 5 and $\bar{5}$ Higgs lead to baryon- and lepton-number-violating dimension-five couplings.

(3) Show how various B-violating four-fermion operators are generated by squark and slepton exchange, starting with the general set of B- and L-violating terms in the superpotential.

13 Supersymmetric dynamics

In the previous chapter, we learned how to build realistic particle physics models based on supersymmetry. There are already significant constraints on such theories, and experiments at the LHC will test whether these sorts of ideas are correct.

If supersymmetry is discovered, the question will become: how is supersymmetry broken? Supersymmetry breaking offers particular promise for explaining large hierarchies. Consider the non-renormalization theorems. Suppose that we have a model consisting of chiral fields and gauge interactions. If the superpotential is such that supersymmetry is unbroken at tree level, the non-renormalization theorems for the superpotential which we proved in Section 9.7 guarantee that supersymmetry is not broken to all orders of perturbation theory. But they do not necessarily guarantee that effects *smaller* than any power of the couplings will not break supersymmetry. So, if we denote the generic coupling constants by g^2, there might be effects of order, say, e^{-c/g^2} which break the symmetry. In the context of a theory like the MSSM, supposing that soft breakings are of this order might account for the wide disparity between the weak scale (correlated with the susy-breaking scale) and the Planck or unification scale.

So, one reason why the dynamics of supersymmetric theories is of interest is its role in aiding our understanding of dynamical supersymmetry breaking and perhaps in studying a whole new class of phenomena in nature. But there are yet other reasons to be interested, as was first clearly appreciated by Seiberg. Supersymmetric Lagrangians are far more tightly constrained than ordinary Lagrangians. It is often possible to make strong statements about the dynamics which would be difficult if not impossible for conventional field theories. We will see this includes phenomena such as electric–magnetic duality and confinement.

13.1 Criteria for supersymmetry breaking: the Witten index

We will consider a variety of theories, some of them strongly coupled. One might imagine that it is a hard problem to decide whether supersymmetry is broken. Even in weakly coupled theories, one might wonder whether one could establish reliably that supersymmetry is *not* broken since, unless one has solved the theory exactly, it would seem hard to assert that there is no tiny non-perturbative effect which does not break the symmetry. One thing we will learn in this chapter is that this is not, however, a particularly difficult problem. We will exploit several tools. One is known as the *Witten index*. Consider the field theory of interest in a finite box. At finite volume the supersymmetry charges are well defined, whether or not supersymmetry is spontaneously broken. Because of the supersymmetry algebra,

$$Q|B\rangle = \sqrt{E}|F\rangle, \quad Q|F\rangle = \sqrt{E}|B\rangle, \tag{13.1}$$

i.e. non-zero-energy states come in Fermi–Bose pairs. Zero-energy states are special; they need not be paired. In the infinite-volume limit, the question of supersymmetry breaking amounts to the question whether there are zero-energy states. To count these, Witten suggested evaluating

$$\Delta = \mathrm{Tr}\,(-1)^F e^{-\beta H}. \tag{13.2}$$

Non-zero-energy states do not contribute to the index. The exponential is present to provide an ultraviolet regulator: the Witten index Δ is *independent of* β. More strikingly, the index is independent of all the parameters of the theory. The only way in which Δ can change as some parameter is changed is by some zero-energy state acquiring non-zero energy or a non-zero-energy state acquiring zero energy. But, because of Eq. (13.1), whenever the number of zero-energy bosonic states changes, the number of zero-energy fermionic states changes by the same amount. The Witten index is thus topological in character, and it is from this that it derives its power as well as its applications in a number of areas of mathematics. What can we learn from this index? If $\Delta \neq 0$ then we can say with confidence that supersymmetry is not broken. If $\Delta = 0$, we do not know whether it is.

Let us consider an example: a supersymmetric gauge theory with gauge group $SU(2)$ and no chiral fields. Since Δ is independent of the parameters, we can consider the theory in a very tiny box, with very small coupling. We can evaluate Δ, somewhat heuristically, as follows. Work in the $A_0 = 0$ gauge. Consider, first, the bosonic degrees of freedom, the A_is, which are matrix valued. In order for the energy to be small, we need the A_is to be constant and to commute. So take A_i to lie in the third dimension in isospin space, and ignore the other bosonic degrees of freedom. One might try to remove these remaining variables by a gauge transformation $g = \exp(iA_i x^i)$, but g is only a sensible gauge transformation if it is single-valued, which means that $A_i^3 = 2\pi n/L$. Thus A_i^3 is a compact variable. This reduces the problem to the quantum mechanics of a rotor. Thus in the lowest state the wave function is a constant. Because the A_i^3s are non-zero, the lowest energy states will only involve the gluinos in the three direction. There are two, λ_1^3 and λ_2^3 (again independent of coordinates).

Now recall that in the $A_0 = 0$ gauge the states must be gauge invariant. One interesting gauge transformation is multiplication by σ_2. This flips the sign of A^3 and λ^3. If we assume that our Fock ground state is even under this transformation, the only invariant states are $|0\rangle$ and $\lambda_1^3\lambda_2^3|0\rangle$. So we find $\Delta = 2$. If we assume that the state is odd then we obtain $\Delta = -2$.

As we indicated, this argument is heuristic. A more detailed, but still heuristic, argument was provided by Witten in his paper on the index Δ. But Witten also provided a more rigorous proof, which yields the same result. For general $SU(N)$, one finds that $\Delta = N$.

This already establishes that a vast array of interesting supersymmetric field theories do not break supersymmetry, not only all the pure gauge theories but any theory with massive matter fields. This follows because Δ is independence of parameters. If the mass is finite, one can take it to be large; if it is sufficiently large we can ignore the matter fields and recover the pure gauge result. Later, we will understand the dynamics of these theories in some detail and will reproduce the result for the index. But we will also see that the limit of zero mass is subtle, and the index calculation is not directly relevant to that case.

13.2 Gaugino condensation in pure gauge theories

Our goal in this section is to understand the dynamics of a pure $SU(N)$ gauge theory with massless fermions in the adjoint representation. Without thinking about supersymmetry one might expect the following, from our experience with real QCD.

1. The theory has a mass gap, i.e. the lowest excitations of the theory are massive.
2. Gauginos, like quarks, condense, i.e.

$$\langle \lambda\lambda \rangle = c\Lambda^3 = ce^{-(8\pi^2/b_0 g^2)}. \tag{13.3}$$

Note that there is no Goldstone boson associated with the gluino (gaugino) condensate. The theory has no continuous global symmetry; the classical symmetry,

$$\lambda \to e^{i\alpha}\lambda, \tag{13.4}$$

is anomalous. However, a discrete subgroup,

$$\lambda \to e^{2\pi i/N}\lambda, \tag{13.5}$$

is free of anomalies. One can see this by considering instantons in this theory. The instanton has $2N$ zero modes; this would appear to preserve a Z_{2N} symmetry. But the transformation $\lambda \to -\lambda$ is actually equivalent to a Lorentz transformation (a rotation by 2π). Multi-instanton solutions also preserve this symmetry, and it is believed to be exact. So the gaugino condensate breaks the Z_N symmetry; there are N degenerate vacua. This neatly accounts for the N value of the index. Later we will show that, even though the theory is strongly coupled, we can demonstrate the existence of the condensate by a controlled semiclassical computation.

Gluino condensation implies a breakdown of the non-renormalization theorems at the non-perturbative level. Recall that the Lagrangian is

$$\mathcal{L} = \int d^2\theta\, SW_\alpha^2, \tag{13.6}$$

so $\langle \lambda\lambda \rangle$ gives rise to a superpotential, i.e.

$$\mathcal{L} = \int d^2\theta\, S\langle \lambda\lambda \rangle. \tag{13.7}$$

This is our first example of a non-perturbative correction to the superpotential. Note, however, that $\langle \lambda\lambda \rangle$ must depend on S, since it depends on g^2:

$$S\langle \lambda\lambda \rangle = e^{-3S/b_0}. \tag{13.8}$$

So we actually have a superpotential for S:

$$W(S) = e^{-S/N}. \tag{13.9}$$

This superpotential violates the continuous shift symmetry which we used to prove the non-renormalization theorem, but it is compatible with the non-anomalous R symmetry,

$$S \to S + i\alpha N, \quad \lambda \to \lambda e^{i\alpha}. \tag{13.10}$$

Under this symmetry the superpotential transforms with charge 2.

13.3 Supersymmetric QCD

A rich set of theories for study is that collectively referred to as *supersymmetric QCD*. These are gauge theories with gauge group $SU(N)$, $N_{\rm f}$ flavor fields Q_f in the N representation and $N_{\rm f}$ flavor fields $\bar{Q}_f$ in the $\bar{N}$ representation; here $f = 1, \ldots, N_{\rm f}$. We will see that the dynamics is quite sensitive to the value of $N_{\rm f}$. First, we will consider the theory without any classical superpotential for the quarks. In this case the theory has a large global symmetry. We can transform the Qs and $\bar{Q}$s by separate $SU(N_{\rm f})$ transformations. We can also multiply the Qs by a common phase and the $\bar{Q}$s by a separate common phase:

$$Q_f \to e^{i\alpha} Q_f, \quad \bar{Q}_f \to e^{i\beta} \bar{Q}_f. \tag{13.11}$$

Finally, the theory possesses an R symmetry, under which the Qs and $\bar{Q}$s are neutral. In terms of component fields, under this symmetry we have

$$\psi_Q \to e^{-i\alpha} \psi_Q, \quad \psi_{\bar{Q}} \to e^{-i\alpha} \psi_{\bar{Q}}, \quad \lambda^a \to e^{i\alpha} \lambda^a. \tag{13.12}$$

Now consider the question of anomalies. The $SU(N_{\rm f})$ symmetries are free of anomalies, as is the vector-like symmetry,

$$Q_f \to e^{i\alpha} Q_f, \quad \bar{Q}_f \to e^{-i\alpha} \bar{Q}_f. \tag{13.13}$$

The R symmetry and the axial $U(1)$ symmetry are both anomalous. But we can define a non-anomalous R by combining the two. The gauginos give a contribution to the anomaly proportional to N, so we need the fermions to carry an R-charge $-N/N_{\rm f}$. Since the bosons (and the chiral multiplets) carry an R-charge that is larger by 1, we have

$$Q_f(x,\theta) \to e^{i\alpha(N_{\rm f}-N)/N_{\rm f}} Q_f(x, \theta e^{-i\alpha}), \quad \bar{Q}_f(x,\theta) \to e^{i\alpha(N_{\rm f}-N)/N_{\rm f}} \bar{Q}_f(x, \theta e^{-i\alpha}). \tag{13.14}$$

So, the symmetry of the quantum theory is $SU(N_{\rm f})_{\rm L} \times SU(N_{\rm f})_{\rm R} \times U(1)_R \times U(1)_V$, where the vector symmetry $U(1)_V$ transforms the Q and $\bar{Q}$ fields by opposite phases.

We have seen that supersymmetric theories often have, classically, a large vacuum degeneracy and this is true of this theory. In the absence of a superpotential, the potential is completely determined by the D terms for the gauge fields. It is helpful to treat D as a matrix-valued field,

$$D = \sum T^a D^a. \tag{13.15}$$

As a matrix, D can be expressed elegantly in terms of the scalar fields. We start with the identity

$$(T^a)_i^j (T^a)_k^l = \delta_i^l \delta_k^j - \frac{1}{N} \delta_i^j \delta_k^l. \tag{13.16}$$

One can derive this result in a number of ways. Consider propagators for fields (such as gauge bosons) in the adjoint representation of the gauge group. Take the group, first, to be $U(N)$. The propagator of the matrix-valued fields satisfies

$$\langle A_i^j A_k^l \rangle \propto \delta_i^l \delta_k^j. \tag{13.17}$$

But this is the same thing as

$$\langle A^a A^b (T^a)_i^j (T^b)_k^l \rangle. \tag{13.18}$$

So we obtain the identity without the $1/N$ terms. Now remembering that A must be traceless, we see that we need to subtract the trace as above. (This identity is important in understanding the $1/N$ expansion in QCD.) Thus a field ϕ in the fundamental representation makes a contribution

$$\delta D_i^j = \phi_i^* \phi^j - \frac{1}{N} \delta_i^j \phi_k^* \phi^k. \tag{13.19}$$

In the antifundamental representation the generators are $-T^{aT}$ (this follows from the fact that these generators are minus the complex conjugates of those in the fundamental representation, and the fact that the T^as are Hermitian). So the full D term is

$$D_i^j = \sum_f Q_i^* Q^j - \bar{Q}_i \bar{Q}^{*j} - \text{Tr terms}. \tag{13.20}$$

In this matrix form it is not difficult to look for supersymmetric solutions, i.e. solutions of $D_i^j = 0$. A simple strategy is first to construct

$$\hat{D}_i^j = \sum_f Q_i^* Q^j - \bar{Q}_i \bar{Q}^{*\,j} \tag{13.21}$$

and demand that $\hat{D}$ either vanish or be proportional to the identity. Let us start with the case $N_{\text{f}} \leq N$. For definiteness, take $N = 3$, $N_{\text{f}} = 2$; the general case is easy to work out. By a sequence of $SU(3)$ transformations, we can bring Q to the following form:

$$Q = \begin{pmatrix} v_{11} & v_{12} \\ 0 & v_{22} \\ 0 & 0 \end{pmatrix}. \tag{13.22}$$

By a sequence of $SU(N_{\text{f}})$ transformations, we can bring this to the simpler form

$$Q = \begin{pmatrix} v_1 & 0 \\ 0 & v_2 \\ 0 & 0 \end{pmatrix}. \tag{13.23}$$

At this point we have used up our freedom to make further symmetry transformations on $\bar{Q}$. But it is easy to find the most general $\bar{Q}$ which makes the D terms vanish. The contribution of Q to D_i^j is simply

$$D = \text{diag}(|v_1|^2, |v_2|^2). \tag{13.24}$$

So, in order that D vanish, $\bar{Q}$ must make an equal and opposite contribution. In order that there be no off-diagonal contributions, $\bar{Q}$ can have entries only on the diagonal, so

$$\bar{Q} = \begin{pmatrix} e^{i\alpha_1} v_1 & 0 \\ 0 & e^{i\alpha_2} v_2 \\ 0 & 0 \end{pmatrix}. \tag{13.25}$$

In general, in these *flat directions* – directions in field space in which the potential is flat – the gauge group is broken to $SU(N - N_\text{f})$. The unbroken flavor group depends on the values of the v_is. We have exhibited N_f complex moduli above, but actually there are more, associated with the generators of the broken flavor symmetries $(SU(N_\text{f}) \times U(1))$. Thus there are $N_\text{f}^2 + 2N_\text{f}$ complex moduli. Note that there are $2NN_\text{f} - N_\text{f}^2$ broken gauge generators, which gain mass by "eating" the components of $Q, \bar{Q}$ that are not moduli. Of the original $2NN_\text{f}$ chiral fields this leaves precisely $N_\text{f}^2 + 2N_\text{f}$ massless fields, so we have correctly identified the number of moduli.

Our discussion, so far, does not look gauge invariant. But this is easily, and elegantly, rectified. The moduli can be written as the gauge-invariant combinations

$$M_{\bar{f}}^f = \bar{Q}_{\bar{f}} Q^f. \tag{13.26}$$

Expanding the fields Q and $\bar{Q}$ about their expectation values gives back the explicit form for the moduli in terms of the underlying gauge-invariant fields. This feature, we will see, is quite general.

The case $N_\text{f} = N$ is similar to the case $N_\text{f} < N$, but there is a significant new feature. In addition to the flat directions with $Q = \bar{Q}$ (up to phases), the potential also vanishes if $Q = v\text{I}$, where I is the identity matrix. This possibility can also be described in a gauge-invariant way since now we have an additional pair of gauge invariant fields, which we will refer to as "baryons":

$$B = \epsilon^{i_1 \dots i_N} \epsilon_{a_1 \dots a_N} Q_{i_1}^{a_1} \cdots Q_{i_N}^{a_N}, \tag{13.27}$$

and similarly for $\bar{B}$.

In the case $N_\text{f} > N$ there is a larger set of baryon-like objects, corresponding to additional flat directions. We will describe them in greater detail later. Before closing this section we should stress that for $N_\text{f} \geq N - 1$ the gauge symmetry is completely broken. For large values of the moduli, the effective coupling of the theory is $g^2(v)$ since infrared physics cuts off at the scale of the gauge field masses. By taking v as sufficiently large that $g^2(v)$ is small, the theories can be analyzed by perturbative and semiclassical methods. Strong coupling is more challenging, but much can be understood. We will see that the dynamics naturally divides into three cases: $N_\text{f} < N$, $N_\text{f} = N$, and $N_\text{f} > N$.

13.4 $N_f < N$: a non-perturbative superpotential

Our problem now is to understand the dynamics of these theories. Away from the origin of the moduli spaces, this turns out to be a tractable problem. We consider first the case $N_{\rm f} < N$. Suppose that the v_is are large and roughly uniform in magnitude. Even here, we have to distinguish two cases. If $N_{\rm f} = N - 1$, the gauge group is completely broken and the low-energy dynamics consists of the set of chiral fields $M_{\bar{f}f}$. If $N_{\rm f} < N - 1$, there is an unbroken gauge group, $SU(N-N_{\rm f})$, with no matter fields (chiral fields) transforming under this group at low energies. The gauge theory is an asymptotically free theory, essentially like ordinary QCD with fermions in the adjoint representation. Such a theory is believed to have a mass gap of order the scale of the theory, $\Lambda_{N-N_{\rm f}}$. Below this scale, again, the only light fields are the moduli $M^f_{\bar{f}}$. In both cases we can try to guess the form of the very-low-energy effective action for these fields from symmetry considerations.

We are particularly interested in whether there is a superpotential in this effective action. If not then the moduli have *exactly* no potential. In other words, even in the full quantum theory, they correspond to an exact, continuous, set of ground states. What features should this superpotential possess? Most important, it should respect the flavor symmetries of the original theory (because the fields M are gauge invariant, it automatically respects the gauge symmetry). Among these symmetries are the $SU(N_{\rm f}) \times SU(N_{\rm f})$ non-Abelian symmetry. The only invariant that we can construct from M is

$$\Phi = \det M. \tag{13.28}$$

The determinant is invariant because it transforms under $M \to VMU$ as $\det V \det U \det M$ and, for $SU(N_{\rm f})$ transformations, the determinant is unity. Under baryon number symmetry, M is invariant. But, under $U(1)_R$ symmetry the its transformation law is more complicated:

$$\Phi \to e^{2i\alpha(N_{\rm f}-N)}. \tag{13.29}$$

Under this R-symmetry, any would-be superpotential must transform with charge 2, so the form of the superpotential is unique:

$$W = \Lambda^{(3N-N_{\rm f})/(N-N_{\rm f})}\Phi^{-1/(N-N_{\rm f})}. \tag{13.30}$$

Here we have inserted a factor Λ, the scale of the theory, on dimensional grounds.

Our goal in the next two sections will be to understand the dynamical origin of this superpotential, known as the Affleck, Dine and Seiberg (ADS) superpotential. We will see that there is a distinct difference between the cases $N_{\rm f} = N - 1$ and $N_{\rm f} < N - 1$. First, though, consider the case $N = N_{\rm f}$. Then the field Φ has R-charge zero, and no superpotential is possible. So, no potential can be generated, perturbatively or non-perturbatively. Similarly, in the case $N_{\rm f} > N$ we cannot construct a gauge-invariant field which is also invariant under the $SU(N_{\rm f}) \times SU(N_{\rm f})$ flavor symmetry. This may not be obvious, since it would seem that we could again construct $\Phi = \det M$. But in this case $\Phi = 0$ in the flat directions.

From the perspective of ordinary, non-supersymmetric, field theories, what we have established here is quite surprising. Normally, we would expect that in an interacting theory, even if the potential vanished classically there would be quantum corrections. For theories with $N \geq N_f$, we have just argued that this is impossible. So this is a new feature of supersymmetric theories: there are often exact moduli spaces, even at the quantum level.

In the next few sections we will demonstrate that non-perturbative effects do indeed generate the superpotential of Eq. (13.30). The presence of the superpotential means that, at least at weak coupling (large v_i), there is no stable vacuum of the theory. At best, we can consider time-dependent, possibly cosmological, solutions. If we add a mass term for the quarks, however, we find an interesting result. If the masses are the same, we expect that all the v_is will be the same, $v_i = v$. Suppose that the mass term is small. Then the full superpotential, at low energies, is

$$W = m\bar{Q}Q + \Lambda^{(3N-N_f)/(N-N_f)}\Phi^{-1/(N-N_f)}. \tag{13.31}$$

Remembering that $\Phi \sim v^{2N_f}$, the equation for a supersymmetric minimum has the form

$$v^{2N/(N-N_f)} = \left(\frac{m}{\Lambda}\right)\Lambda^{2N/(N-N_f)}. \tag{13.32}$$

Note that v is a complex number; this equation has N roots

$$v = e^{2\pi ik/N}\Lambda\left(\frac{m}{\Lambda}\right)^{(N-N_f)/2N}. \tag{13.33}$$

What is the significance of these N solutions? The mass term breaks the $SU(N_f)\times SU(N_f)$ symmetry to the vector sum. It also breaks the $U(1)_R$. But it leaves unbroken a Z_N subgroup of the $U(1)$. In Eqs. (13.14), $\alpha = 2N_f/N$ is a symmetry of the mass term. So these N vacua are precisely those expected from the breaking of the Z_N subgroup. This Z_N is the same as that expected for a pure gauge theory, as one can see by thinking of the case where the mass of the Qs and $\bar{Q}$s is large.

13.4.1 The Λ-dependence of the superpotential

Previously, we proved a non-renormalization theorem for the gauge couplings by thinking of the gauge coupling itself as a background field S. This relied on the shift symmetry

$$S \to S + i\alpha.$$

This symmetry, however, is only a symmetry of perturbation theory. On the one hand, since the imaginary part a of S, couples to $F\tilde{F}$, instanton and other non-perturbative effects violate the symmetry. On the other hand the theory also has an anomalous chiral symmetry, the R symmetry, under which we can take all the scalar fields to be neutral. So the theory is symmetric under this R symmetry combined with a simultaneous shift

$$S \to S + i(N - N_f)\alpha. \tag{13.34}$$

Any superpotential must transform with charge 2 under this symmetry. The field Φ is neutral. But we have, for the Λ parameter,

$$\Lambda = \exp\left(-\frac{8\pi^2}{b_0 g^2}\right) = \exp\left(-\frac{8\pi^2}{3N - N_f}S\right) \tag{13.35}$$

so it transforms as follows:

$$\Lambda^{(3N-N_f)/(N-N_f)} \to e^{2i\alpha}\Lambda^{(3N-N_f)/(N-N_f)}. \tag{13.36}$$

13.5 The superpotential in the case $N_f < N - 1$

Consider first the case $N_f < N-1$. At energies well below the scale v, the theory consists of a pure (supersymmetric) $SU(N-N_f)$ gauge theory and a number of neutral chiral multiplets. The chiral multiplets can couple to the gauge theory only through non-renormalizable operators. Because the moduli are neutral, there are no dimension-four couplings. There are possible dimension-five couplings; they are of the form

$$\delta\phi\, W_\alpha^2, \tag{13.37}$$

where $\delta\phi$ represents the fluctuations of the moduli fields about their expectation values; the coefficient of this operator will be of order $1/v$.

We can be more precise about the form of this coupling by noting that it must respect the various symmetries if it is written in terms of the original, unshifted fields (this is similar to our argument for the form of the superpotential). In particular, a coupling of the form

$$\mathcal{L}_{\text{coup}} = (S + a\ln\Phi)W_\alpha^2 \tag{13.38}$$

respects all the symmetries: it clearly respects the $SU(N_f)$ symmetries, and it also respects the non-anomalous $U(1)_R$ symmetry, for a suitable choice of a, since

$$\ln\Phi \to \ln\Phi + (N-N_f)/N_f\alpha. \tag{13.39}$$

It is not hard to see how this coupling is generated:

$$\Phi \approx v^N + v^{N-1}\phi. \tag{13.40}$$

Thus Im ϕ couples to $F\tilde{F}$ through the anomaly diagram, just like an axion. The real part couples to F^2. One can see this by a direct calculation or by noting that the masses of the heavy fields are proportional to v, so the gauge coupling of the $SU(N-N_f)$ theory depends on v:

$$\alpha_{N-N_f}^{-1}(\mu) = \alpha_N^{-1}(v) + \frac{b_0^{(N-N_f)}}{4\pi}\ln\frac{\mu}{v}. \tag{13.41}$$

Since $\Phi \sim v^{N_f}$, we see that we have precisely the correct coupling. It is easy to see which Feynman graphs generate the couplings to the real and imaginary parts.

But we have seen that in the $SU(N-N_f)$ theory, gaugino condensation gives rise to a superpotential for the coefficient of W_α^2; in this case, it is precisely

$$W = \frac{\Lambda^{(3N-N_f)/(N-N_f)}}{\Phi^{1/(N-N_f)}}. \tag{13.42}$$

So we have understood the origin of the superpotential in these theories.

13.6 $N_f = N - 1$: the instanton-generated superpotential

In the case $N_f = N-1$, the superpotential is generated by a different mechanism: instantons. Before describing the actual computation we give some circumstantial evidence for this fact. Consider the instanton action. This is

$$\exp\left(\frac{-8\pi^2}{g^2(v)}\right). \tag{13.43}$$

Here we have assumed that the coupling is to be evaluated at the scale of the scalar vevs. The gauge group is, after all, completely broken so, provided that the computation is finite, this is the only relevant scale (we are also assuming that all the vevs are of the same order). Thus any superpotential we might compute behaves as

$$W \sim v^3 \left(\frac{\Lambda}{v}\right)^{2N+1} \sim \frac{\Lambda^{2N+1}}{v^{2N-2}}, \tag{13.44}$$

which is the behavior predicted by the symmetry arguments.

To actually compute the instanton contribution to the superpotential, we need to develop further than in Chapter 5 the instanton computation and the structure of the supersymmetry zero modes. The required techniques were developed by 't Hooft, when he computed the baryon-number-violating terms in the effective action of the standard model; 't Hooft started by noting that, in the presence of the Higgs field, *there is no instanton solution*. This can be seen by a simple scaling argument. Here the instanton solution will involve A^μ and ϕ. Suppose one has such a solution. Now simply do a rescaling of all lengths such that

$$x^\mu \to \rho x^\mu, \quad A^\mu \to \frac{1}{\rho} A^\mu, \quad \phi \to \phi \tag{13.45}$$

(because ϕ must tend to its expectation value at ∞, we cannot rescale it). Then the gauge kinetic terms are invariant but the scalar kinetic terms are not; $|D\phi|^2 \to \rho^2 |D\phi|^2$. So the action is changed, and there is no solution.

However, the instanton configuration, while not a solution, is still distinguished by its topology; 't Hooft argued that it makes sense to integrate over solutions of a given topology. This just means that we write down a configuration for each value of ρ, and integrate over ρ. For small ρ we can understand this in the following way. The non-zero modes of the instanton, before turning on the scalar vevs, all have eigenvalues of order $1/\rho$ or larger and can be ignored. There are also zero modes. Those associated with rotations and translations will remain at zero, even in the presence of the scalar, since they correspond to exact symmetries. But this is not the case for the dilatational zero mode; this mode is slightly lifted. The scaling argument above shows that the action is smallest at small ρ; we will see in a moment that the action of the interesting configurations vanishes as $\rho \to 0$. We know from our earlier studies of QCD, however, that renormalization of the coupling tends to make the action large at small ρ. Together, these effects yield a minimum of the action at small but finite ρ, giving a self-consistent justification of the approximation.

To proceed with the computation, we will use 't Hooft's notation for the instanton, which we introduced in Chapter 5. Recall that

$$A^a_\mu(x) = \frac{2\eta_{a\mu\nu}x_\nu}{x^2+\rho^2}. \tag{13.46}$$

It is straightforward to work out $F_{\mu\nu}$ (see the exercises):

$$F^a_{\mu\nu} = \frac{\eta_{a\mu\nu}}{(x^2+\rho^2)^2}. \tag{13.47}$$

We note that F is self-dual, since η is, so this is a solution of the Euclidean equations. A second-rank antisymmetric tensor $F_{\mu\nu}$ is a six-dimensional representation of $SO(4)$; under $SU(2)\times SU(2)$ it decomposes as $(3,1)+(1,3)$, where these are the self-dual and anti-self-dual parts of the tensor. The η symbol is essentially a Clebsch–Gordan coefficient, which describes a mapping of one $SU(2)$ subgroup of $SO(4)$ into $SU(2)$.

At large distances, the instanton is a gauge transformation of "nothing". i.e. vanishing values for the fields. The gauge transformation is just

$$g^i_j = i\bar\sigma^{\mu i}_j \hat{x}^\mu. \tag{13.48}$$

This can be thought of as a mapping of S_3 into $SU(2)$; the winding number of the instanton just counts the number of times space is mapped onto the group.

In this form it is useful to note another way to describe the instanton solution. By an inversion of coordinates one can write

$$A^a_\mu = \frac{2}{g^2}\frac{\rho^2}{x^2+\rho^2}\eta_{a\mu\nu}\frac{x^\nu}{x^2}. \tag{13.49}$$

This *singular gauge* instanton is often useful since it falls off more rapidly at large x than the original instanton solution.

Now, for the doublets we solve the equation

$$D^2 Q = D^2\bar{Q} = 0. \tag{13.50}$$

This has solutions

$$Q^i = \bar{Q}^{i\dagger} = i\bar\sigma^{\mu i}_j\hat{x}^\mu\left(\frac{1}{x^2+\rho^2}\right)^{1/2}\langle Q^j\rangle, \tag{13.51}$$

and similarly for $\bar{Q}$. Like the solution for A^μ, these solutions are "pure gauge" configurations as $r\to\infty$, i.e. they are gauge transformations by g of the constant vev. (Note, here and above, that the σ^μs are the Euclidean versions of the two-component Dirac matrices, $\sigma^\mu = (i,\vec\sigma)$, $\bar\sigma^\mu = (i,-\vec\sigma)$.)

The action of this configuration is

$$S(\rho) = \frac{1}{g^2}(8\pi^2 + 4\pi^2\rho^2 v^2). \tag{13.52}$$

Some features of this result are worth noting.

1. The integral over ρ now converges for large ρ, since it is exponentially damped.
2. Terms in the potential involving $|Q|^4$ make smaller contributions to the action, according to powers of ρ. Rescaling x as ρx, one sees that these terms are of order ρ^4. But ρ is at most of order $gv^{-1} = m_w$ (from item 1 above), so these terms are suppressed. This justifies their neglect in the equations of motion.

Our goal is to compute the instanton contribution to the effective action. We particularly want to see whether the instanton generates the conjectured non-perturbative superpotential. In order to compute the effective action, we need to ask about the fermion zero modes. Before turning on the vevs for the scalars, there are six zero modes. Two of these are generated by supersymmetry transformations of the instanton solution

$$\delta\lambda = \sigma_\alpha^{\mu\nu\beta} F^{\mu\nu} \epsilon_\beta, \tag{13.53}$$

so

$$\lambda_{\alpha a}^{SS[\beta]} = \frac{8\sigma_\alpha^{\mu a\beta}}{(x^2+\rho^2)^2}. \tag{13.54}$$

Note that, because of the anti-self-duality of $\bar\sigma^{\mu\nu}$, two supersymmetry generators annihilate the lowest-order solution, i.e. there are only two supersymmetry zero modes. If we neglect the Higgs, the classical Yang–Mills action has a conformal (scale) symmetry. This is the origin of the zero mode associated with changes in ρ. in the classical solution. In the supersymmetric case, there is, apart from supersymmetry, an additional fermionic symmetry called superconformal invariance. In superspace the corresponding generators are

$$Q^{SC} = \not{x} Q, \tag{13.55}$$

so

$$\lambda_{\alpha a}^{SC[\beta]} = \frac{8\not{x}\sigma_\alpha^{\mu a\beta}}{(x^2+\rho^2)^2}. \tag{13.56}$$

There are also two matter-field zero modes, one for each of the quark doublets:

$$\psi_{Q\alpha}^i = \frac{\delta_\alpha^i}{(x^2+\rho^2)^{3/2}} = \psi_{\bar Q} \tag{13.57}$$

(in the last equation we treated $\bar Q$ as a doublet also; one can instead treat it as a 2^* representation by multiplying by ϵ_{ij}).

When we turn on the scalar vevs these modes are corrected. The superconformal symmetry is broken by the vevs and, not surprisingly, the superconformal zero modes are lifted. In fact, they pair with the two quark zero modes. We can compute this pairing by treating the Yukawa terms in the Lagrangian as a perturbation, replacing the scalar fields by their classical values. Expanding to second order, i.e. including

$$\int d^4x\, Q^* \psi_Q \lambda \int d^4x'\, \bar Q^* \psi_{\bar Q} \lambda \tag{13.58}$$

and expanding the fields in the lowest-order eigenmodes, the superconformal and matter-field zero modes can be absorbed by these terms. Note, in particular, that both Q_{cl} and

$\lambda^{\rm SC}$ are odd under $x \to -x$ while the matter-field zero modes are even, so the integral is non-zero. The supersymmetry zero modes, being even, cannot be soaked up in this way.

The wave functions of the supersymmetry zero modes are altered in the presence of the Higgs fields, and they now have components in the ψ^*_Q and $\psi^*_{\bar Q}$ directions. For ψ_Q, for example, we need to solve the equation

$$D_\mu \bar\sigma^\mu \psi_Q^{\rm SS*} = \lambda^{\rm SS} Q^*. \tag{13.59}$$

This equation is easy to solve, starting with our solution of the scalar equation. If we simply take

$$\psi_Q^{\rm SS} = D_\mu \sigma^\mu Q^*, \tag{13.60}$$

then, substituting back into the left-hand side of Eq. (13.59) we obtain

$$D^2 Q + \sigma_{\mu\nu} F^{\mu\nu} Q; \tag{13.61}$$

the first term vanishes for the classical solution, while the second is indeed just $\lambda^{\rm SS} Q^*$.

With these ingredients we can compute the superpotential terms in the effective action. In particular, the non-perturbative superpotential predicts a non-zero term in the component form of the effective action proportional to

$$\frac{\partial^2 W}{\partial Q \partial \bar Q} = \frac{1}{v^4} \psi_Q \psi_{\bar Q}. \tag{13.62}$$

We can calculate this term by studying the corresponding Green's function. We need to be careful, now, about the various collective coordinates. We want to study the gauge-invariant correlation function

$$\langle \bar Q(x) \psi_Q(x) \psi_{\bar Q}(y) Q(y) \rangle \tag{13.63}$$

in the presence of the instanton. Since we are interested in the low-momentum limit of the effective action, we can take x and y to be widely separated. We need to integrate over the instanton location x_0 and the instanton orientation and scale size. Because the gauge fields are massive, we can take x and y both to be far from the instanton. Then, from our explicit solution for the supersymmetry zero modes, we obtain

$$\psi_Q(x) \propto \not{D} Q \propto \not{D} \frac{i\sigma^\mu (x^\mu - x_0^\mu)}{[(x-x_0)^2 + \rho^2]^{1/2}} \to g(x - x_0) S_F(x - x_0), \tag{13.64}$$

with a similar equation for $\psi_{\bar Q}$. The g and $g^\dagger$ factors are canceled by corresponding factors in Q and $\bar Q$, at large distances. Substituting these expressions into the path integral and integrating over x_0 gives a convolution, $v^2 \int d^4x_0\, S_F(x - x_0) S_F(y - y_0)$. Extracting the external propagators, we obtain the effective action. Integrating over ρ gives a term of precisely the desired form. If we contract the gauge and spinor indices in a gauge and rotationally invariant manner, the integral over rotations just gives a constant factor. It requires some work to do all the bookkeeping correctly. The evaluation of the determinant is greatly facilitated by supersymmetry: there is a precise fermion–boson pairing of all the non-zero modes. In the exercises, you are asked to work out more details of this computation; further details can also be found in the references.

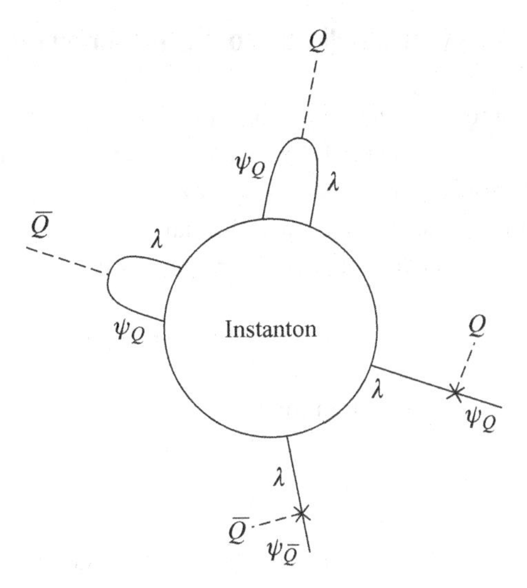

Fig. 13.1 Schematic description of the instanton computation of the superpotential. Four zero modes are tied together by the scalar vevs; two gluino zero modes turn into ψ zero modes as well.

Without working through all the details we can see the main features.

1. The perturbative lifting of the zero modes gives rise to a contribution proportional to v^2 (see Fig. 13.1).
2. The matter-field component of the supersymmetry zero modes studied above gives a contribution to the gauge-invariant correlation function:

$$v^4 \int d^4x_0\, S_f(x - x_0) S_f(y - y_0). \tag{13.65}$$

3. The integral over the gauge collective coordinates (equivalently the rotational collective coordinates) simply gives a constant, since we have computed a gauge- and rotationally invariant quantity.
4. The scale-size collective coordinate integral behaves as

$$W = A \int d\rho\, v^4 \exp - \left[\left(\frac{8\pi^2}{g^2(\rho)} + 4\pi^2 \rho^2 v^2 \right) \right] \tag{13.66}$$

where the power of ρ has been determined from dimensional analysis and A is a constant.
5. Extracting the constant requires careful attention to the normalization of the zero modes and to the Jacobians for the collective coordinates. However, the non-zero modes come in Fermi–Bose pairs, and their contribution to the functional integral cancels.
6. The final ρ integral gives

$$W = A' \frac{\Lambda^5}{v^2}, \tag{13.67}$$

which is consistent with the expectations of the symmetry analysis.

This analysis generalizes straightforwardly to the case of general N_c.

13.6.1 An application of the instanton result: gaugino condensation

The instanton calculation for the case $N_f = N - 1$ is a systematic weak-coupling computation of the superpotential which appears in the low-energy-effective action. Seiberg noted that this result, plus holomorphy, allows systematic study of the strongly coupled regime of other theories. To understand this, take $N = 2$ and add a mass term for the quark. In this case, for very small mass the superpotential is

$$W = m\bar{Q}Q + \frac{\Lambda^{2N+1}}{\bar{Q}Q}. \tag{13.68}$$

We can solve the equation for Q:

$$\bar{Q} = \begin{pmatrix} 0 \\ v \end{pmatrix}, \quad v = \left(\frac{\Lambda^5}{m}\right)^{1/4}. \tag{13.69}$$

Using this we can evaluate the expectation value of the superpotential at the minimum:

$$W(m, \Lambda) = \Lambda^{5/2} m^{1/2}. \tag{13.70}$$

Because W is holomorphic, this result also holds for large m. For large m, the low-energy theory is just a pure $SU(2)$ gauge theory. We expect for large m that the superpotential is $\langle \lambda\lambda \rangle = \Lambda_{\mathrm{le}}^3$. But this is equal to

$$W = \langle \lambda\lambda \rangle = m^3 \exp\left[-\frac{8\pi^2}{2g^2(m)}\right]. \tag{13.71}$$

The right-hand side is simply Λ_{le}^3. We have, in fact, done a systematic, reliable computation of the gluino condensate in a strongly interacting gauge theory!

Suggested reading

Excellent treatments of supersymmetric dynamics appear in the text by Weinberg (1995), and in Michael Peskin's lectures (1997). We have already mentioned 't Hooft's original instanton paper (1976). The instanton computation of the superpotential is described in Affleck *et al.* (1984).

Exercises

(1) Verify that $\sigma_{\mu\nu}$ and $\bar{\sigma}_{\mu\nu}$ are self-dual and anti-self-dual, respectively. This means that $\mathrm{Tr}\,\sigma^a \sigma_{\mu\nu}$ is a self-dual tensor. Verify the connection to η; do the same thing for $\bar{\eta}$.

(2) Verify Eq. (13.47), which shows that F is self-dual and so solves the Euclidean Yang–Mills equations. Check that asymptotically the instanton potential is a gauge transform of "nothing."

(3) Verify the solution Eq. (13.51) of the scalar field equation. Compute the action of this field configuration.

(4) Perform the zero-mode counting for the case of general N_c, $N_f = N_c - 1$. Show that, again, all but two zero modes pair with matter-field zero modes; two supersymmetry zero modes contain matter-field components which can give rise to the expected superpotential.

14 Dynamical supersymmetry breaking

One of the original reasons for the interest in supersymmetry was the possibility of dynamical supersymmetry breaking. So far, however, we have exhibited models in which supersymmetry is unbroken in the true ground state, as in the case of QCD with only massive quarks or models with moduli spaces or approximate moduli spaces. In this chapter, we describe a number of models in which a non-trivial dynamics breaks supersymmetry. We will see that dynamical supersymmetry breaking occurs under special, but readily understood, conditions. In some cases we will be able to exhibit this breaking explicitly, through systematic calculations. In others we will have to invoke more general arguments. Then we will turn to theories in which supersymmetry is preserved in the lowest energy state but in which there exist metastable states with broken supersymmetry. We will argue that this is a generic phenomenon and see that it is even sometimes true in massive QCD.

14.1 Models of dynamical supersymmetry breaking

We might ask why, so far, we have not found supersymmetry to be dynamically broken. In supersymmetric QCD with massive quarks, we might give the Witten index as an explanation. We might also note that there is no promising candidate for a goldstino. With massless quarks we have flat directions and, as the fields get larger, the theory becomes more weakly coupled so that any potential tends to zero.

This suggests two criteria for finding models with dynamical supersymmetry breaking (DSB).

1. The theory should have no flat directions at the classical level.
2. The theory should have a spontaneously broken global symmetry.

The second criterion implies the existence of a Goldstone boson. If the supersymmetry were unbroken, any would-be Goldstone boson must lie in a multiplet with another scalar as well as a Weyl fermion. This other scalar, like the Goldstone particle, has no potential so the theory has a flat direction. But, by assumption, the theory classically (and therefore almost certainly quantum mechanically) has no flat direction. So supersymmetry is likely to be broken. These criteria are heuristic but, in practice, when a systematic analysis is possible, they always turn out to be correct.

Perhaps the simplest model with these features is a supersymmetric $SU(5)$ theory with a single $\bar{5}$ and a single 10 representation. In the exercises, you can show that this theory, in fact, has no flat directions and that it has two non-anomalous $U(1)$ symmetries. One can give arguments showing that these symmetries are broken. So it is likely that this theory breaks supersymmetry.

However, this is a strongly coupled model and it is difficult actually to prove that supersymmetry is broken. In the next section, we will describe a simple weakly coupled theory in which dynamical supersymmetry breaking occurs within a controlled approximation.

14.1.1 The (3, 2) model

A model in which supersymmetry turns out to be broken is the (3, 2) model. This theory has gauge symmetry $SU(3) \times SU(2)$, and matter content

$$Q(3,2), \quad \bar{U}(\bar{3},1), \quad L(1,2), \quad \bar{D}(\bar{3},1). \tag{14.1}$$

This is similar to the field content of a single generation of the standard model, but without the extra $U(1)$ and the positron. The most general renormalizable superpotential consistent with the symmetries is

$$W = \lambda Q L \bar{U}. \tag{14.2}$$

This model admits an R symmetry that is free of anomalies. There is also a conventional $U(1)$ symmetry, under which the charges of the various fields are the same as in the standard model (one can gauge this symmetry if one also adds an e^+ field).

While this model has global symmetries, it is different from supersymmetric QCD in that it does not have classical flat directions. To see this, note that by $SU(3) \times SU(2)$ transformations one can bring Q to the form

$$Q = \begin{pmatrix} a & 0 \\ 0 & b \\ 0 & 0 \end{pmatrix}. \tag{14.3}$$

Now, the vanishing of the $SU(2)$ D term forces

$$L = \left(0, \sqrt{|a^2| - |b^2|}\right). \tag{14.4}$$

The vanishing of the F terms for $\bar{u}$ requires $|a| = |b|$. Then the vanishing of the $SU(3)$ D term forces

$$\bar{U} = \begin{pmatrix} a' \\ 0 \\ 0 \end{pmatrix}, \quad \bar{D} = \begin{pmatrix} 0 \\ a'' \\ 0 \end{pmatrix} \tag{14.5}$$

(up to interchange of the two vevs), with

$$|a'| = |a''| = |a|.$$

Finally, the $\partial W / \partial L$ equations lead to $a = 0$.

To analyze the dynamics of this theory, consider first the case where $\Lambda_3 \gg \Lambda_2$. Ignoring, at first, the superpotential term this is just $SU(3)$ with two flavors. In the flat direction of the D terms there is a non-perturbative superpotential

$$W_{\rm np} = \frac{\Lambda^5}{\det \overline{Q}Q} \sim \frac{1}{v^4}. \tag{14.6}$$

The full superpotential in the low-energy theory is the sum of this term and the perturbative term. It is straightforward to minimize the potential and establish that supersymmetry is broken. One finds

$$a = 1.287\frac{\Lambda}{\lambda^{1/7}}, \quad b = 1.249\frac{\Lambda}{\lambda^{1/7}}, \quad E = 3.593\lambda^{10/7}\Lambda^4. \tag{14.7}$$

If $\Lambda_2 \gg \Lambda_3$, supersymmetry is still broken but the mechanism is different. In this case, before including the classical superpotential the strongly coupled theory is $SU(2)$ with two flavors. This is an example of a model with a *quantum moduli space*. This notion will be explained in the next chapter but it implies that $\langle QL\rangle \neq 0$, so at low energies there is a superpotential (F term) for $\bar{U}$.

There does not exist, at the present time, an algorithm to generate all models which exhibit dynamical supersymmetry breaking, but many classes have been identified. A generalization of the $SU(5)$ model, for example, is provided by an $SU(N)$ model with an antisymmetric tensor field A_{ij} and $N-4$ $\bar{F}$ terms. It is also necessary to include a superpotential,

$$W = \lambda_{ab} A\bar{F}^a\bar{F}^b. \tag{14.8}$$

Other broad classes are known, including generalizations of the $(3,2)$ model. A somewhat different, and particularly interesting, set of models is described in Section 15.4. Catalogs of known models, as well as studies of their dynamics, are given in some references in the suggested reading at the end of this chapter.

14.2 Metastable supersymmetry breaking

In the previous section we established criteria for dynamical supersymmetry breaking and exhibited an example, the $(3,2)$ model, which satisfies the criteria and exhibits dynamic supersymmetry in a stable ground state. But there are a number of ways in which we might view these criteria as limiting. First, while there are many models which satisfy them, they seem exceptional and not particularly generic. Second, it is difficult to build realistic models without spoiling the chiral structure of these theories. Finally, the criteria themselves are troubling, especially the requirement of a continuous global symmetry. We do not expect such symmetries in theories of gravity, so these symmetries must arise as accidents and must hold to some high degree of accuracy. Indeed, these criteria seem less sharp in the framework of supergravity.

If we consider theories with *metastable* ground states, i.e. theories having a stable ground state with unbroken supersymmetry but where supersymmetry is broken in a higher-energy, classically stable, state, the possibilities are greatly enlarged. Indeed, we can consider this question in the O'Raifeartaigh models. Rather than imposing a continuous R symmetry, we can consider discrete symmetries, for example a Z_N subgroup of a continuous R symmetry. For the fields $Z,\ Y,\ A$ we can require, with $\alpha = e^{\frac{2\pi i}{N}}$,

$$Z \to \alpha^2 Z, \quad Y \to \alpha^2 Y, \quad A \to A \tag{14.9}$$

while the superpotential transforms as

$$W \to \alpha^2 W. \tag{14.10}$$

Imposing, for simplicity, an additional symmetry $A \to -A,\ Y \to -Y$, the most general *renormalizable* superpotential takes the form of a simple O'Raifeartaigh model but where, beyond the renormalizable level, additional couplings are allowed:

$$W = Z(A^2 - \mu^2) + mYA + \frac{Z^{N+2}}{M^{N-1}} + \cdots. \tag{14.11}$$

Focusing just on the Z^{N+2} term, there is now a supersymmetric vacuum at

$$(N+2)Z^{N+1} = \mu^2 M^{N-1}. \tag{14.12}$$

For M large (e.g of order the Planck or unification scale) compared with μ this vacuum is far away. Near the origin, the Coleman–Weinberg calculation still leads to a local minimum of the potential. The time required to tunnel from the metastable vacuum to the supersymmetric vacuum grows exponentially with power M/μ (on including effects of general relativity, the time often becomes infinite). So this instability is not a phenomenological concern.

One might imagine that the phenomenon of metastable supersymmetry breaking in theories with discrete R symmetries is rather generic. In models with singlet chiral fields and a continuous R symmetry, if all fields have R charge 0 or 2 then supersymmetry breaking occurs when the number of fields X_i with charge 2 exceeds the number A_α with charge 0. A similar statement holds for the discrete symmetries.

14.2.1 Metastable dynamical supersymmetry breaking: the ISS model

The phenomenon of dynamical metastable supersymmetry breaking appears, then, to be rather generic. Remarkably, this already occurs in supersymmetric QCD with $N_\mathrm{f} > N_\mathrm{c}$, with *massive* quarks, as first pointed out by Intriligator, Shih and Seiberg (ISS). We have already explained that, quite generally, supersymmetric theories with massive, vector-like, fields do not break supersymmetry, in the sense that they possess multiple (typically N, for the gauge group $SU(N)$) supersymmetric ground states. But, consider the case $3N/2 > N_\mathrm{f} \geq N_\mathrm{c} + 1$. Turning off the mass term we will see in Section 16.4 that the theory is dual to a theory with gauge group $SU(N_\mathrm{f} - N_\mathrm{c})$, with:

1. N_f quarks in the fundamental representation, q_f, transforming in the $(1, N_\mathrm{f})$ representation of the flavor symmetry, $SU(N_\mathrm{f})_\mathrm{L} \times SU(N_\mathrm{f})_\mathrm{R}$;
2. N_f in the antifundamental representation, transforming as $(\bar{N}_\mathrm{f}, 1)$ under flavor;
3. a chiral field $\Phi_{f\bar{f}}$, transforming in the $(N_\mathrm{f}, \bar{N}_\mathrm{f})$ representation, which is a singlet of the dual gauge group.

The superpotential of the magnetic theory is

$$W_{\text{mag}} = \mu\bar{q}\Phi q. \tag{14.13}$$

Now turn on a small mass term in the underlying, "electric", theory,

$$\delta W = \bar{Q}mQ. \tag{14.14}$$

We expect the appearance of a small term proportional to m, in the dual, "magnetic", theory. The term $\mu \operatorname{Tr} \Phi\, m$ transforms under the global symmetries (including the anomalous $U(1)$s) in the same way as the original mass term. So we will assume that it is in fact present, i.e. that the full superpotential of the magnetic theory is

$$W_{\text{mag}} = \mu\bar{q}\Phi q + \mu \operatorname{Tr} \Phi\, m. \tag{14.15}$$

Recalling that the fields q, $\bar{q}$ are fundamentals of the dual gauge group and requiring that the D term conditions of this group be satisfied, the vacuum of the dual theory breaks supersymmetry. It is important that $N_{\text{f}} - N < N_{\text{f}}$; the resulting breaking is called "rank breaking". One can see this by using the flavor symmetries to write, for example, for $N_{\text{c}} = 2$, and $N_{\text{f}} = 3$,

$$q = \bar{q} = \begin{pmatrix} v_1 & 0 & 0 & \cdots \\ 0 & v_2 & 0 & \cdots \\ 0 & 0 & v_3 & \cdots \end{pmatrix}. \tag{14.16}$$

With this choice, we can satisfy the equations

$$\frac{\partial W}{\partial \Phi_{f,\bar{f}}} = 0 \tag{14.17}$$

only for $f,\bar{f} = 1, 2, 3$, not for larger f. This generalizes to the other values of $N_{\text{f}}, N_{\text{c}}$ in this class of models.

It still remains to verify that there is a good non-supersymmetric vacuum in the magnetic theory. For this, we need to consider the pseudomoduli of the classical theory. These are components of Φ, essentially those components which cannot gain mass by mixing with the q_f, $\bar{q}_f$ superfields. Clearly, in particular the components of $\Phi_{f,\bar{f}}$ with $f,\bar{f} > N$ are massless at the tree level. A Coleman–Weinberg calculation is necessary to determine the masses of these fields and to establish whether $\Phi = 0$ is a good ground state. The answer turns out to be yes.

We know that in the electric theory there are N supersymmetric ground states. These can be found in the magnetic description; decays to them are highly suppressed for small quark mass.

14.2.2 Retrofitting

A broad class of models exhibiting dynamical metastable supersymmetry breaking can be found by starting with the O'Raifeartaigh models. Again, a simple example is that of Eq. (14.11) above. Now, however, we replace the dimensional parameters m and μ^2 by couplings to a strongly interacting group which generates these scales dynamically.

For simplicity, we will consider μ^2. We introduce an $SU(N)$ gauge group with field strength W_α,

$$W = \lambda Z A^2 + \frac{Z}{M} W_\alpha^2 + mYA \tag{14.18}$$

(we will see that couplings of chiral fields to gauge fields of this type are common in string theory, where M might be the Planck scale or the scale of the string theory). Gaugino condensation in the $SU(N)$ group gives rise to an expectation value Λ^3 for W_α^2,

$$W = Z(A^2 - \mu^2 e^{-Z/(Mb_0)}) + mYA, \tag{14.19}$$

where b_0 is the beta function of the gauge theory.

Near the origin the Coleman–Weinberg calculation is identical to that of the O'Raifeartaigh model, and the potential has a minimum at $Z = 0$. But clearly there are lower energy states at larger fields due to:

1. the exponential term in Eq. (14.19);
2. possible higher-order terms in powers of $Z/M_{\rm p}$.

Models of this type illustrate the fact that metastable dynamical supersymmetry breaking is a generic phenomenon in supersymmetric field theories. They vastly expand the possibilities for supersymmetric model building.

We have seen, in this section, that the dynamical breaking of supersymmetry is common. Flat directions are often lifted and, in many instances the supersymmetry is broken with a stable ground state. So, we are ready to address the question: how might supersymmetry be broken in the real world?

14.3 Particle physics and dynamical supersymmetry breaking

14.3.1 Gravity mediation and dynamical supersymmetry breaking: anomaly mediation

One simple approach to model building which we explored in Chapter 11 was to treat a theory which breaks supersymmetry as a "hidden sector". This construction, as we presented it, was rather artificial. If we replace, say, the Polonyi sector by a sector which breaks supersymmetry dynamically, the situation is dramatically improved. If we suppose that there are some fields transforming under only the Standard Model gauge group and some transforming under only the gauge group responsible for symmetry breaking, the visible/hidden sector division is automatic. As we will see, this sort of division can arise rather naturally in string theory.

In such an approach the scale of supersymmetry breaking is again $m_{3/2}M_{\rm p}$, where we now understand this scale as the exponential of a small coupling at a high-energy scale (presumably the Planck, GUT or string scale). For scalars, soft-supersymmetry-breaking masses and couplings arise just as they did previously. There is no symmetry reason why these masses should exhibit any sort of universality.

One puzzle in this scenario is related to gluino masses. Examining the supergravity Lagrangian, the only term which can lead to gaugino masses is

$$\mathcal{L}_{\lambda\lambda} = e^{K/2} f'_{\alpha\beta k} (D_k W) \lambda^\alpha \lambda^\beta. \tag{14.20}$$

Here f is the gauge coupling function. So, in order to obtain a substantial gaugino mass, it is necessary that there be gauge-singlet fields with non-zero F terms. In most models of stable dynamical supersymmetry breaking there are no scalars which are singlets under all the gauge interactions. In metastable models, such as retrofitted models, it is necessary to suppose that there is some sort of discrete symmetry which accounts for the absence of certain couplings. These symmetries will forbid the coupling of hidden sector fields to visible sector gauge fields through low-dimension operators. In other words, we do not have couplings of the form

$$\frac{S}{M} W_\alpha^2, \tag{14.21}$$

where the F component of S has a non-zero vev. This suggests that gaugino masses would be suppressed relative to squark and slepton masses by powers of $M_{\text{int}}/M_{\text{p}}$.

But this turns out to be not quite correct. This is associated with a phenomenon known as "anomaly mediation". The term is arguably a misnomer; no actual symmetry of the theory is anomalous. The appearance of these terms can be understood, in some cases, as an issue of locality: the gaugino masses are themselves local but the supersymmetric operator which gives rise to them is not (i.e. it includes non-local terms). In other cases, a completely Wilsonian description is not available. Here we simply note that such terms are, in many instances, *required* by supersymmetry. Consider for example a pure $SU(N)$ gauge theory coupled to supergravity, with a small constant W_0 in the superpotential. In this theory, gaugino condensation occurs and gives rise to a non-perturbative correction to the superpotential,

$$W_{\text{np}} = -\frac{N}{32\pi^2} \langle \lambda\lambda \rangle.$$

From $V = -3|W_0 + W_{\text{np}}|^2$, then, we predict the following term in the potential:

$$-\frac{3N}{32\pi^2} W_0^* \langle \lambda\lambda \rangle. \tag{14.22}$$

It is natural to interpret this as resulting from an underlying term in the action,

$$\delta\mathcal{L} = -\frac{b_0}{16\pi^2} m_{3/2} \lambda\lambda. \tag{14.23}$$

One can argue for the presence of such a term for all N and N_{f} in a similar fashion. But the term can be found more directly from the structure of the underlying supergravity theory.

14.3.1.1 Split supersymmetry

The anomaly-mediated expression for the gaugino masses suggests an approach to model building of particular interest, given the large mass scale for squarks suggested by the Higgs mass. Even if one is willing to accept some fine tuning, one might need lighter gauginos to account for WIMP dark matter and to improve the quality of gauge coupling unification. If X denotes the field, with a non-vanishing F component, responsible for supersymmetry breaking, then one might suppose that there is no XW_α^2 coupling. In this case, assuming that the scalar masses are of order $m_{3/2}$, one can contemplate gauginos with masses lighter by a loop factor. So, for example, if squarks are at 30 TeV, one might have gluinos at scales slightly above one TeV and winos (the LSP), according to Eq. (14.23), a factor 3 or so lighter. One can debate how generic a phenomenon this might be.

14.3.2 Low-energy dynamical supersymmetry breaking: gauge mediation

An alternative to the conventional supergravity approach is to suppose that supersymmetry is broken at some much lower energy, with gauge interactions serving as the messengers of supersymmetry breaking. The basic idea is simple. One again supposes that one has some set of new fields and interactions which break supersymmetry. Some of these fields are taken to carry ordinary Standard Model quantum numbers, so that "ordinary" squarks, sleptons and gauginos can couple to them through gauge loops. This approach, which is referred to as *gauge mediated supersymmetry breaking* (GMSB), has a number of virtues.

1. It is highly predictive: as few as two parameters describe all soft breakings.
2. The degeneracies required to suppress flavor-changing neutral currents are automatic.
3. GMSB easily incorporates DSB and so can readily explain the hierarchy.
4. GMSB makes dramatic and distinctive experimental predictions.

The approach, however, also has drawbacks. Perhaps most serious is related to the "μ problem", which we discussed in the context of the MSSM. In theories with high-scale supersymmetry breaking we saw that there is not really a problem at all; a μ term of order the weak scale is quite natural. The μ problem, however, finds a home in the framework of low-energy breaking. The difficulty is that if one is trying to explain the weak scale dynamically then one does not want to introduce the μ term by hand. Various solutions have been offered for this problem. One possibility is that it is protected by symmetries and generated by the same dynamics which generates supersymmetry breaking. In the rest of our discussion we will simply assume that a μ term has been generated in the effective theory and will not worry about its origin.

14.3.2.1 Minimal gauge mediation (MGM)

The simplest model of gauge mediation contains, as messengers, a vector-like set of quarks and leptons, q, $\bar{q}$, ℓ and $\bar{\ell}$. These have the quantum numbers of a 5 and a $\bar{5}$ representation of $SU(5)$. The superpotential is taken to be

$$W_{\text{mgm}} = \lambda_1 q\bar{q} + \lambda_2 S\ell\bar{\ell}. \tag{14.24}$$

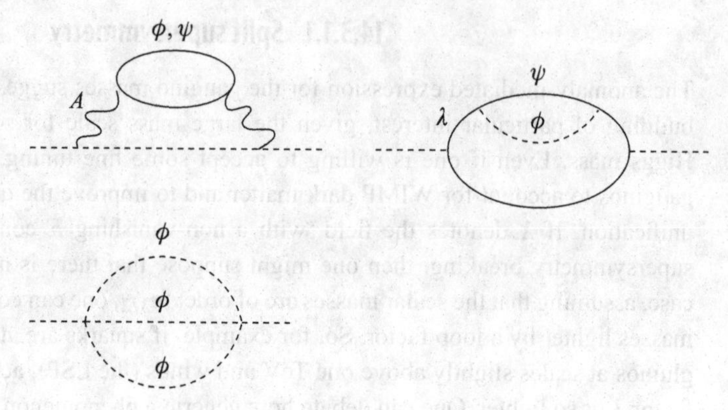

Fig. 14.1 Two-loop diagrams contributing to squark masses in a simple model of gauge mediation.

We will suppose that some dynamics gives rise to non-zero expectation values for S and F_S. We will not provide here a complete microscopic model to explain the origin of the parameters F_S and $\langle S\rangle$ that will figure in our subsequent analysis; retrofitting provides one strategy. To find a compelling model of the underlying dynamics is a good research problem. Instead, we will go ahead and immediately compute the superparticle spectrum for such a model. Ordinary squarks and sleptons gain mass through the two-loop diagrams shown in Fig. 14.1. While the prospect of computing a set of two-loop diagrams may seem intimidating, the computation is actually quite easy. If one treats F_S/S as small then there is only one scale in the integrals. It is a straightforward matter to write down the diagrams, introduce Feynman parameters and perform the calculation. There are also various non-trivial checks. For example, the sum of the diagrams must vanish in the supersymmetric limit. These masses can alternatively be computed by writing down an effective action in terms of spurion fields and computing the wave function renormalization factors as functions of the spurions.

One obtains the following expressions for the scalar masses:

$$\widetilde{m}^2 = 2\Lambda^2 \left[C_3 \left(\frac{\alpha_3}{4\pi}\right)^2 + C_2 \left(\frac{\alpha_2}{4\pi}\right)^2 + \frac{5}{3}\left(\frac{Y}{2}\right)^2 \left(\frac{\alpha_1}{4\pi}\right)^2 \right], \tag{14.25}$$

where $\Lambda = F_S/S$, $C_3 = 4/3$ for color triplets and zero for singlets and $C_2 = 3/4$ for weak doublets and zero for singlets. For the gaugino masses one obtains

$$m_{\lambda_i} = C_i \frac{\alpha_i}{4\pi} \Lambda. \tag{14.26}$$

This expression is valid only to lowest order in Λ. Higher-order corrections have been computed; it is straightforward to compute them exactly in Λ.

All these masses are positive and they are described in terms of a single new parameter, Λ. The lightest new particles are the partners of the $SU(3) \times SU(2)$ singlet leptons. If their masses are of order 100 GeV, we have that $\Lambda \sim 30$ TeV. The spectrum has a high degree of degeneracy. In this approximation the masses of the squarks and sleptons are functions only of their gauge quantum numbers, so flavor-changing processes are suppressed.

Flavor violation arises only through Yukawa couplings, and these can appear only in graphs at high loop order; it is further suppressed because all but the top Yukawa coupling is small.

Apart from the parameter Λ, one has the μ and B_μ parameters (B_μ is the coefficient of the soft-breaking $H_U H_D$ term in the potential; μ and B_μ are both complex), for a total of five. This is three beyond the minimal Standard Model. If the underlying susy-breaking theory conserves CP, this can eliminate the phases, reducing the number of parameters by two.

14.3.2.2 $SU(2)\times U(1)$ breaking

At lowest order, all the squark and slepton masses are positive. The large top quark Yukawa coupling leads to large corrections to $m^2_{H_U}$, however, which tend to drive it negative. The calculation is just a repeat of the one we did in the case of the MSSM. Treating the mass of $\tilde{t}$ as independent of momentum is consistent provided that we cut the integral off at a scale of order Λ (at this scale the calculation leading to Eq. (14.25) breaks down, and the propagator falls rapidly with momentum) and we have

$$m^2_{H_U} = \left(m^2_{H_U}\right)_0 - \frac{6y_t^2}{16\pi^2} \ln \frac{\Lambda^2}{\widetilde{m}_t^2} \left(\widetilde{m}_t^2\right)_0. \tag{14.27}$$

While the loop correction is nominally three-loop in order, because the stop mass arises from gluon loops while the Higgs mass arises at lowest order from W loops we have a substantial effect,

$$\left(\frac{\widetilde{m}_t^2}{m^2_{H_U}}\right)_0 = \frac{16}{9}\left(\frac{\alpha_3}{\alpha_2}\right)^2 \sim 20 \tag{14.28}$$

and the Higgs mass-squared is negative. These contributions are quite large and, given the large value of the Higgs mass, it is again necessary to tune the μ term and other possible contributions to the Higgs mass to a high degree in order to obtain sufficiently small W and Z masses.

14.3.2.3 General gauge mediation

The minimal model of gauge mediation of the previous section makes a quite sharp set of predictions. These predictions, in fact, are referred to as *minimal gauge mediation* (MGM). It is clearly of interest to ask how general they are. It turns out that they are peculiar to our assumption that there is a single set of messengers and that just one singlet is responsible for supersymmetry breaking and R symmetry breaking. Indeed, our messengers have the quantum numbers of a 5 and a $\bar{5}$ representation of $SU(5)$. If, for example, we had considered two singlets, Z_1 and Z_2, with Z_i and F_i non-zero, we could have obtained independent soft-breaking masses for squarks and leptons. Had we allowed different singlets, and taken a 10 and $\overline{10}$ for the messengers, we could have obtained a richer spectrum. Meade, Seiberg and Shih formulated the problem of gauge mediation in a general way and dubbed this formulation *general gauge mediation* (GGM).

They studied the problem in terms of the correlation functions of (gauge) supercurrents. Analyzing the restrictions imposed by Lorentz invariance and supersymmetry on these correlation functions, they found that the general gauge-mediated spectrum is described by three complex parameters and three real parameters. The spectrum can be significantly different from that of the MGM, but the masses are still only functions of the gauge quantum numbers and flavor problems are still mitigated.

The basic structure of the spectrum is readily described. In the formulas for the fermion masses we introduce a separate complex parameter m_i, $i = 1, 2, 3$ for each Majorana gaugino. Similarly, for the scalars we introduce a real parameter Λ_c^2 for the contributions from $SU(3)$ gauge fields Λ_w^2 for those from $SU(2)$ gauge fields and Λ_Y^2 for those from hypercharge gauge fields:

$$\tilde{m}^2 = 2\left[C_3\left(\frac{\alpha_3}{4\pi}\right)^2 \Lambda_c^2 + C_2\left(\frac{\alpha_2}{4\pi}\right)^2 \Lambda_w^2 + \frac{5}{3}\left(\frac{Y}{2}\right)^2\left(\frac{\alpha_1}{4\pi}\right)^2 \Lambda_Y^2\right]. \tag{14.29}$$

One can construct models which exhibit the full set of parameters. In MGM the messengers of each set of quantum numbers each have a supersymmetric contribution to their masses, λM, while the supersymmetry-breaking contribution to the scalar masses goes as λM^2, so in the ratio of these two contributions the coupling cancels out. In GGM model building, additional fields and couplings lead to more complicated relations.

One feature of MGM which is not immediately inherited by GGM is the suppression of new sources of CP violation. Because the gaugino masses are independent parameters, in particular, they introduce additional phases which are inherently CP-violating. Providing a natural explanation of the suppression of these phases is one of the main challenges of GGM model building.

14.3.2.4 Light gravitino phenomenology

There are other striking features of these models. One of the more interesting is that the lightest supersymmetric particle, or LSP, is the gravitino. Its mass is

$$m_{3/2} = 2.5\left(\frac{F}{(100\ \text{TeV})^2}\right)\text{eV}. \tag{14.30}$$

The next-to-lightest supersymmetric particle, or NLSP, can be a neutralino or a charged right-handed slepton. The NLSP will decay to its superpartner plus a gravitino in a time long compared with typical microscopic times but still quite short. The lifetime can be determined from low-energy theorems, in a manner reminiscent of the calculation of the pion lifetime. Just as the chiral currents are linear in the (nearly massless) pion field,

$$j^{\mu 5} = f_\pi \partial^\mu \pi, \quad \partial_\mu j^{\mu 5} = \partial^2 \pi \approx 0, \tag{14.31}$$

so the supersymmetry current is linear in the goldstino G,

$$j_\alpha^\mu = F\gamma^\mu G + \sigma^{\mu\nu}\lambda F_{\mu\nu} + \cdots, \tag{14.32}$$

where F, here, is the goldstino decay constant. From this, if one assumes that the LSP is mostly photino then one can calculate the amplitude for $\tilde{\gamma} \to G + \gamma$ in much the same

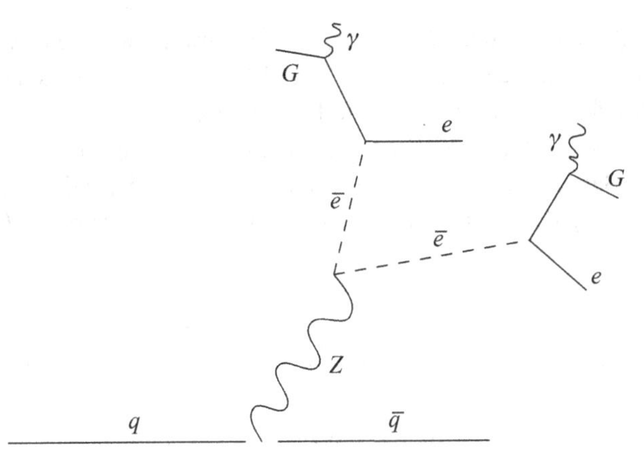

Fig. 14.2 Decay leading to $e^+e^-\gamma\gamma$ events.

way as one considers processes in current algebra. From Eq. (14.32) one sees that $\partial_\mu j^\mu_\alpha$ is an interpolating field for G, so

$$\langle G\gamma|\tilde{\gamma}\rangle = \frac{1}{F}\langle\gamma|\partial_\mu j^\mu_\alpha|\tilde{\gamma}\rangle. \tag{14.33}$$

The matrix element can be evaluated by examining the second term in the current, Eq. (14.32), and noting that $\partial\!\!\!/ \chi = m_\lambda \lambda$.

Given the matrix element, the calculation of the NLSP lifetime is straightforward and yields

$$\Gamma(\tilde{\gamma} \to G\gamma) = \frac{\cos^2\theta_W m_{\tilde{\gamma}}^5}{16\pi F^2}. \tag{14.34}$$

This yields a *decay length*:

$$c\tau = 130\left(\frac{100\ \text{GeV}}{m_{\tilde{\gamma}}}\right)^5\left(\frac{\sqrt{F}}{100\ \text{TeV}}\right)^4\ \mu\text{m}. \tag{14.35}$$

In other words, if F is not too large then the NLSP may decay in the detector. One even has the possibility of measurable displaced vertices. The signatures of such low decay constants would be quite spectacular. Assuming the photino (bino) is the NLSP, one has processes such as $e^+e^- \to \gamma\gamma + \not{E}_t$ and $p\bar{b} \to e^+e^-\gamma\gamma + \not{E}_t$, as indicated in Fig. 14.2, where $\not{E}_t$ is the missing transverse energy.

Suggested reading

There are a number of good reviews of dynamical supersymmetry breaking, including those of Shadmi and Shirman (2000) and Terning (2003). The former includes catalogs of models and mechanisms. The recent interest in metastable supersymmetry breaking was launched by Intriligator *et al.* (2006). There is a large literature on gauge-mediated models and their phenomenology; a good review is provided by Giudice and Rattazzi (1999). The recent development of General Gauge Mediation is described in Meade *et al.* (2008). Models which achieve the full set of parameters are described in Buican *et al.* (2009) and Carpenter *et al.* (2009). A clear exposition of the origin of anomaly mediation is provided in Bagger *et al.* (2000), in Weinberg's text (1995), and in the more recent work of Dine and Seiberg (2007), Dine and Draper (2013), and DiPietro *et al.* (2014).

Exercises

(1) Check that the $SU(N)$ models, with an antisymmetric tensor and $N-4$ antifundamentals, have no flat directions and that they have a non-anomalous $U(1)$ symmetry.
(2) Verify Eq. (14.3) for the case of a $U(1)$ gauge theory with charged field ϕ^+ and ϕ^- introducing a Pauli–Villars regulator field.
(3) Check that Eqn. (14.5) is the most general expression that is consistent with symmetries, at least up to terms linear in m. Verify that there is no supersymmetric vacuum for this superpotential.

15 Theories with more than four conserved supercharges

In theories with more than four conserved supercharges (extended supersymmetry), the supersymmetry generators obey the relations

$$\{Q^I_\alpha, \bar{Q}^J_{\dot\alpha}\} = \not{p}\,\delta^{IJ}, \quad \{Q^I_\alpha, Q^J_\beta\} = Z^{IJ}\epsilon_{\alpha,\beta}. \tag{15.1}$$

The quantities Z^{IJ} are known as central charges. We will see that these can arise in a number of physically interesting ways.

In theories with four supersymmetries, we saw in Chapters 13 and 14 supersymmetry provides powerful constraints on the possible dynamics. Theories with more than four supercharges ($N > 1$ in four dimensions) are not plausible as models of the real world but they do have a number of remarkable features. As in some of our $N = 1$ examples, these theories typically have exact moduli spaces. Gauge theories with $N = 4$ supersymmetry exhibit an exact duality between electricity and magnetism. Theories with $N = 2$ supersymmetry have a rich – and tractable – dynamics, closely related to important problems in mathematics. In all these cases supersymmetry provides remarkable control over the dynamics, allowing one to address questions which are inaccessible in theories without supersymmetry. Supersymmetric theories in higher dimensions generally have more than four supersymmetries, and a number of the features of the theories we study in this chapter will reappear when we come to higher-dimensional field theories and string theory.

15.1 $N = 2$ theories: exact moduli spaces

Theories with $N = 1$ supersymmetry are tightly constrained, but theories with more supersymmetry are even more highly constrained. We have seen that often, in perturbation theory, $N = 1$ theories have moduli; non-perturbatively, sometimes, these moduli are lifted. In theories with $N = 1$ supersymmetry, a detailed analysis is usually required to determine whether the moduli acquire potentials at the quantum level. For theories with more supersymmetries ($N > 1$ in four dimensions; $N \geq 1$ in five or more dimensions), one can show rather easily that the moduli space is exact. Here we consider the case of $N = 2$ supersymmetry in four dimensions. These theories can also be described by a superspace, in this one case built from two Grassmann spinors, θ and $\tilde\theta$. There are two basic types of superfields: vectors and hypermultiplets. The vectors are chiral with respect to both D_α and

$\tilde{D}_\alpha$ and have an expansion, in the case of a $U(1)$ field,

$$\psi = \phi + \tilde{\theta}^\alpha W_\alpha + \tilde{\theta}^2 \bar{D}^2 \phi^\dagger, \tag{15.2}$$

where ϕ is an $N = 1$ chiral multiplet and W_α is an $N = 1$ vector multiplet. The fact that $\phi^\dagger$ appears as the coefficient of the $\tilde{\theta}^2$ term is related to an additional constraint satisfied by ψ. This expression can be generalized to non-Abelian symmetries; the expression for the highest component of ψ is then somewhat more complicated but we will not need it here.

The theory possesses an $SU(2)$ R-symmetry under which θ and $\tilde{\theta}$ form a doublet. Under this symmetry, the scalar component of ϕ and the gauge field are singlets, while ψ and λ form a doublet.

We will not describe the hypermultiplets in detail except to note that, from the perspective of $N = 1$, they consist of two chiral multiplets. The two chiral multiplets transform as a doublet of the $SU(2)$ group. The superspace description of these multiplets is more complicated.

In the case of a non-Abelian theory, the vector field ψ^a is in the adjoint representation of the gauge group. For these fields the Lagrangian has a very simple expression as an integral over half the superspace:

$$\mathcal{L} = \int d^2\theta d^2\tilde{\theta}\ \psi^a \psi^a, \tag{15.3}$$

or, in terms of $N = 1$ components,

$$\mathcal{L} = \int d^2\theta\, W_\alpha^2 + \int d^4\theta \phi^\dagger e^V \phi. \tag{15.4}$$

The theory with vector fields alone has a classical moduli space, given by the values of the fields for which the scalar potential vanishes. Here this just means that the D fields vanish. Written as a matrix we have

$$D = [\phi, \phi^\dagger], \tag{15.5}$$

which vanishes for diagonal ϕ, i.e. for

$$\phi = \frac{a}{2}\begin{pmatrix} 1 & 0 \\ 0 & -1 \end{pmatrix}. \tag{15.6}$$

For many physically interesting questions one can focus on the effective theory for the light fields. In the present case the light field is the vector multiplet ψ. Roughly,

$$\psi \approx \psi^a \psi^a = a^2 + a\delta\psi^3 + \cdots. \tag{15.7}$$

What kind of effective action can we write for ψ? At the level of terms with up to four derivatives, the most general effective Lagrangian has the form[1]

$$\mathcal{L} = \int d^2\theta d^2\tilde{\theta} f(\psi) + \int d^8\theta\, \mathcal{H}(\psi, \psi^\dagger). \tag{15.8}$$

[1] This, and essentially all the effective actions we will discuss, should be thought of as Wilsonian effective actions, obtained by integrating out heavy fields and high-momentum modes.

Terms with covariant derivatives correspond to terms with more than four derivatives when written in terms of ordinary component fields.

The first striking result we can read off from this Lagrangian, with no knowledge of $\mathcal{H}$ and f, is that there is no potential for ϕ, i.e. the moduli space is exact. This statement is true both perturbatively and non-perturbatively.

One can next ask about the function f. This function determines the effective coupling in the low-energy theory and is an object studied by Seiberg and Witten, which we will discuss in Section 15.4.

15.2 A still simpler theory: $N = 4$ Yang–Mills

The $N=4$ Yang–Mills theory is interesting in its own right: it is finite and conformally invariant. It also plays an important role in our current understanding of non-perturbative aspects of string theory. The $N=4$ Yang–Mills has 16 supercharges and is even more tightly constrained than the $N=2$ theories. First, we will describe the theory. In the language of $N=2$ supersymmetry, it consists of one vector multiplet and one hypermultiplet. In terms of $N=1$ superfields, it contains three chiral superfields, ϕ_i and a vector multiplet. The Lagrangian is

$$\mathcal{L} = \int d^2\theta W_\alpha^2 + \int d^4\theta\, \phi_i^\dagger e^V \phi_i + \int d^2\theta\, \phi_i^a \phi_j^b \phi_k^c \epsilon_{ijk} \epsilon^{abc}. \tag{15.9}$$

In the above description there is a manifest $SU(3) \times U(1)$ R-symmetry. Under this symmetry the ϕ_is have $U(1)_R$ charge $2/3$ and form a triplet of $SU(3)$. But the real symmetry is larger – it is $SU(4)$. Under this symmetry, the four Weyl fermions form a 4-dimensional representation, while the six scalars transform in the 6-dimensional representation. Later, our studies of the toroidal compactifications of the heterotic string (Chapter 25) will later give us an heuristic understanding of this $SU(4)$ symmetry: it reflects the $O(6)$ symmetry of the compactified six dimensions. In string theory this symmetry is broken by the compactification manifold; this reflects itself in higher-derivative, symmetry-breaking, operators.

In the $N = 4$ theory there is, again, no modification of the moduli space, perturbatively or non-perturbatively. This can be understood in a variety of ways. We can use the $N = 2$ description of the theory, defining the vector multiplet to contain the $N = 1$ vector and one (arbitrarily chosen) chiral multiplet. Then an identical argument to that given above ensures that there is no superpotential for the chiral multiplet alone. The $SU(3)$ symmetry then ensures that there is no superpotential for any chiral multiplet. Indeed, we can make an argument directly in the language of $N = 1$ supersymmetry. If we tried to construct a superpotential for the low-energy theory in the flat directions, it would have to be an $SU(3)$-invariant holomorphic function of the ϕ_is. But there is no such object.

Similarly, it is easy to see that there are no corrections to the gauge couplings. For example, in the $N = 2$ language, we want to ask what sort of function f is allowed in

$$\mathcal{L} = \int d^2\theta d^2\tilde{\theta}\, f(\psi). \tag{15.10}$$

The theory has a $U(1)$ R-invariance under which

$$\psi \to e^{2/3i\alpha}\psi, \quad \theta \to e^{i\alpha\theta}, \quad \tilde{\theta} \to e^{-i\alpha}\tilde{\theta}. \tag{15.11}$$

Already, then,

$$\int d^2\theta d^2\tilde{\theta}\, \psi\psi \tag{15.12}$$

is the unique structure which respects these symmetries. Now we can introduce a background dilaton field, τ. Classically the theory is invariant under shifts in the real part of τ, $\tau \to \tau + \beta$. This ensures that there are no perturbative corrections to the gauge couplings. With a little more work one can show that there are no non-perturbative corrections either.

One can also show that the quantity $\mathcal{H}$ in Eq. (15.8) is unique in this theory, again using the symmetries. The expression

$$\mathcal{H} = c \ln\psi \ln\psi^\dagger \tag{15.13}$$

respects all the symmetries. At first sight it might appear to violate scale invariance; given that ψ is dimensionful one would expect a scale Λ sitting in the logarithm. However, it is easy to see that if one integrates over the full superspace, any Λ-dependence disappears since ψ is chiral. Similarly, if one considers the $U(1)$ R-transformation, the shift in the Lagrangian vanishes after the integration over superspace. To see that this expression is not renormalized, one merely needs to note that any non-trivial τ-dependence spoils these two properties. As a result, in the case of $SU(2)$ the four derivative terms in the Lagrangian are not renormalized. Note that this argument is non-perturbative. It can be generalized to an even larger class of higher-dimensional operators.

15.3 A deeper understanding of the BPS condition

In our study of monopoles we saw that, under certain circumstances, the complicated second-order non-linear differential equations reduce to first-order differential equations. The main condition is that the potential should vanish. We are now quite used to the idea that supersymmetric theories often have moduli, and we have seen that this is an exact feature of $N = 4$ and many $N = 2$ theories. In the case of an $N = 2$ supersymmetric gauge theory the potential is just that arising from the D term, and one can construct a Prasad–Sommerfield solution. We will now see that the Bogomol'nyi–Prasad–Sommerfield (BPS) condition is not simply magic but is a consequence of the extended supersymmetry of the theory. The resulting mass formula, as a consequence, is *exact*; it is not simply a feature of the classical theory but a property of the full quantum theory. This sort of BPS condition is relevant not only to the study of magnetic monopoles but to topological objects in various dimensions and contexts, particularly in string theory. Here we will give the flavor of the argument without worrying about factors of two. More details can be worked out in the exercises; see also the references.

First, we show that the electric and magnetic charges enter in the supersymmetry algebra of this theory as central charges. Thinking of this as an $N = 1$ theory, we have seen that the supercurrents take the form

$$S^{\mu}_{\alpha} = \sigma^{\mu}_{\alpha\dot{\beta}}(\sigma^{\rho\sigma})^{\dot{\beta}\dot{\gamma}} F_{\rho\sigma}\lambda_{\dot{\gamma}} + \partial_{\rho}\phi^{i}\sigma^{\rho}_{\alpha\dot{\beta}}(\sigma^{\mu})^{\dot{\beta}\gamma}\psi^{i}_{\gamma} + F\text{-term contributions.} \tag{15.14}$$

In this theory, however, there is an $SU(4)$ symmetry and the supercurrents should transform as a 4 representation. It is not hard to guess the other three currents

$$S^{i}_{\mu\alpha} = (\sigma_{\mu})_{\alpha\dot{\beta}}(\sigma^{\rho\sigma})^{\dot{\beta}\dot{\gamma}} F_{\rho\sigma}\psi^{i}_{\dot{\gamma}} + \epsilon_{ijk}\partial_{\rho}\phi^{j}\sigma^{\rho}_{\alpha\dot{\beta}}(\sigma_{\mu})^{\dot{\beta}\gamma}\psi^{k}_{\gamma} + F\text{-term contributions.} \tag{15.15}$$

We are interested in proving bounds on the mass. It is useful to define Hermitian combinations of the charges $Q_{\alpha i} = \int d^3 x S_{\alpha i}$, since we want to study positivity constraints. In this case, it is more convenient to write a four-component expression, using a Majorana (real) basis for the γ matrices. Taking an $N = 2$ subgroup and carefully computing the commutators of the charges, we obtain

$$\{Q_{\alpha i}, Q_{\beta j}\} = \delta_{ij}\gamma^{\mu}_{\alpha\beta}P_{\mu} + \epsilon_{ij}(\delta_{\alpha\beta}U_k + (\gamma_5)_{\alpha\beta}V_k). \tag{15.16}$$

Here

$$\begin{aligned} U_k &= \int d^3x\partial_i(\phi^a_{\text{re }k}E^a_i + \phi^a_{\text{im }k}B^a_i), \\ V_k &= \int d^3x\partial_i(\phi^a_{\text{im }k}E^a_i + \phi^a_{\text{re }k}B^a_i). \end{aligned} \tag{15.17}$$

In the Higgs phase the integrals are, by Gauss's theorem, of electric and magnetic charges multiplied by the Higgs expectation value. From these relations we can derive bounds on masses, using the fact that Q^2_{α} is a positive operator. Taking the expectations of both sides we have, for an electrically neutral system of mass M in its rest frame,

$$M \pm Q_{\text{m}}v \geq 0. \tag{15.18}$$

This bound is saturated when Q annihilates the state. Examining the form of Q_{α}, this is just the BPS condition.

15.3.1 $N = 4$ Yang–Mills theories and electric–magnetic duality

The $N = 4$ theory contains, from the point of view of $N = 1$ supersymmetry, a gauge multiplet and three chiral multiplets in the adjoint representation. In addition to the interactions implied by the gauge symmetry, there is a superpotential

$$W = \frac{1}{6} f_{abc}\epsilon_{ijk}\Phi^a_i\Phi^b_j\Phi^c_k. \tag{15.19}$$

We have normalized the kinetic terms for the fields Φ with a $1/g^2$ factor. So, this interaction has a strength related to the strength of the gauge interactions. This theory has a global $SU(4)$ symmetry. Under this symmetry, the four adjoint fermions transform as a 4, the scalars transform as a 6 and the gauge bosons are invariant. The theory has a large set of flat directions. If we simply take all the Φ fields, regarded as matrices, to be diagonal then the potential vanishes. As a result, this theory has monopoles of the BPS type.

This theory has a symmetry even larger than the Z_2 duality symmetry that we contemplated when we examined Maxwell's equations; the full symmetry is $SL(2, Z)$. We might have guessed this by remembering that the coupling constant is part of the holomorphic variable

$$\tau = \frac{\theta}{2\pi} + \frac{4\pi i}{e^2}. \tag{15.20}$$

Thus in addition to our conjectured $e \to 1/e$ symmetry there is a symmetry $\theta \to \theta + 2\pi$. So, in terms of τ we have the two symmetry transformations

$$\tau \to -\frac{1}{\tau}, \quad \tau \to \tau + 1. \tag{15.21}$$

Together, these transformations generate the group $SL(2, Z)$:

$$\tau \to \frac{a\tau + b}{c\tau + d}, \quad ad - bc = 1. \tag{15.22}$$

Now we can look at our BPS formula. To understand whether it respects the $SL(2, Z)$ symmetry we need to understand how this symmetry acts on the states. Writing

$$M = eQ_e v + \frac{Q_m v}{e}, \tag{15.23}$$

with

$$Q_e = n_e - n_m \frac{\theta}{2\pi}, \quad Q_m = 4\pi \frac{n_m}{e}, \tag{15.24}$$

the spectrum is invariant under the $SL(2, Z)$ transformation of τ accompanied by

$$\begin{pmatrix} n_e \\ n_m \end{pmatrix} \to \begin{pmatrix} d & -b \\ c & -d \end{pmatrix} \begin{pmatrix} n_e \\ n_m \end{pmatrix}. \tag{15.25}$$

Because it follows from the underlying supersymmetry the mass formula is exact, so this duality of the spectrum of BPS objects is a non-perturbative statement about the theory.

15.4 Seiberg–Witten theory

We have seen that $N = 4$ theories are remarkably constrained, and this allowed us, for example, to explore an exact duality between electricity and magnetism. Still, these theories are not nearly as rich as field theories with $N \leq 1$ supersymmetry. The $N = 2$ theories are still quite constrained, but exhibit a much more interesting array of phenomena. They illustrate the power provided by supersymmetry over non-perturbative dynamics. They will also allow us to study phenomena associated with magnetic monopoles in a quite non-trivial way. In this section, we will provide a brief introduction to *Seiberg–Witten theory*. This subject has applications not only in quantum field theory but also for our understanding of string theory and, perhaps most dramatically, in mathematics.

It is convenient to describe the $N = 2$ theories in $N = 1$ language. The basic $N = 2$ multiplets are the vector multiplet and the tensor (or hyper) multiplet. From the point of view of $N = 1$ supersymmetry, the $N = 2$ vector contains an $N = 1$ a vector multiplet and a chiral multiplet. The tensor contains two chiral fields. We will focus mainly on theories with only vector multiplets, with gauge group $SU(2)$. In the $N = 1$ description the fields are a vector multiplet V and a chiral multiplet ϕ, both in the adjoint representation. The Lagrangian density is

$$\mathcal{L} = \int d^4\theta \, \frac{1}{g^2} \phi^\dagger e^V \phi - \frac{i}{16\pi} \int d^2\theta \, \tau W^{a\alpha} W^a_\alpha + \text{h.c.} \tag{15.26}$$

Here

$$\tau = \frac{\theta}{2\pi} + i\frac{4\pi}{g^2}. \tag{15.27}$$

The $1/g^2$ in front of the chiral field kinetic term is somewhat unconventional, but it makes the $N = 2$ supersymmetry more obvious. As we indicated earlier, one way to understand the $N = 2$ supersymmetry is to note that the Lagrangian we have written down has a global $SU(2)$ symmetry. Under this symmetry the scalar fields ϕ^a and the gauge fields A^a_μ are singlets, while the gauginos λ^a and the fermionic components ψ^a of ϕ transform as a doublet. Acting on the conventional $N = 1$ generators, the $SU(2)$ symmetry produces four new generators. So, we have generators Q^A_α, with $A = 1, 2$.

As it stands, the model has flat directions, with

$$\phi = \frac{a}{2} \begin{pmatrix} 1 & 0 \\ 0 & -1 \end{pmatrix}. \tag{15.28}$$

In these directions the spectrum consists of two massive gange bosons and one massless gauge boson, a massive complex scalar that is degenerate with the gauge bosons and a massive Dirac fermion as well as a massless vector and a massless chiral multiplet. The masses of all these particles are

$$M_W = \sqrt{2}a. \tag{15.29}$$

This is precisely the right number of states to fill an $N = 2$ multiplet. Actually, it is a BPS multiplet. It is annihilated by half the supersymmetry generators. The classical theory possesses, in addition to the global $SU(2)$ symmetry, an anomalous $U(1)$ symmetry,

$$\phi \to e^{i\alpha\phi}, \quad \psi \to e^{i\alpha}\psi. \tag{15.30}$$

Under this symmetry, we have

$$\theta \to \theta - 4\alpha \tag{15.31}$$

or

$$\tau \to \tau - 2\pi\alpha. \tag{15.32}$$

Because the physics is periodic in θ with period 2π, $\alpha = \pi/2$ is a symmetry, i.e. the theory has a Z_4 symmetry,

$$\phi \to e^{i\pi/2}\phi. \tag{15.33}$$

Note that ϕ is not gauge invariant. A suitable gauge-invariant variable for the analysis of this theory is

$$u = \langle \mathrm{Tr}\, \phi^2 \rangle. \tag{15.34}$$

Under the discrete symmetry, we have $u \to -u$; at weak coupling

$$u \approx a^2. \tag{15.35}$$

The spectrum of this theory includes magnetic monopoles, in general with electric charges. At the classical level the monopole solutions in this theory are precisely those of Prasad and Sommerfield, with mass

$$M_{\mathrm{M}} = 4\pi\sqrt{2}\frac{a}{g^2}. \tag{15.36}$$

As in the $N = 4$ theory, there is a BPS formula for the masses:

$$m = \sqrt{2}\,|aQ_{\mathrm{e}} + a_{\mathrm{D}}Q_{\mathrm{M}}|. \tag{15.37}$$

At tree level,

$$a_{\mathrm{D}} = \frac{4\pi}{g^2} ia = \tau a, \tag{15.38}$$

where the last equation holds if $\theta = 0$. The appearance of i in this formula is not immediately obvious. To see that it must be present, consider the case of dyonic excitations of monopoles. These should have energy of order the charge, with no factors of $1/g^2$. This is ensured by the relative phase between a and a_{D}. These formulas will receive corrections in perturbation theory and beyond; our goal is to understand the form of these corrections and their (dramatic) physical implications.

Equation (15.38) is not meaningful as it stands; τ is a function of scale. Instead, Seiberg and Witten suggested that

$$\tau = \frac{da_{\mathrm{D}}}{da}. \tag{15.39}$$

They also proposed the existence of a duality symmetry, under which

$$a_{\mathrm{D}} \leftrightarrow a, \quad \tau \to -\frac{1}{\tau}. \tag{15.40}$$

To formulate our questions more precisely and to investigate this proposal, it is helpful, as always, to consider a low-energy effective action. This action should respect the $N = 2$ supersymmetry; in $N = 1$ language this means that the Lagrangian should take the form

$$\mathcal{L} = \int d^4\theta\, K(a, \bar{a}) - \frac{i}{16\pi^2}\int d^2\theta\, \tau(a) W^\alpha W_\alpha. \tag{15.41}$$

The $N = 2$ supersymmetry implies a relation between K and τ; without it these would be independent quantities. Both quantities can be obtained from a holomorphic function called the *prepotential*, $\mathcal{F}(a)$:

$$\tau = \frac{d^2\mathcal{F}}{da^2}, \quad K = \frac{1}{4\pi}\frac{d\mathcal{F}}{da}a^*. \tag{15.42}$$

From

$$\tau = \frac{da_D}{da} = \frac{d}{da}\left(\frac{d\mathcal{F}}{da}\right) \tag{15.43}$$

we have

$$\frac{d\mathcal{F}}{da} = ia_D, \tag{15.44}$$

so that

$$K = \frac{1}{4\pi}\ \mathrm{Im}\ a_D a^*. \tag{15.45}$$

Our goal will be to obtain a non-perturbative description of $\mathcal{F}$. At weak coupling the beta function of this theory is obtained from $b_0 = 3N - N = 2N = 4$, so

$$\tau = \frac{i}{\pi}\ln\frac{u}{\Lambda^2}. \tag{15.46}$$

As a check on this formula note that, under $u \to e^{2i\alpha}u$, $\theta \to \theta - 4\alpha$, we have

$$\tau = \frac{\theta}{2\pi} + 4\pi i g^{-2} \to \tau - \frac{2i\alpha}{\pi}, \tag{15.47}$$

and this is precisely the behavior of the formula Eq. (15.46).

This is similar to phenomena we have seen in $N = 1$ theories. But, when we consider the monopoles of the theory, the situation becomes more interesting. First note that, using the leading-order result for τ,

$$a_D = \frac{2i}{\pi}\left(a\ln\frac{a}{\Lambda} - a\right). \tag{15.48}$$

So, under the transformation $u \to e^{i\alpha u}$ of u,

$$a_D \to e^{i\alpha/4}\left(a_D - \frac{\alpha}{2\pi}a\right). \tag{15.49}$$

Our BPS mass formula transforms to

$$m \to \sqrt{2}\left|a\left(Q_e - \frac{4\alpha}{2\pi}Q_m\right) + a_D Q_M\right|. \tag{15.50}$$

This is the Witten effect, which we discussed earlier: in the presence of θ, the coefficient of $F\tilde{F}$, of (7.39), a magnetic monopole acquires an electric charge. More generally, the spectrum of dyons is altered.

Consider now what happens when we do a full 2π change of θ ($u \to -u$); it should be a symmetry. It is in this case, but in a subtle way: the spectrum of the dyonic excitations of the theory is unchanged but the charges of the dyons have shifted by one fundamental unit. This, in turn, is related to the branched structure of τ.

At the non-perturbative level the structure is even richer. We might expect that

$$\tau(u) = \frac{i}{\pi}\ln\frac{u}{\Lambda^2} + \alpha\exp\left(-\frac{8\pi^2}{g^2}\right) + \beta\exp\left(-\frac{8\pi^2}{g^2}\right) + \cdots. \tag{15.51}$$

Note that, interpreting $\exp(-8\pi^2/g^2)$ as $\exp(2\pi i\tau)$, each term in this series has the correct periodicity in θ. Moreover,

$$\exp(2\pi i\tau) = \frac{\Lambda^2}{u^2}. \tag{15.52}$$

These corrections have precisely the structure required for them to be instanton corrections, and these instanton corrections have been computed. But, following Seiberg and Witten, we can be bolder and consider what happens when g becomes large. Naively, we might expect that some monopoles become light. Associated with this, τ may have a singularity at some point $u_0 = \gamma\Lambda^2$, where Λ is the renormalization-group-invariant mass of the theory. In light of the Z_2 symmetry there must also be a singularity at $-u_0$. Such a singularity arises because a particle is becoming massless. If we think of τ_D as the dual of τ then there is an electrically charged light field of unit charge; more precisely, there must be two particles of opposite charge in order that they can gain mass. So τ_D has the following structure:

$$\tau_D = -\frac{2i}{2\pi}\ln m_M. \tag{15.53}$$

Assuming that a_D has a simple zero,

$$a_D \approx b(u - u_0), \quad m_M = \sqrt{2}a_D, \tag{15.54}$$

then

$$\tau_D = -\frac{i}{\pi}\ln(u - u_0) = -\frac{1}{\tau(u)}. \tag{15.55}$$

Starting with the relation

$$\frac{da}{da_D} = -\tau_D = -\frac{i}{\pi}\ln a_D, \tag{15.56}$$

we have

$$a = \frac{i}{\pi}(a_D \ln a_D - a_D). \tag{15.57}$$

Similarly, we can consider the behavior at the point $-u_0$. This is the mirror image of the previous case, but we must be careful about the relation of a and a_D. They are connected by the symmetry transformation

$$\tilde{a} = ia, \quad \tilde{a}_D = i(a_D - a). \tag{15.58}$$

Now,

$$\tau_D = -\frac{1}{\tau(u)} = \frac{i}{\pi}\ln(u + u_0) \tag{15.59}$$

and

$$\tilde{a} = \frac{1}{\pi}(\tilde{a}_D \ln \tilde{a}_D - \tilde{a}_D). \tag{15.60}$$

Going around the singularities, at u_0 we have

$$a \to a - 2a_D, \quad a_D \to a_D, \tag{15.61}$$

while at $-u_0$

$$a \to 3a - 2a_D, \quad a_D \to 2a - a_D. \tag{15.62}$$

This should be compared with the effect of going around 2π at large u, when $a \to -a$ and $a_D \to -(a_D - a)$. Assuming that these are the only singularities, we can, from this information, reconstruct τ. We will not give the full solution of Seiberg and Witten here, but the basic idea is to note that $\tau(u)$ is the modular parameter of a two-dimensional torus and to reconstruct the torus.

This analysis has allowed us to study the theory deep in the non-perturbative region. Seiberg and Witten uncovered a non-trivial duality, a limit in which monopoles become massless, and they provided insight into confinement. These sorts of ideas have been extended to other theories and to theories in higher dimensions and have provided insight into many phenomena in string theory, quantum gravity and pure mathematics.

Suggested reading

The lectures by Lykken (1996) provide a brief introduction to aspects of $N > 1$ supersymmetry. Olive and Witten (1978) first clarified the connection between the BPS condition and extended supersymmetry, in a short and quite readable paper. Harvey (1996) provides a more extensive introduction to monopoles and the BPS condition. The original paper of Seiberg and Witten (1994) is quite clear; Peskin's lectures, from which we have borrowed extensively here, provide a brief and very clear introduction to the subject.

Exercises

(1) Check the supersymmetry commutators in extended supersymmetry, Eq. (15.16).
(2) Rewrite these supersymmetry commutators in a real basis for the Dirac matrices. Using this, verify the BPS inequality.
(3) Check that the spectrum of monopoles and dyons in Eq. (15.23) is invariant under $SL(2, Z)$ transformations.

16 More supersymmetric dynamics

While motivated in part by the hopes of building phenomenologically successful models of particle physics, we have uncovered in our study of supersymmetric theories a rich treasure trove of field theory phenomena. Supersymmetry provides powerful constraints on the dynamics. In this chapter we will discover more remarkable features of supersymmetric field theories. We will first study classes of (super)conformally invariant field theories. Then we will turn to the dynamics of supersymmetric QCD with $N_f \geq N_c$, where we will encounter new, and rather unfamiliar, types of behavior.

16.1 Conformally invariant field theories

In quantum field theory, theories which are classically scale invariant are typically not scale invariant at the quantum level. Quantum chromodynamics is a familiar example. In the absence of quark masses we believe that the theory predicts confinement and has a mass gap. The CP^N models are an example where we were able to show systematically how a mass gap can arise in a scale-invariant theory. In all these cases the breaking of scale invariance is associated with the need to impose a cutoff on the high-energy behavior of the theory. In a more Wilsonian language one needs to specify a scale where the theory is defined, and this requirement breaks the scale invariance.

There is, however, a subset of field theories which are indeed scale invariant. We have seen this in the case of $N = 4$ supersymmetric field theories in four dimensions. In this section we will see that this phenomenon can occur in $N = 1$ theories and will explore some of its consequences. In the next section we will discuss a set of dualities among $N = 1$ supersymmetric field theories, in which conformal invariance plays a crucial role.

In order that a theory exhibit conformal invariance it is necessary that its beta function vanish. At first sight it would seem difficult to use perturbation theory to find such theories. For example, one might try to choose the number of flavors and colors in such a that the one-loop beta function vanishes. But then the two-loop beta function will generally not vanish. One could try to balance the first term against the second, but this would generally require $g^4 \sim g^2$, and there would not be a good perturbation expansion. Banks and Zaks pointed out that one can find such theories by adopting a different strategy. By taking the number of flavors and colors to be large, one can arrange that the coefficient of the one-loop beta function almost vanishes, and can choose the coupling so that it cancels the two-loop beta function. In this situation one can arrange a cancelation perturbatively, order by order. The small parameter is $1/N$, where N is the number of colors.

We can illustrate this idea in the framework of supersymmetric theories with N colors and N_f flavors. The beta function, through two loops, is given by

$$\beta(g) = -\frac{g^3}{16\pi^2} b_0 - \frac{g^5}{(16\pi^2)^2} b_1, \tag{16.1}$$

where

$$b_0 = 3N - N_f, \quad b_1 = 6N^2 - 2NN_F - 4N_F \frac{(N^2-1)}{2N}. \tag{16.2}$$

In the limit of very large N and N_f, we write $N_f = 3N - \epsilon$, where ϵ is an integer of order one. Then, to leading order in $1/N$, the beta function vanishes for a particular coupling, g_0, given by

$$\frac{g_0^2}{16\pi^2} = \frac{\epsilon}{6N^2}. \tag{16.3}$$

Perturbative diagrams behave as $(g^2N)^n$, and g^2N is small. So, at each order, one can make small adjustments in g^2 so as to make the beta function vanish.

A theory in which the beta function vanishes is genuinely conformally invariant. We will not give a detailed discussion of the conformal group here. The exercises at the end of this chapter guide the reader through some features of the conformal group; good reviews are described in the suggested reading. Here we will just mention a few general features and then perform some computations for our Banks–Zaks fixed point theories to verify these.

Without supersymmetry the generators of the conformal group include the Lorentz generators and the translations,

$$M_{\mu\nu} = -i(x_\mu \partial_\nu - x_\nu \partial_\mu), \quad P_\mu = -i\partial_\mu, \tag{16.4}$$

and the generators of "special conformal transformations" and dilatations,

$$K_\mu = -i(x^2 \partial_\mu - 2x_\mu x_\alpha \partial^\alpha), \quad D = ix_\alpha \partial^\alpha. \tag{16.5}$$

In the presence of supersymmetry the group is enlarged. In addition to the bosonic generators above and the supersymmetry generators, there is a group of *superconformal* generators

$$S_\alpha = X_\mu \sigma^\mu_{\alpha\dot{\alpha}} Q^{\dot{\alpha}}. \tag{16.6}$$

We encountered these in our analysis of the zero modes of the Yang–Mills instanton. The superconformal algebra also includes an R symmetry current.

Conformal invariance implies the vanishing of T^μ_μ. In the superconformal case the superconformal generators and the divergence of the R current also vanish. One can prove a relation between the dimension and the R charge:

$$d \geq \frac{3}{2}|R|. \tag{16.7}$$

States for which the inequality is satisfied are known as *chiral primaries*. An interesting case is provided by the fixed point theories introduced above. For these, the charge of the chiral fields, Q and $\bar{Q}$, under the *non-anomalous* symmetry is

$$R_{Q,\bar{Q}} = \frac{N_f - N}{N_f}. \tag{16.8}$$

Assuming that these fields are chiral primaries, it follows that their dimension d satisfies

$$d - 1 = -\frac{3N - N_f}{2N_f} = -\frac{\epsilon}{6N}. \tag{16.9}$$

At weak coupling, however, the anomalous dimensions of these fields are known:

$$\gamma = -\frac{g^2}{16\pi^2}N = -\frac{\epsilon}{6N}. \tag{16.10}$$

In this chapter we will see that supersymmetric QCD, for a range of N_f and N, exhibits conformal fixed points for which the coupling is not small.

16.2 More supersymmetric QCD

We have studied the dynamics of supersymmetric QCD with $N_f < N$ and observed a range of phenomena: non-perturbative effects which lift the degeneracy among different vacua and non-perturbative supersymmetry breaking. In the case $N_f \geq N_c$ there are exact moduli, even non-perturbatively. In the context of phenomenology such theories are probably of no relevance, but Seiberg realized that, from a theoretical point of view, these theories are a bonanza. The existence of moduli implies a great deal of control over the dynamics. One can understand much about the strongly coupled regimes of these theories, allowing insights into non-perturbative dynamics unavailable in theories without supersymmetry. We will be able to answer questions such as: are there unbroken global symmetries in some region of the moduli space? In regions of strong coupling, are there massless composite particles?

16.3 $N_f = N_c$

The case $N_f = N_c$ already raises issues beyond those of $N_f < N_c$. First, we have seen that there is no invariant superpotential that one can write down. As a result, there is an exact moduli space, perturbatively and non-perturbatively. Yet there is still an interesting quantum modification of the theory, first discussed by Seiberg.

Consider, first, the classical moduli space. Now, in addition to the vacua with $Q = \bar{Q}$ (up-to-flavor transformations) which we found previously, we can also have

$$Q = vI, \quad \bar{Q} = 0 \quad \text{or} \quad Q \leftrightarrow \bar{Q}. \tag{16.11}$$

This is referred to as the "baryonic branch", since now the operator

$$B = \epsilon_{i_1 \ldots i_N} \epsilon^{j_1 \ldots j_N} Q_{j_1}^{i_1} \ldots Q_{j_N}^{i_N} \tag{16.12}$$

is non-vanishing (similarly for the corresponding antibaryon branch).

Classically these two sets of possibilities can be summarized in the condition

$$\det \bar{Q}Q = \bar{B}B. \tag{16.13}$$

Now, this condition is subject to quantum modifications. Both sides are completely neutral under the various flavor symmetries; in principle any function of $B\bar{B}$ or the determinant would be permitted as a modification. But we can use anomalous symmetries (with the anomalies canceled by shifts in S) to constrain any possible corrections. Consider, in particular, possible instanton corrections. These are proportional to

$$v^{2N} e^{-8\pi^2/g^2(v)} \sim \Lambda^{2N} \tag{16.14}$$

and transform just like the left-hand side under the anomalous R symmetry for which

$$Q \to e^{i\alpha Q}. \tag{16.15}$$

So, at the quantum level the moduli space satisfies the condition

$$\det \bar{Q}Q - \bar{B}B = c\Lambda^{2N}. \tag{16.16}$$

This is of just the right form to be generated by a one-instanton correction. We will not do the calculation here; it shows that the right-hand side is indeed generated. We can outline the main features. There are now two superconformal zero modes, two supersymmetry zero modes, $4N-4$ zero modes associated with the gluinos in the $(2, N-2)$ representation of the $SU(2) \times SU(N-2)$ subgroup of $SU(N)$ distinguished by the instanton and $2N$ matter zero modes. We want to compute the expectation value of an operator involving N scalars. To obtain a non-vanishing result it is necessary to replace some fields with their classical values. Others must be contracted with Yukawa terms. The scalar field propagators in the instanton background are known, and the full calculation is reasonably straightforward. Because the classical condition which defines the moduli space is modified, the moduli space of the $N_f = N_c$ theory is referred to as the *quantum moduli space*. This phenomenon appears for other choices of gauge group as well.

16.3.1 Supersymmetry breaking in quantum moduli spaces

We have mentioned that, in the (3, 2) model, in the limit where the $SU(2)$ gauge group is the strong group, supersymmetry breaking can be understood as resulting from an expectation value for QL. The QL vev is non-zero since $N = N_f = 2$. The introduction of a larger class of models, in which a quantum moduli space is responsible for dynamical supersymmetry, is due to Intriligator and Thomas.

Consider a model with gauge group $SU(2)$ and four doublets $Q_I, I = 1-4$ (two "flavors"). Classically, this model has a moduli space labeled by the expectation values of the fields, $M_{IJ} = Q_I Q_J$. These satisfy $\text{Pf}\langle M_{IJ}\rangle = 0$[1] but, as have have just seen, the quantum moduli space is different and satisfies

$$\text{Pf}\langle M_{IJ}\rangle = \Lambda^4. \tag{16.17}$$

[1] In this expression, Pf denotes the Pfaffian. The Pfaffian is defined for $2N \times 2N$ antisymmetric matrices; it is essentially the square root of the determinant of the matrix.

Now add a set of singlets S_{IJ} to the model, with superpotential couplings

$$W = \lambda_{IJ} S_{IJ} Q_I Q_J. \tag{16.18}$$

Unbroken supersymmetry now requires

$$\frac{\partial W}{\partial S_{IJ}} = Q_I Q_J = 0. \tag{16.19}$$

However, this is incompatible with the quantum constraint. So on the one head the supersymmetry is broken.

On the other hand the model, classically, has flat directions in which $S_{IJ} = s_{IJ}$ and all the other fields vanish. So one might worry that there is runaway behavior in these directions, similar to that we saw in supersymmetric QCD. However, for large s it turns out that the energy is growing at infinity. This can be established as follows. Suppose all the components of S are large, $S \sim s \gg \Lambda_2$. In this limit the low-energy theory is a pure $SU(2)$ gauge theory. In this theory gluinos condense,

$$\langle\lambda\lambda\rangle = \Lambda_{\rm LE}^3 = \lambda s \Lambda_2^2. \tag{16.20}$$

Here, $\Lambda_{\rm LE}$ is the Λ parameter of the low-energy theory.

At this level, then, the superpotential of the model behaves as

$$W_{\rm eff} \sim \lambda S \Lambda_2^2, \tag{16.21}$$

and the potential is a constant,

$$V = |\Lambda_2|^4 |\lambda|^2. \tag{16.22}$$

The natural scale for the coupling, λ, which appears here is $\lambda(s)$. This is the correct answer in this case and implies that for large s the potential is growing, since λ is not asymptotically free. So the potential has a minimum in a region of small s.

16.3.2 $N_{\rm f} = N_{\rm c} + 1$

For $N_{\rm f} > N_{\rm c}$ the classical moduli space is exact. But again Seiberg has, pointed out a rich set of phenomena and given a classification of the different theories. As in the case $N_{\rm f} < N_{\rm c}$, different phenomena occur for different values of $N_{\rm f}$.

First, we need to introduce a new tool: the 't Hooft anomaly-matching conditions. 't Hooft was motivated by the following question. When one looks at the repetitive structure of the quark and lepton generations, it is natural to wonder whether the quarks and leptons themselves are bound states of some simpler constituents. 't Hooft pointed out that if this idea were correct then the masses of the quarks and leptons would be far smaller than the scale of the underlying interactions; even at that time it was known that if these particles have any structure then it is on scales shorter than 100 GeV^{-1}. 't Hooft argued that this could only be understood if the underlying interactions left an unbroken chiral symmetry.

One could go on and simply postulate that the symmetry is unbroken, but 't Hooft realized that there are strong – and simple – constraints on such a possibility. Assuming that the mechanism is some strongly interacting non-Abelian gauge theory, 't Hooft imagined gauging the global symmetries of the theory. In general the resulting theory would be anomalous, but one could always cancel the anomalies by adding some "spectator" fields, fields transforming under the gauged flavor symmetries but not the underlying strong interactions. Below the confinement scale of the strong interactions the flavor symmetries might be spontaneously broken, giving rise to Goldstone bosons, or there might be massless fermions. In either case the low-energy theory must be anomaly-free, so the anomalies of either the Goldstone bosons or the massless fermions must be the same as in the original theory. 't Hooft added another condition, which he called the "decoupling" condition: he asked what happened if one added mass terms for some of the constituent fermions. He went on to show that these conditions are quite powerful and that it is difficult to obtain a theory with unbroken chiral symmetries.

As we will see, Seiberg conjectured various patterns of unbroken symmetries for susy QCD. For these the 't Hooft anomaly conditions provide a strong self-consistency check. In the case $N_f = N_c$ there is no point in the moduli space at which the chiral symmetries are all unbroken. So we will move on to the case $N_f = N_c + 1$. The global symmetry of the model is

$$SU(N_f)_L \times SU(N_f)_R \times U(1)_B \times U(1)_R \tag{16.23}$$

where, under $U(1)_R$, the quarks and antiquarks transform as

$$Q_f, \bar{Q}_f \to e^{i\alpha/(N+1)} Q_f, \bar{Q}_f. \tag{16.24}$$

In this theory there two sorts of gauge-invariant objects, the mesons, $M_{\bar{f}f} = \bar{Q}_{\bar{f}} Q$, and the baryons, $B_f = \epsilon_f^{\alpha_1 \cdots \alpha_N} \epsilon_{i_1 \cdots i_n} Q_{\alpha_1}^{i_1} Q_{\alpha_2}^{i_2} \cdots Q_{\alpha_N}^{i_N}$. From these we can build a superpotential that is invariant under all the symmetries:

$$W = (\det M - B_{\bar{f}} M_{\bar{f}f} B_f) \frac{1}{\Lambda^{b_0}}. \tag{16.25}$$

As in all our earlier cases, the power of Λ is determined by dimensional arguments but can also be verified by demanding holomorphy in the gauge coupling.

This superpotential has several interesting features. First, it has flat directions, as we would expect, corresponding to the flat directions of the underlying theory. But also, for the first time, there is a vacuum at the point where all the fields vanish, $B = \bar{B} = M = 0$. At this point all the symmetries are unbroken. The 't Hooft anomaly conditions provide an important consistency check on this whole picture. There are several anomalies to check: ($SU(N_f)_L^3$, $SU(N_f)_R^3$, $SU(N_f)_L^2 U(1)_R$, $\mathrm{Tr}\, U(1)_R$, $U(1)_B^2 U(1)_R$, $U(1)_R^3$ etc.). The cancelations are quite non-trivial. In the exercises, the reader will have the opportunity to check these.

Another test comes from considering decoupling. If we add a mass for one set of fields, the theory should reduce to the $N_f = N$ case. As in examples with smaller numbers of fields we take advantage of holomorphy, writing down expressions for small values of the mass and continuing to large values. So we add to the superpotential a term

$$m\bar{Q}_{N+1} Q_{N+1} = m M_{N+1,N+1}. \tag{16.26}$$

We want to integrate out the massive fields. Because of the global symmetry, it is consistent to set $M_{f,N+1}$ to zero, where $f \leq N$. Similarly, it is consistent to set $B_f = 0, f \leq N$. So we take the M and B fields to have the form

$$M = \begin{pmatrix} M & 0 \\ 0 & m \end{pmatrix}, \quad B_f = \begin{pmatrix} 0 \\ \vdots \\ B \end{pmatrix}, \quad \bar{B}_f = \begin{pmatrix} 0 \\ \vdots \\ \bar{B} \end{pmatrix}. \tag{16.27}$$

Consider the equation $\partial W/\partial m = 0$. This yields

$$(\det M - \bar{B}B) = m\Lambda^{b_0} \tag{16.28}$$

or

$$(\det M - \bar{B}B) = m\Lambda^{b_0} = \Lambda_{N_f}^{2N}. \tag{16.29}$$

In the last line we have used the relation between the Λ parameter of the theory with N_f quarks and that with $N_f + 1$ flavors. This is precisely the expression for the quantum-modified moduli space of the N-flavor theory. Decoupling works perfectly here.

16.4 $N_f > N + 1$

The case $N_f > N + 1$ poses new challenges. We might try to generalize our analysis of the previous section. Take, for example, $N_f = N + 2$. Then the baryons are in the second-rank antisymmetric tensor representations of the $SU(N_f)$ gauge groups, B_{fg} and $\bar{B}_{\bar{f}\bar{g}}$. For a term in the superpotential

$$W \sim B_{fg}\bar{B}_{\bar{k}\bar{l}}M^{f\bar{k}}M^{g\bar{l}}, \tag{16.30}$$

this does not respect the non-anomalous R symmetry.

Seiberg suggested a different equivalence. The baryons, in general, have $\tilde{N} = N_f - N$ indices. So baryons in the same representation of the flavor group can be constructed in a theory with gauge group $SU(\tilde{N})$ and quarks q_f, $\bar{q}_f$ in the fundamental representation. Seiberg postulated that, in the infrared, this theory is dual to the original theory. This is not quite enough. One needs to add a set of gauge-singlet meson fields $M_{\bar{f}f}$, with superpotential

$$W = q^{\bar{f}}M_{\bar{f}f}q^{f}. \tag{16.31}$$

To check this picture we can first check that the symmetries match. There is an obvious $SU(N_f)_L \times SU(N_f)_R \times U(1)_B$. There is also a non-anomalous $U(1)_R$ symmetry. It is important that the dual theory is not asymptotically free, i.e. that it is weakly coupled in the infrared. This is the case for $N > 3N_f/2$. Again, this duality can only apply for a range of N_f and N.

There are a number of checks on the consistency of this picture. Holomorphic decoupling is again one of the most persuasive. Take the case $N_f = N + 2$, so that the dual gauge group is $SU(2)$. In this case, working in the flat directions of the $SU(2)$ theory, one can do an instanton computation. One finds a contribution to the superpotential

$$W_{\text{inst}} = \det M. \tag{16.32}$$

This is consistent with all the symmetries; it is not difficult to see that one can close up all the fermion zero modes with elements of M and q. So one has a superpotential

$$\int d^2\theta (qM\bar{q} - \det M). \tag{16.33}$$

16.5 $N_f \geq 3N/2$

We have noted that Seiberg's duality cannot persist beyond $N_f = 3N/2$. Seiberg also made a proposal for the behavior of the theory in this regime: for $3/2N \leq N_f \leq 3N$ the theories are conformally invariant. Our Banks–Zaks fixed point lies in one corner of this range. As a further piece of evidence, consider the dimension of the operator $\bar{Q}Q$. Under the non-anomalous R symmetry, we have

$$Q \to \exp\left(i\alpha \frac{N_f - N}{2N_f}\right) Q. \tag{16.34}$$

If the theory is superconformal, the dimension of this chiral operator satisfies $d = 3R/2$. As explained in Appendix D, the exact beta function of the theory is given by

$$\beta = -\frac{g^3}{16\pi^2} \frac{3N - N_F + N_f \gamma(g^2)}{1 - N(g^2/8\pi^2)}. \tag{16.35}$$

By assumption this is zero, so

$$\gamma = -\frac{3N - N_f}{N_f}. \tag{16.36}$$

The dimension of $\bar{Q}Q$ is $2 + \gamma$, which is precisely $3R/2$.

We will not pursue this subject further, but there is further evidence that one can provide for all these dualities. They can also be extended to other gauge groups.

Suggested reading

The original papers of Seiberg (1994a,b, 1995a,b; see also Seiberg and Witten 1994) are quite accessible and constitute essential reading on these topics, as the review by Intriligator and Seiberg (1996). Good introductions are provided by the lecture notes of Peskin (1997) and Terning (2003). The use of quantum moduli spaces to break supersymmetry was introduced in Intriligator and Thomas (1996).

Exercises

(1) Discuss the renormalization of the composite operator $\bar{Q}Q$, and verify that the relation $d = 3R/2$ is again satisfied.

(2) Check the anomaly cancelation for the case $N_f = N + 1$. You may want to use an algebraic manipulation program, such as MAPLE or Mathematica, to expedite the algebra.

17 An introduction to general relativity

Even as the evidence for the Standard Model became stronger and stronger in the 1970s and beyond, so the evidence for general relativity grew in the latter half of the twentieth century. Any discussion of the Standard Model and physics beyond it must confront Einstein's theory at two levels. First, general relativity and the Standard Model are very successful at describing the history of the universe and its present behavior on large scales. General relativity gives rise to the big bang theory of cosmology, which, coupled with our understanding of atomic and nuclear physics, explains – indeed predicted – features of the observed universe. But there are features of the observed universe which cannot be accounted for within the Standard Model and general relativity. These include dark matter and dark energy, the origin of the asymmetry between matter and antimatter, the origin of the seeds of cosmic structure (inflation) and more. Apart from these observational difficulties, there are also serious questions of principle. We cannot simply add Einstein's theory onto the Standard Model. The resulting structure is not renormalizable and cannot represent in any sense a complete theory. Black holes, when combined with quantum mechanics, raise further puzzles. In this book we will encounter both these aspects of Einstein's theory. Within extensions of the Standard Model, in the next few chapters we will attempt to explain some features of the observed universe. The second, more theoretical, level is addressed in the third part of this book. String theory, our most promising framework for a comprehensive theory of all interactions, encompasses general relativity in an essential way; some would even argue that what we mean by string theory is the quantum theory of general relativity.

The purpose of this chapter is to introduce some concepts and formulas that are essential to the applications of general relativity in this text. No previous knowledge of general relativity is assumed. We will approach the subject from the perspective of field theory, focusing on the dynamical degrees of freedom and the equations of motion. We will not give as much attention to the beautiful – and conceptually critical – geometric aspects of the subject, though we will return to some of these in the chapters on string theory. Those interested in a more in-depth treatment of general relativity will eventually want to study some of the excellent texts listed in the suggested reading at the end of the chapter.

Einstein put forward his principle of relativity in 1905. At that time, one might quip, half the known laws, those of electricity and magnetism, already satisfied this principle with no modification. The other half, Newton's laws, did not. In considering how one might reconcile gravitation and special relativity, Einstein was guided by the observed equality of gravitational and inertial mass. Inertia has to do with how objects move in space–time in response to forces. Operationally, the way we define space–time, our measurements of length, time, energy and momentum, depends crucially on this notion. The fact that gravity couples to precisely this mass suggests that gravity has a deep connection to the nature of space–time. Considering this equivalence, Einstein noted that an observer in a freely falling elevator (in a uniform gravitational field) would write down the same laws of nature as an observer in an inertial frame without gravity. Consider, for example, an elevator full of particles interacting through a potential $V(\vec{x}_i - \vec{x}_j)$. In the inertial frame,

$$m\frac{d^2\vec{x}_i}{dt^2} = m\vec{g} - \vec{\nabla}_i V(\vec{x}_i - \vec{x}_j). \tag{17.1}$$

The coordinates of the accelerated observer are related to those of the inertial observer by

$$\vec{x}_i = \vec{x}_i' + \frac{1}{2}\vec{g}t^2; \tag{17.2}$$

so, substituting with the equations of motion (17.1), we obtain

$$m\frac{d^2\vec{x}_i'}{dt^2} = -\vec{\nabla}_i V(\vec{x}_i' - \vec{x}_j'). \tag{17.3}$$

Einstein abstracted from this thought experiment a strong version of the *equivalence principle*: the equations of motion should have the same form in any frame, inertial or not. In other words, it should be possible to write down the laws so that in any two coordinate systems, x^μ and $x'^\mu(x)$, they take the same form. This is a strong requirement. We will see that it is similar to gauge invariance, where the requirement that the laws take the same form after gauge transformations determines the dynamics.

17.1 Tensors in general relativity

To implement the equivalence principle, we begin by thinking about the invariant element d_s of distance. In an inertial frame, in special relativity,

$$ds^2 = d\vec{x}^2 - dt^2 = \eta_{\mu\nu}dx^\mu dx^\nu. \tag{17.4}$$

Note here that we have changed the sign of the metric, as we said we would do, from that used earlier in this text. This is the convention of most workers and texts in general relativity and string theory. The above coordinate transformation for the accelerated observer alters the line element. This suggests we consider the generalization

$$ds^2 = g_{\mu\nu}(x)dx^\mu dx^\nu. \tag{17.5}$$

The *metric tensor* $g_{\mu\nu}$ encodes the physical effects of gravitation. We will see that there is a non-trivial gravitational field when we cannot find coordinates which make $g_{\mu\nu} = \eta_{\mu\nu}$ everywhere.

To develop a dynamical theory, we would like to write down invariant actions (which will yield covariant equations). This problem has two parts. We need to couple the fields that we already have to the metric in an invariant way. We also require an analog of the field strength for gravity, which will determine the dynamics of $g_{\mu\nu}$ in much the same way as the field strength $F_{\mu\nu}$ determines the dynamics of the gauge field A_μ. This object is the

Riemann tensor, $\mathcal{R}^{\mu}_{\nu\rho\sigma}$. We will see later that the formal analogy can be made very precise: An object, the spin connection ω_μ, constructed out of the metric tensor plays the role of A^μ. The close analogy will also be seen when we discuss Kaluza–Klein theories, where higher-dimensional general coordinate transformations become lower-dimensional gauge transformations.

We first describe how derivatives and $g_{\mu\nu}$ transform under coordinate transformations. Writing

$$x^\mu = x^\mu(x') \tag{17.6}$$

we have

$$\partial'_\mu \phi(x') = \frac{\partial x^\rho}{\partial x^{\mu'}} \partial_\rho \phi(x) = \Lambda_\mu^{\ \rho}(x) \partial_\rho \phi(x). \tag{17.7}$$

An object which transforms like $\partial_\rho \phi$ is said to be a covariant vector. An object which transforms like $\partial_{\rho_1}\phi \partial_{\rho_2}\phi \cdots \partial_{\rho_n}\phi$ is said to be an nth rank covariant tensor; $g_{\mu\nu}$ is an important example of such a tensor. We can obtain the transformation law for $g_{\mu\nu}$ from the invariance of the line element:

$$g'_{\mu\nu} dx^{\mu'} dx^{\nu'} = g_{\mu\nu} \frac{\partial x^\mu}{\partial x^{\rho'}} \frac{\partial x^\nu}{\partial x^{\sigma'}} dx^{\rho'} dx^{\sigma'}, \tag{17.8}$$

so

$$g'_{\mu\nu} = g_{\rho\sigma} \frac{\partial x^\rho}{\partial x^{\mu'}} \frac{\partial x^\sigma}{\partial x^{\nu'}}. \tag{17.9}$$

Now, dx^μ transforms according to the inverse of Λ:

$$dx'^\mu = \frac{\partial x'^\mu}{\partial x^\rho} dx^\rho, \tag{17.10}$$

where dx^μ is said to be a contravariant vector. Indices can be raised and lowered with $g_{\mu\nu}$; if V^ν is a contravariant vector then $g_{\mu\nu}V^\nu$ transforms as a covariant vector, for example.

Another important object is the volume element, d^4x. Under a coordinate transformation,

$$d^4x = \left| \frac{\partial x}{\partial x'} \right| d^4x'. \tag{17.11}$$

The object in between the vertical lines is the Jacobian of the coordinate transformation, $|\det \Lambda|$. The quantity $\sqrt{-\det g}$ transforms in exactly the opposite fashion. So the four-volume, is invariant.

$$\int d^4x \sqrt{-\det g}. \tag{17.12}$$

We will consider a real scalar field ϕ. The action, before the inclusion of gravity, is

$$S = \int d^4x \frac{1}{2}(-\partial_\mu\phi\, \partial_\nu\phi\, \eta^{\mu\nu} - m^2\phi^2). \tag{17.13}$$

To make this invariant we can replace $\eta^{\mu\nu}$ by $g^{\mu\nu}$ and include a factor $\sqrt{\det(-g)}$ along with d^4x. Then

$$S = \int d^4x \sqrt{\det(-g)} \frac{1}{2}(-\partial_\mu\phi\, \partial_\nu\phi\, g^{\mu\nu} - m^2\phi^2). \tag{17.14}$$

The equations of motion should be *covariant*. They must generalize the equation

$$\partial^2 \phi = -V'(\phi). \tag{17.15}$$

The first derivative of ϕ, we have seen, transforms as a vector, V_μ, under coordinate transformations, but the second derivative does not transform simply:

$$\begin{aligned}\partial_\mu V_\nu &= \partial_\mu \left(\frac{\partial x^{\rho'}}{\partial x^\nu} V'_\rho \right) \\ &= \frac{\partial x^{\rho'}}{\partial x^\nu} \frac{\partial x^{\sigma'}}{\partial x^\mu} \partial'_\sigma V'_\rho + \frac{\partial^2 x^{\rho'}}{\partial x^\mu \partial x^\nu} V_\rho.\end{aligned} \tag{17.16}$$

To compensate for the extra, inhomogeneous, term we need a covariant derivative, as in gauge theories. Rather than look at the equations of motion directly, however, we can integrate the scalar field Lagrangian by parts to obtain *second* derivatives. This yields

$$\sqrt{-g}(g^{\mu\nu}\partial_\mu\partial_\nu\phi + \partial_\mu g^{\mu\nu}\partial_\nu\phi) + g^{\mu\nu}\partial_\mu\sqrt{-g}\phi\,\partial_\nu\phi. \tag{17.17}$$

To bring this into a convenient form, we need a formula for the derivative of a determinant. We can work this out using a trick we have used repeatedly in the case of the path integral. Write

$$\det M = \exp(\mathrm{Tr} \ln M) \tag{17.18}$$

so that

$$\begin{aligned}\det(M + \delta M) &\approx \exp[\mathrm{Tr} \ln M + \ln(1 + M^{-1}\delta M)] \\ &= (\det M)(1 + M^{-1}\delta M).\end{aligned} \tag{17.19}$$

Thus, for example,

$$\frac{d \det M}{dM_{ij}} = M_{ij}^{-1} \det M. \tag{17.20}$$

Putting all this together, we have the quadratic term in the action for a scalar field:

$$\phi \left(g^{\mu\nu}\partial_\mu\partial_\nu\phi + \partial_\mu g^{\mu\nu}\partial_\nu\phi + g^{\mu\nu}\frac{1}{2}g^{\rho\sigma}\partial_\mu g_{\rho\sigma}\partial_\nu\phi \right). \tag{17.21}$$

Writing this as

$$\phi g^{\mu\nu} D_\mu \partial_\nu \phi, \tag{17.22}$$

we have for the covariant derivative

$$D_\mu V_\nu = \partial_\mu V_\nu - \Gamma^\lambda_{\mu\nu} V_\lambda. \tag{17.23}$$

Here

$$\Gamma^\lambda_{\mu\nu} = \frac{1}{2} g^{\lambda\rho}(\partial_\mu g_{\rho\nu} + \partial_\nu g_{\rho\mu} - \partial_\rho g_{\mu\nu}). \tag{17.24}$$

Note that $\Gamma^{\lambda}_{\mu\nu}$ is symmetric in μ, ν. The covariant derivative is often denoted by a semicolon and a Greek letter in the subscript or superscript:

$$D_{\mu} V_{\nu} \equiv V_{\mu;\nu}. \tag{17.25}$$

The reader can check that

$$\Gamma^{\lambda}_{\mu\nu} = \Gamma^{\lambda\prime}_{\mu\nu} - \frac{\partial^2 x^{\lambda}}{\partial x^{\mu} \partial x^{\nu}}, \tag{17.26}$$

which just compensates the extra term in the transformation law (17.16). Here Γ is known as the *affine connection* (the components of Γ are also sometimes referred to as the Christoffel symbols and Γ itself as the Christoffel connection; it is sometimes written as $\left\{ {}_{\nu}{}^{\mu}{}_{\rho} \right\}$). With this definition,

$$D_{\mu} V_{\nu} = \partial_{\mu} V_{\nu} - \Gamma^{\lambda}_{\mu\nu} V_{\lambda} \tag{17.27}$$

transforms like a tensor with two indices, $V_{\mu\nu}$. Similarly, acting on contravariant vectors:

$$D_{\mu} V^{\nu} = \partial_{\mu} V^{\nu} + \Gamma^{\nu}_{\mu\lambda} V^{\lambda} \tag{17.28}$$

transforms correctly. You can also check that $V_{\mu;\nu;\rho}$ transforms as a third-rank covariant tensor, and so on.

To get some practice, and to see how the metric tensor can encode gravity, let us use the covariant derivative to describe the motion of a free particle. In an inertial frame, without gravity,

$$\frac{d^2 x^{\mu}}{d\tau^2} = 0, \tag{17.29}$$

where $\tau = g_{\mu\nu} dx^{\mu} dx^{\nu}$ is the proper time is made covariant by first rewriting it as

$$\frac{dx^{\rho}}{d\tau} \frac{\partial}{\partial x^{\rho}} \left(\frac{dx^{\mu}}{d\tau} \right) = 0. \tag{17.30}$$

We need to replace the derivative $\partial / \partial x^{\rho}$ by a covariant derivative. The covariant version of the left-hand side of Eq. (17.29) is then

$$\frac{dx^{\rho}}{d\tau} D_{\rho} \left(\frac{\partial x^{\mu}}{\partial \tau} \right). \tag{17.31}$$

This becomes, using Eq. (17.28),

$$\frac{\partial x^{\rho}}{\partial \tau} \frac{\partial^2 x^{\mu}}{\partial x^{\rho} \partial \tau} + \Gamma^{\mu}_{\rho\sigma} \frac{\partial x^{\sigma}}{\partial \tau} \frac{\partial x^{\rho}}{\partial \tau}. \tag{17.32}$$

So the equation of motion is

$$\frac{d^2 x^{\mu}}{d\tau^2} + \Gamma^{\mu}_{\rho\sigma} \frac{\partial x^{\sigma}}{\partial \tau} \frac{\partial x^{\rho}}{\partial \tau} = 0. \tag{17.33}$$

This is known as the geodesic equation. Viewed as Euclidean equations, the solutions are geodesics. For a sphere embedded in flat three-dimensional space, for example, the solutions of this equation are easily seen to be great circles. We should be able to recover

Newton's equation for a weak gravitational field. For a weak static gravitational field we might expect that

$$g_{\mu\nu} = \eta_{\mu\nu} + h_{\mu\nu}, \tag{17.34}$$

with $h_{\mu\nu}$ small. Since the gravitational potential in Newton's theory is a scalar, we might further guess that

$$g_{00} = -(1 + 2\phi), \quad g_{ij} = \delta_{ij}. \tag{17.35}$$

Then the non-vanishing components of the affine connection are

$$\begin{aligned} \Gamma^i_{00} &= \frac{1}{2} g^{ij} \left(\partial_0 g_{i0} + \partial_0 g_{0i} - \partial_i g_{00} \right) \\ &= \partial_i \phi \end{aligned} \tag{17.36}$$

and, similarly,

$$\Gamma^0_{0i} = -\partial_i \phi. \tag{17.37}$$

In the non-relativistic limit we can replace τ by t, and we have the equation of motion

$$\frac{d^2 x^i}{dt^2} = -\partial_i \phi. \tag{17.38}$$

17.2 Curvature

Using the covariant derivative we can construct actions for scalars and gauge fields. Fermions require some additional machinery; we will discuss this towards the end of the chapter. Instead, we turn to the problem of finding an action for the gravitational field itself. In the case of gauge fields the crucial object was the field strength, $F_{\mu\nu} = [D_\mu, D_\nu]$. For the gravitational field we will also work with the commutator of covariant derivatives operators. We write

$$[D_\mu, D_\nu] V_\rho = \mathcal{R}^\sigma_{\rho\mu\nu} V_\sigma, \tag{17.39}$$

where $\mathcal{R}$ is known as the *Riemann tensor* or *curvature tensor*. For a Euclidean space it measures what we would naturally call the curvature of the space. It is straightforward to work out an expression for $\mathcal{R}$ in terms of the affine connection:

$$\mathcal{R}^\lambda_{\mu\nu\kappa} = \partial_\kappa \Gamma^\lambda_{\mu\nu} - \partial_\nu \Gamma^\lambda_{\mu\kappa} + \Gamma^\eta_{\mu\nu} \Gamma^\lambda_{\kappa\eta} - \Gamma^\eta_{\mu\kappa} \Gamma^\lambda_{\nu\eta}. \tag{17.40}$$

Unlike F, which is first order in derivatives of A, the Riemann tensor $\mathcal{R}$ is second order in derivatives of g. As a result the gravitational action will be first order in $\mathcal{R}$.

Note that $\mathcal{R}$ transforms as a tensor under coordinate transformations. It has important symmetry and cyclicity properties. These are most conveniently described by lowering the first index on $\mathcal{R}$:

$$\mathcal{R}_{\lambda\mu\nu\kappa} = \mathcal{R}_{\nu\kappa\lambda\mu}, \tag{17.41}$$

$$\mathcal{R}_{\lambda\mu\nu\kappa} = -\mathcal{R}_{\mu\lambda\nu\kappa} = -\mathcal{R}_{\lambda\mu\kappa\nu} = \mathcal{R}_{\mu\lambda\kappa\nu}, \tag{17.42}$$

$$\mathcal{R}_{\lambda\mu\nu\kappa} + \mathcal{R}_{\lambda\kappa\mu\nu} + \mathcal{R}_{\lambda\nu\kappa\mu} = 0. \tag{17.43}$$

Starting with $\mathcal{R}$ we can define other tensors. The most important is the *Ricci tensor*. This has only two indices:

$$\mathcal{R}_{\mu\kappa} = g^{\lambda\nu}\mathcal{R}_{\lambda\mu\nu\kappa}. \tag{17.44}$$

The Ricci tensor is symmetric:

$$\mathcal{R}_{\mu\kappa} = \mathcal{R}_{\kappa\mu}. \tag{17.45}$$

Also very important is the *Ricci scalar*:

$$\mathcal{R}_{\rm s} = g^{\mu\kappa}\mathcal{R}_{\mu\kappa}. \tag{17.46}$$

Note that the Riemann tensor $\mathcal{R}$ also satisfies an important identity, similar to the Bianchi identity for $F^{\mu\nu}$ (which gives the homogeneous Maxwell equations):

$$\mathcal{R}_{\lambda\mu\nu\kappa;\eta} + \mathcal{R}_{\lambda\mu\eta\nu;\kappa} + \mathcal{R}_{\lambda\mu\kappa\eta;\nu} = 0. \tag{17.47}$$

17.3 The gravitational action

Having introduced, through the Riemann tensor $\mathcal{R}$, a description of curvature, we are in a position to write down a generally covariant action for the gravitational field. Terms linear in $\mathcal{R}$, as we noted, will be second order in the derivatives of the metric, so they can provide a suitable action. The action must be a scalar, so we take

$$S_{\rm grav} = \frac{1}{2\kappa^2}\int d^4x\sqrt{-g}\,\mathcal{R}. \tag{17.48}$$

To obtain the equations of motion we need to vary the complete action, including the parts involving matter fields, with respect to $g_{\mu\nu}$. We first consider the variation of the terms involving matter fields. The variation of the matter action with respect to $g_{\mu\nu}$ turns out to be nothing other than the stress–energy tensor, $T^{\mu\nu}$. Once one knows this fact, this gives what is often the easiest way to find the stress–energy tensor for a system. To see that this identification is correct, we first show that $T_{\mu\nu}$ is covariantly conserved, i.e.

$$D_\nu T^{\nu\mu} = T^{\mu\nu}{}_{;\nu} = 0. \tag{17.49}$$

By assumption the fields solve the equations of motion in the gravitational background, so the variation of the action, for any variation of the fields, is zero. Consider, then, a space–time translation:

$$x^{\mu\prime} = x^\mu + \epsilon^\mu. \tag{17.50}$$

Starting with

$$g'_{\mu\nu}(x') = \frac{\partial x^\rho}{\partial x^{\mu\prime}} g_{\rho\sigma} \frac{\partial x^\sigma}{\partial x^{\nu\prime}}, \tag{17.51}$$

we have

$$g'_{\mu\nu}(x+\epsilon) = g_{\mu\nu}(x) - \partial_\mu \epsilon^\rho g_{\rho\nu} - \partial_\nu \epsilon^\sigma g_{\sigma\mu}. \tag{17.52}$$

Thus

$$\delta g_{\mu\nu}(x) = -g_{\mu\lambda}\partial_\nu \epsilon^\lambda - g_{\lambda\nu}\partial_\mu \epsilon^\lambda - \partial_\lambda g_{\mu\nu}\epsilon^\lambda. \tag{17.53}$$

Defining

$$\frac{\delta S_{\text{matt}}}{\delta g_{\mu\nu}} = T^{\mu\nu}, \tag{17.54}$$

under this particular variation of the metric we have

$$\delta S_{\text{matt}} = -\int d^4x \sqrt{-g} T^{\mu\nu} \left(g_{\mu\lambda}\partial_\nu \epsilon^\lambda + g_{\lambda\nu}\partial_\mu \epsilon^\lambda + \partial_\lambda g_{\mu\nu}\epsilon^\lambda\right). \tag{17.55}$$

Integrating the first two terms by parts and using the symmetry of the metric (and consequently the symmetry of $T^{\mu\nu}$), we obtain

$$\delta S_{\text{matt}} = \int d^4x \left[\partial_\mu (T^{\mu\lambda}\sqrt{-g}) - \frac{1}{2}\partial_\lambda g_{\mu\nu} T^{\mu\nu}\sqrt{-g}\right]\epsilon^\lambda. \tag{17.56}$$

The coefficient of ϵ^λ vanishes for fields which obey the equations of motion; this object is $T^{\mu\nu}{}_{;\mu}$. The reader can verify this last identification painstakingly or by noting that

$$\Gamma^\mu_{\mu\lambda} = \frac{1}{\sqrt{-g}}\partial_\lambda \sqrt{g}; \tag{17.57}$$

so, for a general vector, for example, we have

$$V^\mu{}_{;\mu} = \frac{1}{\sqrt{-g}}\partial_\mu \left(\sqrt{-g}V^\mu\right) \tag{17.58}$$

and similarly for higher-rank tensors.

As a check, consider the stress tensor for a free massive scalar field. Once more, the action is

$$S = \int d^4x \sqrt{-g}\left(-\frac{1}{2}g^{\mu\nu}\partial_\mu \phi\, \partial_\nu \phi - \frac{1}{2}m^2\phi^2\right). \tag{17.59}$$

So, recalling our formula for the variation of the determinant,

$$T_{\mu\nu} = \frac{1}{2}\partial_\mu \phi\, \partial_\nu \phi - \frac{1}{4}g_{\mu\nu}(g^{\rho\sigma}\partial_\rho \phi\, \partial_\sigma \phi - m^2\phi^2). \tag{17.60}$$

To find the full gravitational equation – *Einstein's equation* – we need to vary also the gravitational term in the action. This is best done by explicitly constructing the variation of the curvature tensor under a small variation of the field. We leave the details for the exercises, and merely quote the final result:

$$\mathcal{R}_{\mu\nu} - \frac{1}{2}g_{\mu\nu}\mathcal{R}_s = \kappa^2 T_{\mu\nu}. \tag{17.61}$$

We will consider many features of this equation, but it is instructive to see how we obtain Newton's expression for the gravitational field, in the limit where gravity is not too strong. We have already argued that in this case we can write

$$g_{00} = -(1+2\phi), \quad g^{ij} = \delta_{ij}. \tag{17.62}$$

As we have seen, the non-vanishing components of the connection are

$$\Gamma^i_{00} = \partial_i \phi, \quad \Gamma^0_{i0} = -\partial_i \phi. \tag{17.63}$$

Correspondingly, the non-zero components of the Riemann curvature tensor are

$$\mathcal{R}^i_{00j} = \partial_i \partial_j \phi = -\mathcal{R}^i_{0j0} = \mathcal{R}^0_{ij0}, \tag{17.64}$$

where the relations between the various components follow from the symmetries of the curvature tensor. From these we can construct the Ricci tensor and the Ricci scalar:

$$\mathcal{R}_{00} = \nabla^2 \phi, \quad \mathcal{R}_s = -\nabla^2 \phi. \tag{17.65}$$

So, we obtain

$$-\nabla^2 \phi = \kappa^2 T_{00}. \tag{17.66}$$

Note that from this we can identify Newton's gravitational constant in terms of κ,

$$G_N = \frac{\kappa^2}{8\pi}. \tag{17.67}$$

17.4 The Schwarzschild solution

Not long after Einstein wrote down his equations for general relativity, Schwarzschild constructed the solution of the equations for a static isotropic metric. Such a metric can be taken to have the form

$$ds^2 = -B(r)dt^2 + A(r)dr^2 + r^2(d\theta^2 + \sin^2\theta\, d\phi^2). \tag{17.68}$$

Actually, rotational invariance would allow other terms. In terms of vectors $d\vec{x}$ the most general metric has the form

$$-B(r)dt^2 + D(r)\vec{x}\cdot d\vec{x}dt + C(r)d\vec{x}\cdot d\vec{x} + D(r)(\vec{x}\cdot d\vec{x})^2. \tag{17.69}$$

By a sequence of coordinate transformations, however, one can bring the metric to the form (17.68).

We will solve Einstein's equations with $T_{\mu\nu} = 0$. Corresponding to ds^2, we have the non-vanishing metric components

$$g_{rr} = A(r), \quad g_{\phi\phi} = r^2 \sin^2\theta, \quad g_{tt} = -B(r), \quad g_{\theta\theta} = r^2. \tag{17.70}$$

Our goal is to determine A and B. The equations for them follow from Einstein's equations. We first need to evaluate the non-vanishing Christoffel symbols. This is done in the exercises. While straightforward, the calculation of the connection and the curvature is slightly tedious, and this is an opportunity to practise using the computer packages described in the exercises. The non-vanishing components of the affine connection are

$$\Gamma^r_{rr} = \frac{1}{2A(r)}A'(r), \quad \Gamma^r_{\theta\theta} = -\frac{r}{A(r)}, \quad \Gamma^r_{\phi\phi} = -\frac{r\sin^2\theta}{A(r)},$$

$$\Gamma^r_{\phi\phi} = \frac{r\sin^2\theta}{A(r)}, \quad \Gamma^r_{tt} = \frac{1}{2A(r)}B'(r), \tag{17.71}$$

where the primes denote derivatives with respect to r. Similarly,

$$\Gamma^\theta_{r\phi} = \Gamma^\theta_{\theta r} = \frac{1}{r}, \quad \Gamma^\theta_{\phi\phi} = -\sin\theta\,\cos\theta,$$

$$\Gamma^\phi_{\phi r} = \Gamma^\phi_{r\phi} = \frac{1}{r}, \quad \Gamma^\phi_{\phi\theta} = \Gamma^\phi_{\theta\phi} = \cos\theta,$$

$$\Gamma^t_{tr} = \Gamma^t_{rt} = \frac{B'}{2B}. \tag{17.72}$$

The non-vanishing components of the Ricci tensor are

$$\mathcal{R}_{rr} = \frac{B''}{2B} - \frac{1}{4}\frac{B''}{B}\left(\frac{A'}{A} + \frac{B'}{B}\right) - \frac{1}{r}\frac{A'}{A}, \tag{17.73}$$

$$\mathcal{R}_{\theta\theta} = -1 + \frac{r}{2A}\left(-\frac{A'}{A} + \frac{B'}{B}\right) + \frac{1}{A}, \tag{17.74}$$

$$\mathcal{R}_{\phi\phi} = \sin^2\theta\,\mathcal{R}_{\theta\theta}, \quad \mathcal{R}_{tt} = -\frac{B''}{2A} + \frac{1}{4}\frac{B'}{A}\left(\frac{A'}{A} + \frac{B'}{B}\right) - \frac{1}{r}\frac{B'}{A}. \tag{17.75}$$

For empty space, Einstein's equation reduces to

$$\mathcal{R}_{\mu\nu} = 0. \tag{17.76}$$

We will require that, asymptotically, the space–time is just flat Minkowski space, so we will solve these equations with the requirement that

$$A_{r\to\infty} = B_{r\to\infty} = 1. \tag{17.77}$$

Examining the components of the Ricci tensor we see that it is enough to set $\mathcal{R}_{rr} = \mathcal{R}_{\theta\theta} = \mathcal{R}_{tt} = 0$. We can simplify the equations with a little cleverness:

$$\frac{\mathcal{R}_{rr}}{A} + \frac{\mathcal{R}_{tt}}{B} = -\frac{1}{rA}\left(\frac{A'}{A} + \frac{B'}{B}\right). \tag{17.78}$$

From this it follows that $A = 1/B$. Then, from $\mathcal{R}_{\theta\theta} = 0$, we have

$$\frac{d}{dr}(rB) - 1 = 0. \tag{17.79}$$

Thus it follows that

$$rB = r + \text{const}. \tag{17.80}$$

Now $B = -g_{tt}$, so, at a distance far from the origin, where the space–time is nearly flat, $B = 1 + 2\phi$, where ϕ is the gravitational potential. Hence:

$$B(r) = 1 - \frac{2MG}{r}, \quad A(r) = \left(1 - \frac{2MG}{r}\right)^{-1}. \tag{17.81}$$

17.5 Features of the Schwarzschild metric

Finally, then, we have the Schwarzschild metric:

$$ds^2 = -\left(1 - \frac{2MG_N}{r}\right) dt^2 + \left(1 - \frac{2MG_N}{r}\right)^{-1} dr^2 + r^2 d\theta^2 + r^2 \sin^2\theta \, d\phi^2. \tag{17.82}$$

Far from the origin, this clearly describes an object of mass M. While so far we have discussed the energy–momentum tensor for matter, we have not yet discussed the energy of gravitation. The situation is similar to the problem of defining charge in a gauge theory. There, the most straightforward definition involves using the asymptotic behavior of the fields to determine the total charge. In gravity, the energy is similar. There is no invariant local definition of the energy density. But in spaces that are asymptotically flat, one can give a global notion of the energy, known as the Arnowitt, Deser and Misner (ADM) energy. Only the $1/r$ behavior of the fields enters. We will not review this here but, not surprisingly, in the present case this energy P^0 is equal to M.

The curvature of space–time near a star yields observable effects. Einstein, when he first published his theory, proposed two tests of the theory: the bending of light by the Sun and the precession of Mercury's perihelion. In the latter case the theory accounted for a known anomaly in the motion of the planet; the prediction of the bending of light was soon confirmed.

A striking feature of this metric is that it becomes singular at a particular value of r, known as the Schwarzschild radius (the horizon), given by

$$r_{\rm h} = 2MG_N. \tag{17.83}$$

At this point the coefficient of dr^2 diverges, and that of dt^2 vanishes. Both change sign, in some sense reversing the roles of r and t. This singularity is a bit of a fake. No component of the curvature becomes singular. One can exhibit this by choosing coordinates in which the metric is completely non-singular (see the exercises at the end of the chapter).

For most realistic objects, such as planets and typical stars, the $r_{\rm h}$ value is well within the star, where surely it is important to use a more realistic model of $T_{\mu\nu}$. But there are systems in nature where the "material" lies well within the Schwarzschild radius. These systems are known as *black holes*. The known black holes arise from the collapse of very massive stars. It is conceivable that smaller black holes arise from more microscopic processes. These systems have striking properties. Classically, light cannot escape from the region within the horizon; the curvature singularity at the origin is real. Black holes are nearly featureless. Classically, an external observer can only determine the mass, charge and angular momentum of the black hole, however complex the system which may have preceded it.

Bekenstein pointed out that the horizon area has peculiar properties and behaves much like a thermal system. Most importantly, it obeys a relation analogous to the second law of thermodynamics,

$$dA > 0. \tag{17.84}$$

Identifying the area with an entropy suggests that one can associate a temperature T_λ with the black hole, known as the *Hawking temperature*. The black hole horizon is a sphere of area $4\pi r_{\rm h}^2$. So one might guess, on dimensional grounds, that

$$T_{\rm h} = \frac{1}{8\pi G_N M}. \tag{17.85}$$

The precise constant does not follow from this argument. The reader is invited to work through an heuristic path-integral derivation in the exercises.

Quantum mechanically, Hawking showed that this temperature has a microscopic significance. When one studies quantum fields in the gravitational background, one finds that particles do escape from the black hole. These particles have a thermal spectrum with characteristic temperature $T_{\rm h}$. This phenomenon is known as *Hawking radiation*.

These features of black holes raise a number of conceptual questions. For the black hole at the center of the galaxy, for example, with mass millions of times greater than the Sun, the Hawking temperature is ludicrously small. Correspondingly, the Hawking radiation is totally irrelevant. But one can imagine microscopic black holes which would evaporate in much more modest periods of time. This raises a puzzle. The Hawking radiation is strictly thermal. So one could form a black hole, say, in the collapse of a small star. The initial star is a complex system, with many features. The black hole is nearly featureless. Classically, however, one might imagine that some memory of the initial state of the system is hidden behind the horizon; this information would simply be inaccessible to the external observer. But owing to the evaporation, the black hole and its horizon eventually disappear. One is left with just a thermal bath of radiation, with features seemingly determined by the temperature (and therefore the mass). Hawking suggested that this information paradox posed a fundamental challenge for quantum mechanics: it would seem that pure states could evolve into mixed states, through the formation of a black hole. For many years this question was the subject of serious debate. One might respond to Hawking's suggestion by saying that the information is hidden in subtle correlations in the radiation, as would be the case for the burning of, say, a lump of coal initially in a pure state. But more careful consideration indicates that things cannot be quite so simple. Only in relatively recent years has string theory provided at least a partial resolution of this paradox. We will touch on this subject briefly in the chapters on string theory. In the suggested reading the reader will be referred to more thorough treatments.

17.6 Coupling spinors to gravity

In any theory ultimately intended to describe nature, both spinors and general relativity will be present. Coupling spinors to gravity requires some concepts beyond those we have utilized up to now. The usual covariant derivative is constructed for tensors under changes of coordinates. In flat space, spinors are defined by their properties under rotations or more generally, Lorentz transformations. To do the same in general relativity it is necessary, first, to introduce a local Lorentz frame at each point. The basis vectors in this frame are denoted e^a_μ. Here μ is the Lorentz index; we can think of a as labeling the different vectors. The e_μs, in four dimensions are referred to as a tetrad or *vierbein*. In other dimensions they are called *vielbein*.

Requiring that the basis vectors be orthonormal in the Lorentzian sense gives

$$e^a_\mu(x)e_{a\nu}(x) = g_{\mu\nu}(x) \tag{17.86}$$

or, equivalently,

$$e^a_\mu(x)e^{b\mu}(x) = \eta^{ab}. \tag{17.87}$$

The choice of vielbein is not unique. We can multiply e by a Lorentz matrix, $\Lambda^a_b(x)$. Using e we can change indices from space–time (sometimes called "world") indices to tangent space indices:

$$V^a = e^a_\mu V^\mu. \tag{17.88}$$

Using this we can work out the form of the connection which maintains the gauge symmetry. We require that

$$D_\mu V^a = e^{a\nu} D_\mu V_\nu. \tag{17.89}$$

The derivative on the left-hand side is equal to

$$\partial_\mu V^a + (\omega_\mu)^a_b V^b. \tag{17.90}$$

With a bit of work, one can find explicitly the connection between the spin connection and the vielbein:

$$\omega^{ab}_\mu = \frac{1}{2}e^{\nu a}\left(\partial_\mu e^b_\nu - \partial_\nu e^b_\mu\right) - \frac{1}{2}e^{\nu b}\left(\partial_\mu e^a_\nu - \partial_\nu e^a_\mu\right) - \frac{1}{2}e^{\rho a}e^{\sigma b}(\partial_\rho e_{\sigma c} - \partial_\sigma e_{\rho c})e^c_\mu. \tag{17.91}$$

Now we put this together. First, the curvature has a simple expression in terms of the spin connection, which formally is identical to that of a Yang–Mills connection:

$$(\mathcal{R}_{\mu\nu})^a_b = \partial_\mu(\omega_\nu)^a_b - \partial_\nu(\omega_\mu)^a_b + [\omega_\mu, \omega_\nu]^a_b. \tag{17.92}$$

This is connected simply to the Riemann tensor by the basic vectors e^a_σ:

$$(\mathcal{R}_{\mu\nu})^a_b = e^a_\sigma e^\tau_b (\mathcal{R}_{\mu\nu})^\sigma_\tau. \tag{17.93}$$

We can now construct, also, a generally covariant action for spinors:

$$\int d^D x\,\sqrt{g}\, i\bar\psi \Gamma^a e^\mu_a \left(\partial_\mu + \frac{1}{2}\omega^{bc}_\mu \Sigma_{bc}\right)\psi. \tag{17.94}$$

Suggested reading

There are a number of excellent textbooks on general relativity, for example those of Weinberg (1972), Wald (1984), Carroll (2004) and Hartle (2003). Many aspects of general relativity that are important for string theory are discussed in the text of Green *et al.* (1987). A review of black holes in string theory was provided by Peet (2000).

Exercises

(1) Show that $g^{\mu\nu}\partial_\nu$ transforms like dx^μ. Verify explicitly that the covariant derivative of a vector transforms correctly.

(2) Derive Eq. (17.38) by considering the following action for a particle:

$$S = -\int ds = -\int \sqrt{-g_{\mu\nu}\frac{dx^\mu}{d\tau}\frac{dx^\nu}{d\tau}}. \tag{17.95}$$

(3) Verify the formula (17.40) for the Riemann tensor $\mathcal{R}$, its symmetry properties and the Bianchi identities.

(4) Repeat the derivation of the conservation of the stress tensor, being careful with each step. Derive the stress tensor for the Maxwell field of electrodynamics, $F_{\mu\nu}$. Derive Einstein's equations from the action. You will need to show first that

$$\delta R_{\mu\nu} = (\delta\Gamma^\lambda_{\mu\lambda})_{;\nu} - (\delta\Gamma^\lambda_{\mu\nu})_{;\lambda}.$$

(5) Download a package of programs for doing calculations in general relativity in MAPLE, MATHEMATICA or any other program you prefer. A Google search will yield several choices. Practise by computing the components of the affine connection and the curvature for the Schwarzschild solution.

(6) Here is an heuristic derivation of the Hawking temperature. Near the horizon one can choose coordinates such that the metric is almost flat. Check this using

$$\eta = 2\sqrt{r_{\rm h}(r - r_{\rm h})}, \tag{17.96}$$

$$ds^2 = -4r_{\rm h}^2\eta^2 dt^2 + d\eta^2 + r_{\rm h}^2 d\Omega_2^2. \tag{17.97}$$

Now take the time to be Euclidean, $t \to i\phi/(2r_{\rm h})$. Check that now this is the metric of the plane times that of a two-sphere, provided that ϕ is an angle, $0 < \phi < 2\pi$ (otherwise, the space is said to have a conical singularity). Argue that field theory on this sphere is equivalent to field theory at finite temperature $T_{\rm h}$ (you may need to read Appendix C, particularly the discussion of finite-temperature field theory).

18 Cosmology

Very quickly after Einstein published his general theory, a number of researchers attempted to apply Einstein's equations to the universe as a whole. This was a natural, if quite radical, move. In Einstein's theory the distribution of energy and momentum in the universe determines the structure of space–time, and this applies as much to the universe as a whole as to the region of space, say, around a star. To get started, these early researchers made an assumption which, while logical, may seem a bit bizarre. They took the principles enunciated by Copernicus to their logical extreme and assumed that space–time was homogeneous and isotropic, i.e. that there is no special place or direction in the universe. They had virtually no evidence for this hypothesis at the time – definitive observations of galaxies outside of the Milky Way were only made a few years later. It was only decades later that evidence in support of this *cosmological principle* emerged. As we will discuss, we now know that the universe is extremely homogeneous when viewed on sufficiently large scales.

18.1 The cosmological principle and the FRW universe

To implement the principle, just as for the Schwarzschild solution we begin by writing the most general metric consistent with an assumed set of symmetries. In this case the symmetries are homogeneity and isotropy in space. A metric of this form is called a *Friedmann–Robertson–Walker* (FRW) metric. We can derive this metric by imagining our three-dimensional space, at any instant, as a surface in a four-dimensional space. There should be no preferred direction on this surface; in this way we impose both homogeneity and isotropy. The surface will then be one of constant curvature. Consider, first, the mathematics required to describe a $(2+1)$-dimensional space–time of this sort. The three *spatial* coordinates would satisfy

$$x_1^2 + x_2^2 = k\left(R^2 - x_3^2\right), \tag{18.1}$$

where whether $k = \pm 1$ is positive or negative depends on whether the space has positive or negative curvature. Then the line element on the surface is (for positive k):

$$d\vec{x}^2 = dx_1^2 + dx_2^2 + dx_3^2 = dx_1^2 + dx_2^2 + \frac{(x_1 dx_1 + x_2 dx_2)^2}{x_3^2}. \tag{18.2}$$

The equation of the hypersurface gives

$$x_3^2 = R^2 - x_1^2 - x_2^2. \tag{18.3}$$

Setting $x_1 = r' \cos\theta$, $x_2 = r' \sin\theta$, we have

$$d\vec{x}^2 = \frac{R^2 dr'^2}{R^2 - r'^2} + r'^2 d\theta^2. \tag{18.4}$$

It is natural to rescale according to $r' = r/R$. Then the metric takes the form, now for general k,

$$d\vec{x}^2 = \frac{dr'^2}{1 - kr^2} + r^2 d\theta^2. \tag{18.5}$$

Here $k = 1$ for a space of positive curvature; $k = -1$ for a space of negative curvature; $k = 0$ is a spacial case, corresponding to a flat universe.

We can immediately generalize this to three dimensions by allowing the radius R to be a function of time, $R \to a(t)$. In this way we obtain the Friedmann–Robertson–Walker (FRW) metric:

$$ds^2 = -dt^2 + a^2(t)\left(\frac{dr^2}{1 - kr^2} + r^2 d\theta^2 + r^2 \sin^2\theta \, d\phi^2\right). \tag{18.6}$$

By general coordinate transformations this can be written in a number of other convenient and commonly used forms, which we will encounter in the following.

First we will evaluate the connection and the curvature (see Section 17.1). Again, the reader should evaluate a few of these terms by hand and perform the complete calculation using one of the programs mentioned in the exercises in the previous chapter. The non-vanishing components of the Christoffel connection are

$$\Gamma^i_{0j} = \frac{\dot{a}}{a}\delta^i_j, \quad \Gamma^0_{ij} = g_{ij}\frac{\dot{a}}{a}, \quad \Gamma^i_{jk} = \frac{g^{il}}{2}(g_{lj,k} + g_{lk,j} - g_{jk,l}) \tag{18.7}$$

and those of the curvature are

$$\mathcal{R}_{00} = -3\frac{\ddot{a}}{a}, \quad \mathcal{R}_{ij} = g_{ij}\left(\frac{\ddot{a}}{a} + 2H^2 - 2\frac{k}{a^2}\right). \tag{18.8}$$

Here H is known as the Hubble parameter,

$$H = \frac{\dot{a}}{a}, \tag{18.9}$$

and represents the expansion rate of the universe. Today

$$H = 100h \text{ km s}^{-1} \text{ Mpc}^{-1}, \quad h = 0.73 \pm 0.03. \tag{18.10}$$

The assumptions of homogeneity and isotropy greatly restrict the form of the stress tensor: $T_{\mu\nu}$ must take the *perfect fluid* form

$$T_{00} = \rho, \quad T_{ij} = p g_{ij}, \tag{18.11}$$

where ρ and p are the energy density and the pressure and are assumed to be functions only of time. Then the (0, 0) component of the Einstein equation (17.61) gives the *Friedmann equation*,

$$\frac{\dot{a}^2}{a^2} + \frac{k}{a^2} = \frac{8\pi G_N}{3}\rho, \tag{18.12}$$

where G_N is Newton's gravitational constant (see Eq. (17.67)). The i, j components give:

$$\frac{2\ddot{a}}{a} + \frac{\dot{a}^2}{a^2} + \frac{k}{a^2} = -8\pi G_N p. \tag{18.13}$$

There is also an equation which follows from the conservation of the energy momentum tensor, i.e. $T^{\mu\nu}{}_{;\nu} = 0$. This is

$$d(\rho a^3) = -p d(a^3). \tag{18.14}$$

This equation is familiar in thermodynamics as the equation of energy conservation if we interpret a^3 as the volume of a system. Suppose that we have the equation of state $p = w\rho$, where w is a constant. Then Eq. (18.14) says that

$$\rho \propto a^{-3(1+w)}. \tag{18.15}$$

Three special cases are particularly interesting. For non-relativistic matter, the pressure is negligible compared with the energy density, so $w = 0$. For radiation (relativistic matter), $w = 1/3$. For a Lorentz-invariant stress tensor $T_{\mu\nu} = \Lambda g_{\mu\nu}$, we have $p = -\rho$ so $w = -1$. For these cases, it is worth remembering that

$$\text{radiation, } \rho \propto a^{-4}; \quad \text{matter, } \rho \propto a^{-3}; \quad \text{vacuum, } \quad \rho = \text{const.} \tag{18.16}$$

After taking account of the conservation of stress–energy and the Bianchi identities, only one of the two Einstein equations we have written down is independent; and it is conventional to take this as the Friedmann equation. This equation can be rewritten in terms of the Hubble parameter:

$$\frac{k}{H^2 a^2} = \frac{8\pi G_N \rho}{3H^2} - 1. \tag{18.17}$$

Examining the right-hand side of this equation, it is natural to define a *critical density*

$$\rho_c = \frac{3H^2}{8\pi G_N}, \tag{18.18}$$

and to define Ω as the ratio of the density and the critical density,

$$\Omega = \frac{\rho}{\rho_c}. \tag{18.19}$$

So $k = 1$ corresponds to $\Omega > 1$, $k = -1$ to $\Omega < 1$ and $k = 0$, a flat universe, to $\Omega = 1$. It is also natural to break up Ω into various components, such as those due to radiation, matter or cosmological constant. As we will discuss shortly, Ω today is equal to unity within experimental errors; its main components are some unknown form of matter, baryons and dark energy (perhaps a cosmological constant):

$$\Omega_{dm} = 0.267, \quad \Omega_b = 0.049, \quad \Omega_{de} = 0.683. \tag{18.20}$$

The present error bars are of order 3% or less on these quantities (the most recent data is from the Planck satellite). Note that the total is close to unity. The present expansion rate is also known to be at the 2% level.

The history of the universe divides into various eras, in which different forms of energy were dominant. The earliest era for which we have direct observational evidence is a period lasting from a few seconds after the big bang to about 100 000 years, during which the universe was radiation dominated. From the Friedmann equation, setting $k = 0$, we have that

$$a(t) = a(t_0)t^{1/2}, \quad H = \frac{1}{2t}. \tag{18.21}$$

For the period of matter domination, which began about 10^5 years after the big bang and lasted almost to the present:

$$a(t) \propto t^{2/3}, \quad H = \frac{2}{3t}. \tag{18.22}$$

The universe appears today to be passing from an era of matter domination to a phase in which a (positive) cosmological constant dominates. Such a space is called a de Sitter space, with Hubble parameter M_{d}:

$$a(t) \propto e^{H_{\text{d}}t}, \quad H_{\text{d}} = \frac{8\pi G_{\text{N}}}{3}\Lambda. \tag{18.23}$$

In the radiation-dominated and matter-dominated periods, H is, as we can see from the formulas above, roughly a measure of the age of the universe. One can define the age of the universe more formally as:

$$t = \int^{a(t)} \frac{da}{\dot{a}} = \int \frac{da}{aH}. \tag{18.24}$$

The present value of the Hubble constant corresponds to $t \approx 13.8$ billion years. To obtain this correspondence between the age and the measured H_0, it is important to include both the matter and the cosmological constant parts of the energy density. Note, in particular, that the integral is dominated by the most recent times, where H is smallest.

18.2 A history of the universe

As little as 50 years ago, most scientists would have been surprised at just how much we would eventually know about the universe: its present composition, its age and its history, back to times a couple of minutes after the big bang. We have direct evidence of phenomena at much earlier times, though the full implications of this evidence are difficult to interpret. We understand how galaxies formed and the abundance of the light elements. And we have a treasure trove of plausible speculations, some of which we should be able to test over time.

In this section we outline some basic features of this picture. Examining the FRW solution of Einstein's equations, we see that the scale factor $a(t)$ gets monotonically smaller in the past. The Hubble parameter H becomes larger. So, at some time in the past, the universe was much smaller than it is today. More precisely, the objects we see, or their predecessors, were far closer together. Far enough back in time, the material we currently see was highly compressed and hot. So, at some stage, it is likely that the universe was dominated by radiation. Recall that, during a radiation-dominated era,

$$a \sim t^{1/2}, \quad H = \frac{1}{2t}. \tag{18.25}$$

If we suppose that the universe remains radiation dominated as we look further back in time, we face a problem. At $t = 0$ the metric is singular – the curvature diverges. This is a finite time in the past, since

$$\int_0^{\text{today}} dt \sqrt{-g_{00}} \tag{18.26}$$

converges as $t \to 0$. This is not simply a feature of our particular assumptions about the equation of state or the precise form of the metric but a feature of solving Einstein's equations; it is a consequence of the singularity theorems due to Penrose and Hawking. The meaning of this singularity is a subject of much speculation. It might be smoothed out by quantum effects, or it might indicate something else. For now we simply have to accept that extremely early times are inaccessible to us. To start, we will suppose that at time t_0 the universe was extremely hot, with temperature T_0, and reasonably homogeneous and isotropic. We will then allow the universe to evolve, using Einstein's equations, the known particles and their interactions and the basic principles of statistical mechanics. As we will see, we can safely take T_0 to be at least as large as several MeV (corresponding to temperatures larger than 10^{10} K).

To make further progress we need to think about the content of the universe and how it evolves as the universe expands. The universe cannot be precisely in thermal equilibrium but, for much of its history, it has been very nearly so, with matter and radiation evolving adiabatically. To understand why the expansion is adiabatic, note first that H^{-1} is the time scale for the expansion. If the universe is radiation dominated,

$$H \sim \frac{T^2}{M_{\text{p}}}, \tag{18.27}$$

where M_{p} is the Planck mass. The rate for interactions in a gas will scale with T, multiplied perhaps by a few powers of coupling constants. For temperatures well below the Planck scale the reaction rates will be much more rapid than the expansion rate. So, at any given instant, the system will nearly be in equilibrium.

It is worth reviewing a few formulas from statistical mechanics. These formulas can be derived by elementary considerations or by using the methods of finite-temperature field theory, as discussed in Appendix C. For a relativistic weakly coupled Bose gas,

$$\rho = \frac{\pi^2}{30} g T^4, \quad p = \frac{\rho}{3}, \tag{18.28}$$

while, for a similar Fermi gas,

$$\rho = \frac{7}{8}\frac{\pi^2}{30} g T^4, \quad p = \frac{\rho}{3}. \tag{18.29}$$

Here g is a degeneracy factor that counts the number of physical helicity states of each particle type. In the non-relativistic limit, for both bosons and fermions we have

$$n = g\left(\frac{mT}{2\pi}\right)^{3/2} \exp\left[-\frac{(m-\mu)}{T}\right] \tag{18.30}$$

$$\rho = mn, \quad p = nT \ll \rho. \tag{18.31}$$

For temperatures well below m, the density rapidly goes to zero unless $\mu \neq 0$. Note that μ may be non-zero when there is a (possibly approximately) conserved quantum number. Perhaps the most notable example is the baryon number.

We should pause here and discuss an aspect of general relativity which we have not considered up to now. A gravitational field alters the behavior of clocks. This is known as the gravitational red shift. We can understand this in a variety of ways. First, if we have a particle at rest in a gravitational field then the proper time is related to the coordinate time by a factor $\sqrt{g_{00}}$. Consider, alternatively, the equation for a massless scalar field with momentum k in an expanding FRW universe. This is just $D^\mu \partial_\mu \phi = 0$. Using the non-vanishing Christoffel symbols, with $\phi(\vec{x}, t) = e^{i\vec{k}\cdot\vec{x}}\phi(t)$,

$$\ddot{\phi}(k) + 3H\dot{\phi}(k) + \frac{k^2}{a^2(t)}\phi = 0. \tag{18.32}$$

As a result of the last term, the wavelength effectively increases as $a(t)$. A photon red-shifts in precisely the same way.

The implications of this for the statistical mechanical distribution functions are interesting. Consider, first, a massless particle such as the photon. For such a particle, the distribution is

$$\int \frac{d^3k}{(2\pi)^3} \frac{1}{e^{k/T} - 1}. \tag{18.33}$$

The effect of the red shift is to maintain this form of distribution but to change the temperature, $T(t) \propto 1/a(t)$. So even if the particles are not in equilibrium, they maintain an equilibrium distribution appropriate to the red-shifted temperature. This is not the case for massive particles.

Let us imagine, then, starting the clock when the universe is at temperatures well above the scale of QCD but well below the scale of weak interactions, say at 10 GeV. In this regime the density of Ws and Zs is negligible, but the quarks and gluons behave as nearly free particles. So we can take an inventory of the bosons and fermions that are light compared with T. The bosons include the photon and the gluons; the fermions include all the quarks and leptons except the top quark. So the effective g, which we might call g_{10}, is approximately 98. This means, for example, that

$$\rho \approx \frac{g_{10}\pi^2}{30} T^4 \tag{18.34}$$

and the Hubble constant is related to the temperature through

$$H = \left(\frac{8\pi}{3} G_{\rm N} \frac{\pi^2}{30} g_{10} T^4 \right)^{1/2}, \tag{18.35}$$

where $G_{\rm N}$ is Newton's gravitational constant (see Eq. (17.67)). This allows us to write a precise formula for the temperature as a function of time:

$$T = \left(\frac{16\pi}{3} G_{\rm N} \frac{\pi^2}{30} g_{10} \right)^{-1/4} \left(\frac{1}{t} \right)^{1/2}. \tag{18.36}$$

As the universe cools, QCD changes from a phase of nearly free quarks and gluons to a hadronic phase. At temperatures below m_π, the only light species are the electron and the neutrinos. By this time, the antineutrons have annihilated with neutrons and the antiprotons with protons, leaving a small net baryon number, the total number of neutrons and protons. There is, at this time, of order one baryon per billion photons. We will have much more to say about this slight excess later.

At this stage, interactions involving neutrinos maintain an equilibrium distribution of protons and neutrons. We can give a crude, but reasonably accurate, estimate of the temperature at which neutrino interactions drop out of equilibrium by asking when the interaction rate becomes comparable to the expansion rate. The cross section for neutrino–proton interactions is

$$\sigma(\nu + p \to n + e) \approx G_{\rm F}^2 E^2, \tag{18.37}$$

where $G_{\rm F}$ is the Fermi constant (see Eq. (3.3)), and the number density of neutrinos is

$$n_\nu \approx \frac{\pi^2}{30} g_T T^3. \tag{18.38}$$

Combining this with our formula Eq. (18.35) for the Hubble constant as a function of T gives, for the decoupling temperature T_ν,

$$T_\nu^3 \approx G_{\rm F}^{-2} M_p^{-1} \tag{18.39}$$

or

$$T_\nu \approx 2\ {\rm MeV}. \tag{18.40}$$

This corresponds to a time of order 100 s after the big bang. At this point neutron decays are not compensated by the inverse reaction, but many neutrons will pair with protons to form stable light elements such as D and He. At about this time the abundances of the various light elements are fixed.

There is a long history of careful, detailed, calculations of the abundances of the light elements (H, He, D, Li, ...) which result from this period of decoupling. The abundances turn out to be a sensitive function of the ratio of baryon and photons, $n_{\rm B}/n_\gamma$. Astronomers have also made extensive efforts to *measure* this ratio. A comparison of observations and measurements gives, for the baryon to photon ratio,

$$\frac{n_{\rm B}}{n_\gamma} = 6.1^{+0.3}_{-0.2} \times 10^{-10}. \tag{18.41}$$

We will see that this result receives strong corroboration from other sources.

The universe continues to cool in this radiation-dominated phase for a long time. At $t \approx 10^5$ years the temperature drops to about 1 eV. At this time electrons and nuclei can combine to form neutral atoms. As the density of ionized material drops, the universe becomes essentially transparent to photons. This is referred to as *recombination.* The photons now stream freely. As the universe continues to cool the photons red-shift, maintaining a Planck spectrum. Today, these photons behave as if they had a temperature $T \approx 3\,\text{K}$. They constitute the *cosmic microwave background radiation* (CMBR). This radiation was first observed in 1963 by Penzias and Wilson and has since been extensively studied. It is very precisely a black body, with characteristic temperature 2.7 K. We will discuss other features of this radiation shortly.

It is interesting that, given the measured value of the matter density, matter and radiation have comparable energy densities at the recombination stage. At later times matter dominates the energy density, and this continues to be the case to the present time.

In our brief history, another important event occurs at $t \simeq 10^9$ years. If we suppose that initially there were small inhomogeneities, these remain essentially frozen, as we will explain later, until the time of matter–radiation equality. They then grow with time. From observations of the CMBR we know that these inhomogeneities were at the level of one part in 10^5. At about 1 billion years after the big bang, these then grow enough to be non-linear, and their subsequent evolution is believed to give rise to the structure – galaxies, clusters of galaxies, and so on – that we see around us.

One surprising feature of the universe is that most of the energy density is in two forms which we cannot see directly. One is referred to as the *dark matter*. The possibility of dark matter was first noted by astronomers in the 1930s, from observations of the rotation curves of galaxies. Simply using Newton's laws one can calculate the expected rotational velocities and one finds that these do not agree with the observed distribution of stars and dust in the galaxies. This is true for structures on many scales, not only galaxies but clusters and larger structures. Other features of the evolution of the universe are not in agreement with observation unless most of the energy density is in some other form. From a variety of measurements, Ω_{m}, the fraction of the critical energy density (see Eq. (18.18)) in matter, is known to be about 0.3. Nucleosynthesis and the CMBR give a much smaller fraction in baryons, $\Omega_{\text{b}} \approx 0.05$. In support of this picture, direct searches for hidden baryons give results that are compatible with the smaller number; they have failed to find anything like the required density to give Ω_{m}.

Finally, it appears that we are now entering a new era in the history of the universe. For the last 14 billion years, the energy density has been dominated by non-relativistic matter. But, at the present time, there is almost twice as much energy in some new form, with $p < 0$, referred to as dark energy. The dark energy is quite possibly a cosmological constant, Λ. Current measurements are compatible with $w = -1$ $(p = -\rho)$.

The picture we have described has extensive observational support. We have indicated some of this: the light element abundances and the observation of the CMBR. The agreement of these two quite different sets of observations for the baryon to photon ratio is extremely impressive. Observations of supernovae, the age of the universe and features of structure at different scales all support the existence of a cosmological constant (dark energy) constituting about 70% of the total energy.

This is not a book on cosmology, and the overview we have presented is admittedly sketchy; there are many aspects of this picture we have not discussed. Fortunately there are many excellent books on the subject, some of which are mentioned in the suggested reading.

Suggested reading

There are a number of good books and lectures on the aspects of cosmology discussed here. Apart from the text of Weinberg (1972), mentioned earlier, these include the texts of Kolb and Turner (1990), Dodelson (2004) and Weinberg (2008).

Exercises

(1) Compute the Christoffel symbols and the curvature for the FRW metric. Verify the Friedmann equations.
(2) Verify Eq. (18.32).
(3) Consider the case of de Sitter space, $T_{\mu\nu} = -\Lambda g_{\mu\nu}$ with positive Λ. Show that the space expands exponentially rapidly. Compute the horizon, i.e. the largest distance from which light can travel to an observer.

19 Particle astrophysics and inflation

In Chapter 18 we put forward a history of the universe. The picture is extremely simple. Its inputs were Einstein's equations and the assumptions of homogeneity and isotropy. We also used our knowledge of the laws of atomic, nuclear and particle physics. We saw a number of striking confirmations of this basic picture, but there are many puzzles.

1. The most fundamental problem is that we do not know the laws of physics relevant to temperatures greater than about 100 GeV. If there is only a single Higgs doublet at the weak scale, it is possible that we can extend this picture back to far earlier times. If there is, say, supersymmetry or large extra dimensions, the story could change drastically. Even if things are simple at the weak scale, we will not be able to extend the picture all the way back to $t = 0$. We have already seen that the classical gravity analysis breaks down.
2. There are a number of features of the *present* picture we cannot account for within the Standard Model. Specifically, what is dark matter? There is no candidate among the particles of the Standard Model. Is it some new kind of particle? As we will see, there are plausible candidates from the theoretical structures we have proposed, and they are the subject of intense experimental searches.
3. Dark energy is very mysterious. Assuming that it is a cosmological constant, it can be thought of as the vacuum energy of the underlying microphysical theory. As a number, it is totally bizarre. Its natural value should be set by the largest relevant scale, perhaps the Planck or unification scale, or the scale of supersymmetry breaking. Other proposals have been put forward to model dark energy. One which has been extensively investigated is known as quintessence, the possibility that the energy is that of a slowly varying scalar field. Such models typically do not predict $w \neq -1$ (see Section 18.1), and many are already ruled out by observation. But it should be stressed that these models are, if anything, less plausible than the possibility of a cosmological constant. First, one needs to explain why the underlying microphysical theory produces an essentially zero cosmological constant and a potential whose curvature is smaller than the present value of H. Then one needs to understand why the slowly varying field produces the energy density observed today, without disturbing the successes of the cosmological picture for earlier times. It is probably fair to say that no convincing explanation of either aspect of the problem has been forthcoming.
4. The value of the present baryon to photon ratio is puzzling:

$$\frac{n_B}{n_\gamma} = \left(6.1\,^{+0.3}_{-0.2}\right) \times 10^{-10}. \tag{19.1}$$

As we will see, the question can be phrased as why is this so small, or why is it so large? If the universe was always in thermal equilibrium, this number is a constant. So at very early times, there was a very tiny excess of particles over antiparticles. One might imagine that this is simply an initial condition but, as A. Sakharov first pointed out, this is a number that one might hope to explain through cosmology combined with microphysical theory. As we will discuss in detail later, it is necessary that the underlying microphysics violates baryon number and CP and that there is a significant departure from thermal equilibrium. The Standard Model, as we have seen, violates both and can generate a baryon number but, as we will see, it is far too small. So, *modifications of the known physical laws are required to account for the observed density of baryons.*

5. Homogeneity, flatness and topological objects such as monopoles pose puzzles which suggest a phenomenon known as *inflation*. Consider, first, homogeneity. This certainly made the equations simple to solve, but it is puzzling. If we look at the cosmic microwave background, the temperature variation in different directions in the sky is equal to about a part in 10^5. But, as we look out at distances as much as 13.8 billion light years away, points separated by a tiny fraction of a degree were separated, at 100 000 years after the big bang, by an enormous distance compared with the horizon at that time. The problem is that, as we look back, the horizon decreases in size as $\sqrt{t}$. So points separated by a degree were, at that time, separated by about 10^7 light years. But signals could not have traveled more than 10^5 light years by this time. So if these points had not been in causal contact by recombination, why should they have identical temperatures? For nucleosynthesis, which occurs much earlier, the question is even more dramatic.
6. Flatness ($\Omega_{\rm tot} = 1$) may not seem puzzling at first, but consider again the structure of the FRW metric. We have seen that the Friedmann equation can be recast as

$$\frac{8\pi G_N \rho}{H^2} = \Omega - 1. \tag{19.2}$$

Suppose, for example, that $\Omega = 0.999$ today. Then, at recombination, the left-hand side of this equation was more than eight orders of magnitude smaller. So the energy density was equal to the critical density with extraordinary precision. This apparent fine tuning gets more and more extreme as we look further back in time.
7. Monopoles: we have seen that simple grand unified theories predict the existence of magnetic monopoles. Their masses are typically of order the grand unification scale. So unless their density were many orders of magnitude (perhaps 14!) smaller than the density of baryons, their total energy density would be far greater than the observed energy density of the universe. Astrophysical limits turn out to be even smaller; passing through the galaxy, monopoles would deplete the magnetic field. This sets a limit, known as the Parker bound, on the monopole flux in the galaxy:

$$\mathcal{F} < 10^{-16} \left(\frac{M_{\rm mon}}{10^{17}\,{\rm GeV}} \right) {\rm cm}^{-2}\,{\rm s}^{-1}\,{\rm sr}^{-1}. \tag{19.3}$$

However, we might expect, in a grand unified theory, quite extensive monopole production. We have seen that monopoles are topological objects. If there is a phase transition between phases of broken and unbroken $SU(5)$, we would expect twists in the fields on scales of order the Hubble radius at this time, and a density of monopoles of order one per horizon volume. If the transition occurs at $T_0 = 10^{16}$ GeV, the Hubble radius is of order T^2/M_p so the density, in units of the photon density T^3, is of order

$$\frac{n_{\rm mon}}{n_\gamma} = \frac{T^3}{M_p^3} \tag{19.4}$$

and can be *larger* than the baryon density.

In the following sections we discuss these issues. We will study a possible solution to the homogeneity, flatness and monopole problems: inflation, the hypothesis that the universe underwent a period of extremely rapid expansion. We will see that there is some evidence that this phenomenon really occurred. Certainly there is nothing within the Standard Model itself which can give rise to inflation, so this points to the presence of some new phenomena, perhaps fields or perhaps more complicated entities, which are crucial to understanding the universe we see around us. We will describe some simple models of inflation, especially slow-roll inflation and chaotic and hybrid inflation, and some of their successes and the puzzles which they raise. We will discuss inflationary theory's biggest success, that quantum mechanical fluctuations during inflation give rise to the perturbations which grow to give the structure we see around us in the universe. This introduction is not comprehensive but should give the reader some tools to approach the vast literature which exists on this subject.

We next turn to the problem of dark matter. We focus on two candidates: the lightest supersymmetric particle of the MSSM, and the axion. We explain how these particles might rather naturally be produced with the observed energy density and discuss briefly the prospects for their direct detection. Then we turn to baryogenesis. We explain why the Standard Model has all the ingredients to produce an excess of baryons over antibaryons but, given the value of the Higgs mass, this baryon number cannot be nearly as large as is observed. We then turn to baryon production in some of our proposals for physics beyond the Standard Model, focusing on three possibilities: heavy particle decay in grand unified theories, leptogenesis and coherent production by scalar fields.

19.1 Inflation

The underlying idea behind inflation is that the universe behaved for a time as if (or nearly as if) the energy density was dominated by a positive cosmological constant, Λ. During this era the Friedmann equation was that for de Sitter space,

$$H_i^2 = \left(\frac{\dot a}{a}\right)^2 = \frac{8\pi G_N}{3}\Lambda, \tag{19.5}$$

with solution

$$a(t) = e^{H_i t}. \tag{19.6}$$

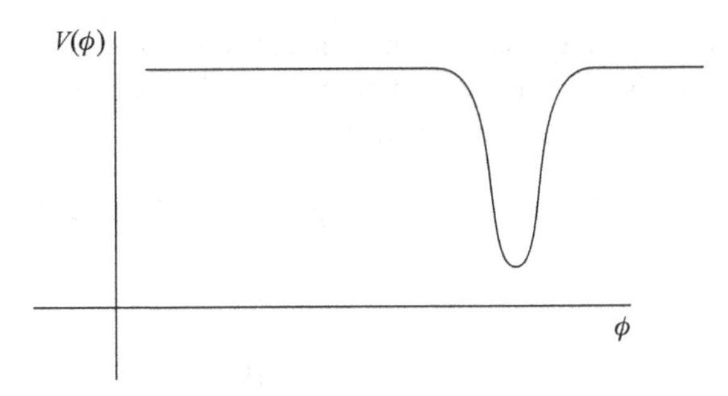

Fig. 19.1 A typical inflationary potential has a region in which $V(\phi)$ varies slowly and then settles into a minimum.

If this situation had held for a time interval such that, say, $\Delta t H_{\rm i} = 60$ then the universe would have expanded by an enormous factor. Suppose, for example, that Λ was 10^{16} GeV; correspondingly $H_{\rm i} \approx 10^{14}$ GeV. Then a patch of size H^{-1} would have grown to be almost a centimeter in size. If, at the end of this period of inflation, the temperature of the universe had been 10^{16} GeV, this patch would have grown, up to the present time, by a factor 10^{29}. This is about the size of our present horizon!

One possibility for how this might have come about is called *slow-roll inflation*. Here one has a scalar field ϕ with potential $V(\phi)$; $V(\phi)$, for some range of ϕ, is slowly varying (Fig. 19.1). What we have called $H_{\rm i}$ is determined by the average value of the potential in the plateau region, V_0. If we assume that we have a patch of initial size a bit larger than $H_{\rm i}^{-1}$, then we can write down an equation of motion for the zero-momentum mode of the field ϕ in this region:

$$g^{\mu\nu} D_\mu \partial_\nu \phi + V'(\phi) = 0. \tag{19.7}$$

Because of our assumptions of homogeneity and isotropy, we can take the metric to have the Robertson–Walker form:

$$\ddot{\phi} + 3h\dot{\phi} + V'(\phi) = 0. \tag{19.8}$$

We assume that the field is moving slowly, so that we can neglect the $\ddot{\phi}$ term. Shortly, we will check whether this assumption is self-consistent. In this limit the equation of motion is first order:

$$\dot{\phi} = -\frac{V'}{3H}. \tag{19.9}$$

We can integrate this equation to get Δt, the time it takes the field to traverse the plateau of the potential; $\Delta t H$ is roughly the number of e-folds of inflation, N, thus

$$N = \frac{1}{M_{\rm p}^2} \int d\phi \frac{V(\phi)}{V'(\phi)}. \tag{19.10}$$

The requirement for obtaining adequate inflation is: that N should be larger than about 60.

Now we can determine the conditions for the validity of the slow-roll approximation. We simply want to check, from our solution, that $\ddot{\phi} \ll 3H\dot{\phi}$ and $V'(\phi)$. Differentiating Eq. (19.9) leads to the conditions

$$\epsilon = \frac{1}{2}M_{\rm p}^2\left(\frac{V'}{V}\right)^2 \ll 1 \tag{19.11}$$

and

$$\eta = M_{\rm p}^2\frac{V''}{V}, \quad |\eta| \ll 1. \tag{19.12}$$

How did inflation end? Near the minimum of the potential we can approximate it as quadratic. So we might try to study an equation of the form

$$\ddot{\phi} + 3H\phi + m^2\phi = 0. \tag{19.13}$$

Were it not for the expansion of the universe, this equation would have a solution

$$\phi = \phi_0 \cos mt. \tag{19.14}$$

In quantum mechanical language this would describe a coherent state of particles, with energy density

$$\rho = \frac{1}{2}m^2\phi_0^2. \tag{19.15}$$

These particles have zero momentum; the pressure, $T_{ij} = p\delta_{ij} = 0$. So, if this field dominates the energy density of the universe, we know that

$$a \sim t^{2/3}, \quad H = \frac{2}{3}t. \tag{19.16}$$

In our toy model we might imagine that $m \sim 10^{16}$ GeV $\gg H$, so we could solve the equation by assuming

$$\phi(t) = f(t)\cos mt \tag{19.17}$$

for a slowly varying function f. Substituting Eqs. (9.1) and (19.16) into Eq. (19.13), one finds

$$f(t) = \frac{1}{t}. \tag{19.18}$$

Note that this means that

$$\rho = \rho_0\left(\frac{t}{t_0}\right)^2 = \rho_0\left(\frac{a}{a_0}\right)^3. \tag{19.19}$$

To summarize, we are describing a system which behaves like pressureless dust – zero-momentum particles – and which is diluted by the expansion of the universe.

This description also gives us a clue as to the fate of the field ϕ. Supposing that the ϕ particles have a finite width Γ, they will decay in time $1/\Gamma$. We can include this in our equation of motion, writing

$$\ddot{\phi} + (3H + \Gamma)\dot{\phi} + V'(\phi) = 0. \tag{19.20}$$

When the particles decay, if their decay products include, for example, ordinary quarks, leptons and gauge fields then their interactions will bring them quickly to equilibrium. We can be at least somewhat quantitative about this. The condensate disappears at a time set by $H \approx \Gamma$. If the universe quickly comes to equilibrium, the temperature must satisfy

$$\frac{\pi^2}{30} g T^4 = H^2 \frac{3}{8\pi G_N}. \tag{19.21}$$

At this temperature we can estimate the rate of interaction. Since the typical particle energy will be of order T, the cross sections will be of order

$$\sigma = \frac{\alpha_i^2}{T^2} \tag{19.22}$$

We can multiply this by the density, $n = (\pi^2/30) g^* T^3$, to obtain a reaction rate. For inflation at the scales we are discussing, this is enormous compared with H. The details by which equilibrium is established have been studied with some care. We can imagine that when a ϕ particle first decays, it produces two very high energy particles. These will have rather small cross sections for scattering with other high-energy decay products, but these interactions degrade the energy, and so the cross sections for subsequent interactions – and for interactions with previously produced particles – are larger. A more careful study leads to a behavior with time where the temperature rises to a maximum and then falls. This maximum temperature is

$$T_{\max} \approx 0.8 g_*^{-1/4} m^{1/2} (\Gamma M_p)^{1/4}, \tag{19.23}$$

where m is the mass of the *inflaton*.

19.1.1 Fluctuations: the formation of structure

One of the most exciting features of inflation is that it predicts that the universe is not exactly homogeneous and isotropic. We cannot do justice to this subject in this short section, but we can at least give the flavor of the analysis and collect the crucial formulae. In order to have inflation we need the metric and fields to be reasonably uniform over a region of size $H_{\rm i}^{-1}$. But, because of quantum fluctuations, the fields and in particular the scalar field ϕ cannot be completely uniform. We can estimate the size of these quantum fluctuations without great difficulty. In order that inflation occurs at all, we need $m_\phi \ll H$. So we will treat ϕ as a massless free field in de Sitter space. As in flat space, we can expand the field ϕ in Fourier modes:

$$\phi(\vec{x}, t) = \int \frac{d^3k}{(2\pi)^3} \left[e^{i\vec{k}\cdot\vec{x}} h(\vec{k}, t) + \text{c.c.} \right]. \tag{19.24}$$

The expansion coefficients h obey the equation $D_\mu \partial_\nu h(\vec{k}, t) e^{i\vec{k}\cdot\vec{x}} = 0$, yielding, in the FRW background,

$$\ddot{h} + 3H\dot{h} + \frac{k^2}{a^2} h = 0. \tag{19.25}$$

Here k/a is the red-shifted momentum. In the case of de Sitter space, a grows exponentially rapidly. As soon as $k/a \sim H$ the system becomes overdamped, and the value of h is essentially frozen. We will see this in a moment when we write down an explicit solution of the equation.

It is convenient to change our choice of time variable. Rather than take the FRW form for the metric, we take a metric more symmetric between space and time:

$$ds^2 = a^2(t)(-d\eta^2 + d\vec{x}^2). \tag{19.26}$$

Here, in terms of our original variables,

$$\frac{d\eta}{dt} = \pm\frac{1}{a}. \tag{19.27}$$

So, choosing the + sign,

$$\eta = \int \frac{da}{(\dot{a}/a)a^2} = \int \frac{da}{Ha^2} \tag{19.28}$$

and

$$\eta = \frac{1}{H_a}. \tag{19.29}$$

In these coordinates, the equation of motion for $h(k, \eta)$ reads:

$$\delta\ddot{\phi} + 2aH\delta\dot{\phi} + k^2\delta\phi = 0. \tag{19.30}$$

This equation is straightforward to solve. The solution can be written in terms of Bessel functions, but more transparently as:

$$\delta\phi_k = \frac{e^{-ik\eta}}{\sqrt{ik}}\left(1 - \frac{i}{k\eta}\right). \tag{19.31}$$

Note that for large times $\eta \to 0$.

Further analysis is required to convert this expression into a fluctuation spectrum. The result is that the fluctuations in the energy density are roughly scale invariant, and

$$\frac{\delta\rho}{\rho} \approx \frac{H^2}{5\pi\dot{\phi}}. \tag{19.32}$$

Using the slow-roll equation

$$3H\dot{\phi} = V' \tag{19.33}$$

gives

$$\frac{\delta\rho}{\rho} = \frac{3H^*}{5\pi V'} = \frac{3V^{3/2}}{5\pi V' M_p^3}. \tag{19.34}$$

Much more detailed discussions of these formulas can be found in the suggested reading. These fluctuations quickly pass out of the horizon during inflation. While outside of the horizon, they are frozen. Subsequently, however, they reenter the horizon and begin to grow. Measurements of the CMBR combined with Eq. (19.34) yield

$$\frac{V^{3/2}}{V'} = 5.15 \times 10^{-4} M_{\rm p}^3 \tag{19.35}$$

on horizon scales. Fluctuations which were within the horizon at the time of matter–radiation equality have grown linearly with time since then. At about 1 billion years after the big bang they became non-linear, and this appears to account adequately for the observed structure in the universe. Precise studies of the CMBR, of the formation of structure and of Type Ia supernovas, as well as of the missing mass in structures on a wide range of scales, has allowed the determination of the composition of the universe.

Other observables of inflation are the spectral index n_s, which measures the deviation of the power spectrum from perfect scale invariance, and r, the ratio of the tensor and scalar fluctuations. The tensor modes arise due to gravitational radiation and are only observable if the scale of inflation is sufficiently high. These quantities, in slow-roll models, are given by

$$n_s = 1 + 2\eta - y\epsilon, \qquad r = 16\epsilon. \tag{19.36}$$

The spectral index has been measured by the Planck collaboration as

$$n_s = 0.9624, \tag{19.37}$$

where the error is about 1%; r is not yet well known. If and when it is measured, it will determine the scale of inflation,

$$\Lambda_{\rm inf} = \left(\frac{r}{0.7}\right)^{1/4} \times (1.8 \times 10^{16})\ {\rm GeV}. \tag{19.38}$$

But, while the inflationary scenario is compelling and has significant observational support, we lack a persuasive microphysical understanding of these phenomena. This is undoubtedly one of the great challenges of theoretical physics. In the next section we describe various classes of models.

19.1.2 Models of inflation

Experiments of the last two decades, and especially WMAP and Planck, have provided strong support for the phenomenon of inflation. This is likely to be information about physics at extraordinarily high energy scales, well above those likely to be accessible to foreseeable accelerators. However, translating the data into a microscopic model is extremely challenging. There are almost as many models of inflation as there are physicists who have thought about the problem, and we cannot possibly sample them all; in this section we survey a few. No existing model is terribly compelling. Essentially, all must be tuned in order to obtain enough e-foldings of inflation and small enough $\delta\rho/\rho$. First, it is known from observations that the Hubble constant during inflation cannot have been much larger than 10^{16} GeV. This means that the scalar mass cannot be comparable to the Planck mass, so we face the usual problem of light scalars. In fact, the difficulties are more severe since the scalar mass must be much lighter than the Hubble scale of inflation, in order to ensure slow roll; as we will see, even with supersymmetry this requires percent-level tunings. Further tunings are typically required to obtain the required fluctuation spectrum.

19.1.2.1 Chaotic inflation

A particularly simple class of models yields what is known as *chaotic inflation*. These models illustrate both possible predictions and the issues of naturalness and tuning. An example is a theory with a single scalar field, with a simple potential

$$V = \frac{1}{2} m^2 \phi^2. \tag{19.39}$$

Requiring 60 *e*-foldings of inflation gives

$$N = \frac{1}{4} \frac{\phi^2}{M_p^2}. \tag{19.40}$$

Correspondingly, $\epsilon = 8.3 \times 10^{-3} = \eta$. One predicts then that the spectral index n_s is approximately 0.967, close to the value measured by Planck, and $r = 0.133$.

While the predictions are interesting, the model is hard to take seriously as a microscopic theory. In particular, solving Eq. (19.34) for m, one obtains $m^2 = 4.6 \times 10^{-12} M_p^2$. There is no symmetry which accounts for this; moreover, we require that the coefficients of all other powers of ϕ be extremely small as well. More generally, the fact that $\phi \gg M_p$ means that we have no control of the physics. Despite these concerns this model has proven useful for considering many aspects of inflation, and it has been argued that some of its features may characterize a larger class of models. Later, we will consider a possible setting for this idea without these problems.

19.1.2.2 Inflation with supersymmetry: hybrid inflation

Given that supersymmetry naturally produces light scalars, supersymmetry would seem a natural context in which to construct models of inflation. We have mentioned that in supersymmetric field theories and in string theory one often encounters moduli, i.e. scalar fields whose potentials vanish in the some limit. Banks suggested that, for such fields, a potential of the form

$$V = \mu^4 f(\phi/M_p) \tag{19.41}$$

will often arise. Here μ is an energy scale determined by some dynamical phenomenon such as the scale of supersymmetry breaking. For such a potential, assuming that typical field values are of order M_p we have from Eq. (19.34),

$$\frac{\delta\rho}{\rho} \approx \frac{\mu^2}{M_p^2}. \tag{19.42}$$

From this we have $\mu \approx 10^{15.5}$ GeV. The number of *e*-foldings is generically of order one; the potential must be tuned to the level of 1%, for example, if one is to obtain sufficient inflation. Still, this may seem less troubling than having many couplings less than 10^{-12}.

Note that μ, the energy scale in a supersymmetric model of this kind with a single field, is far larger than those we have considered previously for low-energy supersymmetry breaking. Banks proposed that, at the minimum of the potential, supersymmetry is unbroken with vanishing $\langle W \rangle$ as a result of an R symmetry.

Another class of models of some interest are known as hybrid models. These involve at least two fields. They are particularly interesting in the context of supersymmetry, where such models have been dubbed "supernatural" by Guth and Randall since the presence of light scalars is again natural.

Hybrid inflation is often described in terms of fields and potentials with rather detailed, special, features but it can be characterized in a conceptual way. Inflation occurs in all such models on a pseudomoduli space, in a region where supersymmetry is badly broken (possibly by a larger amount than in the present universe) and on which the potential is slowly varying. We have seen that moduli are common in supersymmetric theories. We will find that they are ubiquitous in string theory. The simplest (supersymmetric) hybrid model has two fields, I and ϕ:

$$W = I(\kappa\phi^2 - \mu^2). \tag{19.43}$$

The field ϕ is known as the waterfall field. Classically, for large I the potential is independent of I,

$$V_{\rm cl} = \mu^4 \quad (\phi = 0).$$

This is the regime of inflation. Quantum mechanical effects generate a potential for I such that it rolls slowly from larger to smaller values. Inflation ends either when the slow-roll conditions are not satisfied or when I is small enough that the ϕ curvature is negative. In any case, at this point ϕ moves quickly towards its minimum. As it oscillates about the minimum, reheating occurs.

The quantum mechanical corrections control the dynamics of the inflaton. These involve a Coleman–Weinberg calculation of a type with which we are now familiar:

$$V(I) = \mu^4\left(1 + \frac{\kappa^2}{16\pi^2}\log\frac{|I|^2}{\mu^2}\right). \tag{19.44}$$

Here κ is constrained to be extremely small in order that the fluctuation spectrum be of the correct size; κ is proportional, in fact, to V_I, the energy during inflation. The quantum corrections determine the slow-roll parameters. We have

$$\kappa = 0.17 \times \left(\frac{\mu}{10^{15}\,\text{GeV}}\right)^2 = 7.1 \times 10^5 \times \left(\frac{\mu}{M_{\rm p}}\right)^2. \tag{19.45}$$

The problem of fine tuning in these models can be readily characterized. For example, Planck-suppressed terms in the Kahler potential K can spoil slow roll;

$$K = \frac{\alpha}{M_{\rm p}^2} I^\dagger I I^\dagger I \tag{19.46}$$

gives too large an η value unless $\alpha \sim 10^{-2}$. This is an irreducible tuning of (supersymmetric) hybrid models. However, the very small value required of κ is arguably a more severe tuning issue. In any case the model as it stands predicts $n_s > 1$ and is ruled out by the results from the Planck satellite, Eq. (19.3). Modifications are possible which avoid this prediction. Indeed, the moduli space of the simplest model does not closely resemble those we will encounter in string theory; broadening these considerations leads to different possibilities.

19.1.3 Constraints on reheating: the gravitino problem

In the context of supersymmetric theories, it is thought that there may be an upper bound on the reheating temperature. This is the problem of producing too many gravitinos. The gravitino lifetime is quite long,

$$\Gamma_{3/2} \approx \frac{m_{3/2}^3}{M_p^2}; \tag{19.47}$$

gravitinos might even be stable. As a minimal requirement we need to suppose that gravitinos did not dominate the energy density at the time of nucleosynthesis. Otherwise the expansion rate at the time of nucleosynthesis is not consistent with the observed abundances of the light elements but, even more dramatically, their decay products would break up He^4 and other nuclei. Even though gravitinos are very weakly interacting, there is a danger that they would be overproduced during the period of reheating that follows inflation. A natural estimate for their production rate per unit volume is obtained by assuming that they are produced in two-body scattering, by light particles with densities of order T^3, and that their cross sections behave as $1/M_p^2$:

$$n^2\langle\sigma v\rangle \approx T^6\langle\sigma v\rangle \approx \frac{T^6}{M_p^2}. \tag{19.48}$$

Integrating this over a Hubble time M_p/T^2 and dividing by the photon density, of order T^3, gives a rough estimate:

$$\frac{n_{3/2}}{s} \sim \frac{T}{M_p}. \tag{19.49}$$

Assuming 1 TeV for the gravitino mass, the requirement that gravitinos do not dominate before nucleosynthesis gives $T < 10^{12}$ GeV. But this is too crude. Considering the destruction of deuterium and lithium gives $T < 10^9$ GeV or possibly a much smaller value. This is a strong constraint on the nature of reheating after inflation, but it is not a problem for the low-scale hybrid models we discussed in the previous subsection.

19.2 The axion as the dark matter

Within the set of ideas we have discussed for physics beyond the Standard Model, there are two promising candidates for dark matter. One is the axion, which we discussed in Chapter 5 as a possible solution to the strong CP problem. A second is the lightest supersymmetric

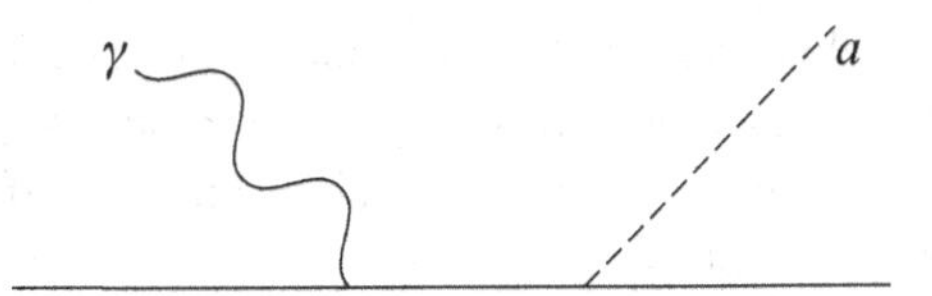

Fig. 19.2 In a Bremsstrahlung-like process, a lepton or nucleon can emit an axion when struck by a photon.

particle in models with an unbroken R-parity. We first discuss the axion, mainly because the theory is particularly simple. To begin, we need to consider the astrophysics of the axion a little further. There is a lower bound on the axion decay constant or, equivalently, an upper bound on its mass, arising from processes in stars. Axions can be produced by collisions deep within a star. Then, because of their small cross section, most axions will escape, carrying off energy. This has the potential to disrupt the star. We can set a limit by requiring that the flux of energy from the stars is not more than a modest fraction of the total energy flux.

To estimate these effects, we can first ask what sorts of processes might be problematic. A pair of photons can collide and produce an axion (using the $aF\tilde{F}$ coupling of the axion to the photon). Axions can be produced from nuclei or electrons in Compton-like and Bremsstrahlung processes (Fig. 19.2). The typical energies will be of order T.

For the Compton-like process of Fig. 19.2, the cross section is of order given by

$$\sigma_a \approx \frac{\alpha}{f_a^2}. \tag{19.50}$$

The total rate per unit time for a given electron to scatter off a photon in this way will be proportional to the photon density, which we will simply approximate as T^3. To obtain the total emission per unit volume we need to multiply, as well, by the electron density in the star. In the Sun this number is of order the total number of protons or electrons, 1.16×10^{57}, divided by the cube of the solar radius (in particle physics units, 3.5×10^{25} GeV^{-1}). This corresponds to

$$n_e \approx 3 \times 10^{-16}\ (\text{GeV})^3 \text{ electrons.} \tag{19.51}$$

Rather than calculate the absolute rate, we will compare it with the rate for neutrino production. We would expect that if axions carry off far more energy than neutrinos, this would be problematic. For neutrino production we might take n_e^2 and multiply by a typical weak cross section:

$$\sigma_\nu = G_F^2 E_\nu^2. \tag{19.52}$$

where E_ν is a typical neutrino energy. Finally, we take the temperature in the core of the star to be of order 1 MeV. Taking $f_a = 10^9$ gives, for the axion production rate,

$$R_a = 10^{-47}\ \text{GeV}^{-4} \tag{19.53}$$

while

$$R_\nu = 10^{-47}\ \text{GeV}^{-4}. \tag{19.54}$$

Clearly this analysis is crude; much more care is required in enumerating the different processes, evaluating their cross sections and integrating over particle momentum distributions. But this rough calculation indicates that 10^9 GeV is a plausible lower limit on the axion decay constant.

So, we have a lower bound on the axion decay constant. An upper bound arises from cosmology. Suppose that the Peccei–Quinn symmetry breaks before inflation. Then, throughout what will become the observable universe, the axion field is essentially constant. But, at early times the axion potential is negligible. To be more precise, consider the equation of motion of the axion field:

$$\ddot{a} + 3H\dot{a} + V'(a) = 0. \tag{19.55}$$

At very early times $H \gg m$ and the system is overdamped. The axion simply does not move. If the universe were very hot, the axion mass would actually have been much smaller than its current value. This is explained in Appendix C, but is easy to understand: at very high temperatures, the leading contribution in QCD to the axion potential comes from instantons. Instanton corrections are suppressed by $\exp\left[\frac{-8\pi^2}{g^2(T)}\right] = (\Lambda/T)^{b_0}$. They are also suppressed by powers of the quark masses. In other words, they behave as

$$V(a) = \prod_f m_f \Lambda^{b_0} T^{-b_0+n_f-4} \cos\theta, \tag{19.56}$$

where $\theta = a/f_a$ and n_f is the number of flavors with mass $\ll T$. This goes very rapidly to zero at temperatures above the QCD scale.

So the value of the axion field – the θ-angle – at early times, is most likely to be simply a random variable. Let us consider, then, the subsequent evolution of the system. The equation of motion for such a scalar field in an FRW background is by now quite familiar:

$$\ddot{a} + H\dot{a} + V'(a) = 0. \tag{19.57}$$

The potential $V(a)$ also depends on $T(t)$, which complicates the solution slightly, so let us first solve the problem with just the zero-temperature axion potential. In this case, the axion will start to oscillate when $H \sim m_a$. After this, the axions on the one hand dilute like matter, i.e. as $1/a^3$. The energy in radiation, on the other hand, dilutes like $a^4 \propto T^{-4}$. Assuming radiation domination when the axion starts to oscillate, we can determine the temperature at that time. Using our standard formula for the energy density,

$$\rho = \frac{\pi^2}{30} g^* T^4, \tag{19.58}$$

we have, just above the QCD phase transition, $g^* \approx 48$ (with the gluons, three quark flavors, three light neutrinos and the photon). Just below it we do not have the quarks or gluons but we should include the pions, so $g^* \approx 30$. Taking the larger value,

$$T_a = 10^2\ \text{GeV} \left(\frac{10^{11}\,\text{GeV}}{f_a}\right)^{1/2}. \tag{19.59}$$

At this time, the fraction of the energy density in axions is approximately

$$\frac{\rho_a}{\rho} = \frac{\frac{1}{2}f_a^2 m_a^2}{\rho} \approx \frac{1}{6}\frac{f_a^2}{M_p^2}. \tag{19.60}$$

So, if $f_a = 10^{11}$ GeV, axions come to dominate the energy density quite late, at $T \approx 10^{-3}$ eV. The temperature of axion domination scales with f_a, so 10^{16} GeV axions would dominate the energy density at 100 eV, which would be problematic.

However, the axion potential, as we have seen, is highly suppressed at temperatures above a few hundred MeV. So oscillation, sets in much later, in fact. We can make another crude estimate by simply supposing that the axion potential turns on at $T = 100$ TeV. In this case the axion fraction is large, of order $1/g^*$. So, if the axion density is to be compatible with the observed dark matter density for any value of f_a, we need to allow for the detailed temperature dependence of the axion mass. Using our formula for the axion potential as a function of temperature we can ask when the associated mass becomes of order the Hubble constant. After that time, axion oscillations are more rapid than the Hubble expansion so we might expect that the axion density will damp, subsequently, like matter. Let us take, specifically, $f_a = 10^{11}$ GeV. For the axion mass we can take

$$m_a(T) \approx 0.1 m_a(T=0)\left(\frac{\Lambda_{\rm QCD}}{T}\right)^{3.7}. \tag{19.61}$$

The axion then starts to oscillate when $T \approx 1.5$ GeV. At this time, axions represent about 10^{-9} of the energy density. One needs to do a bit more work to show that, in the subsequent evolution, the energy in axions relative to the energy in radiation falls roughly as $1/T$ but, for this decay constant, the axion and radiation energies become equal at roughly 1 eV. If the decay constant is significantly higher than 10^{11} GeV then the axions start to oscillate too late and dominate the energy density too early. If the decay constant is significantly smaller then the axions cannot constitute the presently observed dark matter.

So, on the one hand it is remarkable that there is a rather narrow range of axion decay constants that are consistent with observation. On the other hand, some assumptions that we have made in this section are open to question. In particular, as we will see when we discuss the problem of moduli in cosmology, there are reasons to suspect that the universe may never have been hotter than tens of MeV. In this case the upper limit on the axion decay constant, as we will discuss further below, can be much weaker.

19.3 The LSP as the dark matter

Stability is one criterion for a dark matter candidate; a suitable production rate is another. We can make a crude calculation, which indicates that with susy breaking in the TeV range the LSP density is in a suitable range for the LSP to be the dark matter. Consider particles X with mass of order 100 GeV and interacting with weak interaction strength.

Their annihilation and production cross sections go as $G_F^2 E^2$. So, in the early universe, the corresponding interaction rate is of order

$$\Gamma \approx \rho_X G_F^2 E^2 \approx \rho_X G_F^2 T^2. \tag{19.62}$$

These interactions will drop out of equilibrium when the mass of the particle X is small compared with the temperature, so that there is a large Boltzmann suppression of their production. This will occur when this rate is of order the expansion rate, or

$$T^3 e^{-M_X/T} \langle v\sigma \rangle \sim \frac{T^2}{M_p}. \tag{19.63}$$

Since the exponent is very small, once $T \sim 10 M_X$ we can get a rough estimate of the density by saying that

$$e^{-M_X/T} T^3 \sim \frac{G_F^{-2}}{M_p}. \tag{19.64}$$

The ratio of the X particle density and the total entropy, n_X/s, is then given by

$$\frac{n_X}{s} \approx \frac{G_F^{-2}}{(M_p T^3)}. \tag{19.65}$$

Assuming that $M_X \sim 100$ GeV and $T \sim 10$ GeV, this gives about 10^{-9} for the right-hand side. Since the energy density in radiation damps as T^{-4} while that for matter damps as T^{-3}, this gives matter–radiation equality at temperatures of order an electronvolt, as in the standard big bang cosmology.

Needless to say, this calculation is quite crude. Extensive, and far more sophisticated calculations have been done to find the regions of parameter space in different supersymmetric models which are compatible with the observed dark matter density. The basic starting point for these analyses is the Boltzmann equation. If the basic process is of the form $1 + 2 \leftrightarrow 3 + 4$ then

$$\begin{aligned} a^{-3}\frac{d}{dt}(n_1 a^3) = &\int \frac{d^3p_1}{(2\pi)^3 2E_1} \int \frac{d^3p_2}{(2\pi)^3 2E_2} \int \frac{d^3p_3}{(2\pi)^3 2E_3} \int \frac{d^3p_4}{(2\pi)^3 2E_4} \\ &\times [f_3 f_4 (1 \pm f_1)(1 \pm f_2) - f_1 f_2 (1 \pm f_3)(1 \pm f_4)] \\ &\times (2\pi)^4 \delta^4(p_1 + p_2 - p_3 - p_4)|\mathcal{M}|^2, \end{aligned} \tag{19.66}$$

where $\mathcal{M}$ is the invariant matrix element for the scattering process under consideration. The functions $f_1, \ldots, f_4$ are the distribution functions for the different species. These equations can be simplified in the high-temperature limit using Boltzmann statistics:

$$f(E) \rightarrow e^{\mu/T} e^{-E/T}. \tag{19.67}$$

Interactions are still fast enough at this time to maintain the equilibrium of the X momentum distributions (i.e. kinetic equilibrium) but not that of the X number. So it is the limiting value of the X chemical potential, μ_X, which we seek. In this limit, we have

$$f_3 f_4(1 \pm f_1)(1 \pm f_2) - f_1 f_2(1 \pm f_3)(1 \pm f_4)$$
$$\to e^{-(E_1+E_2)/T}\left(e^{(\mu_1+\mu_2)/T} - e^{(\mu_3+\mu_4)/T}\right). \tag{19.68}$$

Here we have used $E_1 + E_2 = E_3 + E_4$.

Things simplify further since all but the X particle (particle 1) are light and are nearly in equilibrium. Defining $n_i^{(0)}$ as the distributions in the absence of a chemical potential, and defining the thermally averaged cross section

$$\langle \sigma v \rangle = \frac{1}{n_1^{(0)} n_2^{(0)}} \int \frac{d^3p_1}{(2\pi)^3 2E_1} \int \frac{d^3p_2}{(2\pi)^3 2E_2} \int \frac{d^3p_3}{(2\pi)^3 2E_3} \int \frac{d^3p_4}{(2\pi)^3 2E_4} |\mathcal{M}|^2, \tag{19.69}$$

we have

$$a^{-3}\frac{d(n_x a^3)}{dt} = n_X^{(0)} n_2 \langle \sigma v \rangle \left(1 - \frac{n_X}{n_X^{(0)}}\right). \tag{19.70}$$

Detailed solutions of these equations (often without some of these simplifications) reveal, as one would expect, a range of parameters in the MSSM that are compatible with the observed dark matter density (Fig. 19.3).

So, while it is disturbing that we need to impose additional symmetries in the MSSM in order to avoid proton decay, it is also exciting that this leads to a possible solution of one of the most critical problems of cosmology: the identity of the dark matter.

Having contemplated stable, weakly interacting particles as the dark matter, it is clear that this is a possibility that one can consider without invoking supersymmetry. One can

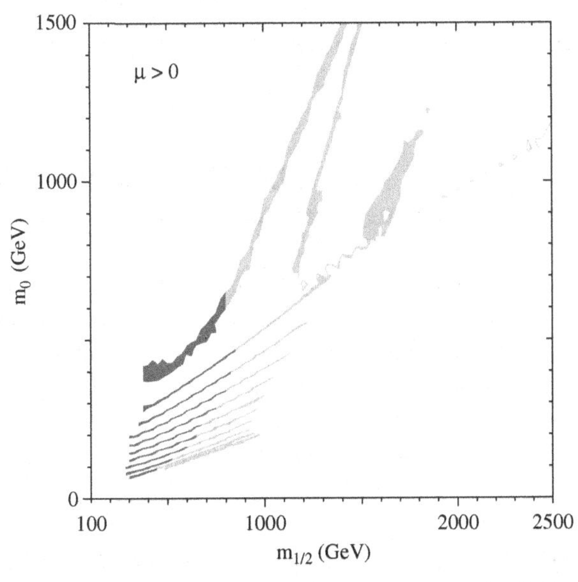

Fig. 19.3 Reprinted from J. R. Ellis *et al.*, Supersymmetric dark matter, *Phys. Lett. B*, **565**, 176 (2003), Figure 9. Copyright 2003, with permission from Elsevier.

simply postulate the existence of a massive stable particle with weak interactions with Standard Model fields. The fact that such a particle automatically leads to more or less the observed dark matter density is referred to as the "wimp miracle". One can also suppose that this particle has interactions with other particles, possibly lighter than it. Indeed, partly in response to potential signals, physicists have explored a broad range of possibilities.

19.3.1 The search for wimp dark matter

There is a variety of strategies that one can contemplate to search for weakly interacting masssive particle (*wimp*) dark matter, and this has become a major area of experimental and theoretical activity. There is no space here for an extensive review, so we will just mention the main strategies.

1. *Direct detection of dark matter* Here one searches for the scattering of dark matter particles off a target. Typically the targets are heavy nuclei, and one searches for the energy transferred to the recoiling nucleus. Such experiments must be conducted deep underground. Detectors must be sensitive to tiny energy depositions.
2. *Indirect detection of dark matter* Here one looks for the annihilation of dark matter particles against each other with the production of pairs of photons or neutrinos, for example. The galactic center, which is believed to contain a high concentration of dark matter particles, is a particularly interesting potential source for such events. A variety of experiments, particularly involving satellites such as PAMELA, Fermi and AMS, have been engaged in such searches.
3. *Accelerator searches* Many models of dark matter predict observable accelerator signals. Clearly direct observation in accelerators, complemented by discovery in either direct or indirect searches, would have the potential to provide a convincing discovery.

Among direct and indirect searches, significant exclusions have been achieved, There have also been tantalizing hints of possible signals.

19.4 The moduli problem

We have seen that in supersymmetric theories there are frequently light moduli (we have invoked this idea in our discussion of hybrid inflation). In string models we will find that such fields are ubiquitous. Such moduli, if they exist, pose a cosmological problem with some resemblance to the problems of axion cosmology.

In this section we will formulate the problem as it arises in gravity-mediated supersymmetry breaking. The potential for a modulus ϕ would be expected to take the form

$$V(\phi) = m_{3/2}^2 M_p^2 f(\phi/M_p). \tag{19.71}$$

By assumption f has a minimum at some value ϕ of order M_p. In the early universe, when the Hubble parameter is much greater than $m_{3/2}$, this potential is effectively quite small and there is in general no obvious reason why the field should sit at its minimum. So, when

$H \sim m_{3/2}$, the field is likely to lie at a distance M_p in field space from the minimum and to store an energy of order $m_{3/2}^2 M_p^2$. Like the axion, after this time, assuming it is within the domain of attraction of the minimum, it will oscillate, behaving like pressureless dust. Almost immediately, given our assumptions about scales, it comes to dominate the energy density of the universe and continues to do so until it decays. The problem is that the decay occurs quite late and the temperature after the decay is likely to be quite low.

We can estimate this temperature after ϕ decay, T_r, by first considering the lifetime of the ϕ particle. We might expect this to be

$$\Gamma_\phi = \frac{m_{3/2}^3}{M_p^2}, \tag{19.72}$$

assuming that the couplings of the ϕ field to other light fields are suppressed by a single power of M_p. Assuming that the decay products quickly thermalize, and noting that Γ_ϕ is the Hubble constant at the time of ϕ decay, gives

$$T_r^4 \approx \frac{m_{3/2}^6}{M_p^2} \approx (10 \text{ keV})^4. \tag{19.73}$$

Here we are assuming that $m_{3/2} \approx 1$ TeV. This is a temperature well below the temperature at which nucleosynthesis occurs. So, in such a picture, the universe is matter dominated during nucleosynthesis. But the situation is actually far worse: the decay products almost certainly destroy deuterium and the other light nuclei.

Two plausible resolutions for this puzzle have been put forward. The first is the obvious one, that there may simply be no moduli. Related to this, it is possible that, at the minimum of their potential, all the moduli may be charged under unbroken symmetries; these might be new discrete symmetries, for example beyond those of the MSSM. Furthermore, they may be much more strongly interacting than suggested above. In models with some degree of low-energy supersymmetry, there is a problem with this proposal. Assuming that the strong CP problem is solved by an axion, this field is accompanied by another scalar. This scalar must acquire mass in a supersymmetry-violating fashion, otherwise it would be quite heavy. Conceivably, of course, either there is no axion or supersymmetry is broken at an extremely high energy scale.

Alternatively, the moduli might be significantly more massive than 1 TeV. Note that T_r scales, like the moduli masses, to the $3/2$ power, so if the moduli masses are of order 30 - TeV or more, this temperature could be sufficiently high (10 MeV) that nucleosynthesis occurs (again).

Such a scenario raises interesting questions. First, one could well imagine that one or more of these moduli play the role of the inflaton. In this case the reheating temperature would be much lower than usually contemplated. Indeed, in effect the universe was never very hot. The conventional picture of the thermal production of dark matter cannot be operative. Even if the late-decaying moduli are not connected with inflation, these decays will dilute whatever dark matter might have been produced earlier. This dilution factor can easily be a factor of $10^9 - 10^{12}$. Any baryon number produced before these decays is also diluted by this factor. One can hope that the baryons are produced in the decays of these moduli, but this requires one to understand why such low-energy baryon-number violation does not cause difficulties for proton decay. In the rest of this section, we will consider non-thermal mechanisms to produce the dark matter; in the next section we will discuss possible mechanisms to produce the baryon asymmetry, and we will see that there is one mechanism which is capable of producing a large enough asymmetry to survive moduli decays.

19.4.1 The axion as dark matter again

If moduli dominated the energy density of the universe for some period, then the cosmological constraints on the axion mass and decay constant are appreciably modified. These can be formulated quite simply. If the axion initially has amplitude f_a then, when the axion begins to oscillate and decay, at $H \approx m_a$, the fraction of the energy density stored in axions is of order f_a^2/M_p^2. If, when the moduli decay, they reheat the universe to 10 MeV, the ratio of axions (dark matter) to radiation is

$$r_a = \frac{f_a^2}{M_p^2} \frac{10\,\text{MeV}}{T}. \tag{19.74}$$

In order that this fraction be of order unity only when the temperature is of order 1 eV, we require $f_a < 10^{14.5}$ GeV. This is close to, say, the unification scale.

19.4.2 Moduli and wimp dark matter

As we have noted, moduli domination followed by reheating to nucleosynthesis temperatures does not permit the usual thermal production of wimps. One possibility which has been widely considered is that dark matter might be produced in moduli decays. The problem with this is that typically dark matter is then overproduced. In an approximately supersymmetric limit, moduli decays to particles and their superpartners have equal branching ratios. This means that, when the moduli decay, an order-one fraction goes into each accessible state. If the LSP is one of these decay products, it will likely be overproduced (typically, subsequent annihilations are not strong enough to avoid this difficulty). There may be special ranges of parameters where dark matter production in this way is possible; alternatively, one might argue that a picture with moduli favors axions, or some other coherently produced particle, as the dark matter.

19.5 Baryogenesis

The baryon to photon ratio, n_B/n_γ, is quite small. At early times, when QCD was nearly a free theory, this slight excess would have been extremely unimportant. But, for the structure of our present universe, it is terribly important. One might imagine that n_B/n_γ is simply an initial condition, but it would be more satisfying if we could have some microphysical understanding of this asymmetry between matter and antimatter. Andrei

Sakharov, after the experimental discovery of CP violation, was the first to state precisely the conditions under which the laws of physics could lead to a prediction for the asymmetry.

1. *The underlying laws must violate baryon number* This condition is obvious; if there is, for example, no net baryon number initially, and if baryon number is conserved, the baryon number will always be zero.
2. *The laws of nature must violate CP* Otherwise, for every particle produced in interactions, an antiparticle will be produced as well.
3. *The universe, in its history, must have experienced a departure from thermal equilibrium* Otherwise, the CPT theorem ensures that the numbers of baryons and of antibaryons at equilibrium are zero. This can be proven with various levels of rigor, but one way to understand it is to observe that CPT ensures that the masses of the baryons and antibaryons are identical, so at equilibrium their distributions should be the same.

Subsequently, there have been many proposals for how the asymmetry might arise. In the next sections, we will describe several. *Leptogenesis* relies on lepton-number violation, something we know is true in nature but of whose underlying microphysics we are ignorant. *Baryogenesis through coherent scalar fields* (Affleck–Dine baryogenesis) also seems plausible. It is only operative if supersymmetry is unbroken up to comparatively low energies, but it can operate quite late in the evolution of the universe and can be extremely efficient. This could be important in situations like moduli decay or hybrid inflation where the entropy of the universe is produced very late, after the baryon number.

19.5.1 Baryogenesis through heavy particle decays

One well-motivated framework in which to consider baryogenesis is grand unification. Here one can satisfy all the requirements for baryogenesis. Baryon-number violation is one of the hallmarks of GUTs, and these models possess various sources of CP violation. As far as departure from equilibrium is concerned, the decays of massive gauge bosons X provide good candidates for a mechanism. To understand in a little more detail how the asymmetry can come about, note that CPT requires that the total decay rate of X is the same as that of its antiparticle $\bar{X}$. But it does not require equality of the number of decays to particular final states (partial widths). So, starting with equal numbers of X and $\bar{X}$ particles, there can be a slight asymmetry between processes such as

$$X \to d + e^-, \quad X \to d + u^c \tag{19.75}$$

and

$$\bar{X} \to d^c + e^+, \quad \bar{X} \to d^c + u, \tag{19.76}$$

where the superscript c denotes an antiparticle. The tree graphs for these processes are necessarily equal; any CP-violating phase simply cancels out when we take the absolute square of the amplitude (see Fig. 19.4). This is not true in higher order, where additional phases associated with real intermediate states can appear. Actually computing the baryon asymmetry requires an analysis of the Boltzmann equations, of the kind that we have encountered in our discussion of dark matter.

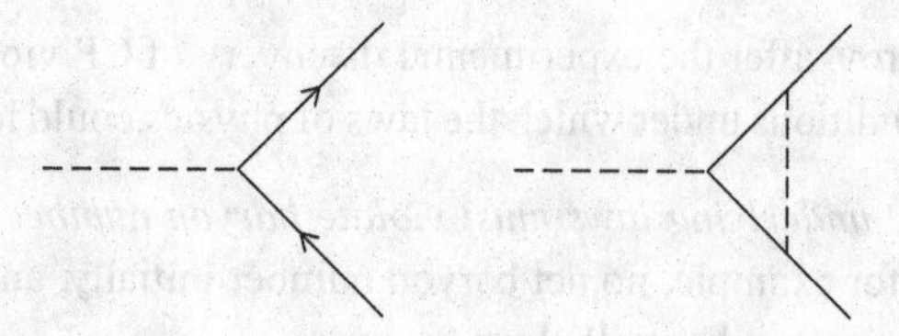

Fig. 19.4 Tree and loop diagrams whose interference can lead to an asymmetry in heavy particle decay.

There are reasons to believe, however, that GUT baryogenesis is not the origin of the observed baryon asymmetry. Perhaps the most compelling of these has to do with inflation. Assuming that there was a period of inflation, any pre-existing baryon number was greatly diluted by this. So, in order that one produces baryons through X boson decay, it is necessary that the reheating temperature after inflation be at least comparable with the X boson mass; but, we have seen that the scale of inflation is constrained to be less than 10^{16} GeV so we would require very efficient conversion of the energy density during inflation to radiation for this mechanism to be operative. Also, as we have explained, in supersymmetric theories a reheating temperature greater than 10^9 GeV leads to the overproduction of gravitinos.

19.5.2 Electroweak baryogenesis

The Standard Model, for some range of parameters, can satisfy all the conditions for baryogenesis. We saw in our discussion of instantons that the Standard Model violates baryon number. This violation is extremely small at low temperatures, so small that it is unlikely that in the history of the universe a single baryon has decayed in this way. The rate is so small because baryon-number violation is a tunneling process. If one could excite the system to high energies, one might expect that the rate would be enhanced. At high enough energies the system might simply be above the barrier. One can find the configuration which corresponds to sitting on top of the barrier by looking for static but unstable solutions of the equations of motion. Such a solution is known. It is called a *sphaleron* (from the Greek, meaning "ready to fall"). The barrier is quite high – from familiar scaling arguments, the sphaleron energy is of order $E_{\rm sp} = 1/(\alpha M_W)$. But this configuration is large compared with its energy; its size is of order M_W. As a result, it is difficult to produce in high-energy scattering. Two particles with enough energy to produce the sphaleron need to have momenta much higher than M_W. As a result, their overlap with the sphaleron configuration is exponentially suppressed.

At high temperatures one might expect that the sphaleron rate would be controlled by a Boltzmann factor, $e^{-E_{\rm sp}/T}$. Then, as the temperature increases, the rate would grow significantly. This turns out to be the case. In fact the rate is even larger than one might expect from this estimate, because $E_{\rm sp}$ itself is a function of T. At very high temperatures the rate has no exponential suppression at all and behaves as:

$$\Gamma = (\alpha_w T)^4. \tag{19.77}$$

These phenomena are discussed in Appendix C.

If the Higgs mass is not too large, the Standard Model can produce a significant departure from equilibrium. As the temperature rises, a simple calculation, described in Appendix C, shows that the Higgs mass increases (the mass-squared value becomes less negative) with temperature. At very high temperatures, the $SU(2) \times U(1)$ symmetry is restored. The phase transition between these two phases, for a sufficiently light Higgs, is first order. It proceeds by the formation of bubbles of the unbroken phase. The surfaces of these bubbles can be sites for baryon number production. These phenomena are also discussed in Appendix C. So, the third of Sakharov's conditions can be satisfied.

Finally, we know that the Standard Model violates CP. We also know, however, that it is crucial that there are three generations and that this CP violation vanishes if any quark masses are zero. As a result, even if the Higgs mass is small enough that the transition is strongly first order, any baryon number produced is suppressed by several powers of Yukawa couplings and is far too small to account for the observed matter–antimatter asymmetry.

In the MSSM the situation is somewhat better. There is a larger region of the parameter space in which the transition is first order and, as we have seen, there are many new sources of CP violation. As a result there is, as of the time of writing, a small range of parameters where the observed asymmetry could be produced in this way.

19.5.3 Leptogenesis

There is compelling evidence that neutrinos have mass. The most economical explanation of these masses is that they arise from a seesaw, involving gauge singlet fermions N_a. These couplings violate lepton number. So, according to Sakharov's principles, we might hope to produce a lepton asymmetry in their decays. Because the electroweak interactions violate baryon and lepton number at high temperatures, the production of a lepton number leads to the production of baryon number.

In general, there may be several N_a fields, with couplings of the form

$$\mathcal{L}_N = M_{ab} N_a N_b + h_{ai} H L_i N_a + \text{c.c.} \tag{19.78}$$

In a model with three Ns, there are CP-violating phases in the Yukawa couplings of the Ns to the light Higgs. The heaviest of the right-handed neutrinos, say N_1, can decay to ℓ and a Higgs, or to $\bar{\ell}$ and a Higgs. At tree level, as in the case of GUT baryogenesis, the rates for production of leptons and antileptons are equal, even though there are CP-violating phases in the couplings. It is necessary, again, to look at quantum corrections, since in these dynamical phases can appear in the amplitudes. At one loop the decay amplitude for N has a discontinuity associated with the fact that the intermediate N_1 and N_2 can be on-shell (a similar situation to that in Fig. 19.4). So, one obtains an asymmetry ϵ proportional to the imaginary parts of the Yukawa couplings of the Ns to the Higgs:

$$\epsilon = \frac{\Gamma(N_1 \to \ell H_2) - \Gamma(N_1 \to \bar{\ell}\bar{H}_2)}{\Gamma(N_1 \to \ell H_2) + \Gamma(N_1 \to \bar{\ell}\bar{H}_2)} = \frac{1}{8\pi} \frac{1}{hh^\dagger} \sum_{i=2,3} \text{Im}[(h_\nu h_\nu^\dagger)_{1i}]^2 f\left(\frac{M_i^2}{M_1^2}\right), \tag{19.79}$$

where f is a function that represents radiative corrections. For example, in the Standard Model, $f = \sqrt{x}[(x - 2)/(x - 1) + (x + 1)\ln(1 + 1/x)]$ while in the MSSM, $f = \sqrt{x}[2/(x - 1) + \ln(1 + 1/x)]$. Here we have allowed for the possibility of multiple Higgs fields, with H_2 coupling to the leptons. The rough order of magnitude here is readily understood by simply counting loop factors. It need not be very small.

Now, as the universe cools through temperatures of order the masses of the Ns, they drop out of equilibrium and their decays can lead to an excess of neutrinos over antineutrinos. Detailed predictions can be obtained by integrating a suitable set of Boltzmann equations. However, a rough estimate can be obtained by noting that the N_as drop out of equilibrium when their production rate becomes comparable with the expansion rate of the universe. If α represents a typical coupling, this occurs roughly when

$$\pi\alpha^2 T e^{-M_N/T} \approx \frac{T^2}{M_{\rm p}}. \tag{19.80}$$

Assuming that, in the polynomial terms, $T \sim M_N/10$ gives a density at this time of order

$$\frac{\rho_N}{\rho_{\rm tot}} \sim \frac{\pi T}{M_{\rm p}\alpha^2}. \tag{19.81}$$

Multiplying by ϵ, the average asymmetry in N decays, this estimate suggests a lepton number – and hence a baryon number – of order

$$\frac{\rho_{\rm B}}{\rho_{\rm tot}} \approx \epsilon \frac{M_N}{10\pi\alpha^2 M_{\rm p}}. \tag{19.82}$$

We have seen that ϵ is suppressed by a loop factor and by Yukawa couplings. So the above number can easily be compatible with observations, or even somewhat larger, depending on a variety of unknown parameters.

These decays, then, produce a net lepton number but not a net baryon number (hence they produce a net $B - L$). The resulting lepton number will be further processed by sphaleron interactions, yielding a net lepton and baryon number (recall that sphaleron interactions preserve $B - L$ but violate B and L separately). One can determine the resulting asymmetry by an elementary thermodynamics exercise: one introduces chemical potentials for each neutrino, quark and charged lepton species and then considers the various interactions between the species at equilibrium. For any allowed chemical reaction, the sum of the chemical potentials on each side of the reaction must be equal. For neutrinos, the relations come from:

1. the sphaleron interactions themselves,

$$\sum_i (3\mu_{q_i} + \mu_{\ell_i}) = 0; \tag{19.83}$$

2. a similar relation for QCD sphalerons,

$$\sum_i (2\mu_{q_i} - \mu_{u_i} - \mu_{d_i}) = 0; \tag{19.84}$$

3. the vanishing of the total hypercharge of the universe,

$$\sum_i \left(\mu_{q_i} - 2\mu_{\bar{u}_i} + \mu_{\bar{d}_i} - \mu_{\ell_i} + \mu_{\bar{e}_i}\right) + \frac{2}{N}\mu_{\rm H} = 0; \tag{19.85}$$

4. the quark and lepton Yukawa coupling relations

$$\mu_{q_i} - \mu_\phi - \mu_{d_j} = 0, \quad \mu_{q_i} - \mu_\phi - \mu_{u_j} = 0, \quad \mu_{\ell_i} - \mu_\phi - \mu_{e_j} = 0. \tag{19.86}$$

The number of equations here is the same as the number of unknowns. Combining these, one can solve for the chemical potentials in terms of the lepton chemical potential and, finally, in terms of the initial $B - L$. With N generations we obtain,

$$B = \frac{8N + 4}{22N + 13}(B - L). \tag{19.87}$$

Reasonable values of the neutrino parameters give asymmetries of the order we seek to explain. Note the sources of small numbers:

1. the phases in the couplings;
2. the loop factor;
3. the small density of the N particles when they drop out of equilibrium; parametrically, one has, e.g., for production,

$$\Gamma \sim e^{(-M/T)} g^2 T, \tag{19.88}$$

which is much less than $H \sim T^2/M_{\rm p}$ once the density is suppressed by $T/M_{\rm p}$, i.e. Γ is of order 10^{-6} for a 10^{13} GeV particle.

It should be noted that implementing this mechanism requires a high reheating temperature after inflation, of order the mass of the right-handed neutrinos. It is conceivable, as we have seen, that the reheating temperature is this high. It is also possible that the right-handed neutrinos are light. If the reheating temperatures (after inflation or moduli decay) are *very* low, some other mechanism to produce the dark matter is required.

It is interesting to ask, assuming that these processes are the source of the observed asymmetry, how many parameters which enter into the computation can be measured? It is likely that, over time, many parameters of the light neutrino mass matrices, including possible CP-violating phases, will be measured. But, while these measurements determine some N_i couplings and masses, they are not in general enough. In order to give a precise calculation, analogous to nucleosynthesis calculations, of the baryon number density one needs additional information about the masses of the fields N_i. One either requires some other (currently unforeseen) experimental access to this higher-scale physics or a compelling theory of neutrino mass in which symmetries, perhaps, reduce the number of parameters.

19.5.4 Baryogenesis through coherent scalar fields

In supersymmetric theories the ordinary quarks and leptons are accompanied by scalar fields. These scalar fields carry baryon and lepton number. A coherent field, i.e. a large classical value of such a field, can in principle carry a large amount of baryon number. As we will see, it is quite plausible that such fields were excited in the early universe, and this could have led to a baryon asymmetry.

To understand the basics of the mechanism, consider first a model with a single complex scalar field. Take the Lagrangian to be

$$\mathcal{L} = |\partial_\mu \phi|^2 - m^2|\phi|^2. \tag{19.89}$$

This Lagrangian has a symmetry, $\phi \to e^{i\alpha}\phi$, and a corresponding conserved current, which we will refer to as the baryon number:

$$j^\mu_{\rm B} = i(\phi^* \partial^\mu \phi - \phi \partial^\mu \phi^*). \tag{19.90}$$

It also possesses a CP symmetry:

$$\phi \leftrightarrow \phi^*. \tag{19.91}$$

With supersymmetry in mind we will think of m as of order M_W.

If we focus on the behavior of spatially constant fields, $\phi(\vec{x}, t) = \phi(t)$, this system is equivalent to an isotropic harmonic oscillator in two dimensions. In field theory, however, we expect that higher-dimensional terms will break the symmetry. In the isotropic oscillator analogy, this corresponds to anharmonic terms which break the rotational invariance. With a general initial condition the system will develop some non-zero angular momentum. If the motion is damped, so that the amplitude of the oscillations decreases, these rotationally non-invariant terms will become less important with time.

In the supersymmetric field theories of interest, supersymmetry will be broken by small quartic and higher-order couplings as well as by soft masses for the scalars. So, as a simple model, take:

$$\mathcal{L}_I = \lambda|\phi|^4 + \epsilon\phi^3\phi^* + \sigma\phi^4 + \text{c.c.} \tag{19.92}$$

These interactions clearly violate the conservation of B. For general complex ϵ and σ, they also violate CP. As we will shortly see, once supersymmetry is broken, quartic and higher-order couplings can be generated but these couplings $\lambda, \epsilon, \sigma \ldots$ will be extremely small, $\mathcal{O}(m_{3/2}^2/M_{\rm p}^2)$ or $\mathcal{O}(m_{3/2}^2/M_{\rm GUT}^2)$.

In order that these tiny couplings could have led to an appreciable baryon number, it is necessary that the fields, at some stage, were very large. To see how the cosmic evolution of this system can lead to a non-zero baryon number, first note that at very early times, when the Hubble constant $H \gg m$ (see Eq. (19.89)), the mass of the field is irrelevant. It is thus reasonable to suppose that at this early time $\phi = \phi_0 \gg 0$; later we will describe some specific suggestions as to how this might come about. This system then evolves like the axion or moduli. In the radiation- and matter-dominated eras, respectively, one has that

$$\phi = \frac{\phi_0}{(mt)^{3/2}} \sin mt \qquad \text{(radiation)} \tag{19.93}$$

$$\phi = \frac{\phi_0}{mt} \sin mt \qquad \text{(matter)}. \tag{19.94}$$

In either case the energy behaves, in terms of the scale factor $R(t)$, as

$$E \approx m^2 \phi_0^2 \left(\frac{R_0}{R}\right)^3, \tag{19.95}$$

i.e. it decreases as R^3, as would the energy of pressureless dust. One can think of this oscillating field as a coherent state of ϕ particles with $\vec{p} = 0$.

Now let us consider the effects of the quartic couplings. Since the field amplitude damps with time, their significance will decrease with time. Suppose, initially, that $\phi = \phi_0$ is real. Then the real and imaginary parts of ϕ satisfy, in the approximation that ϵ and δ are small,

$$\ddot{\phi}_i + 3H\dot{\phi}_i + m^2\phi_i \approx \text{Im}(\epsilon + \delta)\phi_r^3. \tag{19.96}$$

For large times, the right-hand side falls off as $t^{-9/2}$ whereas the left-hand side falls off only as $t^{-3/2}$. As a result, just as in our mechanical analogy, baryon number (angular momentum) violation becomes negligible. Equation (19.96) goes over to the free equation, with a solution of the form

$$\phi_i = a_\text{r}\frac{\text{Im}(\epsilon + \delta)\phi_0^3}{m^2(mt)^{3/4}}\sin(mt + \delta_\text{r}) \qquad \text{(radiation)}, \tag{19.97}$$

$$\phi_i = a_\text{m}\frac{\text{Im}(\epsilon + \delta)\phi_0^3}{m^3 t}\sin(mt + \delta_\text{m}) \qquad \text{(matter)}, \tag{19.98}$$

in the radiation- and matter-dominated cases, respectively. The constants δ_m, δ_r, a_m and a_r can easily be obtained numerically, and are of order unity:

$$a_\text{r} = 0.85, \quad a_\text{m} = 0.85, \quad \delta_\text{r} = -0.91, \quad \delta_\text{m} = 1.54. \tag{19.99}$$

However, now we have a non-zero baryon number; substituting in the expression for the current,

$$n_\text{B} = 2a_\text{r}\,\text{Im}(\epsilon + \delta)\frac{\phi_0^2}{m(mt)^2}\sin(\delta_\text{r} + \pi/8) \qquad \text{(radiation)} \tag{19.100}$$

$$n_\text{B} = 2a_\text{m} \quad \text{Im}(\epsilon + \delta)\frac{\phi_0^2}{m(mt)^2}\sin\delta_\text{m} \qquad \text{(matter)}. \tag{19.101}$$

Note that CP violation can be provided here by phases in the couplings and/or the initial fields. Note also that as expected, n_B is conserved at late times, in the sense that the baryon number per comoving volume is constant.

This mechanism for generating baryon number could be considered without supersymmetry. In that case, several questions would be begged.

- What are the scalar fields carrying baryon number?
- Why are the ϕ^4 terms so small?
- How are the scalars in the condensate (see Section 19.8) converted to more familiar particles?

In the context of supersymmetry there is a natural answer to each question. First, as we have stressed, there are scalar fields carrying baryon and lepton number. As we will see, in the limit in which supersymmetry is unbroken, there are typically approximate flat directions in the field space in which the quartic terms in the potential vanish. Finally, the scalar quarks and leptons can decay (in a baryon- and lepton-number-conserving fashion) to ordinary quarks.

19.6 Flat directions and baryogenesis

To discuss the problem of baryon number generation, we first want to examine the theory in a limit in which we ignore the soft susy-breaking terms. After all, at very early times, $H \gg M_W$ and these terms were irrelevant. We are now quite familiar with the fact that supersymmetric theories often exhibit flat directions. At the renormalizable level the MSSM possesses many flat directions. A simple example is

$$H_u = \begin{pmatrix} 0 \\ v \end{pmatrix}, \quad L_1 = \begin{pmatrix} v \\ 0 \end{pmatrix}, \tag{19.102}$$

where L_1 denotes the first-generation lepton doublet and v is an (at this point arbitrary) expectation value. This direction is characterized by a modulus which carries lepton number. Written in a gauge-invariant fashion, $\Phi = H_u L$. As we have seen, producing a lepton number is for all intents and purposes like producing a baryon number. Non-renormalizable, higher-dimensional terms, with more fields, can lift the flat direction. For example, the quartic term in the superpotential,

$$\mathcal{L}_4 = \frac{1}{M}(H_u L)^2 \tag{19.103}$$

respects all the gauge symmetries and is invariant under R-parity. Here M denotes some very large scale, perhaps the planck mass M_p. The term (19.103) gives rise to a potential

$$V_\text{lift} = \frac{|v|^6}{M^2}. \tag{19.104}$$

There are many more flat directions, and many of them do carry baryon or lepton number. A flat direction with both baryon and lepton numbers excited is the following:

$$\text{first generation, } Q_1^1 = b, \quad \bar{u}_2 = a, \quad L_2 = b; \tag{19.105}$$

$$\text{second generation, } \bar{d}_1 = \sqrt{|b|^2 + |a|^2}, \tag{19.106}$$

$$\text{third generation, } \bar{d}_3 = a. \tag{19.107}$$

(On Q the upper index is a color index and the lower index is an $SU(2)$ index; we have suppressed the generation indices.)

Higher-dimensional operators can again lift this flat direction. In this case the leading term is

$$\mathcal{L}_7 = \frac{1}{M^3}[Q^1 \bar{d}^2 L^1][\bar{u}^1 \bar{d}^2 \bar{d}^3]. \tag{19.108}$$

Here the superscripts denote flavor. We have suppressed the color and $SU(2)$ indices, but the brackets indicate sets of fields which are contracted in $SU(3)$- and $SU(2)$-invariant ways. In addition to being completely gauge invariant, this operator is invariant under ordinary R-parity. (There are lower-dimensional operators, including operators of dimension four, which violate R-parity.) It gives rise to a term in the potential

$$V_{\text{lift}} = \frac{\Phi^{10}}{M^6}. \tag{19.109}$$

Here Φ refers in a generic way to the fields whose vevs parameterize the flat directions (a, b).

19.7 Supersymmetry breaking in the early universe

We have indicated that higher-dimensional, supersymmetric operators give rise to potentials in the flat directions. To fully understand the behavior of the fields in the early universe, we need to consider supersymmetry breaking, which gives rise to additional potential terms.

In the early universe, we expect that supersymmetry was much more badly broken than it is in the present era. For example, during inflation, the non-zero energy density (the cosmological constant) breaks supersymmetry. Suppose that I is the inflaton field and that the inflaton potential arises because of a non-zero value of the auxiliary field for I, $F_I = \partial W/\partial I$. So, during inflation, supersymmetry is broken by a large amount. Not surprisingly, as a result there can be an appreciable supersymmetry-breaking potential for the field Φ. These contributions to the potential have the form

$$V_H = H^2\Phi^2 f(\Phi^2/M_{\text{p}}^2). \tag{19.110}$$

It is perfectly possible for the second derivative of the potential near the origin to be negative. In this case, writing our higher-dimensional term as

$$W_n = \frac{1}{M^n}\Phi^{n+3}, \tag{19.111}$$

the potential takes the form

$$V = -H^2|\Phi|^2 + \frac{1}{M^{2n}}|\Phi|^{2n+4}. \tag{19.112}$$

The minimum of the potential then lies at

$$\Phi_0 \approx M\left(\frac{H}{M}\right)^{1/(n+1)}. \tag{19.113}$$

More generally, one can see that the higher the dimension of the operator that raises the flat direction, the larger the starting value of the field and the larger the ultimate value of the baryon number. Typically, there is plenty of time for the field to find its minimum during inflation. After inflation, H decreases and the field Φ evolves adiabatically, oscillating slowly about the local minimum for some time.

Our examples illustrate that, in models with R-parity, the value of n and hence the size of the initial field can vary appreciably. Which flat direction is most important depends on the form of the mass matrix (i.e. it depends on in which directions the curvature of the potential is negative). With further symmetries, it is possible that n is larger and even that all operators which might lift the flat direction are forbidden. For the rest of this section, however, we will continue to assume that the flat directions are lifted by terms in the superpotential. If they are not, the required analysis is different since the lifting of a flat direction is entirely associated with supersymmetry breaking.

19.7.1 Appearance of the baryon number

The term $|\partial W/\partial \Phi|^2$ in the potential does not break either baryon number or CP. In most models it turns out that the leading sources of B and CP violation come from supersymmetry-breaking terms associated with F_I. These have the form

$$am_{3/2}W + bHW. \tag{19.114}$$

Here a and b are complex dimensionless constants. The relative phase in these two terms, δ, violates CP. This is crucial; if the two terms carry the same phase then this phase can be eliminated by a field redefinition, and we have to look elsewhere for possible CP-violating effects. Examining Eqs. (19.103) and (19.108), one sees that the term proportional to W violates B and/or L. In following the evolution of the field Φ, the important era occurs when $H \sim m_{3/2}$. At this point the phase misalignment of the two terms, along with the B-violating coupling, leads to the appearance of a baryon number. From the equations of motion, the equation for the time rate of change of the baryon number is

$$\frac{dn_{\rm B}}{dt} = \frac{\sin\delta\, m_{3/2}}{M^n}\phi^{n+3}. \tag{19.115}$$

Assuming that the relevant time is H^{-1}, one is led to the estimate (supported by numerical studies)

$$n_{\rm B} = \frac{1}{M^n}\sin\delta\,\Phi_0^{n+3}. \tag{19.116}$$

Here, Φ_0 is determined by $H \approx m_{3/2}$, i.e. $\Phi_0^{2n+2} = m_{3/2}^2 M^{2n}$.

19.8 The fate of the condensate

Of course, we do not live in a universe dominated by a coherent scalar field. In this section we consider the fate of a homogeneous condensate in the early universe, ignoring possible inhomogeneities. The following section will deal with the inhomogeneities and the interesting array of phenomena to which they might give rise.

We will adopt the following model for inflation. The features of this picture are true of many models of inflation but by no means all. We will suppose that the energy scale of inflation is $E \sim 10^{15}$ GeV and that inflation is due to a field, the inflaton I. We will take the amplitude of the inflaton, just after inflation, to be of order $M \approx 10^{18}$ GeV (the usual reduced Planck mass). Correspondingly, we will take the mass of the inflaton to be $m_I = 10^{12}$ GeV (so that $m_I^2 M_p^2 \approx E^4$). Correspondingly, the Hubble constant during inflation is of order $H_I \approx E^2/M_p \approx 10^{12}$ GeV.

After inflation ends the inflaton oscillates about the minimum of its potential, much like the field Φ, until it decays. We will suppose that the inflaton couples to ordinary particles with a rate suppressed by a single power of the Planck mass. Dimensional analysis then gives, for a rough value of the inflaton lifetime,

$$\Gamma_I = \frac{m_I^3}{M^2} \sim 1 \text{ GeV}. \tag{19.117}$$

The reheating temperature can then be obtained by equating the energy density at time, $H \approx \Gamma(\rho = 3H^2M^2)$, to the energy density of the final plasma:

$$T_R = T(t = \Gamma_I^{-1}) \sim (\Gamma_I M_p)^{1/2} \sim 10^9 \text{ GeV}. \tag{19.118}$$

The decay of the inflaton is not sudden, however, but leads to a gradual reheating of the universe, as described, for example, in the book by Kolb and Turner (1990). As a function of time,

$$T \approx \left(T_R^2 H(t) M_p\right)^{1/4}. \tag{19.119}$$

where $H(t)$ is the Hubble parameter as a function of time. For the field Φ our basic assumption is that during inflation it obtains a large value, in accord with Eq. (19.113). When inflation ends the inflaton, by assumption, still dominates the energy density for a time, oscillating about its minimum; the universe is matter dominated during this period. The field Φ now oscillates about a time-dependent minimum, given by Eq. (19.113). The minimum decreases in value with time, dropping to zero when $H \sim m_{3/2}$. During this evolution, a baryon number develops classically. This number is frozen once $H \sim m_{3/2}$.

Eventually the condensate will decay, through a variety of processes. As we have stressed, the condensate can be thought of as a coherent state of Φ particles. These particles – linear combinations of the squark and slepton fields – are unstable and will decay. However, for $H \leq m_{3/2}$ these lifetimes are much longer than those in the absence of the condensate. The reason is that the fields to which Φ couples have mass of order Φ, and Φ is large. Particles which are light in the presence of large ϕ form an ambient thermal bath. In most cases, the most important process which destroys the condensate is what we might call evaporation: particles in the ambient thermal bath can scatter off the particles in the condensate.

We can make a crude estimate for the reaction rate as follows. Because the particles which couple directly to Φ are heavy, interactions of Φ particles with light particles must involve loops. So we include a loop factor in the amplitude, of order α_2^2, the weak coupling squared. Because of the large masses, the amplitude is suppressed by Φ. Squaring and multiplying by the thermal density of the scattered particles gives a crude estimate for the reaction rate.

$$\Gamma_p \sim \alpha_2^2 \pi \frac{1}{\Phi^2} \left(T_R^2 H M\right)^{3/4}. \tag{19.120}$$

The condensate will evaporate when this quantity is of order H. Since we know the time dependence of Φ, this allows us to solve for this time. One finds that equality occurs, in the case $n = 1$, for $H_I \sim 10^2$–10^3 GeV. For $n > 1$ it occurs significantly later; for $n < 4$ it occurs before the decay of the inflaton and for $n \geq 4$ a slightly different analysis is required from that which follows. In other words, for the case $n = 1$, the condensate evaporates shortly after the baryon number is created but for larger n, it evaporates significantly later.

The expansion of the universe is unaffected by the condensate as long as the energy density in the condensate, $\rho_\Phi \sim m_\Phi^2 \Phi^2$, is much smaller than that of the inflaton, $\rho_I \sim H^2 M^2$. Assuming that $m_\Phi \sim m_{3/2} \sim 0.1$–$1$ TeV, a typical supersymmetry-breaking scale, one can estimate the ratio of the two densities at the time when $H \sim m_{3/2}$ as

$$\frac{\rho_\Phi}{\rho_I} \sim \left(\frac{m_{3/2}}{M_\text{p}}\right)^{2/(n+1)}. \tag{19.121}$$

We are now in a position to calculate the baryon to photon ratio in this model. Given our estimate of the inflaton lifetime, the coherent motion of the inflaton still dominates the energy density when the condensate evaporates. The baryon number equals the Φ density just before evaporation divided by the Φ mass (assumed to be of order $m_{3/2}$), while the inflaton number is ρ_I/M_I. So the baryon to inflaton ratio follows from Eq. (19.121). With the assumption that the inflaton energy density is converted to radiation at the reheating temperature, T_R, we obtain

$$\frac{n_\text{B}}{n_\gamma} \sim \frac{n_\text{B}}{\rho_I/T_\text{R}} \sim \frac{n_\text{B}}{n_\Phi}\frac{T_\text{R}}{m_\Phi}\frac{\rho_\Phi}{\rho_I} \sim 10^{-10}\left(\frac{T_\text{R}}{10^9\,\text{GeV}}\right)\left(\frac{M_\text{p}}{m_{3/2}}\right)^{(n-1)/(n+1)}. \tag{19.122}$$

Clearly the precise result depends on factors beyond those indicated here explicitly, such as the precise mass of the Φ particle(s). But as a rough estimate it is rather robust. For $n = 1$, it is in *precisely the right range* to explain the observed baryon asymmetry. For larger n, it can be significantly larger. In light of our discussion of the late decays of moduli this is potentially quite interesting. These decays produce a huge amount of entropy, typically increasing it by a factor 10^7 or so. The baryon density is diluted by a corresponding factor. But we see that coherent production can readily yield, prior to moduli decay, baryon to photon densities of the needed size.

There are many other issues which can be studied, both in leptogenesis and in Affleck–Dine baryogenesis, but it appears that both types of process might well account for the observed baryon asymmetry. The discovery (or not) of low-energy supersymmetry, and further studies of neutrino masses, might make one or the other picture more persuasive. Both pose challenges, as they involve couplings which we are not likely to be able to measure directly.

19.9 Dark energy

It has long been recognized that any cosmological constant in nature is far smaller than the scales of particle physics. Before the discovery of dark energy, many physicists conjectured that for some reason of principle this energy was zero. However, as we have seen, we now know that the dark energy is non-zero and in fact that it is the largest component of the energy density of the universe. Present data is compatible with the idea that this energy density represents a cosmological constant (for a cosmological constant, in the equation of state, we have $p = w\rho$ and $w = -1$; the Planck satellite, for example, gives $w = -1.10^{+0.08}_{-0.07}$) but other suggestions, typically involving time-varying scalar fields, have been offered and future surveys will improve the measurement of w.

Apart from its smallness, another puzzle surrounding the cosmological constant is simply one of coincidence: why is the dark energy density today comparable to the dark matter density? Weinberg has argued that it could not be much different from this in a universe containing stars and galaxies, provided that all the other laws of nature are as we observe. The basic point is that if the dark energy were, say, 10^3 times more dense than we observe, it would have come to dominate the energy density when the universe was much younger than it is today, at a time prior to the formation of galaxies and stars. The rapid acceleration after that time would have prevented the formation of structure. More refined versions of the argument give estimates for the dark energy within a factor ten of the measured value.

Weinberg speculated that perhaps the universe is much larger than we see (i.e. than our current horizon) and that in other regions it has different values of the cosmological constant. Only in those regions where Λ is very small would stars – and hence observers – form. Weinberg called this possible explanation (actually a prediction) of Λ the *weak anthropic argument*. We will return to this question in our studies of string theory, where we will see that such a *landscape* of ground states may exist.

Suggested reading

Seminal papers on inflation include that of Guth (1981), which proposed a version of inflation now often referred to as "old inflation," and those of Linde (1982) and Albrecht and Steinhardt (1982), which contain the germ of the slow-roll inflation idea stressed in this work. The ideas of hybrid inflation were developed by Linde (1994); those specifically discussed here were introduced by Randall *et al.* (1996) and Berkooz *et al.* (2004). There are a number of good texts on inflation and related issues, some of which we have mentioned in the previous chapter. These texts include those of Dodelson (2004), Kolb and Turner (1990) and Linde (1990). Dodelson provide a particularly up-to-date discussion of dark matter, including more detailed calculations than those presented here, and dark energy, including surveys of observational results. For a review of axions and their cosmology and astrophysics, see Turner (1990). For more recent papers which raise questions about the cosmological axion limits see, for example, Banks *et al.* (2003). The cosmological moduli problem, and possible solutions, were first discussed by Banks *et al.* (1994) and de Carlos *et al.* (1993). A general review of electroweak baryogenesis, including detailed discussions of phenomena at the bubble walls, appears in Cohen *et al.* (1993). A discussion of electroweak baryogenesis within the MSSM is given in Carena *et al.* (2003). A detailed review of baryogenesis is to be found in Buchmuller *et al.* (2005), while Enqvist and Mazumdar (2003) focuses on Affleck–Dine baryogenesis. A more comprehensive review of baryogenesis mechanisms appears in Dine and Kusenko (2003). Aspects of the cosmological constant, and especially Weinberg's anthropic prediction of Λ, are explained clearly in Weinberg (1989), with more recent additions in Vilenkin (1995) and Weinberg (2000).

Exercises

(1) Verify the slow-roll conditions, Eqs. (19.11) and (19.12). Determine the number of e-foldings and the size of $\delta\rho/\rho$ as a function of N.

(2) Work through the limits on the axion in more detail. Attempt to analyze the behavior of the axion energy in the high-temperature regime.

(3) Construct a discrete R symmetry which guarantees that the $H_U L$ flat direction is exactly flat. Assuming that the universe reheats to 100 MeV when a modulus decays, estimate the final baryon number of the universe in this case.

(4) Suppose that the characteristic scale of supersymmetry breaking is much higher than 1 TeV, say 10^9 GeV. Discuss baryogenesis by coherent scalar fields in such a situation.

PART 3

STRING THEORY

20 Introduction

String theory was stumbled on, more or less, by accident. In the late 1960s, string theories were first proposed as theories of the strong interactions. It was quickly realized, however, that, while hadronic physics has a number of string-like features, string theories were not suitable for a detailed description. In their simplest form, string theories have massless spin-2 particles and more than four dimensions of space–time, hardly features of the strong interactions. But a small group of theorists appreciated that the presence of a spin-2 particle implied that these theories were generally covariant and explored them through the 1970s and early 1980s, as possible theories of quantum gravity. Like field theories, the number of possible string theories seemed to be infinite, while, unlike field theories, there was reason to believe that these theories did not suffer from ultraviolet divergences. In the 1980s, however, studies of anomalies in higher dimensions suggested that all string theories with chiral fermions and gauge interactions suffered from quantum anomalies. But in 1984 it was shown that the anomalies cancel for two choices of gauge group. It was quickly recognized that the non-anomalous string theories do come close to unifying gravity and the Standard Model of particle physics. Many questions remained. Beginning in 1995, great progress was made in understanding the deeper structure of these theories. All the known string theories were understood to be different limits of some larger structure. As string theories still provide the only framework in which one can do systematic computations of quantum gravity effects, many workers use the term "string theory" to refer to some underlying structure which unifies quantum mechanics, gravity and gauge interactions.

String theory has provided us with many insights into what a fundamental theory of gravity and gauge interactions might look like, but there is still much we do not understand. We cannot really begin a course of action by enunciating some great principle and seeing what follows. We might, for example, have imagined that the underlying theory would be a string field theory, whose basic objects would create and annihilate strings. Some set of organizing principles would determine the action for this system, and the rest would be a problem of working out the consequences. But there are good reasons to believe that string theory is not like this. Instead, we can at best provide a collection of facts, organized according to the teacher/author/professor's view of the subject at any given moment. As a result, it is perhaps useful first to give at least some historical perspective as to how these facts were accumulated, if only to show that there are, as of yet, no canonical texts or sacred principles in the subject. In the next section we review a little of the remarkable history of string theory. In the following section we will attempt to survey what is known as of the time of writing: the various string theories, with their spectra and interactions, and the connections between them.

20.1 The peculiar history of string theory

For electrodynamics, the passage from classical to quantum mechanics is reasonably straightforward. But general relativity and quantum mechanics seem fundamentally incompatible. Viewed as a quantum field theory, Einstein's theory of general relativity is a non-renormalizable theory; its four-dimensional coupling constant has dimensions of inverse mass-squared. As a result, quantum corrections are very divergent. From the point of view developed in Part 1, these divergences should be thought of as cut off at some scale associated with new physics: general relativity is an incomplete theory. Hawking has discussed another sense in which gravity and quantum mechanics seem to clash. Hawking's paradox appears to be associated with phenomena at arbitrarily large distances – in particular, with the event horizons of large black holes. Because black holes emit a thermal spectrum of radiation, it seems possible for a pure state – a large black hole – to evolve into a mixed state. Such puzzles suggest that reconciling quantum mechanics and gravity will require a radical rethinking of our understanding of very short-distance physics.

Apart from its potential to reconcile quantum mechanics and general relativity, there is another reason that string theory has attracted so much attention: it is finite and free of the ultraviolet divergences that plague ordinary quantum field theories. In the previous chapters of this book we have adopted the point of view that our theories of nature should be viewed as effective theories; it is not clear that they can be complete in any sense. One might wonder whether some sort of structure exists where the process stops; where some finite, fundamental, theory accounts for the features of our present, more tentative, constructions. Many physicists have speculated through the years that these two questions are related; that an understanding of quantum general relativity would provide a fundamental length scale. The finiteness of string theory suggests it might play this role.

As mentioned above, string theory was discovered by accident in the 1960s, as physicists tried to understand certain regularities of the hadronic S-matrix. In particular, hadronic scattering amplitudes exhibited a feature then referred to as *duality*. Scattering amplitudes with two incoming and two outgoing particles (so-called $2 \rightarrow 2$ processes) could be described equally well by an exchange of mesons in the s channel or in the t channel (but not both simultaneously). This is not a property, at least perturbatively, of conventional quantum field theories. Veneziano succeeded in writing down an expression for an S-matrix with just the required properties. Veneziano's result was extended in a variety of ways and it was soon recognized, by Nambu, Susskind and others, that this model was equivalent to a theory of strings.

One could well imagine coming to string theory by a different route. Quantum field theory describes point particles. Apart from properties like mass and charge, no additional features (size, shape) are assigned to the basic entities. One could well imagine that this is naive but, in describing nature, quantum field theory is extraordinarily successful. In fact, there is no evidence for any size of the electron or the quarks down to distances of order 10^{-17} cm (energy scales of order several TeV). Still, it is natural to try to go beyond the idea of particles as points. The simplest possibility is to consider entities with a one-dimensional extent, strings. In the next few chapters we will discuss the features of theories of string. Here we just note that a straightforward analysis yields some remarkable results. A relativistic quantum string theory is necessarily:

1. a theory of general relativity;
2. a theory with gauge interactions;
3. finite; string world sheets are smooth. String interactions do not occur at space–time points but are spread out. As a result, in perturbation theory one does not have the usual ultraviolet divergences of quantum theories of relativistic particles.

These features are not postulated; they are inevitable. Other, seemingly less desirable, features also emerge: the space–time dimension has to be 26 or 10. Many string theories also contain tachyons in their spectrum, whose interpretation is not immediately clear.

As theories of hadronic physics, string theories had only limited success. Their spectra and S-matrices did share some features in common with those of the real strong interactions. But, as a result of the features described above – massless particles and unphysical space–time dimensions as well as the presence of tachyons in many cases – strings were quickly eclipsed by QCD as a theory of the strong interactions.

Despite these setbacks, string theory remained an intriguing topic. String theories were recognized to have short-distance behavior very different – and better – from that of quantum field theories. There was reason to think that such theories were free of ultraviolet divergences altogether. Scherk and Schwarz, and also Yoneya, made the bold proposal that string theories might well be sensible theories of quantum gravity. At the time, any concrete realization of this suggestion seemed to face enormous hurdles. The first string theories contained bosons only. But string theories with fermions were soon studied and were discovered to have another remarkable, and until then totally unfamiliar, property: supersymmetry. We have already learned a great deal about supersymmetry, but at this early stage its possible role in nature was completely unclear. In their early formulations, string theories only made sense in special, and at first sight uninteresting, space–time dimensions. But it had been conjectured since the work of Kaluza and Klein that higher-dimensional space-times might be "compactified", leaving theories which appear four-dimensional; Scherk and Schwarz hypothesized that this might be the case for string theories. Over a decade, Green and Schwarz studied supersymmetric string theories further, developing a set of calculational tools in which supersymmetry was manifest and which were suitable for tree level and one-loop computations.Witten and Alvarez-Gaume, however, pointed out that higher-dimensional theories in general suffer from anomalies, which render them inconsistent. They argued that almost all the then-known chiral string theories suffered from just such anomalies. It appeared that the string program was doomed; only two known string theories, theories without gauge interactions, seemed to make sense. Green and Schwarz, however, persisted. By a direct string computation they discovered that, while it was true that almost all would-be string theories with gauge symmetries are inconsistent, there was one exception among the then-known theories, with a gauge group $O(32)$. They reviewed the standard anomaly analysis and realized why $O(32)$ is special; this work raised the possibility that there might be one more consistent string theory, based on the gauge group $E_8 \times E_8$. The corresponding string theory, as well as another with gauge group $O(32)$ (known as the heterotic string theories), was promptly constructed.

This work stimulated widespread interest in string theory as a unified theory of all interactions, for now these theories appeared to be not only finite theories of gravity but also nearly unique. Compactification of the heterotic string on six-dimensional manifolds known as Calabi–Yau spaces were quickly shown to lead to theories which at low energies closely resemble the Standard Model, with similar gauge groups, particle content and other features such as repetitive generations, low-energy supersymmetry and dynamical supersymmetry breaking. The various string theories have since been shown to be part of a larger theory, suggesting that one is studying some unique structure which describes quantum gravity. Some basic questions about quantum gravity theories, such as Hawking's puzzle, have been at least partially resolved.

Many questions remain, however. There is still no detailed understanding of how string theory can make contact with experiment. There are a number of reasons for this. String theory, as we will see, is a theory with no dimensionless parameters. This is a promising beginning for a possible unified theory. But it is not clear how a small expansion parameter can actually emerge, allowing systematic computation. String theory provides no simple resolution of the cosmological-constant puzzle. Finally, while there are solutions which resemble nature, there are vastly more which do not. A principle, or dynamics, which might explain the selection of one vacuum or another has not emerged.

Yet string theory is the only model we have for a quantum theory of gravity. More than that, it is the only model we have for a finite theory which could be viewed as some sort of ultimate theory. At the same time, string theory addresses almost all of the deficiencies we have seen in the Standard Model and has the potential to encompass all the solutions we have proposed. The following are some examples.

1. The theory unifies gravity and gauge interactions in a consistent, quantum mechanical, framework.
2. The theory is completely finite. It has no free parameters. The constants of nature must be determined by the dynamics or other features internal to the theory.
3. The theory possesses solutions in which space–time is four-dimensional, with gauge groups close to the Standard Model and repetitive generations. It is in principle possible to compute the parameters of the Standard Model.
4. Many solutions exhibit low-energy supersymmetry, of the sort we have considered in the first part of this book.
5. Other solutions exhibit large dimensions, technicolor-like structures and the like.
6. The theory does not have continuous global symmetries but often possesses discrete symmetries, of the sort we have considered.

While these are certainly encouraging signs, we are a long way from a detailed understanding of how string theory might describe nature. We will see that there are fundamental obstacles to such an understanding. At the same time we will see that string theory provides a useful framework in which to assess proposals for Beyond the Standard Model physics.

The third part of this book is intended to provide the reader with an overview of superstring theory, with a view to connecting string theory with nature. In the next chapter we will study the bosonic string. We will understand how to find the spectra of string theories. We will also understand string interactions. The reason that string theories are so constrained is that strings can only interact in a limited set of ways, essentially by splitting and joining. We will explain how to translate this into concrete computations of scattering amplitudes.

In subsequent chapters we will turn to superstring theories obtain their spectra and understand their interactions. We will then turn to the compactification of string theories, focusing mainly on compactifications to four dimensions. We first consider toroidal compactifications of strings, whose features can be worked out quite explicitly. We also discuss orbifolds, simple string models which can exhibit varying amounts of supersymmetry. Then we devote a great deal of attention to compactifications on Calabi–Yau spaces. These are smooth spaces; superstring theories compactified on these spaces exhibit varying amounts of supersymmetry. Many look quite close to the real world.

Finally, we will turn to the question of developing a realistic string phenomenology. Having seen the many intriguing features of string models, we will point out some of the challenges. Among these are the following.

1. There is a proliferation of classes of string vacua.
2. Within different classes, moduli exist.
3. Mechanisms which generate potentials for moduli are known but, in regimes where calculations can be performed systematically, they tend not to produce stable minima. The question of supersymmetry breaking is closely related to the question of stabilizing moduli.
4. There are detailed issues, such as proton decay, features of quark and lepton masses and many others.

We will touch on some proposed solutions to these puzzles. Much string model building simply posits that moduli have been fixed in some way and a vacuum with desirable properties has somehow been selected by some (unknown) overarching principle. This is often backed up by calculations which, while not systematic, are at least suggestive that moduli are stabilized. An alternative viewpoint is provided by the *landscape*. This refers to the possibility that the theory possesses a huge array of stable and/or metastable ground states. We have already discussed such a hypothesis in the context of the cosmological-constant problem. It is conceivable that string theory provides a realization of this possibility. In particular, string theories possess various tensor fields which, when compactified, support quantized fluxes. The possible choices of flux vastly increase the possible array of (metastable) string ground states. If one simply accepts that there is such a landscape of states, and that the universe samples ("scans") many of these states in some way, then one is led to think about the distributions of parameters of low-energy physics. This applies not merely to the coupling constants but also to the gauge groups, particle content, scale of supersymmetry breaking and value of the cosmological constant. For better or worse, this is in some sense the ultimate realization of the notions of naturalness which so concerned us in Part 1. The question is: why is the universe we see around us the likely outcome of a distribution of this sort? We will leave it for the readers – and for experiment – to sort out which, if any, of these viewpoints may be correct.

This is not a string theory textbook. The reader will not emerge from these few chapters with the level of technical proficiency in weakly coupled strings provided by Polchinski's text, or with the expertise in Calabi–Yau spaces provided by the book of Green *et al.* (1987). In order to obtain quickly the spectra of various string theories, the following chapters heavily emphasize light cone techniques. While some aspects of the covariant treatment are developed in order to explain the rules for computing the *S*-matrix, many important topics, especially the Polyakov path integral approach and Becchi, Rouet, Stora and Tyutin (BRST) quantization, are given only a cursory treatment. Similarly, the introduction to *D*-brane physics provides some basic tools but does not touch on much of the well-developed machinery of the subject.

Suggested reading

The introduction of the book by Green *et al.* (1987) provides a particularly good overview of the history of string theory and some of its basic structure. The introductory chapter of Polchinski's text (1998) provides a good introduction to more recent developments and a perspective on why strings might be important in the description of nature. The reader who wishes a more thorough grounding in the physics of *D*-branes will want to consult the texts of Polchinski (1998), Johnson (2003) and Becker *et al.* (2007).

21 The bosonic string

A particle moving through space sweeps out a path called a *world line*. The action of the particle is just the integral of the invariant length element along the path, up to a constant.

Suppose we want to describe the motion of a string. A string, as it moves, sweeps out a two-dimensional surface in space–time called a *world sheet*. We can parameterize the path in terms of two coordinates, one time-like and one space-like, denoted τ and σ or σ_0 and σ_1. The action should not depend on the coordinates we use to parameterize the surface. Polyakov stressed that this can be achieved by using the formalism of general relativity. Introduce a two-dimensional metric $\gamma_{\alpha\beta}$. Then an invariant action is

$$S = \frac{T}{2} \int d^2\sigma \sqrt{-\gamma}\, \gamma^{\alpha\beta} \partial_\alpha X^\mu \, \partial_\beta X^\nu \, \eta_{\mu\nu}. \tag{21.1}$$

Here our conventions are such that, for a flat space,

$$\gamma = \eta = \begin{pmatrix} -1 & 0 \\ 0 & 1 \end{pmatrix} \tag{21.2}$$

(similarly, our D-dimensional space–time metric is $ds^2 = -dt^2 + d\vec{x}^2$).

This action has a large symmetry group. There are, first, general coordinate transformations of the two-dimensional surface. For a simple topology (plane or sphere), these permit us to bring the metric to the form

$$\gamma = e^\phi \eta. \tag{21.3}$$

In this gauge (the *conformal gauge*) the action is independent of the angle ϕ:

$$S = -\frac{T}{2} \int d^2\sigma \, \eta^{\alpha\beta} \, \partial_\alpha X^\mu \, \partial_\beta X^\nu \, \eta_{\mu\nu}. \tag{21.4}$$

It is possible to fix this symmetry further. To motivate this gauge choice, we consider an analogous problem in field theory. In a gauge theory such as QED we can fix a covariant gauge, $\partial \cdot A = 0$. This gauge fixing, while manifestly Lorentz invariant, is not manifestly unitary. We might try to quantize covariantly by introducing creation and annihilation operators a^μ. These would obey

$$[a^\mu, a^{\dagger\nu}] = g^{\mu\nu}, \tag{21.5}$$

so that some states would seem to have a negative norm. If one proceeds in this way, it is necessary to prove that states with negative (or vanishing) norm cannot be produced in scattering amplitudes.

One way to deal with this is to choose a non-covariant gauge. The Coulomb gauge is a familiar example, but a particularly useful description of gauge theories is obtained by choosing the *light cone gauge*. First, define light cone coordinates

$$x^{\pm} = \frac{1}{\sqrt{2}}(x^0 \pm x^{D-1}). \tag{21.6}$$

We will simply denote as $\vec{X}$ the remaining, transverse, coordinates. Correspondingly, one defines the light cone momenta

$$p^{\pm} = \frac{1}{\sqrt{2}}(p^0 \pm p^{D-1}), \qquad \vec{p}. \tag{21.7}$$

Note that

$$A \cdot B = -(A^+B^- + A^-B^+) + \vec{A} \cdot \vec{B}. \tag{21.8}$$

Now we will think of x^+ as our time variable. The "Hamiltonian" generates translations in x^+; it is in fact p^-. Note that for a particle,

$$p^2 = -2p^+p^- + \vec{p}^{\,2} \tag{21.9}$$

and the Hamiltonian is

$$H = \frac{1}{p^+}p^-. \tag{21.10}$$

Having made this choice of variables, one can then make the gauge choice $A^+ = 0$. In the Lagrangian there are no terms involving $\partial_+ A^-$, so A^- is not a dynamical field; only the $D-2$ A^is are dynamical. So we have the correct number of physical degrees of freedom. One simply solves for A^- by using its equations of motion. In the early days of QCD this description proved useful in understanding very high energy scattering. In practice, similar algebraic gauges are still very useful.

Light cone coordinates, more generally, are very helpful for identifying physical degrees of freedom. Consider the problem of counting the degrees of freedom associated with some tensor field $A^{\mu\nu\rho}$. For a massive field, one counts by going to the rest frame and restricting the indices μ, ν, ρ to be $(D-1)$-dimensional. For a massless field, the relevant group is the "little group" of the Lorentz group, $SO(D-2)$. Correspondingly, one restricts the indices to be $(D-2)$-dimensional. So, for example, for a massless vector, there are $D-2$ degrees of freedom; for a symmetric traceless tensor (the graviton), there are $[(D-2)(D-1)/2]-1$. Light cone coordinates and the light cone gauge, provide an immediate realization of this counting.

For many questions in quantum field theory, covariant methods are much more powerful than use of the light cone. Quantum field theorists are familiar with techniques for coping with covariant gauges. These involve the introduction of additional fictitious degrees of freedom (Faddeev–Popov ghosts). It is probably fair to say that most quantum field theorists do not know much about gauges such as the light cone gauge (there is almost no treatment of these topics in standard texts). But we will see in string theory that the light cone gauge is quite useful in isolating the physical degrees of freedom of strings. It lacks some of the elegance of covariant treatments but avoids the need to introduce an intricate ghost structure and, as in field theory, the physical degrees of freedom are manifest. The differences between the covariant and light cone treatments, as we will see, are most dramatic when we consider supersymmetric strings. In the light cone approach of Green and Schwarz, space–time supersymmetry is manifest. In the covariant approach, it is not at all apparent. However, for the discussion of interactions the light cone treatment tends to be rather awkward. In this chapter we will first introduce the light cone gauge and then go on to discuss aspects of the covariant formulation. The suggested readings should satisfy the reader interested in more details of the covariant treatment.

21.1 The light cone gauge in string theory

21.1.1 Open strings

In the conformal gauge, (see Eq. (21.3)) we can use our coordinate freedom to choose $X^+ = \tau$. We also can choose the coordinates such that the momentum density $\mathcal{P}^+$ is constant on the string. In this gauge, in D dimensions the independent degrees of freedom of a single string are the coordinates $X^I(\sigma, \tau)$, $I = 1, \ldots, D-2$. They are each described by the Lagrangian of a free two-dimensional field,

$$S = \frac{T}{2} \int d^2\sigma [(\partial_\tau X^I)^2 - (\partial_\sigma X^I)^2]. \tag{21.11}$$

It is customary to define another quantity, α' (the *Regge slope*), with dimensions of length-squared:

$$\alpha' = \frac{1}{2\pi T}. \tag{21.12}$$

We will generally take a step further and use units with $\alpha' = 1/2$. In this case, the action is:

$$S = \frac{1}{2\pi} \int d^2\sigma [(\partial_\tau X^I)^2 - (\partial_\sigma X^I)^2]. \tag{21.13}$$

The reader should be alerted to the fact that there is another common choice of units, $\alpha' = 2$, and we will encounter this later. In this case, the action has a factor $1/(8\pi)$ out front.

In order to write down the equations of motion, we need to specify boundary conditions in σ. Consider, first, open strings, i.e. strings with two free ends. We want to choose boundary conditions such that when we vary the action we can ignore surface terms. There are two possible choices:

1. *Neumann boundary conditions*,

$$\partial_\sigma X^I(\tau, 0) = \partial_\sigma X^I(\tau, \pi) = 0; \tag{21.14}$$

2. *Dirichlet boundary conditions,*

$$X^I(\tau, 0) = X^I(\tau, \pi) = \text{ const.} \tag{21.15}$$

It is tempting to discard the second possibility, as it appears to violate translation invariance. So, for now, we will consider only Neumann boundary conditions but will return later to the Dirichlet conditions.

We want to write down a Fourier expansion for the X^Is. The normalization of the coefficients is conventionally taken to be somewhat different from that of relativistic quantum field theories:

$$X^I = x^I + p^I\tau + i\sum_{n\neq 0} \frac{1}{n}\alpha_n^I e^{-in\tau} \cos n\sigma. \tag{21.16}$$

The α_n^μs are, up to constants, ordinary creation and annihilation operators:

$$\alpha_n^I = \sqrt{n}a_n, \quad \alpha_{-n}^I = \sqrt{n}a_n^\dagger. \tag{21.17}$$

Because we are working at finite volume (in the two-dimensional sense) there are normalizable zero modes, the x^Is and p^Is. They correspond to the coordinate and momentum of the center of mass of a string. From our experience in field theory, we know how to quantize this system. We impose the commutation relation

$$[\partial_\tau X^I(\sigma, \tau), X^J(\sigma', \tau)] = \frac{-i}{\pi}\delta^{IJ}(\sigma - \sigma'). \tag{21.18}$$

This is satisfied by

$$[x^I, p^J] = i\delta^{IJ}, \quad [\alpha_n^I, \alpha_{n'}^J] = n\delta_{n+n',0}\delta^{IJ}. \tag{21.19}$$

The states of this theory can be labeled by their transverse momenta $\vec{p}$ and by integers corresponding to the occupation numbers of the infinite set of oscillator modes. It is helpful to keep in mind that this is just the quantization of a set of free two-dimensional fields in a finite volume.

We can write down a Hamiltonian for this system. With normal ordering this is

$$H = \vec{p}^{\,2} + N + a, \tag{21.20}$$

where

$$N = \sum_{n=1}^{\infty} \alpha_{-n}^I \alpha_n^I \tag{21.21}$$

and a is a normal ordering constant. States can be labeled by the occupation numbers for each mode, N_{n_i} and their momentum p^I:

$$|p^I, \{N_{n_i}\}\rangle \tag{21.22}$$

The light cone Hamiltonian H generates translations in τ. It is convenient to refine the gauge choice as follows:

$$X^+ = p^+\tau.$$

Since p^- is conjugate to the light cone time x^+, we have

$$p^- = H/p^+ \tag{21.23}$$

or

$$p^+p^- = \vec{p}^{\,2} + N + a, \quad M^2 = N + a. \tag{21.24}$$

So the quantum string describes a tower of states, of arbitrarily large mass. The constant a is not arbitrary; we will see shortly that

$$a = -1. \tag{21.25}$$

This means that the lowest state is a tachyon. We can label this state simply as

$$|T(\vec{p})\rangle = |\vec{p}, \{0\}\rangle \equiv |\vec{p}\rangle. \tag{21.26}$$

The state carries transverse momentum $\vec{p}$ and longitudinal momenta p^+ and p^- and is annihilated by the infinite tower of oscillators. The significance of this instability is not immediately clear; we will close our eyes to it for now and proceed to look at other states in the spectrum. When we study the superstring, we will often find that there are no tachyons.

The first excited state is

$$|A^I\rangle = \alpha^I_{-1}|\vec{p}\rangle. \tag{21.27}$$

Its mass is given by

$$m_A^2 = 1 + a. \tag{21.28}$$

Now we can see why $a = -1$. Here, $\vec{A}$ is a vector field with $D-2$ components. In D dimensions, a massive vector field has $D-1$ degrees of freedom; a massless vector has $D-2$ degrees of freedom. So $\vec{A}$ must be massless and $a = 1$ if the theory is Lorentz invariant. Later, we will give a fancier argument for the value of a but the content is equivalent.

At level 2 we have a number of states,

$$\alpha^I_{-2}|\vec{p}\rangle, \quad \alpha^I_{-1}\alpha^J_{-1}|\vec{p}\rangle. \tag{21.29}$$

These include a vector, a scalar and a symmetric tensor. We will not attempt here to group them into representations of the Lorentz group.

It turns out that the value of D is fixed: $D = 26$. In the light cone formulation the issue is that the light cone theory is not manifestly Lorentz invariant. To establish that the theory is Poincaré invariant, it is necessary to construct the full set of Lorentz generators and carefully check their commutators. This analysis yields the conditions $D = 26$ and $a = -1$. Later, we will discuss further the derivation of this result. In a manifestly covariant formulation such as the conformal gauge: the issue is one of unitarity, as in gauge field theories. The decoupling of negative- and zero-norm states yields, again, the condition $D = 26$.

Turning to the gauge boson, it is natural to ask: what are the fields charged under the gauge symmetry? The answer is suggested by a picture of a meson as a quark and antiquark connected by a string. We can allow the ends of the strings to carry various types of charge. These are known as Chan–Paton factors. In the case of the bosonic string these can be, for example, a fundamental and antifundamental of $SU(N)$. Then the string itself transforms as a tensor product of vector representations. Because the open strings include massless gauge bosons, this product must lie in the adjoint representation of the group. In bosonic string theory one can also have $SO(N)$ and $Sp(N)$ groups. In the case of a superstring we will see that the group structure is highly restricted. The theory will make sense only in ten flat dimensions, and then only if the group is $O(32)$.

21.2 Closed strings

We have begun with open strings, since these are in some ways the simplest, but theories of open strings by themselves are incomplete. There are always processes which will produce closed strings. For closed strings, we again have a set of fields $X^I(\sigma, \tau)$. Their action is identical to what we wrote down before, but they now obey the boundary conditions

$$X^I(\sigma + \pi, \tau) = X^I(\sigma, \tau). \tag{21.30}$$

Again, we can write a mode expansion:

$$X^I = x^I + p^I\tau + \frac{i}{2}\sum_{n\neq 0}\frac{1}{n}\left(\alpha_n^I e^{-2in(\tau-\sigma)} + \tilde{\alpha}_n^I e^{-2in(\tau+\sigma)}\right). \tag{21.31}$$

The exponential terms are the familiar solutions to the two-dimensional wave equation. One can speak of modes moving to the left ("left movers") and to the right ("right movers") on the string. Again we have commutation relations:

$$[x^I, p^J] = i\delta^{IJ}, \quad [\alpha_n^I, \alpha_{n'}^J] = n\delta_{n+n'}\delta^{IJ}, \quad [\tilde{\alpha}_n^I, \tilde{\alpha}_{n'}^J] = n\delta_{n+n'}\delta^{IJ}. \tag{21.32}$$

Now the Hamiltonian is

$$H = \vec{p}^{\,2} + N + \tilde{N} + b, \tag{21.33}$$

where

$$N = \sum_{n=1}^{\infty} \alpha_{-n}^I \alpha_n^I, \quad \tilde{N} = \sum_{n=1}^{\infty} \tilde{\alpha}_{-n}^I \tilde{\alpha}_n^I. \tag{21.34}$$

In working out the spectrum there is an important constraint. There should be no special point on the string, i.e. translations in the σ direction should leave states alone. The generator of constant shifts of σ can be found by the Noether procedure:

$$\mathcal{P}_\sigma = \int d\sigma\, \partial_\tau X^I \partial_\sigma X^I = N - \tilde{N}. \tag{21.35}$$

So we need to impose the constraint $N = \tilde{N}$ on the states.

Once more, the lowest state is a scalar,

$$|T\rangle = |\vec{p}\rangle, \quad m_T^2 = b. \tag{21.36}$$

Because of the constraint, the first excited state is

$$|\Psi_{IJ}\rangle = \tilde{\alpha}^{I}_{-1}\alpha^{J}_{-1}|\vec{p}\rangle. \tag{21.37}$$

We can immediately decompose these states into irreducible representations of the little group; there is a symmetric traceless tensor, a scalar (the trace) and an antisymmetric tensor. A symmetric, traceless, tensor should have, if massive, $D^2 - D - 1$ states. Here, however, we have only $D^2 - 3D + 1$ states. This is precisely the correct number of states for a massless, spin-2 particle – a graviton. The remaining states are precisely the number for a massless antisymmetric tensor field and a scalar. So we learn that $b = -2$.

This is a remarkable result. General arguments, going back to Feynman, Weinberg and others, show that a massless spin-2 particle, in a relativistic theory, necessarily couples like a graviton in Einstein's theory. So string theory is a theory of general relativity. This bosonic string is clearly unrealistic, but the presence of the graviton will be a feature of all string theories, including the more realistic ones.

21.3 String interactions

The light cone formulation is very useful for determining the spectrum of string theories, but it is somewhat more awkward for the discussion of interactions. As explained in the introduction to this chapter, string interactions are determined geometrically, by the nature of the string world sheet. Actually turning drawings of world sheets into a practical computational method is surprisingly straightforward. This is most easily done using the conformal symmetry of the string theory. So we return to the conformal gauge. There are close similarities between the treatment of open and closed strings. We will start with closed strings, for which the Green's functions are somewhat simpler. At the end of this chapter we will return to open strings.

21.3.1 String theory in conformal gauge

In conformal gauge the action is

$$S = \frac{1}{\pi}\int d^2\sigma[(\partial_\tau X^\mu)^2 - (\partial_\sigma X^\mu)^2]. \tag{21.38}$$

Introducing the two-dimensional light cone coordinates

$$\sigma_\pm = \sigma_0 \pm \sigma_1, \tag{21.39}$$

the flat world-sheet metric takes the form

$$\eta_{+-} = \eta_{-+} = -\frac{1}{2} \tag{21.40}$$

and the action can be written as

$$S = \frac{1}{8\pi}\int d\sigma_+ d\sigma_-\, \partial_{\sigma_+}X^\mu\, \partial_{\sigma_-}X^\mu. \tag{21.41}$$

At the classical level this action is invariant under a conformal rescaling of the coordinates. If we introduce light cone coordinates on the world sheet then the action is invariant under the transformations

$$\sigma_{\pm} \to f_{\pm}(\sigma_{\pm}). \tag{21.42}$$

Later, we will Wick-rotate and work with complex coordinates; these conformal transformations will then be the conformal transformations familiar in complex variable theory. It is well known that, by a conformal transformation, one can map the plane into a sphere, for example. In this case the regions at infinity with incoming or outgoing strings are mapped to points. The creation or destruction of strings at these points is described by local operators in the two-dimensional world-sheet theory. In order to respect the conformal symmetry these operators must, like the action, be integrals over the world sheet of local dimension-two operators. These operators are known as vertex operators, $V(\sigma, \tau)$.

In conformal gauge the action also contains Faddeev–Popov ghost terms, associated with fixing the world-sheet general coordinate invariance. We will discuss some of their features later. But we will focus on the fields X^{μ} first. If we simply write down mode expansions for the fields (taking closed strings, for definiteness),

$$X^{\mu} = x^{\mu} + p^{\mu}\tau + i\sum_{n\neq 0} \frac{1}{n}\left(\alpha_n^{\mu} e^{-2in(\tau-\sigma)} + \tilde{\alpha}_n^{\mu} e^{-2in(\tau+\sigma)}\right), \tag{21.43}$$

then we will encounter difficulties. The α^{μ}s will now obey the commutation relations

$$[x^{\mu}, p^{\nu}] = ig^{\mu\nu}, \quad \left[\alpha_n^{\mu}, \alpha_{n'}^{\nu}\right] = \left[\tilde{\alpha}_n^{\mu}, \tilde{\alpha}_{n'}^{\nu}\right] = n\delta_{n+n'}g^{\mu\nu}. \tag{21.44}$$

If we proceed naively, for $\mu = \nu = 0$ the minus sign from g^{00} means that we will have states in the spectrum of negative or zero norm.

The appearance of negative-norm states is familiar in gauge field theory. The resolution of the problem, there, is gauge invariance. One can either choose a gauge in which there are no states with negative norm or one can work in a covariant gauge in which the negative-norm states are projected out. In a modern language, this projection is implemented by the BRST procedure. But it is not hard to check that, in a covariant gauge, low-order diagrams in QED, for example, give vanishing amplitudes to produce negative- or zero-norm states (i.e. photons with time-like or light-like polarization vectors). In gauge theories it is precisely the gauge symmetry which accounts for this. In string theory it is another symmetry, the residual conformal symmetry of the conformal gauge.

In Chapter 17 on general relativity we learned that differentiation of the matter action with respect to the metric gives the energy–momentum tensor. In Einstein's theory, differentiating the Einstein term as well gives Einstein's equations. In the string case the world-sheet metric has no dynamics (the Einstein action in two dimensions is a total derivative), and the Euler–Lagrange equation for γ yields an equation starting that the energy–momentum tensor vanishes. Quantum mechanically, these become constraint equations. The components of the energy–momentum tensor are

$$T_{10} = T_{01} = \partial_0 X \cdot \partial_1 X, \quad T_{00} = T_{11} = \frac{1}{2}[(\partial_0 X)^2 + (\partial_1 X)^2]. \tag{21.45}$$

The energy–momentum tensor is traceless. This is a consequence of conformal invariance; you can show that the trace is the generator of conformal transformations. In terms of the light cone coordinates, the non-vanishing components of the stress tensor are

$$T_{++} = \partial_+ X \cdot \partial_+ X, \quad T_{--} = \partial_- X \cdot \partial_- X. \tag{21.46}$$

Note that $T_{+-} = T_{-+} = 0$. Energy–momentum conservation then says that

$$\partial_- T_{++} = 0, \quad \partial_+ T_{--} = 0. \tag{21.47}$$

As a result, any quantity of the form $f(x^+)T_{++}$ or $f(x^-)T_{--}$ is also conserved. Integrating over the world sheet, this gives an infinite number of conserved charges.

We want to impose the condition of vanishing stress tensor as a condition on states. There is an obstacle, however, and this leads to one way of understanding the origin of the critical dimension, 26. The obstacle is an anomaly, similar to the anomalies we encountered in the first part of this text. One can see the problem if one takes the mode expansions for the X^μs and works out the commutators for the Ts. We will show in the next section that

$$\begin{aligned}&[T_{++}(\sigma), T_{++}(\sigma')]\\ &\quad = \frac{i}{24}(26 - D)\delta'''(\sigma - \sigma') + i[T_{++}(\sigma) + T_{++}(\sigma')]\delta'(\sigma - \sigma'),\end{aligned} \tag{21.48}$$

and a similar equation holds for T_{--}. The first term in Eq. (21.48) is clearly an obstruction to imposing the constraint unless $D = 26$. The number 26 arises from the energy–momentum tensor of the Faddeev–Popov ghosts. Were it not for the ghosts, strings would *never* make sense quantum mechanically. One can calculate this commutator painstakingly by decomposing in modes. But there are simpler methods, which also provide important insights into string theory and which we will develop in the next section.

21.4 Conformal invariance

The analysis of conformal invariance is enormously simplified by passing to Euclidean space. Define

$$w = \tau + i\sigma, \quad \bar{w} = \tau - i\sigma. \tag{21.49}$$

The ws describe a cylinder. Again, in this section $\alpha' = 2$. This choice will make the coordinate space Green's functions for the X^μs very simple. The Euclidean action is now

$$S = \frac{1}{8\pi}\int d^2w\, \partial_w X^\mu\, \partial_{\bar{w}} X^\mu. \tag{21.50}$$

In complex coordinates the non-vanishing components of the energy–momentum tensor are

$$T_{ww} = -\frac{1}{2}\partial_w X \cdot \partial_w X, \quad T_{\bar{w}\bar{w}} = -\frac{1}{2}\partial_{\bar{w}} X \cdot \partial_{\bar{w}} X. \tag{21.51}$$

We saw in the previous section that the string action, in Minkowski coordinates, is invariant under the transformations

$$\sigma^+ \to f(\sigma^+), \quad \sigma_- \to g(\sigma^-). \tag{21.52}$$

In terms of the complex coordinates this becomes invariance under the transformations

$$w \to f(w), \; \bar{w} \to f^*(\bar{w}). \tag{21.53}$$

These are conformal transformations of the complex variable and, as a result of this symmetry, we are able to bring all the machinery of complex analysis to bear on this problem. One particularly useful conformal transformation is the mapping of the cylinder onto the complex plane

$$z = e^w, \quad \bar{z} = e^{\bar{w}}. \tag{21.54}$$

Under this mapping, surfaces of constant τ on the cylinder are mapped into circles in the complex plane; $\tau \to -\infty$ is mapped into the origin and $\tau \to \infty$ is mapped to ∞. Surfaces of constant τ are mapped into circles.

It is convenient to write our previous expression for X^μ in terms of the variable z. First, we write down our previous expressions again:

$$\begin{aligned} X^\mu &= x^\mu + p^\mu \tau + i \sum_{n\neq 0} \frac{1}{n} (\alpha_n^\mu e^{-2in(\tau-\sigma)} + \tilde{\alpha}_n^\mu e^{-2in(\tau+\sigma)}) \\ &= X_{\rm L}^\mu + X_{\rm R}^\mu, \end{aligned} \tag{21.55}$$

where

$$X_{\rm L}^\mu = \frac{1}{2} x^\mu + \frac{1}{2} p^\mu (\tau - \sigma) + i \sum_{n\neq 0} \frac{1}{n} \alpha_n^\mu e^{-in(\tau-\sigma)}, \tag{21.56}$$

$$X_{\rm R}^\mu = \frac{1}{2} x^\mu + \frac{1}{2} p^\mu (\tau + \sigma) + i \sum_{n\neq 0} \frac{1}{n} \alpha_n^\mu e^{-in(\tau+\sigma)}. \tag{21.57}$$

Here $X_{\rm L}$ is holomorphic (analytic) in z and $X_{\rm R}$ is antiholomorphic:

$$\partial X_{\rm L} = -i\alpha_n^\mu z^{-n-1}, \quad \bar{\partial} X_{\rm R} = -i\tilde{\alpha}_n^\mu \bar{z}^{-n-1}, \tag{21.58}$$

where $\partial \equiv \partial/\partial z, \sim \bar{\partial} \equiv \partial/\partial \bar{z}$ and $\alpha_0^\mu = \tilde{\alpha}_0^\mu = \frac{1}{2} p^\mu$.

Let us evaluate the propagator of the xs in coordinate space. The Xs are just two-dimensional quantum fields. Their kinetic term, however, is somewhat unconventional. Because we are working with units $\alpha' = 2$, the action has a factor $1/(8\pi)$ out front. Accounting for the extra 4π, the coordinate-space propagator is (in Euclidean space)

$$\langle X^\mu(\sigma) X^\nu(0) \rangle = 4\pi \delta^{\mu\nu} \int \frac{d^2k}{(2\pi)^2} \frac{e^{i\sigma \cdot k}}{k^2}. \tag{21.59}$$

The right-hand side is logarithmically divergent in the infrared. We can use this fact to our advantage, cutting off the integral at scale μ and isolating the $\ln(\mu|z-z'|)$ factor. The logarithmic dependence can be seen almost by inspection of the integral:

$$\langle X^\mu(z) X^\nu(z') \rangle = 2g^{\mu\nu} \ln(|z-z'|\mu) = g^{\mu\nu} \left[\ln(z-z') + \ln(\bar{z}-\bar{z}') + \ln \mu^2 \right]. \tag{21.60}$$

As we will see shortly, the infrared cutoff drops out of the physically interesting quantities, so we will suppress it in the following.

In the covariant formulation, conformal invariance is crucial to the quantum theory of strings. To understand the workings of two-dimensional conformal invariance, we can use techniques of complex variable theory and the operator product expansion (OPE). We have discussed the OPE previously, in the context of two-dimensional gauge anomalies. It is also important in QCD in the analysis of various short-distance phenomena. The basic idea is that, for two operators, $\mathcal{O}(z_1)$ and $\mathcal{O}(z_2)$, when $z_1 \to z_2$ we have

$$\mathcal{O}_i(z_1)\mathcal{O}_j(z_2) \underset{z_1 \to z_2}{\to} \sum_k C_{ijk}(z_1 - z_2)\mathcal{O}_k(z_1). \tag{21.61}$$

The coefficients C_{ijk} are, in general, singular as $z_1 \to z_2$. The singularity is determined by the conformal dimension of $\mathcal{O}_i$ defined below (Eq. (21.75)).

To implement this rather abstract statement one can insert the above two operators into a Green's function with other operators located at some distance from z_1. In other words, one studies

$$\langle \mathcal{O}_i(z_1)\mathcal{O}_j(z_2)\Psi(z_3)\Psi(z_4)\cdots\rangle. \tag{21.62}$$

The operators in $\mathcal{O}(z_1)$ can be contracted with those in $\mathcal{O}(z_2)$, giving expressions which are singular as $z_1 \to z_2$, or with the other operators, giving non-singular expressions. The leading term in the OPE comes from the term with the maximum number of operators at z_1 contracted with operators at z_2; less singular operators arise when we contract fewer operators.

As an example which will be useful shortly, consider the product $\partial X^\mu(z)\partial X^\nu(w)$. If this appears in a Green's function, the most singular term as $z \to w$ will be that where we contract $\partial X(z)$ with $\partial X(w)$. The result will be equivalent to the insertion of the unit operator at a point times the singular function $1/(z-w)^2$, so we can write:

$$\partial X^\mu(z)\partial X^\nu(\omega) \sim \frac{g^{\mu\nu}}{(z-w)^2} + \cdots. \tag{21.63}$$

A somewhat more non-trivial, and important, set of operator product expansions is provided by the stress tensor and derivatives of X:

$$T(z)\partial X^\nu(w) = \partial X^\mu(z)\partial X^\mu(z)\partial X^\nu(w). \tag{21.64}$$

Now the most singular term arises when we contract the $\partial X(w)$ factor with one of the $\partial X(z)$ factors in $T(z)$. The other $\partial X(z)$ is left alone; in Green's functions, it must be contracted with other away operators that are further away. So we are left with

$$T(z)\partial X(w) \approx \frac{1}{(z-w)^2}\partial X(w) + \frac{1}{z-w}\partial^2 X(w) + \cdots. \tag{21.65}$$

Another important set of operators will turn out to be exponentials of x:

$$T(z)e^{ik\cdot x} = \frac{k^2}{(z-w)^2}e^{ik\cdot x} + \cdots. \tag{21.66}$$

To get some sense of the utility of conformal invariance and OPEs, we will compute the commutators of the α^μs. Start with

$$\alpha_n^\mu = \oint \frac{dz}{2\pi} z^n \partial X^\mu, \tag{21.67}$$

where the contour is taken about the origin. Now use the fact that, on the complex plane, time ordering becomes radial ordering, So, for $|z| > |w|$,

$$T\langle \partial X^\mu(z) \partial X^\nu(w)\rangle = \langle \partial X^\mu(z) \partial X^\nu(w)\rangle. \tag{21.68}$$

For $|z| < |w|$,

$$T\langle \partial X^\mu(z) \partial X^\nu(w)\rangle = \langle \partial X^\nu(w) \partial X^\mu(z)\rangle. \tag{21.69}$$

Thus we have

$$[\alpha_m^\mu, \alpha_n^\nu] = \left(\oint \frac{dz}{2\pi} z^m \oint \frac{dw}{2\pi} w^n - \oint \frac{dw}{2\pi} w^n \oint \frac{dz}{2\pi} z^m \right) T\left(\partial X^\mu(z) \partial X^\nu(w)\right), \tag{21.70}$$

where the contour can be taken to be a circle about the origin. In the first term, we take $|z| > |w|$, and in the second, $|w| > |z|$. Now, to evaluate the integral, we do, the z integral first, say. For fixed w, deform the z contour so that it encircles w (Fig. 21.1). Then

$$\begin{aligned}[\alpha_m^\mu, \alpha_n^\nu] &= \oint \frac{dw}{2\pi} w^n \oint \frac{dz}{2\pi} z^m \frac{1}{(z-w)^2} g^{\mu\nu} \\ &= m\delta_{m+n} g^{\mu\nu}.\end{aligned}$$

Let us now return to the stress tensor. We expect that the stress tensor is the generator of conformal transformations and that its commutators should contain information about the dimensions of operators. What we have just learned, by example, is that the operator products of operators encode the commutators. We could show by the Noether procedure

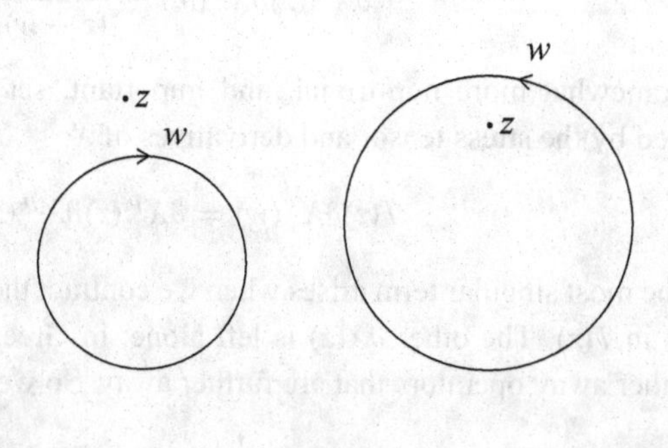

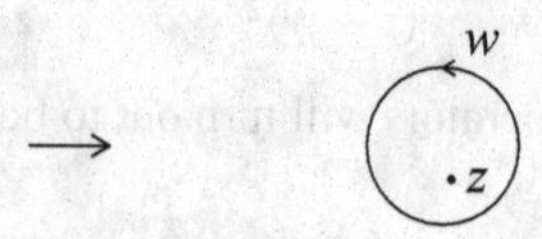

Fig. 21.1 Contour integral manipulations used to evaluate commutators in conformal field theory.

that the stress tensor is the generator of conformal transformations. But we can verify this directly. Consider the transformation

$$z \to z + \epsilon(z). \tag{21.71}$$

We expect that the generator of this transformation is

$$\oint dz\, T(z)\epsilon(z). \tag{21.72}$$

Let us take the special case of an overall conformal rescaling:

$$\epsilon(z) = \lambda z. \tag{21.73}$$

Now suppose that we have an operator $\mathcal{O}(w)$ and that

$$T(z)\mathcal{O}(w) = \frac{h}{(z-w)^2}\mathcal{O}(w) + \text{ less singular terms.} \tag{21.74}$$

Then

$$\begin{aligned}\left[\frac{1}{2\pi i}\oint T(z)\epsilon(z), \mathcal{O}(w)\right] &= \frac{1}{2\pi i}\oint dz \frac{\lambda z h \mathcal{O}(w)}{(z-w)^2} \\ &= \lambda h \mathcal{O}(w).\end{aligned} \tag{21.75}$$

This means that, under the conformal rescaling, we have $\mathcal{O} \to h\mathcal{O}$, just as we would expect for an operator of dimension h. As an example, consider $\mathcal{O} = (\partial)^n X$. This should have dimension n, and the leading term in its OPE is just of the form of Eq. (21.74), with $h = n$.

More precisely, an operator is called a primary field of dimension d if

$$T(z)\mathcal{O}(w) = \frac{d\mathcal{O}}{(z-w)^2} + \frac{\partial \mathcal{O}}{z-w}. \tag{21.76}$$

Note that $\partial X(z)$ is an example; $e^{ip\cdot x}$ is another. However, $(\partial)^n X$ is not, in general, as the $1/(z-w)$ term does not have quite the right form. A particularly interesting operator is the stress tensor itself. Naively, this has dimension two, but it is not a primary field. In the OPE, the most singular term arises from the contraction of all the derivative terms. This is proportional to the unit operator. The first subleading term, where one contracts just one pair of derivatives, gives a contribution proportional to the stress tensor itself:

$$T(z)T(w) = \frac{D}{(z-w)^4} + \frac{1}{(z-w)^2}T(w). \tag{21.77}$$

When one includes the Faddeev–Popov ghosts, one finds that they give an additional contribution, changing D to $D - 26$.

The algebra of the Fourier modes of T is known as the Virasoro algebra, and is important in string theory, conformal field theory and mathematics. In string theory it provides important constraints on states. Define the operators

$$L_n = \frac{1}{2\pi i}\oint dz\, z^{n+1} T(z). \tag{21.78}$$

In terms of these we have

$$T(z) = \sum_{m=-\infty}^{\infty} \frac{L_m}{z^{m+2}}, \tag{21.79}$$

and similarly for $\bar{z}$. Because the stress tensor is conserved, we are free to choose any time (i.e. radius for the circle). The operator product (21.77) is equivalent to the commutation relations above. Proceeding as we did for the commutators of the αs gives

$$[L_n, L_m] = (m-n)L_{m+n} + \frac{D}{12}(m^3 - m)\delta_{m+n}. \tag{21.80}$$

Using expression (21.16) we can construct the L_ns:

$$L_m = \frac{1}{2}\sum : \alpha^{\mu}_{m-n}\alpha_{\mu\, n} :, \quad \tilde{L}_m = \frac{1}{2}\sum : \tilde{\alpha}^{\mu}_{m-n}\tilde{\alpha}_{\mu\, n} :, \tag{21.81}$$

where the colons indicate normal ordering. Only when $m = 0$ is this significant. In this case we have to allow for the possibility of a normal-ordering constant. This constant is related to the constant we found in the Hamiltonian in light cone gauge,

$$L_0 = \sum_{n=0}^{\infty} \alpha^{\mu}_{-n}\alpha_{\mu n} - a, \quad \tilde{L}_0 = \sum_{n=0}^{\infty} \tilde{\alpha}^{\mu}_{-n}\tilde{\alpha}_{\mu n} - a. \tag{21.82}$$

Now we want to consider the constraint on states corresponding to the classical vanishing of the stress tensor. Because of the commutation relations, we cannot require all of Ls annihilate physical states. We require instead that

$$L_m|\Psi\rangle = 0 \tag{21.83}$$

for $m \geq 0$. Since $L^{\dagger}_m = L_{-m}$, this ensures that

$$\langle\Psi|L_n|\Psi\rangle = 0 \quad \forall\, n. \tag{21.84}$$

The constraint (21.35) in the light cone of invariance under translations along the string now becomes the condition $L_0 = \tilde{L}_0$. At the first excited level we have the state:

$$|\epsilon\rangle = \epsilon_{\mu\nu}\alpha^{\mu}_{-1}\tilde{\alpha}^{\nu}_{-1}|p^{\mu}\rangle. \tag{21.85}$$

The L_ns, for $n > 1$, trivially annihilate the state. For $n = 1$ we have

$$L_1|\epsilon\rangle = \alpha^{\mu}_0 \epsilon_{\mu\nu}|p^{\nu}\rangle. \tag{21.86}$$

Taking into account also $\tilde{L}_1$, we have the conditions

$$p_{\mu}\epsilon^{\mu\nu} = 0 = p_{\nu}\epsilon^{\mu\nu}. \tag{21.87}$$

This is similar to the condition $k_{\mu}\epsilon^{\mu}$ familiar in covariant gauge electrodynamics and it eliminates the negative-norm states. Consider, now, L_0:

$$L_0|\epsilon\rangle = (p^2 - a + 1)|\epsilon\rangle. \tag{21.88}$$

So, if $a = 1$ then the constraint is $p^2 = 0$, as we expect from Lorentz invariance. For open strings there is an analogous construction.

21.5 Vertex operators and the S-matrix

We have argued that, when the cylinder is mapped to the plane, the creation or destruction of states is described by local operators known as *vertex operators*. In this section we discuss the properties of these operators and their construction. We explain how the space–time S-matrix is obtained from correlation functions of these operators, and compute a famous example.

21.5.1 Vertex operators

There is a close correspondence between states and operators: $z \to 0$ corresponds to $t \to -\infty$. So consider, for example,

$$\partial_z X^\mu |0\rangle; \tag{21.89}$$

as $z \to 0$ we have

$$\partial_z X(z \to 0)|0\rangle = -i \sum_{m=-1}^{\infty} \frac{\alpha_m^\mu}{z^{m+1}} |0\rangle. \tag{21.90}$$

All terms but the term $m = -1$ annihilate the state to the right. Combining this with a similar left-moving operator creates a single-particle state.

More generally, in conformal field theories there is a one-to-one correspondence between states and operators. This is the realization of the picture discussed in the introduction. By mapping the string world sheet to the plane the incoming and outgoing states have been mapped to points, and the production or annihilation of particles at these points is described by local operators.

The construction of the S-matrix in string theory relies on this connection between states and operators. The operators which create and annihilate states are known as vertex operators. What properties should a vertex operator possess? The production of the particle should be represented as an integral over the string world sheet (so that there is no special point along the string). The expression

$$\int d^2 z\, V(z, \bar{z}) \tag{21.91}$$

should be invariant under conformal transformations. This means that the operator should possess dimension two; more precisely, it should possess dimension one with respect to both the left- and the right-moving stress tensors, so that

$$T(z)V(w, \bar{w}) = \frac{1}{(z-w)^2} V(w, \bar{w}) + \frac{1}{z-w} \partial_w V(w, \bar{w}) + \cdots \tag{21.92}$$

and similarly for $\bar{T}$. An operator with this property is called a $(1, 1)$ operator.

A particularly important operator in two-dimensional free-field theory (i.e. the string theories we have been describing up to now) is constructed from the exponential of the scalar field:

$$\mathcal{O}_p = e^{ip\cdot x}. \tag{21.93}$$

This has dimension

$$d = p^2 \tag{21.94}$$

with respect to the left-moving stress tensor, and similarly for the right-moving part.

With these ingredients, we can construct operators of dimension $(1, 1)$. These are in one-to-one correspondence with the states we found in the light cone construction, as follows.

1. *The tachyon*:

$$e^{ip\cdot x}, \quad p^2 = 1. \tag{21.95}$$

2. *The graviton, antisymmetric tensor, and dilaton*:

$$\epsilon_{\mu\nu}\partial X^\mu \bar{\partial} X^\nu e^{ip\cdot x}, \quad p^2 = 0. \tag{21.96}$$

The operator product

$$\partial X^\rho(z)\partial X_\rho(z)\epsilon_{\mu\nu}(p)\bar{\partial}X^\mu(w)\partial X^\nu(w)e^{ip\cdot x}(w) \tag{21.97}$$

contains terms which go as $1/(z-w)^3$ and have come from contracting one derivative in the stress tensor with $e^{ip\cdot x}$ and one with ∂X^μ. Examining Eq. (21.92), this leads to the requirement

$$p^\mu \epsilon_{\mu\nu}(p) = 0, \tag{21.98}$$

which we expect for massless spin-2 states. In our earlier operator discussion, this was one of the Virasoro conditions.

3. *Massive states*:

$$\epsilon_{\mu_1\cdots\mu_n}(p)\partial X^{\mu_1}\partial X^{\mu_2}\cdots\partial X^{\mu_n}e^{ip\cdot x}, \quad p^2 = 1-n. \tag{21.99}$$

Obtaining the correct OPE with the stress tensor now gives a set of constraints on the polarization tensor; again these are just the Virasoro constraints. Without worrying about degeneracies, we have a formula for the masses:

$$M_n^2 = n - 1. \tag{21.100}$$

This is what we found in the light cone gauge. Traditionally, the states were organized in terms of their spins. States of a given spin all lie on straight lines, known as *Regge trajectories*. These results are all in agreement with the light cone spectra we found earlier.

21.5.2 The S-matrix

Now we will make a guess as to how to construct an S-matrix. Our vertex operators, integrated over the world-sheet, are invariant under reparameterizations and conformal transformation of the world-sheet coordinates. We have seen that they correspond to the creation and annihilation of states in the far past and far future. We will normalize the vertex operators in such a way that

$$V_i(z)V_j(w) \sim \frac{\delta_{ij}}{|z-w|^4}. \tag{21.101}$$

So, we need to study correlation functions of the form

$$\mathcal{A} = \int d^2z_1 \cdots d^2z_n \langle V_1(z_1,p_1) \cdots V_n(z_n p_n)\rangle. \tag{21.102}$$

We will include a coupling constant g with each vertex operator.

Before evaluating this expression in special cases, let us consider the problem of evaluating the correlation functions of exponentials

$$\left\langle \exp\left(i \sum p_i \cdot X(z_i)\right)\right\rangle. \tag{21.103}$$

An easy way to evaluate this expression is to work in the path integral framework. Then the exponential has the structure

$$\int d^2z J_\mu(z)X(z), \tag{21.104}$$

where

$$J_\mu(z) = \sum_i p_{i\mu}\delta^2(z-z_i). \tag{21.105}$$

But we know that the result of such a path integral is

$$\exp\left(i\int d^2zd^2z' J_\mu(z)J^\mu(z')\Delta(z-z')\right) = \exp\left(\sum p_i \cdot p_j \ln|(z_i-z_j)|^2\mu^2\right), \tag{21.106}$$

where we have made a point of restoring the infrared cutoff.

We will consider the infrared cutoff first. Overall, we have a factor:

$$\mu^{(\sum p_i)^2}. \tag{21.107}$$

This vanishes as $\mu \to 0$ unless $\sum p_i = 0$, i.e. *unless momentum is conserved*. This result is related to the Mermin–Wagner–Coleman theorem, which states that there is no spontaneous breaking of global symmetries in two dimensions. Translational invariance is a global symmetry of the two-dimensional field theory; $e^{ip\cdot x}$ transforms under this symmetry. The only non-vanishing correlation functions are translationally invariant.

This correlation function also has an ultraviolet problem, coming from the $i=j$ terms in the sum. Eliminating these corresponds to the normal ordering of the vertex operators, and we will do this in what follows (we can, if we like, introduce an explicit ultraviolet cutoff; this gives a factor which can be absorbed into the definition of the vertex operators).

There is one more set of divergences with which we need to deal. These are associated with a part of the conformal invariance that we have not yet fixed. The operators L_0, L_1 and L_{-1} form a closed algebra. On the plane they generate overall rescalings (L_0), translations (L_1) and more general transformations (L_{-1}) which can be unified in $SL(2, C)$, the Möbius group. It transforms coordinates z to coordinates z^1, where

$$z = \frac{\alpha z' + \beta}{\gamma z' + \delta}. \tag{21.108}$$

Such transformations have the feature that they map the plane once into itself. It is necessary to fix this symmetry and divide by the volume of the corresponding gauge group. We can choose the location of three of the vertex operators, say z_1, z_2, z_3. These location are conventionally taken to be $0, 1, \infty$. It is necessary also to divide by the volume of this group; the corresponding factor is

$$\Omega_M = |z_1 - z_2|^2 |z_1 - z_3|^2 |z_2 - z_3|^2. \tag{21.109}$$

One can simply accept that this factor emerges from a Faddeev–Popov condition or it can be derived in the exercises at the end of the chapter. Finally, it is necessary to divide by g_s^2. This ensures that a three-particle process is proportional to g_s, a four-particle process to g_s^2 and so on.

Using these results we can construct particular scattering amplitudes. While it is physically somewhat uninteresting, the easiest case to examine is simply the scattering of tachyons. Let us specialize to the case of two incoming and two outgoing particles. Putting together our results above we have (remembering that $z_3 \to \infty$) the amplitute for particle scattering takes the form

$$\begin{aligned}\mathcal{A} = \frac{1}{\Omega_M} \int d^2 z_4\, |z_1 - z_2|^2 |z_1 - z_3|^2 |z_2 - z_3|^2 \\ |z_3|^{p_3 \cdot (p_1 + p_2 + p_3)} |z_1 - z_2|^{p_1 \cdot p_2} |z_4|^{p_4 \cdot p_1} |z_4 - 1|^{p_4 \cdot p_2}.\end{aligned} \tag{21.110}$$

Using momentum conservation, the z_4-independent contributions cancel out in the limit and we are left with

$$\mathcal{A} = \int d^2 z\, |z|^{2 p_1 \cdot p_4} |z - 1|^{2 p_2 \cdot p_4}. \tag{21.111}$$

Now we need an integral table to obtain

$$\begin{aligned} I &= \int d^2 z\, |z|^{-A} |1 - z|^{-B} \\ &= \beta\left(1 - \frac{A}{2}, 1 - \frac{B}{2}, \frac{A + B}{2} - 1\right). \end{aligned} \tag{21.112}$$

The beta function is defined by

$$\beta = \pi \frac{\Gamma(a)\Gamma(b)\Gamma(c)}{\Gamma(a + b)\Gamma(b + c)\Gamma(c + a)}. \tag{21.113}$$

We can express this result in terms of the Mandelstam invariants for $2 \to 2$ scattering, $s = -(p_1 + p_2)^2$, $t = -(p_2 - p_3)^2$ and $u = -(p_1 - p_4)^2$. Using the mass shell conditions,

$$p_4 \cdot p_1 = \frac{1}{2}\left[u + (p_1^2 - p_4^2)\right],$$

$$p_4 \cdot p_2 = -(p_3 + p_2 + p_1) \cdot p_2 = \frac{1}{2}(-s - t + 2m^2), \tag{21.114}$$

gives

$$\mathcal{A} = \frac{\kappa^2}{4\pi}\beta(-4s+1, -4t+1, -4u+1). \tag{21.115}$$

This is the Virasoro–Shapiro amplitude. There are a number of interesting features of this amplitude. It has singularities at precisely the locations of the masses of the string states. It should be noted, also, that we have obtained this result by an analytic continuation. The original integral is only convergent for a range of momenta, corresponding, essentially, to rules sitting below the threshold for the tachyon in the intermediate states.

We will not develop the machinery of open-string amplitudes here, but it is similar. One again needs to compute correlation functions of vertex operators. The vertex operators are somewhat different. Also, the boundary conditions for the two-dimensional fields, and thus the Green's functions, are different. The scattering amplitude for open-string tachyons is known as the Veneziano formula (see Section 21.6).

21.5.3 Factorization

The appearance of poles in the S-matrix at the masses of the string states is no accident. We can understand it in terms of our vertex operator and OPE analysis. Suppose that particles one and two, with momenta p_1 and p_2, have $s = (p_1 + p_2)^2 = -m_n^2$, the mass-squared of a physical state of the system. Consider the OPE of their vertex operators:

$$e^{ip_1 \cdot X(z_1)} e^{ip_2 \cdot X(z_2)} \approx e^{i(p_1 + p_2)\cdot X(z_2)} |z_1 - z_2|^{2p_1 \cdot p_2}. \tag{21.116}$$

So, in the S-matrix, fixing $z_2 = 0$, $z_3 = 1$ and $z_4 = \infty$, we encounter:

$$\int d^2z\, |z_1|^{2p_1 \cdot p_2} \langle e^{i(p_1+p_2)\cdot X(z_2)} e^{ip_3 \cdot X(z_3)} e^{ip_4 \cdot X(z_4)} \rangle. \tag{21.117}$$

Using momentum conservation and the on-shell conditions for p_1 and p_2 we obtain

$$2p_2 \cdot p_1 = q^2 - 8, \tag{21.118}$$

where $q = p_1 + p_2$. So the z-integral gives a pole,

$$\mathcal{A} \sim \frac{1}{4 - q^2} \tag{21.119}$$

i.e. it vanishes when the intermediate state is an on-shell tachyon.

This is general. Poles appear in the scattering amplitude when intermediate states go on-shell. The coefficients are precisely the couplings of the external states to the (nearly) on-shell physical state; this follows from the OPE.

21.6 The S-matrix versus the effective action

The Virasoro–Shapiro and Veneziano amplitudes are beautiful formulas. Analogous formulas for the case of massless particles can be obtained. These are particularly important for the superstring. For many of the questions which interest us, we are not directly interested in the *S*-matrix. One feature of the string *S*-matrix construction is that it involves on-shell states; the momenta appearing in the exponential factors satisfy $p^2 = -m^2$, where m is the mass of the state. So one cannot calculate, for example, the effective potential for the tachyon, since this requires that all momenta vanish. For massless particles things are better, since $p = 0$ is the limiting case of an on-shell process. But the *S*-matrix is not precisely the effective action. Instead, given the *S*-matrix, it is usually a straightforward matter to determine a low-energy effective action which will reproduce it. At tree level, one just needs to subtract massless particle exchanges. In loops, one must be more careful.

It is particularly easy to extract three-point couplings of massless particles at tree level. One just needs to study an "*S*-matrix" for three particles (one could also be a little could also more careful and study a four-particle amplitude, isolating the coefficient of the massless propagator). From our previous analysis, we need

$$\mathcal{A} = \frac{1}{\Omega_M}\langle V_1(z_1)V_2(z_2)V_3(z_3)\rangle, \tag{21.120}$$

where we do not integrate over the locations of the vertex operators. We are free to take z_1 and z_2 arbitrarily close to one another. Then the operator product will involve

$$V_1(z_1)V_2(z_2) \approx C_{123}\frac{1}{|z_1 - z_2|^2}V_3(z_2). \tag{21.121}$$

The final correlation function follows from the normalization of the vertex operators and cancels the Möbius volume. So the net result is that $g_s C_{123}$ is the coupling.

As an example, consider the coupling of two gravitons in the bosonic string. The vertex operator is

$$V_1 = \epsilon_{\mu\nu}(k_1)\,\partial X^\mu(z)\bar{\partial}X^\nu(z)\,e^{ik_1\cdot X(z)}, \tag{21.122}$$

and similarly for V_2 and V_3. So the operator product has the following structure:

$$V_1(z)V_2(w)$$
$$= \frac{1}{|z-w|^4} + \epsilon_{\mu\nu}(k_1)\epsilon_{\rho\sigma}(k_2)e^{i(k_1+k_2)\cdot X(z)}\left(k_1^\nu k_2^\sigma\frac{1}{|z-w|^2}\partial X^\mu(z)\bar{\partial}X^\rho(z) + \cdots\right). \tag{21.123}$$

Here the first term arises from the contraction of all the ∂X terms with each other. Loosely speaking, it is related to the production of off-shell tachyons. We will ignore it. The second term that we have indicated explicitly comes from contracting the first $\bar{\partial}X$ factor with the second exponential and the second ∂X factor with the first exponential. The ellipses indicate a long set of contractions. The complete vertex is precisely the on-shell coupling of three gravitons in Einstein's theory, along with couplings to the antisymmetric tensor and dilaton. We will not worry with the details here. When we discuss the heterotic string, we will show that the theory completely reproduces the Yang–Mills vertex in much the same way. We should not be surprised that it is difficult to define off-shell Green functions. In gravity, apart from the S-matrix it is in general hard to define coordinate-invariant observables.

21.7 Loop amplitudes

So far, we have considered tree amplitudes. Closed or open strings interact by splitting and joining. Once we allow for quantum fluctuations, strings in intermediate states can split and join too. Because of conformal invariance, the only invariant characteristic of these diagrams is their topology (for closed-strings, the tree level world sheet has the topology of a sphere). In the closed-string case, each additional loop adds a handle to the world sheet. In general, the theory of string loops is complicated, but the description of one-loop diagrams is rather simple and exposes important features of the theory not apparent in tree diagrams. In the case of closed strings, requiring that the one-loop amplitude be sensible places strong constraints on the theory. Invariance under certain (global) two-dimensional general coordinate transformations, known as modular transformations, accounts for many features of both the bosonic and superstring theories. In space–time, satisfying these constraints is a necessary condition for the unitarity of the scattering amplitude. In this section we provide only a brief introduction. We will leave for later the discussion of open-string loops.

The one-loop amplitude has the topology of a donut, or torus. A simple representation of a torus is as indicated in Fig. 21.2. In this figure, the world sheet is flat and of finite size. We can think of this torus as living in the complex plane. It is (up to conformal transformations) the world sheet appearing in the Euclidean path integral. The two possible periods of the torus are translated into two complex periods, λ_1 and λ_2. We require that the fields are periodic under

$$z \to z + m\lambda_1 + n\lambda_2. \tag{21.124}$$

We can transform λ_1 and λ_2 by a transformation in the modular group, $SL(2, Z)$,

$$\begin{pmatrix} \lambda_1 \\ \lambda_2 \end{pmatrix} = \begin{pmatrix} a & b \\ c & d \end{pmatrix} \begin{pmatrix} \lambda_1' \\ \lambda_2' \end{pmatrix} \tag{21.125}$$

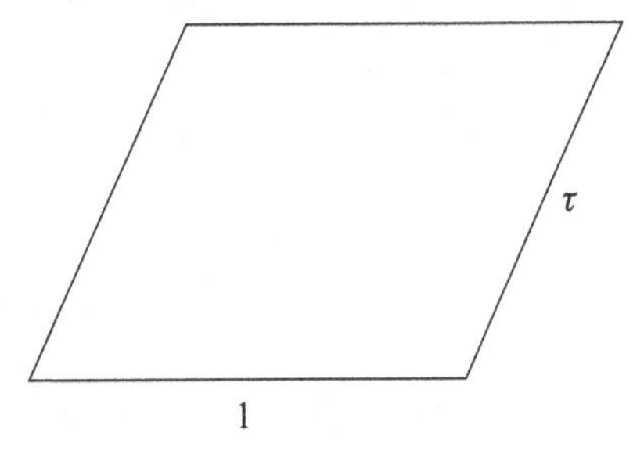

Fig. 21.2 A simple representation of a torus.

with a, b, c and d integers satisfying $ad - bc = 1$, provided that we also transform the integers n and m by the inverse matrix,

$$\begin{pmatrix} m \\ n \end{pmatrix} = \begin{pmatrix} d & -b \\ -c & a \end{pmatrix} \begin{pmatrix} m' \\ n' \end{pmatrix} \tag{21.126}$$

Now rescale z by λ_1, and set $\tau = \lambda_2/\lambda_1$. Then z has the periodicities 1 and τ. Under modular transformations, τ transforms as follows:

$$\tau \to \frac{a\tau + b}{c\tau + d}. \tag{21.127}$$

The modular transformations are general coordinate transformations of the world-sheet theory, but they are not continuously connected to the identity. In order that one-loop string amplitudes make sense, we require that they be invariant under this transformation. The general amplitude will be a correlation function

$$\langle V(z_1)V(z_2)\cdots\rangle_{\text{torus}}, \tag{21.128}$$

evaluated on the torus, as indicated. The simplest amplitude is that with no vertex operators inserted. (At tree level this amplitude vanishes owing to the division by the infinite Möbius volume.) For the bosonic string, we can evaluate the amplitude in light cone gauge. We simply need to evaluate the functional determinant. As these are free fields on a flat space, this is not too difficult. It is helpful to remember some basic field theory facts. The path integral, with initial configuration $\phi_{\rm i}(x)$ and final configuration $\phi_{\rm f}(x)$, computes the quantum mechanical matrix element:

$$\langle\phi_{\rm f}|e^{-iHT}|\phi_{\rm i}\rangle. \tag{21.129}$$

If we take the time to be Euclidean, impose periodic boundary conditions and sum (integrate) over all possible $\phi_{\rm i}$, we will have computed

$$\mathrm{Tr}\, e^{-HT} \tag{21.130}$$

i.e. the quantum mechanical partition function. As described in Appendix C, this observation is the basis of the standard treatments of finite-temperature phenomena in quantum field theory. In the present case the periodicity is in the τ direction. So we compute

$$\mathrm{Tr}\, e^{-H_{\rm lc}\tau}. \tag{21.131}$$

It is convenient to rewrite the light cone Hamiltonian, $H_{\rm lc}$, in terms of L_0 and $\bar{L}_0$. Introducing

$$q = e^{2\pi i\tau}, \quad \bar{q} = e^{-2\pi i\bar{\tau}} \tag{21.132}$$

we want to evaluate

$$\mathrm{Tr}\left(q^{L_0}, \bar{q}^{\bar{L}_0}\right). \tag{21.133}$$

From any oscillator with oscillator number n, just as in quantum mechanics we obtain $(1 - q^n)^{-1}$; so, allowing for the different values of n and the $D - 2$ transverse directions, we have

$$\prod q^{D/24}\bar{q}^{D/24}(1-q^n)^{2-D}(1-\bar{q}^n)^{2-D}. \tag{21.134}$$

This is conveniently expressed in terms of a standard function, the Dedekind η-function,

$$\eta(q) = q^{1/24}\prod_{n=1}^{\infty}(1-q^n). \tag{21.135}$$

We also need the contribution of the zero modes. This is

$$\int \frac{d^{D-2}p}{(2\pi)^{D-2}}e^{-\tau_2 p^2} \propto \tau_2^{D-2}. \tag{21.136}$$

In the final expression, we need to integrate over τ. The measure for this can be derived from the Faddeev–Popov ghost procedure, but it can be guessed from the requirement of modular invariance. It is easy to check that

$$\int \frac{d^2\tau}{\tau_2^2} \tag{21.137}$$

is invariant. So, in 26 dimensions, we finally have

$$Z \propto \int \frac{d^2\tau}{\tau_2^2}\tau_2^{-12}|\eta(\tau)|^{-48}. \tag{21.138}$$

Now, to check that this is modular invariant we note, first, that the full modular group is generated by the transformations

$$\tau \to \tau + 1, \quad \tau \to -1/\tau. \tag{21.139}$$

Under these transformations, as we said, the measure is invariant. The Dedekind η function transforms as

$$\eta(\tau+1) = e^{i\pi/12}\eta(\tau), \quad \eta(-1/\tau) = (-i\tau)^{1/2}\eta(\tau). \tag{21.140}$$

Since $\tau_2 \to \tau_2/\tau_1^2 + \tau_2^2$, under $\tau \to -1/\tau$ we have that Z is invariant. Here we see that the bosonic string makes sense only in 26 dimensions.

Suggested reading

More detail on the material in this chapter can be found in Green *et al.* (1987) and in Polchinski (1998). The light cone treatment described here is nicely developed in Peskin (1985).

Exercises

(1) Enumerate the states of the bosonic closed string at the first level with positive mass-squared. Don't worry about organizing them into irreducible representations, but list their spins.

(2) OPEs: explain why X^μ and X^ν do not have a sensible operator product expansion. Work out the OPE of ∂X^μ and ∂X^ν as in the text. Verify the commutator of α^μ and α^ν, as in the text.

(3) Work out the Virasoro algebra, starting with the operator product expansion for the stress tensor and using the contour method.

(4) The Mermin–Wagner–Coleman theorem: consider a free two-dimensional quantum field theory with a single, massless, complex field ϕ. Describe the conserved $U(1)$ symmetry. Show that correlation functions of the form

$$\langle e^{iq_1\phi(x_1)} \dots e^{iq_n\phi(x_n)} \rangle \tag{21.141}$$

are non-vanishing only if $\sum q_i = 0$. Argue that this means that the global symmetry is not broken. From this construct an argument that global symmetries are never broken in two dimensions.

(5) Show that the factor Ω_M of Eq. (21.109) is invariant under the Möbius group. You might want to proceed by analogy with the Faddeev–Popov procedure in gauge theories.

(6) Show that the factorization of tree level S-matrix elements is general, i.e. that if the kinematics are correctly chosen for two incoming particles 1 and 2, so that $(p_1+p_2)^2 \approx m_n^2$, that the amplitude is approximately a product of the coupling of particles 1 and 2 to particle n, times a nearly on-shell propagator for the n.

22 The superstring

The theories we have described were motivated by thinking of a picture of a string moving in space–time. We arrived in this way at a description of strings in terms of two-dimensional quantum fields. The theories, so far, are theories of bosons only. But, in this more abstract picture, we can imagine adding two-dimensional fermionic fields as well. This possibility was first considered by Ramond, Neveu and Schwarz and leads to the superstring theories, Type I, Types IIA and IIB and the two heterotic string theories. We first develop the theories in the light cone gauge, where their spectra are readily exhibited. Then we discuss interactions.

22.1 Open superstrings

A priori there appears to be a great deal of freedom in how we introduce fermions: their number, their representations under the (space–time) Lorentz group and possibly other options. Various consistency conditions restrict these choices. In the case of open strings we have to introduce one fermion ψ^I for each coordinate X^I. For the action of the fermions we take

$$S_\psi = \frac{1}{2\pi}\int d^2\sigma\, i\bar{\psi}^I(\partial_\alpha \gamma^\alpha)\psi^I. \tag{22.1}$$

In two dimensions, a particularly simple choice for the γ-matrices is

$$\gamma^0 = \sigma_2, \quad \gamma^1 = i\sigma_1 \tag{22.2}$$

and the analog of γ_5 in four dimensions is

$$\gamma_3 = \sigma_3. \tag{22.3}$$

The Dirac equation in this basis is purely imaginary, so we can take the fermions to be real (Majorana). We can work with eigenfunctions of σ_3:

$$\psi^I = \begin{pmatrix} \psi^I_- \\ \psi^I_+ \end{pmatrix}. \tag{22.4}$$

In this way, if we again introduce light cone coordinates on the world sheet,

$$\sigma^\pm = \tau \pm \sigma, \tag{22.5}$$

the action becomes

$$S_\psi = \frac{1}{2\pi}\int d^2\sigma\,(\psi^I_+\partial_-\psi^I_+ + \psi^I_-\partial_+\psi^I_-). \tag{22.6}$$

We need to impose boundary conditions at the string end points. To determine suitable boundary conditions, we vary the Lagrangian to obtain the Euler–Lagrange equations. The surface terms which arise in the variation involve $\psi_+\delta\psi_+ - \psi_-\delta\psi_-$. So the boundary terms vanish if $\psi_+ = \pm\psi_-$. An overall sign doesn't matter, so we can take the plus sign at $\sigma = 0$:

$$\psi^I_+(0,\tau) = \psi^I_-(0,\tau) \tag{22.7}$$

This leaves two choices for the boundary conditions at $\sigma = \pi$:

$$\psi^I_+(\pi,\tau) = \pm\psi^I_-(\pi,\tau). \tag{22.8}$$

Fermions which obey the boundary condition with the plus sign are called Ramond fermions; those with the minus sign are called Neveu–Schwarz (NS) fermions. Corresponding to the Ramond case are the mode expansions

$$\psi^I_- = \frac{1}{\sqrt{2}}\sum_{n\in Z} d^I_n e^{-in(\tau-\sigma)}, \quad \psi^I_+ = \frac{1}{\sqrt{2}}\sum_{n\in Z} d^I_n e^{-in(\tau+\sigma)}. \tag{22.9}$$

In the NS case we have

$$\psi^I_- = \frac{1}{\sqrt{2}}\sum_{r\in Z+1/2} b^I_r e^{-ir(\tau-\sigma)}, \quad \psi^I_- = \frac{1}{\sqrt{2}}\sum_{r\in Z+1/2} b^I_r e^{-ir(\tau-\sigma)}. \tag{22.10}$$

Now we quantize these fields:

$$\{\psi^I(\sigma,\tau)_\pm, \psi^J(\sigma',\tau)_\pm\} = \pi\delta(\sigma-\sigma')\delta^{IJ}\delta_{\pm\pm} \tag{22.11}$$

This gives, for the modes:

$$\{b^I_r, b^J_s\} = \delta^{IJ}\delta_{r+s}, \quad \{d^I_m, d^J_n\} = \delta^{IJ}\delta_{m+n}. \tag{22.12}$$

The Hamiltonian in light cone gauge, for the Ramond sector, is

$$H = \vec{p}^{\,2} + N_\alpha + N_d. \tag{22.13}$$

Here the Ns are the various number operators:

$$N_\alpha = \sum_{m=1}^{\infty}\alpha^I_{-m}\alpha^I_m, \quad N_d = \sum_{m=1}^{\infty} m d^I_{-m} d^I_m. \tag{22.14}$$

For the NS sector, N_d is replaced by N_b:

$$N_b = \sum_{r=1/2}^{\infty} m b_{-r} b_r. \tag{22.15}$$

Each of these Hamiltonian contributions has a normal-ordering constant. We will determine these shortly. The states of the theory are the eigenstates of the fermion number operators

$b_n^\dagger b_n$, $d_n^\dagger d_n$ etc. for non-zero n. The eigenvalues can take the values 0 or 1 in each case. The zero modes, which arise in the Ramond sector, are special. They give rise to space–time fermions.

22.2 Quantization in the Ramond sector: the appearance of space–time fermions

Usually, we do field theory at infinite volume but here we are considering field theory at a finite volume ($0 < \sigma < \pi$), and this has introduced some new features. For the bosonic fields X^I we have already seen that there are zero modes, which gave rise to the coordinates and momenta of space–time. For the fermions we now have the new feature that there are two sectors, with two independent Hilbert spaces. It is tempting to simply keep one sector, but it turns out that when we consider string interactions it is necessary to include both: even if we attempted to exclude, say, the Ramond states, they would appear in string loop diagrams.

There is another feature: the appearance of fermion zero modes d_0^I in the Ramond sector. These are not conventional creation and annihilation operators. They obey the commutation relations

$$\left\{d_0^I, d_0^J\right\} = \delta^{IJ}. \tag{22.16}$$

These are, up to a factor 2, the anticommutation relations of the Dirac gamma matrices for a $(D-2)$-dimensional space, i.e. they are associated with the group $O(D-2)$. Anticipating the fact that $D=10$, we are interested in the Dirac matrices of $O(8)$. Before giving a construction of the spinor representations of $O(8)$, let us first simply state the basic result: $O(8)$ has two spinor representations, 8_s and $8_s'$, and a vector representation, 8_v, all eight-dimensional. So we can realize the commutation relations, not on a Fock space, but on a space corresponding to one of the eight-dimensional representations of $O(8)$. Labeling these states a, $\dot{a}$, then

$$\langle \dot{a}|d_0^I|a\rangle = \frac{1}{\sqrt{2}}\gamma_{\dot{a}a}^I. \tag{22.17}$$

We can construct an explicit representation for these matrices in various ways. A simple and easy to remember construction is to think of $O(8)$ as acting on eight coordinates x^I. Group these into complex coordinates:

$$z^1 = x^1 + ix^2, \quad z^2 = x^3 + ix^4, \quad z^3 = x^5 + ix^6, \quad z^4 = x^7 + ix^8 \tag{22.18}$$

and their complex conjugates. This defines an embedding of $U(4)$ in O(8). Correspondingly, we define

$$a^1 = d_0^1 + id_0^2, \tag{22.19}$$

etc. The a^is obey the commutation relations

$$\{a^i, a^{j\dagger}\} = \delta^{ij}, \tag{22.20}$$

all others vanishing. These are just the conventional anticommutation relations of fermion creation and annihilation operators (but remember that for this discussion they are just matrices and should not be confused with the d_ns, which are genuinely creation and annihilation operators). Among products of these operators we can distinguish two classes: those built from an even number of as and those built from an odd number. In four dimensions the analogous distinction corresponds to the eigenvalue (± 1) of γ^5.

Now we define a state, $|0\rangle$, annihilated by the a^is. We can then form two sets of states, those with even fermion number and those with odd fermion number. The even states are

$$|0\rangle, \quad a^{i\dagger} a^{j\dagger}|0\rangle, \quad a^{1\dagger} a^{2\dagger} a^{3\dagger} a^{4\dagger}|0\rangle. \tag{22.21}$$

These states form one of the eight representations, say 8_s. The second is formed by the states of odd fermion number. States are now labeled $|p^I, a, \{\text{oscillators}\}\rangle$.

What we have learned is that the states in the Ramond sector are space–time *fermions*; the states in the NS sector are space–time bosons.

22.3 Type II theory

For closed strings we still have two-component fields ψ, but the possible choices of boundary conditions are somewhat different. We still require that the fermion surface terms vanish, but we also require that currents such as $\psi_+^I \psi_+^J$ be periodic. (These currents are part of the generators of rotations in space–time.) So we impose the Ramond and Neveu–Schwarz boundary conditions independently on the left and right movers. Recalling that the Lagrangian for the fermions breaks up into left- and right-moving parts, we treat the left- and right-moving fermions as independent fields. The fermions have the mode expansions

$$\psi^I = \sum_{n \epsilon Z} d_n^I e^{-2in(\tau-\sigma)}, \quad \psi^I = \sum_{n \epsilon Z+1/2} b_r^I e^{-2ir(\tau-\sigma)} \tag{22.22}$$

in the Ramond and NS sectors, respectively, and

$$\tilde{\psi}^I = \sum \tilde{d}_n^I e^{-2in(\tau+\sigma)} \tilde{\psi}^I = \sum \tilde{b}_r^I e^{-2ir(\tau+\sigma)}. \tag{22.23}$$

The light cone Hamiltonian is now

$$H = p^2 + N_\alpha + \tilde{N}_\alpha + N_d + \tilde{N}_d - a. \tag{22.24}$$

In constructing the spectrum, this must be supplemented with the condition of invariance under shifts in σ; in the covariant formulation this was the $L_0 = \tilde{L}_0$ constraint (see the discussion after Eq. (21.84)).

22.4 World-sheet supersymmetry

Before considering the spectra of superstring theories, we consider the question of supersymmetry. The theory we are considering is supersymmetric in two dimensions. Just as we decomposed the fermions into left and right movers, we can introduce a two-component anticommuting parameter θ:

$$\theta = \begin{pmatrix} \theta_- \\ \theta_+ \end{pmatrix}. \tag{22.25}$$

Then we define the superfield

$$Y^I = X^I + \bar{\theta}\psi^I + \frac{1}{2}\bar{\theta}\theta B^I. \tag{22.26}$$

We will see shortly that B^I is an auxiliary field, which in the case of strings in flat space we can set to zero by its equations of motion. The supersymmetry generators are

$$Q_A = \frac{\partial}{\partial \bar{\theta}_A} + i(\gamma^\alpha \theta)_A \partial_\alpha \tag{22.27}$$

(we are using the capital letter A for two-dimensional spinor indices here to distinguish them from lower case a, which we used for $O(8)$ spinor indices, and from α, which we used for two-dimensional vector indices). As in four dimensions, we can introduce a covariant derivative operator which anticommutes with the supersymmetry generators:

$$D = \frac{\partial}{\partial \bar{\theta}} - i\gamma^\alpha \theta \partial_\alpha. \tag{22.28}$$

In terms of the superfields, the action may be written in a manifestly invariant way:

$$\begin{aligned} S &= \frac{i}{4\pi} \int d^2\sigma d^2\theta\, \bar{D}Y^\mu DY_\mu \\ &= \frac{-1}{2\pi} \int d^2\sigma \, (\partial_\alpha X^I \partial^\alpha X^I - i\bar{\psi}^I \gamma^\alpha \partial_\alpha \psi^I_\mu - B^I B^I). \end{aligned} \tag{22.29}$$

Note that B^I vanishes by its equations of motion.

Finally, note that, in the NS sectors, the boundary conditions explicitly break the world-sheet supersymmetry; they map bosonic fields into fermionic fields and vice versa, and these fields obey different boundary conditions. The Ramond sector is supersymmetric.

In the covariant formulation, this supersymmetry is essential to an understanding of the full set of constraints on the states. But it is important to stress that it is a symmetry of the world-sheet theory; its implications for the theory in space–time are subtle.

22.5 The spectra of the superstrings

We have, so far, considered first the world-sheet structure of the superstring theories. We have not yet explored their spectra in detail. As in the case of the bosonic string, we will see that these theories possess a massless graviton. We will also find that they have a massless spin-3/2 particle, the gravitino. For the couplings of such a particle to be consistent requires that the space–time theory is supersymmetric.

22.5.1 The normal-ordering constants

First, we give a general formula for the normal-ordering constant. This is related to the algebra of the energy–momentum tensor we discussed in Section 21.4. For a left- or right-moving boson, with modes which differ from an integer by η (e.g. the modes are $1-\eta$, $2-\eta$ etc.), the contribution to the normal-ordering constant is

$$\Delta = -\frac{1}{24} + \frac{1}{4}\eta(1-\eta). \tag{22.30}$$

For fermions, the contribution is the opposite. So we can recover some familiar results. In the bosonic string, with 24 transverse degrees of freedom, we see that the normal-ordering constant is -1. For the superstring, in the NS–NS sector (see below) we have a contribution of $-1/24$ for each boson and $1/24 - 1/16$ for each of the eight fermions on the left (and similarly on the right). So the normal-ordering constant is $-1/2$. For the RR sector, the normal-ordering constant vanishes.

There are simple derivations of the above formula, whose justification requires careful consideration of conformal field theory. The normal-ordering constant is just the vacuum energy of the corresponding two-dimensional free-field theory. So we need

$$f(\eta) = \frac{1}{2}\sum_{1}^{\infty}(n+\eta). \tag{22.31}$$

Ignoring the fact that the sum is ill-defined, we can shift n by one and compensate by a change in η:

$$f(\eta) = f(\eta+1) + \frac{\eta}{2}. \tag{22.32}$$

If we assume that the result is quadratic in η, we recover the formula above, up to a constant. We can "calculate" this constant by the following trick, known as zeta function regularization. For $\eta = 0$ we need

$$\sum_{n=1}^{\infty} n = \lim_{s\to-1}\sum_{n=1}^{\infty} n^{-s} \tag{22.33}$$

The object on the right-hand side of this equation is $\zeta(s)$, the *Riemann zeta function*. The analytic structure of this function is something of great interest to mathematicians, but one well-known fact is that its singularities lie off the real axis. Using integral representations one can derive a standard result: $\zeta(-1) = -1/12$. This fixes the constant as $-1/24$. This argument may (or should) appear questionable to the reader. The real justification comes from considering questions in conformal field theory.

22.5.2 The different sectors of the Type II theory

In the Type II theory there are four possible choices of boundary condition: NS for both left and right movers, Ramond for both left and right movers, Ramond for left and NS for right and NS for left and R for right. We will refer to these as the NS–NS, R–R, R–NS and NS–R sectors. Consider, first, the NS–NS sector. There are no zero-mode fermions, so we just have a normal (unique) ground state for the oscillators. From our computation of the normal ordering constants in the previous section, we see that $a = -1/2$ for both left and right movers. The lowest state is simply the state $|\vec{p}\rangle$. It has mass-squared -1 (in units with $\alpha' = 2$). Since no oscillators are excited, the $L_0 = \tilde{L}_0$ condition is satisfied. Now consider the first excited states; again, we must have invariance under σ translations, so these are the states

$$\psi^I_{-1/2}\tilde{\psi}^J_{-1/2}|\vec{p}\rangle. \tag{22.34}$$

Because $a = -1/2$ for both left and right movers, these states are massless. The symmetric combination here contains a scalar and a massless spin-2 particle, the graviton; the antisymmetric combination is an antisymmetric tensor field. At the next level we can create massive states using four space–time fermions or two bosons or two fermions and one boson.

Now let us turn to the other sectors. Consider, first, the R–NS sector, where ψ is Ramond and $\tilde{\psi}$ is NS. Now, the left-moving normal-ordering constant is zero, while the right-moving constant is $-1/2$. So we can satisfy the level-matching condition (invariance under σ translations) if we take the left movers to be in their ground state and take the right-moving NS state to be an excitation with a single fermion operator above the ground state, i.e.

$$|\Psi^I_a\rangle = \tilde{\psi}^I_{-1/2}|a\,\vec{p}\rangle. \tag{22.35}$$

From the space–time viewpoint, these are particles of spin-3/2 and 1/2. In the NS–R sector, we have another spin-3/2 particle.

Just as a massless spin-2 particle requires that the underlying theory be generally covariant, a massless spin-3/2 particle, as we discussed in the context of four-dimensional field theories, requires *space–time* supersymmetry. But now we seem to have a paradox. With space–time supersymmetry we cannot have tachyons, yet our lowest state in the NS–NS sector, $|\vec{p}\rangle$, is a tachyon.

The solution to this paradox was discovered by Gliozzi, Scherk and Olive, who argued that it is necessary to project out states, i.e. to keep only states in the spectrum which satisfy a particular condition. This projection, which yields a consistent supersymmetric theory, is known as the GSO projection. Note, first, that we have been a bit sloppy with the fermion indices on the ground states. We have two types of fermion indices, a and $\dot{a}$, corresponding to the two spinor representations of $O(8)$. So we do the following. We keep only states on the left which are odd under the left-moving world-sheet fermion number; we do the same on the right but we include in the definition of the world-sheet fermion number the chirality of the zero-mode states. We take

$$(-1)^F = \exp(i\pi\gamma^9)\exp\left(i\pi\sum_{1/2}^{\infty}\psi_n\psi_{-n}\right). \tag{22.36}$$

In the R–NS sector we make a similar set of projections. Here we have a choice, however, in which chirality we take. If we project on states of opposite $(-1)^F$ then we get the Type IIA theory; if we take the same chirality, we get the Type IIB theory.

Returning to the NS–NS sector we make a similar projection, keeping only states which are odd under both left- and right-moving fermion number. In this way we eliminate the would-be tachyon in this sector.

Somewhat more puzzling is the R–R sector in each theory. Here both the left- and right-moving ground states are spinors. So, in space–time the states are bosons. We can organize them as tensors by constructing antisymmetric products of γ-matrices, $\gamma^{ijk\cdots}$. As we know from our experience in four dimensions these form irreducible representations, in this case of the little group $O(8)$. Thinking of our construction of the γ-matrices in terms of the as, we can see that γs with even numbers of indices connect states of opposite chirality while those with odd numbers of indices connect states with the same chirality. Which tensors appear depends on whether we consider the IIA or IIB theories. In the IIA case, only the tensors of even rank are non-vanishing. These tensors correspond to field strengths (one can consider an analogy with the magnetic moment coupling in electrodynamics, $\bar{\psi}\gamma^{\mu\nu}\psi$). So, in the IIA theory one has second- and fourth-rank tensors; the sixth- and eighth-rank field strengths are dual to these. In terms of gauge fields there are a one-index tensor (a vector) and a third-rank antisymmetric tensor. In the IIB theory, there are a scalar, a second-rank tensor and a fourth-rank tensor. In string perturbation theory, because the couplings are through the field strengths, there are no objects carrying the fundamental charge. Later we will see that there are non-perturbative objects, *D-branes*, which do carry these charges.

22.5.3 Other possibilities: modular invariance and the GSO projection

The reader may feel that the choices of projections, and for that matter the choices of representations for the two-dimensional fermions, seem rather arbitrary. It turns out that the possible choices, at least for flat background space–times, are highly restricted. There are only a few consistent theories. Those we have described are the only ones without tachyons but with both left- and right-moving supersymmetries on the world sheet.

In the bosonic string theory, we saw that it is crucial that the theory be formulated in 26 dimensions. One problem with the theory outside 26 dimensions is that it is not modular invariant. This means that it is not invariant under certain global two-dimensional general coordinate transformations. This world-sheet anomaly is correlated with anomalies in space–time. As for the gauge anomalies in field theories, these lead to breakdown of unitarity, Lorentz invariance or both.

For the superstring theories we will now explain why modular invariance demands a projection like the GSO projection. The point is that modular transformations relate sectors with different choices of boundary condition.

In our discussion of string theories up to this point, path integrals have appeared only occasionally, but they are extremely useful in discussing string perturbation theory. The propagation of strings can be described by a two-dimensional path integral over the string coordinates, $X_\mu(\sigma,\tau)$, weighted by e^{-S}, where S is the string action. At tree level the closed-string world sheet has the topology of a sphere. At one loop it has the topology of a torus. So, at one loop, string amplitudes can be described as path integrals of a two-dimensional field theory on a torus. Note that we need here the full path integral, not simply the generator of the Green's function for the field theory. The path integral on the torus, with no insertion of vertex operators, yields the partition function of the two-dimensional field theory. To understand this, let us consider the fermion partition function. Actually, there are several fermion partition functions. We begin with a single right-moving Majorana fermion and take, first, Neveu–Schwarz boundary conditions. There are two sorts of partition function we might define. First,

$$\mathrm{Tr}\, q^{L_0} = \prod_{r=1/2}^{\infty} (1 + q^r). \tag{22.37}$$

Alternatively,

$$\mathrm{Tr}\, (-1)^F q^{L_0} = \prod_{r=1/2}^{\infty} (1 - q^r). \tag{22.38}$$

From a path integral point of view, the first expression is like a standard thermal partition function. It can be represented as a path integral with antiperiodic boundary conditions in the time direction. The second integral corresponds to a path integral with even boundary conditions for fermions in the time direction. We can represent the torus as in Fig. 21.2. Taking the vertical direction to be the time direction and the horizontal direction the space direction, we can indicate the boundary conditions with plus and minus signs along the sides of the square. Recalling the action of modular transformations on the torus, however, we see that the modular group mixes up the various boundary conditions. Not only does it mix the temporal boundary conditions, it mixes the spatial boundary conditions as well.

It will be convenient for much of our later analysis to group the fermions in complex pairs. In the present case this grouping is rather arbitrary, say $\Psi^1 = \psi^1 + i\psi^2$ and so on. Then the partition functions can be conveniently written in terms of ϑ functions. These functions, which have been extensively studied by mathematicians, transform nicely under modular transformations:

$$\vartheta \begin{bmatrix} \theta \\ \phi \end{bmatrix} (0,\tau) = \eta(\tau) e^{2\pi i \theta\phi} q^{\theta^2/2 - 1/24} \prod_{m=1}^{\infty} \left(1 + e^{2\pi i \phi} q^{m+\theta-1/2}\right) \times \left(1 + e^{-2\pi i \phi} q^{m-\theta-1/2}\right). \tag{22.39}$$

Under $\tau \to \tau + 1$,

$$\vartheta \begin{bmatrix} \theta \\ \phi \end{bmatrix} (0,\tau+1) = e^{i\theta^2 - \theta - \theta\phi} \vartheta \begin{bmatrix} \theta \\ \phi - \theta \end{bmatrix} (0,\tau), \tag{22.40}$$

while, under $\tau \to -1/\tau$,

$$\vartheta \begin{bmatrix} \theta \\ \phi \end{bmatrix} (0, 1/\tau) = e^{2\pi i \theta\phi} \vartheta \begin{bmatrix} -\phi \\ \theta \end{bmatrix} (0, \tau). \tag{22.41}$$

These transformation properties have a physical interpretation. Returning to Eqs. (21.125)–(21.127), the transformation $\tau \to -1/\tau$ exchanges the time and space directions of the torus. So these transformations interchange sectors with a given projection (the multiplication of states by a given phase) with states with a twist in the space direction. This is precisely what one would expect from a path integral, where boundary conditions in the time direction correspond to the weighting of states with (symmetry) phases.

Setting

$$Z^{\alpha}_{\beta}(\tau) = \frac{1}{\eta(\tau)} \vartheta \begin{bmatrix} \alpha/2 \\ \beta/2 \end{bmatrix} (0, \tau), \tag{22.42}$$

the partition function for the eight fermions in the NS sector is $(Z^1_0)^4$, for example. If we include a factor $(-1)^F$, this is replaced by $(Z^1_1)^4$. We can work out similar expressions for the Ramond sector. From our expression for the transformation of the ϑ functions, it is clear that none of these is modular invariant by itself, as we would expect from our path integral arguments. So it is necessary to combine them and include also the eight bosons. When we do, we have the possibility of including minus signs (in more general situations, as we will see later, we will have more complicated possible phase choices). There are a finite number of possible choices. Two that work are

$$Z^{\pm} = \frac{1}{2}\left[Z^0_0(\tau)^4 - Z^0_1(\tau)^4 + Z^1_0(\tau)^4 \mp Z^1_1(\tau)^4\right]. \tag{22.43}$$

These transform simply under the modular transformations; all the terms transform to each other, up to an overall factor. There is a similar factor from the left-moving fermions (where one need not, a priori, take the same phase). Recall that the bosonic partition function is

$$Z_X(\tau) = (4\pi\alpha'\tau_2)^{-1/2}|\eta(q)|^{-2}. \tag{22.44}$$

Here the η function comes from the oscillators. The τ_2 factors come from the integration over the momenta. There are two additional such factors, coming from the integrals over the two light cone momenta. So the full partition function is

$$Z = C \int \frac{d^2\tau}{\tau_2^2} Z^8_X Z^+(\tau) Z^{\pm}(\tau)^*. \tag{22.45}$$

It is not hard to check that this expression is modular invariant.

If we examine the partition function carefully, we see that we have uncovered the GSO projection. Consider the first two terms in $Z^{\pm}$. They amount to just

$$\mathrm{Tr}[1 - (-1)^F]_{\mathrm{NS}}, \tag{22.46}$$

i.e. the physical states of the theory, in the NS sector, are only those of odd fermion number. There is a similar projector in the Ramond sector. The two possible choices of left- relative to right-moving Zs correspond precisely to the two possible supersymmetric string theories. Our original argument for the GSO projector was consistency in space–time, but here we have a more direct, world-sheet, consistency argument.

These are the only choices of phases which lead to supersymmetric strings in ten dimensions. However, there are other choices which lead to non-supersymmetric strings. These give what has come to be known as the Type 0 superstring. We will leave consideration of these theories to the exercises.

22.5.4 More on the Type I theory: gauge groups

In our discussion of the bosonic string theory, we mentioned that one can obtain non-Abelian gauge groups by allowing charges at the ends of the strings. There is an infinite set of possibilities, which we have not explore, as all these theories have other problematic features if one is trying to describe nature.

In the case of open superstrings, it turns out that the possible structures are quite constrained. First, it is necessary to include closed strings as well, in order to obtain a unitary theory. This can be seen by considering the scattering of four open strings. By stretching the diagram of Fig. 22.1 one can see that closed strings appear in intermediate states. These strings cannot be oriented. This leads to a different structure in the closed string sector from what we saw in the IIA or IIB theories. It is necessary to require that states be symmetric under the exchange of left- and right-moving quantum numbers. We will discuss the required projection later when we talk about D-branes and orientifold planes.

Second, it turns out that the absence of anomalies fixes uniquely the gauge symmetry as $O(32)$. From the point of view of our experience with four-dimensional anomalies this is somewhat surprising, but it turns out that in ten dimensions supergravity by itself can be anomalous, and this is the case for the open string. Allowing for charges at the end of the string leads to a set of additional mixed gauge and gravitational anomalies. Almost miraculously, if one takes the ends of the string to lie in the vector representation of $O(32)$, all anomalies cancel.

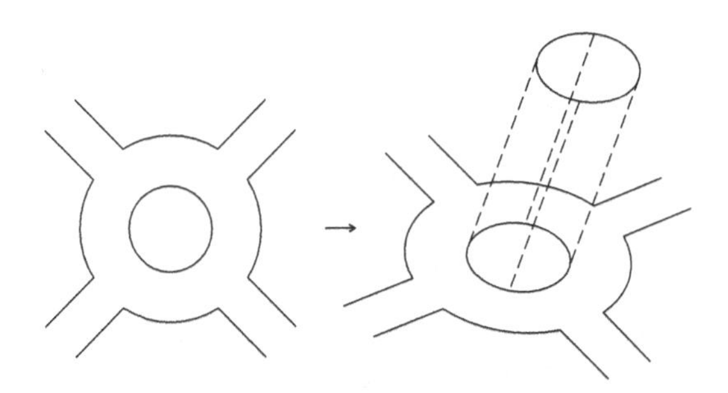

Fig. 22.1 Deforming the diagram for open-string scattering reveals an intermediate closed-string state.

22.6 Manifest space–time supersymmetry: the Green–Schwarz formalism

In the Ramond–Neveu–Schwarz formalism, space–time supersymmetry is obscure. It only arises after imposing the GSO projector. The supersymmetry operators must connect the different sectors, which are essentially different two-dimensional field theories. Such operators can be constructed, although we will not do that in this text. Instead, we consider in this section a different formalism, the *Green–Schwarz formalism*, in which the space–time supersymmetry is manifest. This formalism is best understood in the light cone gauge.

In the Green–Schwarz formalism one still has the bosonic coordinates X^I, but the eight fermionic coordinates ψ^I in the vector representation of $O(8)$ are replaced by eight fermionic coordinates in a spinor representation of $O(8)$ (we have already seen that $O(8)$ possesses two spinor representations, of opposite chirality). These are usually written as $S^a(\sigma, \tau)$. Their Lagrangian is

$$\mathcal{L}_{\text{gs}} = \frac{i}{2\pi} \bar{S}^a \rho^\alpha \partial_\alpha S^a, \tag{22.47}$$

where we have written the Ss as two component fermions and ρ^α denotes the two-dimensional γ-matrices. The S_as can be taken as real (Majorana). They can be decomposed into left and right movers, $S_\pm$. Unlike the case of RNS fermions, for both closed and open strings one has only one boundary condition. As in the case of the RNS fermions, for open strings the boundary conditions relates the left and right movers:

$$S_+^a(0, \tau) = S_-^a(0, \tau), \quad S_+^a(\pi, \tau) = S_-^a(\pi, \tau). \tag{22.48}$$

For closed strings one simply has a periodicity condition,

$$S_\pm^a(\sigma + \pi, \tau) = S_\pm^a(\sigma, \tau). \tag{22.49}$$

The mode expansions, in the case of closed strings, are

$$S_+^a = \sum_{-\infty}^{\infty} S_n^a e^{-2in(\tau-\sigma)},$$

$$S_-^a = \sum_{-\infty}^{\infty} \tilde{S}_n^a e^{-2in(\tau+\sigma)}. \tag{22.50}$$

The S_ns obey the anticommutation relations

$$\{S_n^a, S_m^b\} = \delta^{ab} \delta_{m+n}, \quad \{\tilde{S}_n^a, \tilde{S}_m^b\} = \delta^{ab} \delta_{m+n}. \tag{22.51}$$

For non-zero n these are canonical fermion creation-and-annihilation-operator anticommutation relations. Because of their quantum numbers, the Ss, acting on space–time bosonic states, produce fermionic states and vice versa.

The light cone Hamiltonian, in terms of these fields, takes the form:

$$H = \frac{1}{2p^+}[(p^I)^2 + N + \tilde{N}], \tag{22.52}$$

where

$$N = \sum_{m=1}^{\infty} (\alpha^I_{-m}\alpha^I_m + mS^a_{-m}S^a_m), \quad \tilde{N} = \sum_{m=1}^{\infty} (\tilde{\alpha}^I_{-m}\tilde{\alpha}^I_m + m\tilde{S}^a_{-m}\tilde{S}^a_m). \tag{22.53}$$

Note that there is no normal-ordering constant; more precisely, the normal-ordering constants associated with the left- and right-moving fields vanish, because the contributions of the bosonic and fermionic fields cancel (as they do in the Ramond sector of the superstring).

As in the Ramond sectors of the superstring theories, the anticommutation relations of the zero modes are important and interesting:

$$\{S^a_0, S^b_0\} = \delta^{ab}. \tag{22.54}$$

Again they are similar to the anticommutation relations of Dirac γ-matrices, but now the indices are different from the RNS case. The solution is to allow S_0 to act on 16 states, eight of which carry spinor labels, $\dot{b}$, and eight of which carry $O(8)$ vector labels, I. Then

$$\langle I|S^a_0|\dot{b}\rangle = \gamma^I_{a\dot{b}}. \tag{22.55}$$

We will leave the verification of this relation for the exercises and proceed directly to the identification of the massless states of the closed-string theories. The IIA and IIB theories are distinguished by the relative helicities of the S and $\tilde{S}$ fields. In the IIA case they are opposite; in the IIB case, the same. The massless fields are obtained just by tensoring the left and right states of the zero modes. The states

$$\epsilon_{IJ}|I\rangle|J\rangle \tag{22.56}$$

are the graviton, B-field and dilaton; the states where $I \to a$ or $J \to a$ are the two gravitini of the theory; those where both I and J are replaced by spinor indices are the states that we discovered in the Ramond–Ramond sector of the superstring theories.

In this formalism the space–time supersymmetry is manifest. There are two sets of supersymmetry generators. One generates not only space–time supersymmetries, but world-sheet supersymmetries as well. This is as it should be; the world-sheet Hamiltonian in the light cone gauge is also the space–time Hamiltonian,

$$Q^{\dot{a}} = \frac{1}{\sqrt{P^+}}\gamma^I_{a,\dot{a}} \sum_{-\infty}^{\infty} S^a_{-n}\alpha^I_n. \tag{22.57}$$

The second set is built of the zero modes alone:

$$Q^a = \sqrt{2P^+}S^a_0. \tag{22.58}$$

The supersymmetry generators obey the commutation relations:

$$\{Q^a, Q^b\} = 2P^+\delta^{ab}, \tag{22.59}$$

$$\{Q^a, Q^{\dot{a}}\} = \sqrt{2}\gamma^I_{a\dot{a}}P^I, \tag{22.60}$$

$$\{Q^{\dot{a}}, Q^{\dot{b}}\} = 2H\delta^{\dot{a}\dot{b}}. \tag{22.61}$$

The manifest supersymmetry and the close connection between world-sheet and space–time supersymmetries make the Green–Schwarz formalism a powerful tool, both conceptually and computationally, despite its lack of manifest Lorentz invariance.

22.7 Vertex operators

Because there are more world-sheet fields in the superstring than in the bosonic string, the vertex operators are more complicated. In the RNS formalism, the supersymmetry on the world sheet is a relic of a larger, local, supersymmetry, much as conformal invariance is a relic of the general coordinate invariance of the two-dimensional supersymmetry. The resulting superconformal symmetry provides constraints on vertex operators beyond those of the Virasoro algebra. These constraints can be implemented in a variety of ways, depending on how one treats the superconformal ghosts. In the simplest version, the vertex operators must be supersymmetric. In the case of the Type II theories, the vertex operators must respect both the left- and right-moving supersymmetries. For the massless fields of the Type II theory, for example,

$$V = \epsilon_{\mu\nu}(\bar{\partial}X^\mu - ik_\rho\psi^\rho\psi^\mu)(\bar{\partial}X^\nu - ik_\sigma\tilde{\psi}^\sigma\tilde{\psi}^\nu)e^{ik\cdot x}. \tag{22.62}$$

Here ϵ is subject to the constraint $k^\mu\epsilon_{\mu\nu} = 0$. Depending on the symmetries of ϵ, the vertex operator describes the production of gravitons, dilatons or antisymmetric tensor fields. It is straightforward to check that the coupling of three gravitons is that expected from the Einstein Lagrangian.

In the Green–Schwarz formalism, it is Lorentz invariance which governs the form of the vertex operators. As in the covariant formulation, the vertex operators in the Type II theory are products of separate vertex operators for the left and the right movers, with $e^{ik\cdot x}$ factors. These products have the structure

$$V_B = \zeta_{\mu\nu}B^\mu\tilde{B}^\nu e^{ik\cdot X}, \tag{22.63}$$

where

$$B^I = \partial X^I - R^{IJ}k^J, \quad B^+ = p^+ \tag{22.64}$$

and, from the light cone gauge condition, $\zeta^{\mu+} = 0$. Here

$$R^{IJ} = \frac{1}{4}\gamma^{IJ}_{ab}S^aS^b. \tag{22.65}$$

In the Green–Schwarz approach, it is no more difficult to deal with vertex operators for fermions than those for bosons. The polarizations $\zeta_{\mu\nu}$ are replaced by polarizations with one or two spinor indices. Then, as appropriate, one replaces the B^μs with fermionic operators, F^a and $F^{\dot{a}}$. We will not give these here, as we will not need them in the text, but they can be found in the references. In the covariant approach, more conformal field theory machinery is required to construct fermion emission operators.

Suggested reading

The superstring is well treated in various textbooks. Green *et al.* (1987) focuses heavily on the light cone formulation; Polchinski (1998) focuses on the RNS formulation. Both provide a great deal of additional detail, including the construction of vertex operators and S-matrices in the two formalisms. A concise and quite readable introduction to the problem of fermion vertex operators in the RNS formulation is provided by the lectures of Peskin (1987).

Exercises

(1) Consider the R–R sectors of the IIA and IIB theories, and study the objects

$$\bar{u}\gamma^{IJK\cdots}u.$$

Show that, in the IIA case, only even-rank tensors are non-vanishing while in the IIB theory only the odd-rank tensors are non-vanishing. Phrase this in the language of ten dimensions rather than the eight light cone dimensions. To do this consider a particle moving along the direction x^9, and show that the Dirac equation correlates chirality in ten dimensions with chirality in eight. To do this, you may want to make the following choice of Γ-matrices:

$$\Gamma^0 = \sigma_2 \otimes I_{16}, \quad \Gamma^i = i\sigma_1 \otimes \gamma^i, \quad \Gamma^9 = i\sigma_3 \otimes I_{16}. \tag{22.66}$$

(2) Write down the Green–Schwarz Lagrangian in a superspace formulation. Show that $Q^{\dot{a}}$ is the supersymmetry generator expected in this approach. Construct the symmetry generated by Q^a, and show that this has the structure of a non-linearly realized (spontaneously broken) supersymmetry. Can you offer some interpretation?

(3) Verify that, with the choice of Eq. (22.55), the zero modes of the Green–Schwarz operators S^a obey the correct anticommutation relations.

(4) Verify the expression for the partition function for the Type II theories. Show that it is modular invariant. Consider a different choice, which defines the type-0 superstring,

$$|Z_0^0|^8 + |Z_1^0|^8 + |Z_0^1|^8 \mp |Z_1^1|^8. \tag{22.67}$$

Attempt to verify that this is also modular invariant, but at least show that the spectrum does not include a spin-3/2 particle.

(5) Verify that the operator product of two graviton vertex operators in the RNS formalism yields the correct on-shell coupling of three gravitons. Remember the gauge condition in this analysis. The three-graviton vertex in Einstein's theory can be found, for example, in Sannan (1986).

23 The heterotic string

In the Type II theory we have seen that the left and right movers are essentially independent. At the level of the two-dimensional Lagrangian, there is a reflection symmetry between left and right movers; however, this symmetry does not hold sector by sector and is broken by boundary conditions and projectors.

In the *heterotic theory* this independence is taken further, and the degrees of freedom of the left and right movers are taken to be independent – and different. There are two convenient world-sheet realizations of this theory, known as the fermionic and bosonic formulations. In both there are eight left-moving and eight right-moving X^Is, associated with ten flat coordinates in space–time. There are eight right-moving two-dimensional fermions, ψ^I. There is a right-moving supersymmetry but no left-moving supersymmetry. In the fermionic formulation there are, in addition, 32 left-moving fermions which have no obvious connection with space–time, λ^A. In the bosonic description there are an additional 16 left-moving bosons. In other words, there are 24 left-moving bosonic degrees of freedom. There are actually several heterotic string theories in ten dimensions. Rather than attempt a systematic construction, we will describe the two supersymmetric examples. These have gauge groups $O(32)$ and $E_8 \times E_8$. The group E_8, one of the exceptional groups in Cartan's classification, is not very familiar to most physicists. However, it is in this theory that we can most easily find solutions which resemble the Standard Model. We will introduce certain features of E_8 group theory as we need them. More detail can be found in the suggested reading. In this chapter we will work principally in the fermionic formulation. We will develop some features of the bosonic formulation in later chapters, once we have introduced the compactification of strings.

23.1 The $O(32)$ theory

The $O(32)$ ($SO(32)$) theory is somewhat simpler to write down, so we will develop it first. In this theory the 32 λ^A fields are taken to be on an equal footing. The GSO projector, for the right movers, is as in the superstring theory. In the RNS formalism, in the NS sector we keep only states of odd fermion number and similarly in the Ramond sector, where fermion number includes a factor $e^{i\Gamma_{11}}$. For the left movers, the conditions are different. Again, we have a Ramond and an NS sector. In the NS sector we keep only states of even fermion number. In the R sector the ground state is a spinor of $SO(32)$. The spinor representation can be constructed just as we constructed the spinor representation of $O(8)$. Again, there are two inequivalent irreducible representations. There is a chirality, which we can call Γ_{33}.

The lowest spinor representation of definite chirality is the 32. Again, in the Ramond sector we project (by convention) onto states of even fermion number.

As for the superstring, there is a different light cone Hamiltonian for each sector. The right-moving part is just as in the superstring. The left-moving part includes a contribution from the bosonic operators and a contribution from the fermions, λ^A. As for the superstring, in the Ramond sector the λ^As are integer moded; they are half-integer moded in the NS sector. From our formula, the left-moving normal-ordering constant is -1 in the NS sector and zero in the R sector.

Now, we can consider the spectrum. Take, first, the NS–NS sector, i.e. the sector with NS boundary conditions for both the left and the right movers. The states are space–time bosons. The left-moving normal-ordering constant is -1. Without λ^As, the lowest mass states we can form are

$$\tilde{\alpha}^I_{-1}\psi^J_{-1/2}|0\rangle. \tag{23.1}$$

From our discussion of the normal-ordering constants, we see that these states are massless. They have the quantum numbers of a graviton, antisymmetric tensor and scalar field.

Using the left-moving fermion operators, we can construct additional massless states in this sector:

$$\lambda^A_{-1/2}\lambda^B_{-1/2}\psi^J_{-1/2}|0\rangle. \tag{23.2}$$

These are vectors in space–time. Because the λ^As are fermions, they are antisymmetric under $A \leftrightarrow B$. So, they are naturally identified as gauge bosons of the gauge group $SO(32)$. We will show shortly that they have the couplings of $O(32)$ Yang–Mills theories.

Let's first consider the other sectors. In the NS–R sector, the right-moving states, $\psi^J_{-1/2}|\vec{p}\rangle$, are replaced by the states we labeled $|a\rangle$. Again these must be massless, so we now have particles with the quantum numbers of the gravitino, one additional fermion and the gauginos of $O(32)$. In the NS–R and R–R sectors, however, it turns out that there are no massless states, as can be seen by computing the normal-ordering constants. It is necessary to include the R sector for the left movers. Here the normal-ordering constant is $+1$, and there are no massless states.

23.2 The $E_8 \times E_8$ theory

The E_8 group is unfamiliar to many physicists, and one might wonder how one could obtain two such groups from a string theory. To begin, it is useful to note that E_8 has an $O(16)$ subgroup. Under this group the adjoint of E_8, which is 248-dimensional, decomposes as a 120 – the adjoint of $O(16)$ – and a 128, a spinor of $O(16)$.

In ten dimensions we have seen we can build a sensible string theory with eight left-moving bosons and 32 left-moving fermions. So the strategy is to break the fermions into two groups of 16, λ^A and $\lambda^{\tilde{A}}$, and to treat these as independent. This gives a manifest $O(16) \times O(16)$ symmetry, similar to the symmetry of the $O(32)$ theory. There are now NS and R sectors for each set of fermions separately. The right-moving GSO projectors are as before. For the left movers, in each NS sector the action of the left-moving projector is onto states of even fermion number. With a suitable convention for the Γ_{11} chirality, this is also true of the R sectors. So, consider again the spectrum. In the NS–NS–NS sector, just as before, there are a graviton, antisymmetric tensor and scalar field. We can also construct gauge bosons in the adjoint of each of the two $O(16)$s,

$$\lambda^A \lambda^B \psi^J_{-1/2}|0\rangle, \quad \lambda^{\tilde{A}} \lambda^{\tilde{B}} \psi^J_{-1/2}|0\rangle. \tag{23.3}$$

Note that, because of the projectors, there are no massless states carrying quantum numbers of both $O(16)$ groups simultaneously. In the NS–NS–R sector we find the superpartners of these fields.

Now consider the R–NS–NS sector. Here the ground state is a spinor of the first $O(16)$. So now we have a set of gauge bosons in the spinor 128-dimensional representation. Similarly, in the NS–R–NS sector we have a spinor of the other $O(16)$. These are the correct set of states to form the *adjoints* of two E_8s. Again, establishing that the group is actually $E_8 \times E_8$ requires showing that the gauge bosons interact correctly. We will do that in the following section.

Finally, in the R–R–NS and R–R–R sectors there are no massless states.

23.3 Heterotic string interactions

We would like to show that the states we have identified as gauge bosons in the heterotic string interact at low energies, as required by Yang–Mills gauge invariance. To do this we work in the covariant formulation and construct vertex operators corresponding to the various states. Consider the $O(32)$ theory first. With our putative gauge bosons we associate the vertex operators

$$\int d^2z\, V^{AB\mu} = \int d^2z\, \lambda^A(\bar{z})\lambda^B(\bar{z}) \left[\partial_z X^\mu(z) - ik_\nu \psi^\mu \psi^\nu(z)\right] e^{ik\cdot x}. \tag{23.4}$$

For the right movers, as in the Type II theories we have required invariance under the right-moving world-sheet supersymmetry. For the left-moving vertex operators we have simply required that the operators have dimension one, so that overall the vertex operator has dimension one with respect to the left- and right-moving conformal symmetry (the operator is said to be $(1,1)$, just like those of the Type II theory). To determine their interactions, we will study the operator product of two such operators. The left-moving part of the vertex operator is a current,

$$j^{AB}(\bar{z}) = \lambda^A(\bar{z})\lambda^B(\bar{z}). \tag{23.5}$$

The operator product of two of these currents is

$$j^{AB}(\bar{z})j^{CD}(\bar{w}) = \frac{\delta^{AC}\delta^{BD} + \cdots}{(\bar{z}-\bar{w})^2} + \frac{\delta^{AC}\lambda^B(\bar{z})\lambda^D(\bar{w}) + \cdots}{\bar{z}-\bar{w}}. \tag{23.6}$$

An algebra of currents of this kind is called a *Kac–Moody algebra*. It has the general form

$$j^a(\bar{z})j^b(\bar{w}) = \frac{k\delta^{ab}}{(\bar{z}-\bar{w})^2} + \frac{f^{abc}j^c(\bar{w})}{\bar{z}-\bar{w}}, \tag{23.7}$$

where k is called the central extension of the algebra. In our case $k = 1$. The f^{abc}s are the structure constants of the group. They can be found from Eq. (23.6).

To see the Yang–Mills structure it is helpful to use the general Kac–Moody form, denoting the currents and the corresponding vertex operators by a subscript a. Regarding the operator product, we have seen from our discussion of factorization that the interaction is proportional to the coefficient of $1/|z-w|^2$. In the product $V_a(z)V_b(w)$ the term $1/(\bar{z}-\bar{w})$ is proportional to f_{abc}, just what is needed for the Yang–Mills vertex. The momentum and $g_{\mu\nu}$ contributions arise from the right-moving operator product. In

$$[\partial X^\mu(z) + k_{1\rho}\psi^\rho(z)\psi^\mu(z)]e^{ik_1\cdot X(z)}[\partial X^\nu(w) + k_{2\sigma}\psi^\sigma(w)\psi^\nu(w)]e^{ik_2\cdot X(w)} \tag{23.8}$$

the $1/(z-w)$ terms arise from various sources. One can contract the ∂X factors in each vertex with the exponential factors. This gives

$$V_a^\mu V_b^\nu \sim \frac{f^{abc}V^{c\nu}\left(k_2^\mu - k_1^\mu\right)}{|z-w|^2}. \tag{23.9}$$

Contracting the two ∂X factors with each other gives two factors of $z-w$ in the denominator. These can be compensated by Taylor-expanding $X(z)$ about w. Additional terms arise from contracting the fermions with each other. The details of collecting all the terms and comparing with the three-gauge-boson vertex are left for the exercises.

23.4 A non-supersymmetric heterotic string theory

One can verify the modular invariance of the heterotic string theory, with the GSO projections we have used, in precisely the same way as we did for the superstring theories. This raises the question: are there other ten-dimensional heterotic theories, obtained by combining the partition functions of the separate sectors in different ways? The answer is definitely yes. Several of these have tachyons, but one does not. Its gauge group is $O(16) \times O(16)$. It is most readily described in the Green–Schwarz formalism. It will also provide us with our first example of "modding out", i.e. obtaining a new string theory by making various projections.

On the other hand, in order to obtain the smaller gauge group we need to get rid of the gauge bosons from E_8 which lie in the spinor representation. On the other hand there is no harm in having the corresponding gauginos, if supersymmetry is broken. So we take the original $E_8 \times E_8$ theory and keep only states which are even under the symmetry $(-1)^F$ in space–time and a corresponding symmetry in the gauge group (i.e. spinorial representations are odd, and non-spinorial are even). This immediately gets rid of:

1. the gravitinos, and
2. the gauge bosons which are in spinorial representations of the group.

However, we have seen that, for consistency, it is important that string theories be modular invariant. Simply throwing away states spoils modular invariance; it is necessary to add in additional states. In the present case one has to add a sector with different, twisted, boundary conditions for the fields, as follows:

$$S_a(\sigma + \pi, \tau) = -S_a(\sigma, \tau). \tag{23.10}$$

For the gauge fermions there is a related boundary condition but this is more easily described in the bosonic formulation which we will discuss in Chapter 25 on compactification.

Suggested reading

The original heterotic string papers by Gross *et al.* (1985, 1986) are remarkably clear. Polchinski's book (1998) provides a quite thorough overview of these theories. For example, for those who are not enamored of the Green–Schwarz formalism, it develops the non-supersymmetric $O(32)$ in the RNS formalism in some detail. The absence of global symmetries in the heterotic string is demonstrated in Banks and Dixon (1988).

Exercises

(1) Construct the states corresponding to the gauge bosons of $E_8 \times E_8$. In particular, use the creation–annihilation operator construction of $O(2N)$ spinor representations to build the 128-dimensional representations of $O(16)$.

(2) Verify that the algebra of $O(32)$ currents is of the Kac–Moody form. To work out the structure constants, remember that the generators of O groups are just the antisymmetric matrices

$$(\omega^{AB})_{CD} = \delta^{AC}\delta^{BD} - \delta^{AD}\delta^{BC}. \tag{23.11}$$

(3) Verify that, on-shell, the three-gluon vertex has the correct form. In addition to carefully evaluating the terms in the operator product expansion, it may be necessary to use momentum conservation and the transversality of the polarization vectors.

24 Effective actions in ten dimensions

In ten dimensions, supersymmetry greatly restricts the allowed particle content and effective actions of theories with massless fields. Without gauge interactions there are only two consistent possibilities. These correspond to the low-energy limits of the IIA and IIB theories. These have $N = 2$ supersymmetry (they have 32 conserved supercharges). Because the symmetry is so restrictive, we can understand a great deal about the low-energy limits of these theories without making any detailed computations. We can even make exact statements about the non-perturbative behavior of these theories. This is familiar from our studies of field theories in four dimensions with more than four supercharges. In ten dimensions, supersymmetric gauge theories have $N = 1$ supersymmetry (16 supercharges). Classically, specification of the gauge group completely specifies the terms in the effective action with up to two derivatives. Quantum mechanically, only the gauge groups $O(32)$ and $E_8 \times E_8$ are possible.

24.1 Eleven-dimensional supergravity

Rather than start with these ten-dimensional theories, it is instructive to start in eleven dimensions. Eleven is the highest dimension where one can write a supersymmetric action (in higher dimensions, spins higher than 2 are required). This fact by itself has focused much attention on this theory. But it is also known that the theory in eleven dimensions has a connection with string theory. As we will see later, if one takes the strong coupling limit of the Type IIA string theory, one obtains a theory whose low-energy limit is eleven-dimensional supergravity.

The particle content of the eleven-dimensional theory is simple: there is a graviton, g_{MN} (44 degrees of freedom) and a three-index antisymmetric tensor field, C_{MNO} (84 degrees of freedom); here $M, N, O = 0, \ldots, 9$ are space–time indices. There is also a gravitino, ψ_M. This has 16×8 degrees of freedom. We have, as usual, counted degrees of freedom by considering a theory in nine dimensions, remembering that g_{MN} is symmetric and traceless and that the basic spinor representation in nine dimensions is sixteen-dimensional (it combines the two eight-dimensional spinors of $O(8)$).

The Lagrangian for the eleven-dimensional theory, in addition to the Ricci scalar, involves a field strength for the three-index field, C_{MNO}. The corresponding field strength F_{MNOP} is completely antisymmetric in *its* indices, similar to the field strength of electrodynamics:

$$F_{MNOP} = \frac{3!}{4!}(\partial_M C_{NOP} - \partial_N C_{MOP} + \cdots)$$
$$= \frac{3!}{4!}\sum_P (-1)^P \partial_M C_{NOP}, \tag{24.1}$$

where the sum is over all permutations and the factor $(-1)^P$ is ± 1 depending on whether the permutation is even or odd. It is convenient to describe such antisymmetric tensor fields in the language of differential forms. For the reader unfamiliar with these, an introduction is provided later, in Section 26.1. For now we note that antisymmetric tensors with p indices are p-forms. The operation of taking the curl, as in Eq. (24.1), takes a p-form to a $(p+1)$-form. It is denoted by the symbol d and is called the *exterior derivative*. In terms of forms, Eq. (24.1) can be written compactly as

$$F = dC. \tag{24.2}$$

The theory has a gauge invariance:

$$C \to C + d\Lambda, \quad C_{MNO} \to \frac{2}{3!}\sum_P (-1)^P \partial_M \Lambda_{NO} \tag{24.3}$$

where Λ is a two-form.

We will not need the complete form of the action. The bosonic terms are

$$\mathcal{L}_{\text{bos}} = -\frac{1}{2\kappa^2}\sqrt{g}R - \frac{1}{48}\sqrt{g}F^2_{MNPQ} - \frac{\sqrt{2}\kappa}{3456}\epsilon^{M_1 \ldots M_{11}} F_{M_1 \ldots M_4} F_{M_5 \ldots M_8} C_{M_9 \cdots M_{11}}. \tag{24.4}$$

The last of these is a Chern–Simons term. It respects the gauge invariance of Eq. (24.3) if one integrates by parts. Such terms can arise in field theories with odd dimensions; in (2+1)-dimensional electrodynamics, for example, they play an interesting role. The fermionic terms include covariant derivative terms for the gravitino as well as couplings to F and various four-fermion terms. The supersymmetry transformation laws have the structure

$$\delta e^A_M = \frac{\kappa}{2}\bar{\eta}\Gamma^A \psi_M, \tag{24.5}$$

$$\delta A_{MNP} = -\frac{\sqrt{2}}{8}\bar{\eta}\Gamma_{[MN}\psi_{P]}, \tag{24.6}$$

$$\delta \psi_M = \frac{1}{\kappa} D_M \eta + (F\eta \text{ terms}). \tag{24.7}$$

Here e^A_M is the *vielbein* field and the covariant derivative is constructed from the spin connection (discussed in Section 17.6).

24.2 The IIA and IIB supergravity theories

The eleven-dimensional fields are functions of the coordinates $x_0, \ldots, x_{10}$. We obtain the IIA supergravity theory (the low-energy limit of the Type IIA string) if we truncate the eleven-dimensional supergravity theory to ten dimensions, i.e. if we simply eliminate the dependence on x_{10}. We need to relabel the fields as well, since it is not appropriate to have a 10 index. So we take the components of g with ten-dimensional indices to be the ten-dimensional metric. Then $g_{10\,10}$ is a ten-dimensional scalar, which we call ϕ, and $g_{10\,\mu}$ is a ten-dimensional vector, which corresponds to the Ramond–Ramond vector of the IIA string theory. Note that $C_{10\,\mu\nu} = B_{\mu\nu}$ is a two-index antisymmetric tensor field in ten dimensions (corresponding to the two-index tensor we found in the NS–NS sector). The gravitino decomposes into two ten-dimensional gravitinos, and two spin-1/2 particles. With $H = dB$, the bosonic terms in the ten-dimensional action for the NS–NS fields are

$$\mathcal{L}_{\text{bos}} = -\frac{1}{2\kappa^2}R - \frac{3}{4}\phi^{-3/2}H^2_{\mu\nu\rho} - \frac{9}{16\kappa^2}\left(\frac{\partial_\mu \phi}{\phi}\right)^2. \tag{24.8}$$

The IIB theory is not obtained in this way. But, from string theory, we can see that the NS–NS action must be the same as in the Type IIA theory. The reason is that in the NS–NS sector the vertex operators of the IIA and IIB theories are the same, so the scattering amplitudes – and hence the effective action – are the same as well.

24.3 Ten-dimensional supersymmetric Yang–Mills theory

From our studies of the heterotic string we know the field content of this theory. There is a metric, an antisymmetric tensor field (which we again call $B_{\mu\nu}$), a scalar ϕ and the gauge fields, A^a_μ. The Lagrangian for g, B and ϕ is the same as in the Type II theories. The gauge terms are

$$\mathcal{L}_{\text{YM}} = -\frac{\phi^{-3/4}}{4g^2}F^2_{\mu\nu} - \frac{1}{2}\bar{\chi}^a(D_M\chi)^a. \tag{24.9}$$

It turns out that there is another crucial modification in the Yang–Mills case. The field strength H_{MNO} is not simply the curl of B_{MN} but contains an additional contribution, which closely resembles the Chern–Simons term we encountered in our study of instantons in four-dimensional Yang–Mills theory:

$$H = dB - \frac{\kappa}{\sqrt{2}}\omega_3 \tag{24.10}$$

(the notation will be thoroughly explained in Chapter 26), with

$$\omega_3 = A^aF^a - \frac{1}{3}gf_{abc}A^aA^bA^c = A^a dA^a + \frac{2}{3}gf_{abc}A^aA^bA^c. \tag{24.11}$$

There is also a gravitational term, with a similar form. This extra term plays an important role in understanding anomaly cancellation. In four dimensions we will see that it leads to the appearance of axions in the low-energy theory.

24.4 Coupling constants in string theory

The Standard Model is defined, in part, by specifying a set of coupling constants. The fact that there are so many parameters is one of the reasons we have given that the model is not satisfactory as some sort of ultimate description of nature. In our discussion of string interactions we introduced a coupling constant g_s. There is one such constant for each of the string theories we have introduced, bosonic, Type I, Types IIA and IIB and heterotic, as well as for non-supersymmetric strings. But the idea that string theory possesses a free parameter is, it turns out, an illusion. By changing the expectation value of the dilaton field, we can change the value of the coupling. This is similar to phenomena we observed in four-dimensional supersymmetric gauge theories. In situations with a great deal of supersymmetry there will be no potential, perturbative or non-perturbative, for this field and the choice of coupling will correspond to a choice of vacuum. But, in vacua in which supersymmetry is broken, we would expect that dynamical effects would fix the value of this and any other moduli. The coupling constants of the low-energy theory would then be determined fully in ways which, in principle, one could understand and eventually hope to calculate. In the next few sections we explain this connection between coupling constants and fields.

24.4.1 Couplings in closed-string theories

When we constructed vertex operators we saw that we could include a coupling constant g_s in the definition of the vertex operator. In the heterotic string the same coupling enters in all vertices. This is a consequence of unitarity. At tree level, for example, we saw that scattering amplitudes factorize near poles of the S-matrix; if one introduced independent couplings for each vertex operator, the amplitudes would not factorize correctly. As a result all amplitudes can be expressed in terms of a single parameter. In the heterotic string theory this means that there is a calculable relation between the gravitational constant and the Yang–Mills coupling. To work out this coupling, one needs to calculate the three-point interactions for three gravitons and for three gauge bosons carefully (see the exercises at the end of the chapter). The results are necessarily of the form

$$\kappa_{10}^2 = a g^2 (2\alpha')^4, \quad g_{\rm YM}^2 = b g^2 (2\alpha')^3. \tag{24.12}$$

The calculation yields $a = 1/4$, $b = 1$.

A similar analysis in the Type I theory gives a relation between the open-string and closed-string couplings and a relation between the gauge and gravitational couplings.

In both theories we see that the string scale is smaller than the Planck scale:

$$M_s = (g_s)^{1/4} M_p. \tag{24.13}$$

This is a satisfying result. It means that if we think of M_s as the cutoff on the gravity theory, gravitational loops are suppressed by powers of g_s.

24.4.2 The coupling is not a parameter in string theory

So far, in all the string theories it would appear that there is an adjustable, dimensionless, parameter. As we said earlier this is not really the case; the reason can be traced to the *dilaton*. Classically, in all the string theories we have studied the dilaton has no potential, so its expectation value is not fixed. In the next two short subsections we will demonstrate that changing the expectation value of the dilaton changes the effective coupling. In four dimensions with $N > 1$ supersymmetry (and automatically in dimensions greater than five) there is no potential for the dilaton, so the question of the value of the coupling is equivalent to a choice among degenerate vacuum states. Without supersymmetry (or with $N \leq 1$ supersymmetry in four dimensions), one expects quantum mechanical effects to generate a potential for the dilaton, and the value of the coupling is then a *dynamical* question.

24.4.3 Effective Lagrangian argument

Perhaps the simplest way to understand the role of the dilaton is to examine the ten-dimensional effective action. We start with the case of the heterotic string in ten dimensions. We can redefine ϕ as $g^{-2}\kappa^{3/2}\phi'$, eliminating g everywhere in the action. Note that since $\kappa \propto g$ this means that $\phi' \sim g^{1/2}$. Then we can do a Weyl rescaling

$$g_{\mu\nu} = \phi^{-1} g_{\mu\nu}. \tag{24.14}$$

This puts a common power of ϕ in front of the action, ϕ^{-4} and is consistent with g being the string loop parameter, since effectively we have a factor g^{-2} at the front.

With this rescaling it is the string scale which is fundamental. Remember that $M_\mathrm{p}^2 = M_\mathrm{s}^2/(g^2)^{1/4}$. By rescaling the metric we have rescaled lengths which were originally expressed in units of M_p in terms of M_s. So we have a consistent picture. The cutoff for the effective Lagrangian is M_s. All dimensional parameters in the Lagrangian are of order M_s, and loops are accompanied by $g^2 \sim \phi^4$.

24.4.4 World-sheet coupling of the dilaton

As we will discuss further in the next chapter, we can define a generating functional for the *S*-matrix by taking the two-dimensional field theory and adding space–time fields weighted by vertex operators. So, for example, for the bosonic string we would add terms to the action of the form

$$(\eta_{\mu\nu} + h_{\mu\nu})e^{ik\cdot x}\partial x^\mu \bar{\partial} x^\nu. \tag{24.15}$$

We can generalize this to a background metric, yielding a two-dimensional non-linear sigma model,

$$g_{\mu\nu}(x)\partial x^\mu \bar{\partial} x^\nu. \tag{24.16}$$

Just as we can couple the graviton to the world sheet, we can also couple the dilaton to it. The dilaton turns out to couple to the two-dimensional curvature:

$$\mathcal{L}_\Phi = \frac{1}{4\pi}\int d^2\sigma\sqrt{h}\Phi(X)R^{(2)}. \tag{24.17}$$

In two dimensions, however, the dynamics of gravity is trivial. Indeed, if we use our usual counting rules, the graviton has less than a single degree of freedom. So, the $R^{(2)}$ factor should not generate any sensible graviton dynamics. If we go to the conformal gauge,

$$h_{\alpha\beta} = e^\phi \eta_{\alpha\beta}, \tag{24.18}$$

the curvature is a total divergence:

$$R^{(2)} = \partial^2\phi. \tag{24.19}$$

Thus at most this factor in the action is topological. To get some feeling for this, let us evaluate the integral in the case of a sphere. We have seen that one representation for the sphere is provided by the space CP^1. This space has one complex coordinate. It is Kahler, which means that the only non-vanishing component of g is

$$g_{z\bar{z}} = \partial_z\partial_{\bar{z}}K(z,\bar{z}), \tag{24.20}$$

where, in this case,

$$K = \ln(1+\bar{z}z). \tag{24.21}$$

So we have

$$g = \left(\frac{1}{1+\bar{z}z}\right)^2. \tag{24.22}$$

From this, we can read off ϕ,

$$\phi = 2\ln(1+\bar{z}z) = -2\ln\left(1+\sigma_x^2+\sigma_y^2\right), \tag{24.23}$$

and the integral over the curvature is

$$\frac{1}{4\pi}\int d^2\sigma\,\partial^2\left[-2\ln\left(1+\sigma_x^2+\sigma_y^2\right)\right] = 2. \tag{24.24}$$

Note that this is invariant under a constant Weyl rescaling; it is topological. It is known as the Euler character of the surface and satisfies

$$\chi = \frac{1}{4\pi}\int d^2\sigma\sqrt{h}R^{(2)} \tag{24.25}$$

and

$$\chi = 2(1-g). \tag{24.26}$$

In this expression, χ is known as the Euler character of the manifold and g is the genus. For the sphere, $g = 0$; for the torus, $g = 1$; and so on for higher-genus string amplitudes. So string amplitudes, for constant Φ, come with a factor

$$e^{-2\Phi(1-g)}. \tag{24.27}$$

Thus we can identify e^Φ with the string coupling constant.

Suggested reading

Ten-dimensional effective actions were described in some detail by Green *et al.* (1987). The couplings of the dilaton in string theory were discussed in detail by Polchinski (1998).

Exercise

(1) By studying the OPEs of the appropriate vertex operators, verify Eq. (24.12). To avoid making this calculation too involved, you may want to isolate particular terms in the gravitational and Yang–Mills couplings. The required vertices in general relativity can be found in Sannan (1986).

25 Compactification of string theory I. Tori and orbifolds

We do not live in a ten-dimensional world, and certainly not in a 26-dimensional world without fermions. But if we don't insist on Lorentz invariance in all directions then there are other possible ways to construct consistent string theories. In this chapter we will uncover many consistent string theories in four dimensions (and in others). If anything, our problem will shortly be an "embarrassment of riches:" we will see that there are vast numbers of possible string constructions. The connection of these various constructions to one another is not always clear. Many of them can be obtained from others by varying the expectation values of the light fields (i.e. the moduli). One might imagine that others could be obtained by exciting massive fields as well. In general, though, this is not known and in any case the meaning of such connections in a theory of gravity is obscure. But, before exploring these deep and difficult questions, we need to acquire some experience with constructing strings in different dimensions.

25.1 Compactification in field theory: the Kaluza–Klein program

The idea that space–time might be more than four-dimensional was first put forward by Kaluza and Klein shortly after Einstein published his general theory of relativity. They argued that *five-dimensional* general coordinate invariance would give rise to both four-dimensional general coordinate invariance and a $U(1)$ gauge invariance, unifying electromagnetism and gravity. In modern language they considered the possibility that space–time is five-dimensional, with the structure $M^4 \times S^1$. This is, on first exposure, a bizarre concept but its implications are readily understood by considering a toy model. Take a single scalar field Φ in five dimensions. Denote the coordinates of M^4 by x^μ as usual and that of the fifth dimension by y,

$$0 \leq y < 2\pi R. \tag{25.1}$$

Because y is a periodic variable, we can expand the field Φ in Fourier modes:

$$\Phi(x,y) = \sum_n \frac{1}{\sqrt{2\pi R}} \phi_n(x) e^{ip_n y}, \quad p_n = \frac{n}{R}. \tag{25.2}$$

Taking a simple free-field Lagrangian for Φ in five dimensions, the Lagrangian, written in terms of the Fourier modes, takes the form

$$\int d^4xdy\,\mathcal{L} = -\int d^4xdy\frac{1}{2}\left[(\partial\phi)^2 + M^2\phi^2)\right]$$
$$= -\int d^4x\sum_n\frac{1}{2}\left[\partial_\mu\phi^2 + (M^2 + p_n^2)\phi^2\right]. \tag{25.3}$$

So, from a four-dimensional perspective, this theory describes an infinite number of fields, with ever increasing mass. In the gravitational case, symmetry considerations will force $M = 0$. If we set $M = 0$ in our scalar model, we obtain one massless state in four dimensions ($n = 0$) and an infinite tower – the Kaluza–Klein tower – of massive states. If R is very small, say $R \approx M_{\rm p}^{-1}$, the massive states are all extremely heavy. For the physics of the everyday world we can integrate out these massive fields and obtain an effective Lagrangian for the massless field. The effects of the infinite set of massive fields – the signature of extra dimensions – will show up only in tiny, higher-dimensional operators. So, in the end, finding evidence for these extra dimensions is likely to be extremely difficult.

Having understood this simple model, we can turn to Kaluza and Klein's theory of gravitation and electromagnetism. The five-dimensional theory has the Lagrangian

$$\mathcal{L} = \frac{1}{2\kappa^2}\sqrt{g}R. \tag{25.4}$$

Now there is an infinite tower of massive states corresponding to modes of the five-dimensional metric: $g_{\mu\nu}$, $g_{\mu 4}$ and g_{44}. Our principal interest is in the massless states, which arise from modes that are independent of y (we will need to refine this identification shortly). We expect to find a four-dimensional metric tensor, $g_{\mu\nu}$, a field which transforms as a vector of the four-dimensional Lorentz group, $g_{4\mu}$, and a scalar, g_{44}. There are various ways in which we can rewrite the five-dimensional fields in terms of four-dimensional fields. The physics is independent of this choice, but clearly some choices will be more helpful than others. The most sensitive choice is that of the gauge field; we would like to choose this field in such a way that its gauge transformation properties are simple. The general coordinate invariance associated with transformations of the fifth dimension, $x_4 = x_4 + \epsilon_4(x)$, is given by

$$g_{\mu 4} = g_{\mu 4} + \partial_\mu\epsilon_4(x). \tag{25.5}$$

This looks just like the transformation of a gauge field. So, we adopt the conventions

$$g_{\mu 4} = A_\mu, \quad g_{44}(x) = e^{2\sigma(x)}, \quad g_{\mu\nu} = g_{\mu\nu}. \tag{25.6}$$

Note we are defining, here, a reference metric and are measuring distances relative to that; we can take the basic distance to be the Planck length. Substituting this ansatz back into the five-dimensional action, one can proceed very straightforwardly, working out the Christoffel symbols and from these the various components of the curvature. Gauge invariance significantly constrains the possible terms. One obtains

$$\mathcal{L} = \frac{2\pi R}{2\kappa^2}\sqrt{g}e^\sigma R + \frac{1}{4}e^{-\sigma}F_{\mu\nu}^2. \tag{25.7}$$

So the theory at low energies consists of a $U(1)$ gauge field, the graviton and a scalar. The Lagrangian is not quite in the canonical form; usually one writes the action for general relativity in a form where the coefficient of the Ricci scalar (the "Einstein term") is field independent. One can achieve this by performing an overall rescaling of the metric, known as a Weyl rescaling,

$$g_{\mu\nu} \to e^{-\sigma} g_{\mu\nu}. \tag{25.8}$$

This introduces a kinetic term for the scalar:

$$\mathcal{L} = \frac{1}{2\kappa^2}\left[R + \frac{3}{2}(\partial\phi)^2\right]. \tag{25.9}$$

The scalar field here is particularly significant. As it corresponds to g_{55}, giving it an expectation value amounts to changing the radius of the internal space. In the Lagrangian there is no potential for σ so, at this level, nothing determines this expectation value. As in our four-dimensional examples, σ is said to be a modulus. We now show that quantum mechanical effects generate a potential for σ even at one loop. This potential falls to zero rapidly as the radius becomes large. If there is a minimum of the potential, it occurs at radii of order one, where the computation is certainly not reliable.

The calculation is equivalent to a Casimir energy computation in quantum field theory; one can think of the system as sitting in a periodic box of size $2\pi R$ and can ask how the energy depends on the size of the box. We can guess the form of the answer before doing any calculation. Since this is a one-loop computation, the result is independent of the coupling. On dimensional grounds the energy density is proportional to $1/R^4$.

To simplify matters, we will treat the gravitational field as a scalar field. At one loop,

$$\Gamma = \mathrm{Tr}\ln\left(-\partial^2 + \frac{n^2}{R^2}\right), \tag{25.10}$$

where we can do the calculation in Euclidean space. We can obtain a more manageable expression by differentiating with respect to R. The trace can be interpreted now as a sum over the possible momentum states in four Euclidean dimensions, in a box of volume VT. Replacing the sum by an integral gives an explicit factor of VT; the coefficient is the energy per unit volume:

$$\frac{\partial V}{\partial R} = \int \frac{d^4p}{(2\pi)^4 R^3}\sum_n \frac{n^2}{p^2 + (n^2/R^2)}. \tag{25.11}$$

This can be evaluated using the same trick as one uses to compute the partition function in finite-temperature field theory (this is described in Appendix C). One first converts the sum into a contour integral, by introducing a function with simple poles located at the integers:

$$\frac{\partial V}{\partial R} = \int \frac{d^4p}{(2\pi)^3}\oint \frac{dz}{2\pi i}\frac{1}{z^2 + p^2}\frac{1}{1 - e^{2\pi i R z}}\frac{z^2}{R}. \tag{25.12}$$

The contour consists of one line running slightly above the real axis and one line running slightly below it. Now deform the contour in such a way that the upper line encircles the pole at $z = ip$ and the lower line encircles the pole at $z = -ip$. The resulting expression is divergent, but we can separate off a term independent of R and a convergent, R-dependent, term:

$$\frac{\partial V}{\partial R} = \frac{1}{R}\int \frac{d^4p}{(2\pi)^4}\frac{p^2}{2p}\left(1+\frac{1}{e^{2\pi pR}-1}\right)$$
$$= \frac{24\zeta[5]}{(2\pi)^4R^5} + R\text{-independent, term;} \tag{25.13}$$

the zeta function was defined in Eq. (22.33).

25.1.1 Generalizations and limitations of the Kaluza–Klein program

So far we have considered the compactification of a five-dimensional theory on a circle, but one can clearly consider compactifications of more dimensions on more complicated manifolds. It is possible to obtain, in this way, non-Abelian groups. So, one might hope to understand the interactions of the Standard Model. The principal obstacle to such a program turns out to be obtaining chiral fermions in suitable representations. The existence of chiral fermions in a particular compactification is a topological question, as can readily be seen. As one smoothly varies the size and shape of the manifold, it is possible that some fields will become massless; equivalently, massless fields can become massive. However, fields which gain mass must come in vector-like pairs; the chiral structure of a theory will not change as one continuously changes the parameters of the compactification.

That chirality is special follows from the observation that spinors in higher dimensions decompose as left–right symmetric pairs with respect to four dimensions. For compactification manifolds with non-trivial topology, it is indeed possible to obtain chiral fermions. However, it turns out to be impossible to obtain chiral fermions in the required representations of the Standard Model group. We will see, though, that string theory can generate both gauge groups and chiral fermions upon compactification.

25.2 Closed strings on tori

So far we have considered compactifications of field theories in higher dimensions, but general higher-dimensional field theories are non-renormalizable and must be viewed as low-energy limits of some other structure. The only sensible structure we know in higher dimensions is string theory. At the same time, if string theory is to have anything to do with the world around us then it must be compactified to four dimensions.

It is not complicated to repeat the field theory analysis for the case of closed strings on circles, or more generally on tori. Consider first compactifying one dimension, X^9, on a circle of radius $2\pi R$. We require that states be invariant under translations by $2\pi R$. This means that the momenta, as in the field theory case, are quantized,

$$p^9 = \frac{n}{R}. \tag{25.14}$$

But now there is a new feature. Because of the identification of points, the string fields themselves (X^9) need not be strictly periodic. Instead, we now have the mode expansion

$$X^9 = x^9 + p^9\tau + 2mR\sigma + \frac{i}{2}\sum_{n\neq 0}\frac{1}{n}\left(\alpha_n^9 e^{-in(\tau-\sigma)} + \tilde{\alpha}_n^9 e^{-in(\tau+\sigma)}\right), \tag{25.15}$$

where m is an integer. The states with non-zero m are called *winding modes*. They correspond to the possibility of a string winding around, or wrapping, the extra dimension. Now the mass operator, in addition to including a contribution $(p^9)^2 = n^2/R^2$, includes a contribution from the windings, m^2R^2 (if there is no momentum). If R is large compared with the string scale, these states are very heavy. At small R, however, they become light while the momentum (Kaluza–Klein) states become heavy. This reciprocity often corresponds, as we will see, to a symmetry between compactification at large and at small radius.

Let us focus on the various superstring theories. It is convenient to break up X^9 in terms of left- and right-moving fields:

$$X_{\rm L}^9 = \frac{x^9}{2} + \left(\frac{n}{2R} + mR\right)(\tau - \sigma) + \frac{i}{2}\sum_{n\neq 0}\frac{1}{n}\alpha_n^9 e^{-in(\tau-\sigma)}, \tag{25.16}$$

$$X_{\rm R}^9 = \frac{x^9}{2} + \left(\frac{n}{2R} - mR\right)(\tau + \sigma) + \frac{i}{2}\sum_{n\neq 0}\frac{1}{n}\tilde{\alpha}_n^9 e^{-in(\tau+\sigma)}. \tag{25.17}$$

It is then natural to define left- and right-moving momenta:

$$p_{\rm L} = \frac{n}{2R} + mR, \quad p_{\rm R} = \frac{n}{2R} - mR. \tag{25.18}$$

The world-sheet fermions are untouched by this compactification. The mass operators are essentially as before, with p replaced by $p_{\rm L}$ for the left movers and by $p_{\rm R}$ for the right movers:

$$L_0 = \frac{1}{2}p_{\rm L}^2 + N, \quad \tilde{L}_0 = \frac{1}{2}p_{\rm R}^2 + \tilde{N}. \tag{25.19}$$

Suppose we compactify on a simple product of circles, whose coordinates are labeled X^I. The left- and right-moving momenta form a lattice:

$$p_{\rm L}^I = \frac{n^I}{R^I} + 2m^I R^I, \quad p_{\rm R}^I = \frac{n}{R^I} - 2m^I R^I. \tag{25.20}$$

Now we will determine the spectrum, focusing on the light states. Consider, first, the heterotic string; to simplify the formulas, we take the $O(32)$ case. The $O(32)$ symmetry is unbroken. The original ten-dimensional gauge bosons,

$$|A_M^{AB}\rangle = \lambda_{-1/2}^A \lambda_{-1/2}^B \psi_{M\,-1/2}|p\rangle, \tag{25.21}$$

now decompose into a set of four-dimensional gauge bosons, corresponding (in light cone gauge) to $M = 2$ and 3, and six scalars corresponding to $M = I$. The graviton, scalar and antisymmetric tensor field now decompose as a set of scalars, g_{IJ}, B_{IJ}, vectors $g_{\mu i}$, $B_{\mu I}$, a four-dimensional graviton $g_{\mu\nu}$, an antisymmetric tensor, $b_{\mu\nu}$, and a scalar, ϕ.

In order to understand space–time fermions, we will work in light cone gauge and return to our description of $O(8)$ spinors. Group the γ-matrices into a set associated with the internal six dimensions and a set associated with the (transverse) Minkowski directions. In other words, instead of the four creation and annihilation operators $a^i, a^{\bar{i}}$, we group these into one set of three (labeled a^i, where now $i = 1, 2, 3$) and b, together with their conjugates). So the 8_s, which previously consisted of the states

$$|0\rangle, \quad a^{i\dagger} a^{j\dagger} |0\rangle, \quad a^{1\dagger} a^{2\dagger} a^{3\dagger} a^{4\dagger} |0\rangle, \tag{25.22}$$

now decomposes as

$$|0\rangle, \quad a^{i\dagger} a^{j\dagger} |0\rangle, \quad b^{\dagger} a^{j\dagger} |0\rangle, \quad b^{\dagger} a^{1\dagger} a^{2\dagger} a^{3\dagger} |0\rangle. \tag{25.23}$$

There are four states with no bs and four with one b. These groups have opposite *four-dimensional* helicity. They can also be classified according to their transformation properties under $O(6)$. The group $O(6)$ is isomorphic to $SU(4)$. We have just seen that $8_s = 4 + \bar{4}$. We can also see that, under the $SU(3)$ subgroup of $SU(4)$, the spinor decomposes as

$$8 = 3 + \bar{3} + 1 + 1. \tag{25.24}$$

Now consider how the gravitino in ten dimensions decomposes under $O(3, 1) \times SU(4)$. We see that it consists of a set of spin-3/2 particles in the four-dimensional representation of $SU(4)$ and their antiparticles. So, from the perspective of four dimensions, this is a theory with $N = 4$ supersymmetry. This is not really surprising since the ten-dimensional theory is a theory with 16 supercharges, and none of these is touched by this reduction to four dimensions.

Because of the high degree of susy, one cannot write a potential for the scalar fields g_{IJ}, b_{IJ} etc.; they are exactly flat directions. If we redo our Casimir energy calculation then we will find that, because there is a fermionic state degenerate with every bosonic state, there are cancelations.

To what do these moduli correspond? Those which arise from the diagonal components of the metric correspond to the fact that the radii are not fixed. There is a string solution for any value of the R^I. The off-diagonal components are related to the fact that the general torus in six dimensions is not simply a product of circles; there can be non-trivial angles.

The massless scalars arising from the gauge bosons A^I are also moduli. For constant values of these fields there is no associated field strength, so they carry zero energy. But there are non-trivial Wilson lines:

$$U_I = \exp\left(i \int_0^{2\pi R_I} dx^I A_I \right). \tag{25.25}$$

Because of the periodicity these are gauge invariant and correspond to distinct physical states. These moduli are often themselves called Wilson lines.

The periodicities of a general N-dimensional torus can be characterized in terms of N basis vectors e_a^I, $a = 1, \ldots, N$. The theory is defined by the identifications

$$X^I = X^I + 2\pi n^a e_a^I. \tag{25.26}$$

The set of integers defines a lattice. To determine the allowed momenta we define the dual lattice, with unit vector $\tilde{e}^I_a$, satisfying

$$\tilde{e}^I_a e^I_b = \delta_{a,b}. \tag{25.27}$$

In terms of these, we can write the momenta for the general torus:

$$p^I = n^a \tilde{e}^I_a, \tag{25.28}$$

while the windings are

$$w^I = m^a e^I_a. \tag{25.29}$$

We can break these into left-moving and right-moving parts:

$$p^I_{\rm L} = (p^I/2 + w^I), \quad p^I_{\rm R} = (p^I/2 - w^I). \tag{25.30}$$

The lattice of left- and right-moving momenta $(p_{\rm L}, p_{\rm R})$ has some interesting features. Considered as a Lorentzian lattice, it is even and self-dual. The term "even" refers to the fact that the inner product of a vector with itself,

$$p^2_{\rm L} - p^2_{\rm R} = 2nm, \tag{25.31}$$

is even. The self-duality means that the basis vectors of the lattice and the dual are the same (Eq. (25.27)).

In bosonic or Type II theories, these are are the most general four-dimensional compactifications with $N = 8$ supersymmetry. The different possible choices of torus define a moduli space of such theories. These moduli space correspond to varying the metric and antisymmetric tensor fields. In the heterotic case, the four dimensional theory has $N = 4$ supersymmetry. Additional moduli arise from Wilson lines. As in the case of the simple compactification on a circle, these are essentially constant gauge fields. A constant gauge field is almost a pure gauge transformation (take I fixed, for simplicity),

$$A^I = ie^{ix_I A^I} \partial^I e^{-ix_I A^I} = ig\, \partial^I g^\dagger, \tag{25.32}$$

but the gauge transformation is only periodic if $A^I = 1/R_I$. In this case the Wilson line is unity. But we can do a redefinition of all the charged fields which eliminates the A^Is,

$$\phi = g\phi'. \tag{25.33}$$

With this choice, the charged fields are no longer periodic but obey boundary conditions

$$\phi'(X^I) = e^{2\pi i R_I A^I} \phi'. \tag{25.34}$$

This means that the momenta are shifted:

$$p^I = \frac{n}{R_I} + A^I. \tag{25.35}$$

Shortly, we will see how all the different momentum lattices can be understood in terms of constant background fields.

25.3 Enhanced symmetries and T-duality

For large radius, the spectrum of the toroidally compactified string theory is very similar to that expected from Kaluza–Klein field theories. The principal new feature, the winding states, is not important. At smaller radius, however, these states introduce startling new phenomena. We focus first on the compactification of just one dimension. Examining the momenta

$$p_L = \frac{m}{2R} + nR, \quad p_R = \frac{m}{2R} - nR \tag{25.36}$$

we see that these are symmetric under $R \to 1/(2R)$. This symmetry is often called *T-duality*. It means that there is no sense in which one can take the compactification radius to be arbitrarily small; it is our first indication that there is some sort of fundamental length scale in the theory. The T-duality symmetry is *not* a feature of the compactification of field theory; the string windings are critical.

What is the physical significance of this symmetry? The answer depends on which string theory we study. Consider the heterotic string. We first ask whether duality is truly a symmetry or just a feature of the spectrum of that theory. To settle this we can check that it has a well-defined action on all vertex operators. Alternatively, we can note that there is a self-dual point, $R_{sd} = 1/\sqrt{2}$. Examining Eq. (25.19) we see that, at this radius, various states can become massless. These include both scalars (from the point of view of the non-compact dimensions) and gauge bosons:

$$\psi^{I,\mu}_{-1/2}|n = \pm 1, m = \mp 1\rangle. \tag{25.37}$$

Together with the $U(1)$ gauge boson, the spin-1 particles form the adjoint of an $SU(2)$ group. We can check this by studying the operator product expansions of the associated vertex operators (see the exercises at the end of this chapter).

Now we can understand the $R \to 1/R$ symmetry. At the fixed point the symmetry is an unbroken symmetry. It transforms as follows:

$$p_L \to -p_L, \quad p_R \to p_R. \tag{25.38}$$

In world-sheet terms this corresponds to a change of sign of ∂X_L,

$$\partial X_L \to -\partial X_L, \quad \partial X_R \to \partial X_R. \tag{25.39}$$

From (25.37) X_L is the third component of isospin, T_3, so $T_3 \to -T_3$ under T-duality.

This transformation corresponds to a 90° rotation about the 1 or 2 axis in the $SU(2)$ space, i.e. it is a gauge transformation! This means that the large and small radii do not merely exhibit the same physics, they *are* the same. It also means that, provided the theory makes sense, the symmetry is an exact symmetry of the theory, in perturbation theory and beyond. As for any gauge symmetry, any violation of this symmetry would signal an inconsistency.

Returning to the self-dual point, the momentum lattice at this point can be thought of as a group lattice, with the p_Ls labeling the $SU(2)$ charges. Much larger symmetry groups can be obtained by making special choices of the torus, Wilson lines and antisymmetric tensor fields.

In other string theories the symmetry has a different significance. Consider the Type II theories; take the case of IIA for definiteness. Then, since $\psi_R^9 \to -\psi_R^9$, the GSO projection in the right-moving Ramond sectors is flipped. So this transformation takes the Type IIA theory to the Type IIB theory. In other words, *the IIA theory at large R is equivalent to the IIB theory at small R.*

25.4 Strings in background fields

The possibilities for string compactification are not limited to tori, they are much richer. We will explore them in this and the next chapter. We can approach the problem in two ways, each of which is very useful. First, we can examine the low-energy effective field theory which describes the massless modes of the string in ten dimensions and look for solutions corresponding to large compactified (i.e. internal) spaces. The effective action can be organized into terms with more and more derivatives. The spaces must be large in order that this use of the low-energy effective action makes sense. Alternatively, we can look for more direct ways to construct classical solutions in string theory. Both approaches have turned out to have great value.

We will first formulate the string problem in a more general way. We want to ask: how do we describe a string propagating in a background which is not flat? The background might be described by a metric, G_{MN}, but it might also include an antisymmetric tensor, B_{MN}, a dilaton, ϕ, and, in the case of the heterotic string, gauge fields. We first focus on the metric. Start with the bosonic string. It is natural, as we saw in the previous chapter, to generalize the string action

$$\frac{1}{2\pi}\int d^2\sigma\, \partial_\alpha X^M \partial^\alpha X^N \eta_{MN} \tag{25.40}$$

to

$$\frac{1}{2\pi}\int d^2\sigma\, \partial_\alpha X^M \partial^\alpha X^N G(X)_{MN}. \tag{25.41}$$

From a world-sheet point of view, we have replaced a simple free-field theory with a non-trivial, interacting-field, theory: a two-dimensional non-linear sigma model. We can think of the X^Ms as fields which propagate on a manifold with metric G_{MN}. Often this space is called the *target space* of the theory; the Xs then provide a mapping from two-dimensional space–time to this target space.

This looks plausible, and we can give some evidence that in fact it is the correct prescription. Suppose, in particular, we consider a metric which is nearly that of flat space:

$$G_{MN} = \eta_{MN} + h_{MN}. \tag{25.42}$$

Substitute this form in the action, and examine the path integral for the field theory:

$$Z[h] = \int [dX^M] \exp\left(iS_0 + \frac{1}{2\pi}\int d^2\sigma\, \partial_\alpha X^M \partial^\alpha X^N h(X)_{MN}\right). \tag{25.43}$$

Differentiating with respect to h brings down a vertex operator for the graviton. In other words, the path integral for this action is the generating functional for the graviton S-matrix.

This observation suggests a general treatment for backgrounds for the massless particles

$$I = \frac{1}{2\pi}\int d\tau \int_0^\pi d\sigma\, (g_{IJ}\partial_\alpha X^I + \epsilon^{\alpha\beta} B_{IJ}\partial_\alpha X^I \partial_\beta X^J). \tag{25.44}$$

The corresponding path integral generates the S-matrix elements for both the graviton and the antisymmetric tensor field. But we would like to consider configurations which are not close to the flat metric with vanishing B_{MN}. We can ask: what are acceptable backgrounds for string propagation? To answer this question, we need to remember that, for the free string, *conformal invariance* was the crucial feature to the consistency of the picture. It was conformal invariance which guaranteed Lorentz invariance and unitarity. So we need to look for interacting two-dimensional field theories which are conformally invariant.

25.4.1 The beta function

Field theories of the type we have just encountered are called non-linear sigma models. In 1+1 dimensions these are renormalizable theories: g_{IJ}, B_{IJ} etc. are dimensionless. A priori, however, they are general functions of the fields, and there are an infinite – continuously infinite – set of possible couplings.

Physically, the statement that these theories must be conformally invariant is equivalent to the statement that their beta functions must vanish. To get some feeling for what this means, let us consider a special situation. Suppose that B_{IJ} vanishes and that the metric is close to the flat-space metric η_{MN}:

$$g_{MN} = \eta_{MN} + \int d^D k\, h_{MN}(k) e^{ik\cdot x}. \tag{25.45}$$

The action is then

$$I = \frac{1}{2\pi}\int d^2\sigma \left(\eta_{IJ}\,\partial_\alpha X^I \partial^\alpha X^J + \sum_k h_{IJ}(k) e^{ik\cdot x}\, \partial_\alpha X^I \partial^\alpha X^J\right). \tag{25.46}$$

We can treat the term involving h as a perturbation. Working to second order, we have

$$\left\langle \int d^2 z_1 \left[h_{\mu\nu}(k) e^{ik\cdot X(z_1)}\, \partial X(z_1)^\mu \partial X(z_1)^\nu\right] \right.$$
$$\left. \times \int d^2 z_2 \left[h_{\rho\sigma}(k') e^{ik'\cdot X(z_2)}\, \partial X(z_2)^\rho \partial X(z_2)^\sigma\right]\right\rangle. \tag{25.47}$$

We will write this simply as

$$\int d^2z \int d^2z'\, h_1 \mathcal{O}_1(z_1) h_2 \mathcal{O}_2(z_2). \tag{25.48}$$

Ultraviolet divergences will arise in this integral when $z_1 \to z_2$. In this limit, we can use the operator product expansion

$$\mathcal{O}_1(z_1)\mathcal{O}_2(z_2) = \frac{c_{12j}}{|z_1 - z_2|^2} \mathcal{O}_j(z_2) + \cdots. \tag{25.49}$$

The integral over z_2 is ultraviolet divergent. If we cut it off at scale Λ^{-1} then we have the correction to the world-sheet Lagrangian:

$$\int d^2z\, h_1 h_2 c_{12j}\, \mathcal{O}_j \,\ln \Lambda. \tag{25.50}$$

There is another divergence associated with the couplings h_1 and h_2; this comes from normal ordering. In the case of the graviton vertex operator, if we simply expand the exponential factors and contract the xs, we obtain

$$\int d^2z\, h_1(x) k^2 \ln \Lambda. \tag{25.51}$$

Requiring, then, that the beta function for the coupling h_1 should vanish gives

$$k^2 h_1 + h_2 h_3\, c_{123} = 0. \tag{25.52}$$

Recall now that c_{ijk} is the three-point coupling for the three fields. So this is just the equation of motion to quadratic order in the fields.

This result is general. At higher orders, one encounters divergences of two types. First, there are terms involving a single logarithm of the cutoff times more powers of the fields. Second, there are terms involving higher powers of logarithms. These higher powers are, from a renormalization perspective, associated with iterations of lowest-order divergences, and they are systematically subtracted in computing the beta functions. From a space–time point of view, these correspond to the appearance of massless intermediate states, which must be subtracted in constructing the effective action or equations of motion.

This procedure can be used to recover Einstein's equations. A more elegant and efficient approach is to apply the background-field method. For a general gravitational background, one can view X as a fixed background which solves the two-dimensional equations of motion and study fluctuations about it. For a suitable choice of coordinates, the metric is second order in the fluctuations. One can include in this analysis the background antisymmetric tensor fields and a background dilaton. The antisymmetric tensor can be analyzed along the lines of our analysis of $h_{\mu\nu}$. The dilaton Φ is more subtle. In our action above we omitted one possible coupling: the two-dimensional curvature. The dilaton couples to the world-sheet fields through

$$\int d^2\sigma\, \Phi \mathcal{R}^{(2)}; \tag{25.53}$$

here $\mathcal{R}^{(2)}$ is the two-dimensional curvature scalar.

The full analysis leads to the equations of motion:

$$\beta_{\mu\nu} = 0 = \alpha' R_{\mu\nu} + 2\alpha' \nabla_\mu \nabla_\nu \Phi - \frac{\alpha'}{4} H_{\mu\lambda\omega} H_\nu^{\lambda\omega}, \tag{25.54}$$

$$\beta^B_{\mu\nu} = -\frac{\alpha'}{2} \nabla^\omega H_{\omega\mu\nu} + \alpha' \nabla^\omega \Phi H_{\omega\mu\nu} + \mathcal{O}(\alpha')^2, \tag{25.55}$$

$$\beta^\Phi = \frac{D-26}{6} - \frac{\alpha'}{2} \nabla^2 \Phi + \alpha' \nabla_\omega \Phi \nabla^\omega \Phi - \frac{\alpha'}{24} H^{\mu\nu\lambda} H_{\mu\nu\lambda}. \tag{25.56}$$

It is possible to extend these methods to describe quantum corrections to the equations, at least in the case of supersymmetric compactifications.

25.4.2 More general toroidal compactification

As a first application, we consider the heterotic string theory in the case of more general toroidal compactification.

For general metrics and backgrounds for both the antisymmetric tensor and gauge fields, one obtains a somewhat more involved expression for the momenta. A particularly elegant way to derive this is to argue that constant background fields should affect only slow modes of the string. In the presence of a background, a constant metric and antisymmetric tensor fields, the action is

$$I = \frac{1}{2\pi} \int d\tau d\sigma \int_0^\pi \left(g_{IJ}\, \partial_\alpha X^I \partial^\alpha X^J + \epsilon^{\alpha\beta} B_{IJ}\, \partial_\alpha X^I \partial_\beta X^J \right). \tag{25.57}$$

To realize the notion of slowly varying fields, one makes the ansatz

$$X^I = q^I(\tau) + 2\sigma m^I, \tag{25.58}$$

where the second term allows for the possibility of winding. Substituting this back in the action and performing the integral over σ:

$$I = \int d\tau \left(\frac{1}{2} g_{IJ} \dot{q}^I \dot{q}^J + 2B_{IJ} \dot{q}^I m^J - 2g_{IJ} m^I n^J \right). \tag{25.59}$$

Now we can read off the canonical momenta:

$$P_I = g_{IJ} \dot{q}^J + 2B_{IJ} m^J. \tag{25.60}$$

In quantum mechanics it is the canonical momenta which act by differentiation on wave functions, so it is the canonical momenta which must be quantized for a periodic system:

$$P_I = n_I, \tag{25.61}$$

where n_I is an integer. In terms of q^I this gives

$$\dot{q}^I = g^{IJ} m_J - 2B^I_J n^J. \tag{25.62}$$

Finally, integrating this equation and substituting back into X^I:

$$X^I = q^I + 2\sigma m^I + \tau \left(g^{IJ} n_J - 2B^I_J m^J \right). \tag{25.63}$$

From this, we can read off the left- and right-moving momenta:

$$p_L^I = m^I + \frac{1}{2} g^{IJ} n_J - g^{IJ} B_{JK} m^K,$$
$$p_R^I = -m^I + \frac{1}{2} g^{IJ} n_J - g^{IJ} B_{JK} m^K. \tag{25.64}$$

Once again, $p_L p_L' - p_R p_R'$ is an integer; the lattice, thought of as a Lorentzian lattice, is even and self-dual.

Including Wilson lines is slightly more subtle, because of their asymmetric coupling between left and right movers. For small A, the modification is essentially what we guessed above. There is also a modification of the internal, E_8-charge, lattice.

25.5 Bosonic formulation of the heterotic string

We have seen that, in toroidal compactifications of string theory, new unbroken gauge symmetries can arise at particular radii. We have also seen that a toroidal compactification can be described by a lattice. So far, in describing the heterotic string we have worked in what is known as the fermionic formulation. There is an alternative formulation, in which the 32 left-moving fermions are replaced by 16 left-moving bosons.

It is an old result that two-dimensional fermions are equivalent to bosons; more precisely, two real left-moving fermions are equivalent to a single real boson, and vice versa. The correspondence, for a complex fermion, λ, is

$$\lambda(z) = e^{i\phi(z)}, \tag{25.65}$$

where ϕ is a left-moving boson. The equal sign here is subtle; at finite volume care is required with the zero modes, as we will see. To be convinced that this equivalence is plausible, consider correlation functions at infinite volume. From our previous analyses of two-dimensional Green's functions, we have

$$\langle \lambda(z)\lambda(w) \rangle = \left\langle e^{i\phi(z)} e^{i\phi(w)} \right\rangle \sim \frac{1}{z-w}. \tag{25.66}$$

This suggests that in the case of, say, the $SO(32)$ heterotic string, we can replace the 32 left-moving fermions by 16 left-moving bosons. Note that this means, loosely, that we have 26 left-moving coordinates, as in the bosonic string (but still only 10 right-moving bosons). At finite volume (i.e. $0 < \sigma < \pi$), we can write the usual mode expansions for these fields:

$$X_L^A = \frac{1}{2} p_L^A + \frac{i}{2} \sum_n \frac{1}{n} \tilde{\alpha}_n^A e^{-in(\tau+\sigma)}. \tag{25.67}$$

Now the p_Ls are elements of the group lattice. Modular invariance requires that the lattice be even and self-dual. In 16 dimensions there are two such lattices, those of $O(32)$ and $E_8 \times E_8$.

The bosonization of fermions which we have described here is useful for the right-moving fields as well, and also for the fermions of the Type II theories. We have avoided discussing space–time supersymmetry in the RNS formalism because the fermion vertex operators and the supersymmetry generators must change the boundary conditions on two-dimensional fields. But, in this bosonized form, this problem is simpler. Once again, we have relations of the form

$$\psi_i \sim e^{i\phi_i}. \tag{25.68}$$

The ϕs live on a torus, whose "momenta" describe both N and RS states. Operators of the form $e^{i\phi/2}$ change NS to R states, i.e. they connect bosons to fermions. This connection allows the construction of fermion vertex operators and supersymmetry generators.

25.6 Orbifolds

Toroidal compactifications of string theory are simple; they involve free two-dimensional field theories. But they are also unrealistic. Even in the case of the heterotic string they have far too much supersymmetry, and their spectra are not chiral. There is a simple construction which reduces the amount of supersymmetry, yielding models with interesting gauge groups and a chiral structure. The corresponding world-sheet theories are still free, so explicit computations are straightforward. These constructions are also interesting in other ways. They correspond to particular submanifolds of the moduli space of larger classes of solutions. They exhibit interesting features such as discrete symmetries and subtle cancelations of four-dimensional anomalies. At low orders it is a simple matter to work out their low-energy effective actions. Through a combination of world-sheet and space–time methods, one can understand their perturbative, and in some cases non-perturbative, dynamics.

Here, we will work out one example in some detail. Other examples can be studied in a similar way. We will also mention some other free-field constructions of interesting string solutions.

We start with a toroidal compactification on a particular lattice, a product of three tori as shown in Fig. 25.1. It is convenient to introduce complex coordinates,

$$z^1 = x^1 + ix^2, \quad z^2 = x^3 + ix^4, \quad z^3 = x^5 + ix^6. \tag{25.69}$$

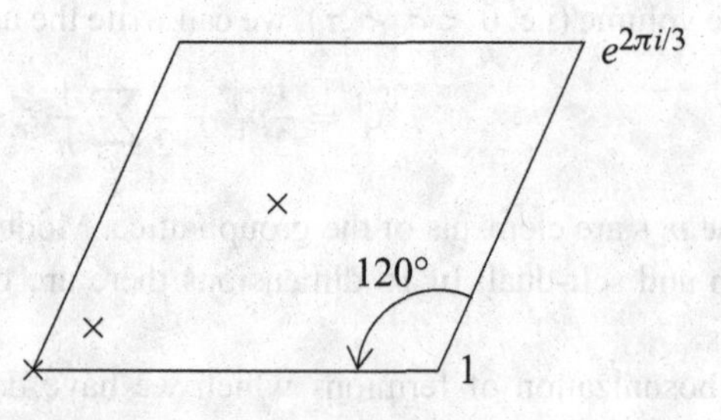

Fig. 25.1 A torus that admits a Z_3 symmetry, allowing an orbifold construction.

This lattice is invariant under the Z_3 symmetry

$$z^i \rightarrow e^{2\pi i/3} z^i. \tag{25.70}$$

This can be seen by examining the figure carefully. The lattice vector $(1, 0)$, for example, in the original Cartesian coordinates is rotated into the lattice vector $(-1/2, 1/\sqrt{2})$. This is related by a lattice vector translation to $(1/2, 1/\sqrt{2})$.

Now we identify points under the symmetry, i.e. two points related by a symmetry transformation are considered to be the same point. The result is almost a manifold, but not quite. There are three particular points which are invariant under the symmetry. These are called *fixed points*. They are the points

$$(0, 0), \quad (1/2, \sqrt{3}/2), \quad (1, \sqrt{3}). \tag{25.71}$$

The geometry near each of these points is singular. If one parallel-transports about, say, the point at the origin then after 120° one has returned to where one started. The space is said to have a deficit angle (a *conical singularity*). It is as if there were an infinite amount of curvature located at each of the points. Such a space is called an *orbifold*.

In quantum mechanics, requiring such an identification of points under a symmetry means requiring that states be invariant under the quantum mechanical operator which implements the symmetry. Consider the various states of the original ten-dimensional theory. In the Type II theory, for example, in the NS–NS sector we have the following states, before making any identifications:

$$\tilde{\psi}^{\mu}_{-1/2}\psi^{\nu}_{-1/2}|0\rangle, \quad \tilde{\psi}^{\bar{j}}_{-1/2}\psi^{i}_{-1/2}|0\rangle, \quad \tilde{\psi}^{j}_{-1/2}\psi^{\bar{i}}_{-1/2}|0\rangle, \tag{25.72}$$

$$\tilde{\psi}^{j}_{-1/2}\psi^{i}_{-1/2}|0\rangle, \quad \tilde{\psi}^{\bar{i}}_{-1/2}\psi^{\bar{j}}_{-1/2}|0\rangle. \tag{25.73}$$

After the identifications, the first set of states is invariant; the latter two are not. These states all have simple interpretations. The first three are the four-dimensional graviton, the antisymmetric tensor and the dilaton. The second two are the moduli of the torus. The parts symmetric under $i \rightarrow \bar{j}$ correspond to the metric components $g_{i\bar{j}}$ in the original theory. The antisymmetric parts correspond to the corresponding components of $B_{i\bar{j}}$.

The diagonal components, $g_{i\bar{i}}$, are easily understood. Changing the value of these components slightly corresponds to changing the overall radius of the ith torus. This does not change the symmetry properties. The off-diagonal components, $g_{1\bar{2}}$ etc., correspond to deformations which mix up the three planes but leave a lattice with an overall Z_3 symmetry.

To understand what happens to the supersymmetries, we will focus on the gravitino. It is convenient to work in light cone gauge and to decompose the spinors as we did earlier. To determine how the spinors transform under the Z_3, we need to decide how the state we called $|0\rangle$ transforms under the symmetry. Consider a rotation, say in the 12 plane, by 120°. The rotation generator is

$$S_{12} = \frac{i}{4}(\gamma_1\gamma_2 - \gamma_2\gamma_1) = a^{1\dagger}a^1 + \frac{1}{2}. \tag{25.74}$$

So, the rotation of the state $|0\rangle$ is described by

$$e^{(2\pi i/6)s_{12}}|0\rangle = e^{2\pi i/6}|0\rangle. \tag{25.75}$$

The transformations of the other states can then be read off from the transformation laws of the a^is:

$$|0\rangle \rightarrow e^{-2\pi i/6}|0\rangle, \quad a^{\bar{i}}|0\rangle \rightarrow e^{2\pi i/6} a^{\bar{i}}|0\rangle. \tag{25.76}$$

Now we have to be a bit more precise about the orbifold action. This is a product of Z_3s for each plane. But we see that, acting on fermions, the separate transformations are Z_6s. In order that the group action be a sensible Z_3 we need to take, for example:

$$Z^1 \rightarrow e^{2\pi i/3} Z^1, \quad Z^2 \rightarrow e^{2\pi i/3} Z^2, \quad Z^3 \rightarrow e^{-4\pi i/3} Z^3. \tag{25.77}$$

With this definition the fermion component 0, which we will write as $|0\rangle$, is invariant under the orbifold projection. The components i, which we will write as $a^{\bar{i}}|0\rangle$, are not.

We can label the gravitinos

$$\psi^{\mu}_{0,\alpha}, \quad \tilde{\psi}^{\mu}_{\tilde{0},\alpha}, \quad \psi^{\mu}_{i,\alpha}, \quad \tilde{\psi}^{\mu}_{i,\alpha}. \tag{25.78}$$

After the projection, instead of eight gravitinos, as in the toroidal case, there are only two; we have $N = 2$ supersymmetry in four dimensions.

In addition to projecting out states we need to consider a new class of states. We can consider closed strings which sit at the fixed points. More precisely, in addition to the strict periodic boundary condition we can consider strings which satisfy

$$X^i(\sigma + \pi) = e^{2\pi i/3} X^i(\sigma). \tag{25.79}$$

These boundary conditions do not permit the usual bosonic zero modes. Instead, we have a mode expansion

$$X^i = x^i_{(a)} + \frac{i}{2} \sum_n \left(\alpha^i_{n-1/3} e^{2i(n-1/3)(\sigma-\tau)} + \tilde{\alpha}^i_{n-1/3} e^{2i(n-1/3)(\sigma+\tau)} \right). \tag{25.80}$$

The mode numbers are now fractional; the absence of a momentum term indicates that the strings sit at fixed points (labeled by a). In this case there are 27 fixed points. For the fermions, we again have to distinguish the Ramond and Neveu–Schwarz sectors. In the NS sectors the fermions have modes which differ from integers by multiples of $1/2 - 1/3 = 1/6$:

$$\psi^i = \sum \psi_{n-1/6} e^{-2i(n-1/6)(\tau-\sigma)}, \tag{25.81}$$

with a similar expansion for $\tilde{\psi}$.

We can readily work out the normal-ordering constant, using a formula that we wrote down earlier (Eq. (22.30)). We have, in the NS–NS sector:

$$a = 6 \times \frac{1}{4}\left(\frac{1}{3} \times \frac{2}{3}\right) - 6 \times \frac{1}{4}\left(\frac{1}{6}\right) \times \left(\frac{5}{6}\right) - 4 \times \frac{1}{4}\left(\frac{1}{2}\right) \times \left(\frac{1}{2}\right) = 0. \tag{25.82}$$

So, the ground state is massless in the twisted sectors. Again, because of the $N = 2$ supersymmetry there can be no potential for this field. So there is a modulus in each twisted sector. Unlike the moduli in the untwisted sector, this modulus does not correspond to a simple change in the features of the torus which defines the orbifold. Instead, it represents a deformation which, from a space–time viewpoint, smooths out the orbifold singularity.

The resulting smooth space is an example of a Calabi–Yau manifold, of a type that we will discuss in the next chapter.

We now turn to the heterotic string theory on this orbifold. We will take the same projector on the spatial coordinates X^i as before. As a result there is only one gravitino; the four-dimensional theory has $N = 1$ supersymmetry. The moduli are in one-to-one correspondence with the scalars of the NS–NS sector of the $N = 2$ theory: $g_{i\bar{j}}, B_{i\bar{j}}, \phi$. We can also make a projection on to the world-sheet gauge degrees of freedom. There are many possible choices of this gauge transformation; the principal restriction comes from the requirement of modular invariance. A particularly simple one is almost symmetrical between the left and right movers. In the fermionic formulation it works as follows. Take $E_8 \times E_8$ for definiteness. Of the 16 fermions in the first E_8, single out six, and rewrite them in terms of three complex fermions, λ^i. Call the remaining ten fermions λ^a. Now, in the projection, require invariance under

$$Z^i \to e^{2\pi i/3} Z^i, \quad \psi^i \to e^{2\pi i/3} \psi^i, \quad \lambda^i \to e^{2\pi i/3} \lambda^i. \tag{25.83}$$

In the untwisted sector this projection has no effect on the graviton or the moduli which we have identified previously. But consider the various gauge fields. In ten dimensions these were vectors in the adjoint of the two E_8s and their fermionic partners. The fields with space–time indices in the internal dimensions now appear as four-dimensional scalars. In order that they be invariant under the full projection, it is necessary to choose their gauge quantum numbers appropriately. In the NS sector, for each E_8, the invariant states include the following.

1. *A set of fields in the adjoints of E_6 and E_8 and an $SU(3)$* Of these, an $O(10)$ subgroup of the E_6 is manifest in the NS–NS–NS sector, as well as an $O(16)$ subgroup of E_8. Correspondingly, the gauge bosons are

$$\lambda^a_{-1/2} \lambda^b_{-1/2} \psi^\mu_{-1/2} |0\rangle \tag{25.84}$$

in $O(10)$,

$$\lambda^A_{-1/2} \lambda^B_{-1/2} \psi^\mu_{-1/2} |0\rangle \tag{25.85}$$

in $O(16)$ and, in $SU(3) \times U(1)$,

$$\lambda^i_{-1/2} \lambda^{\bar{i}}_{-1/2} \psi^\mu_{-1/2} |0\rangle. \tag{25.86}$$

Note that all these states are invariant. The $U(1)$ is actually an E_6 generator. The group E_6 has an $O(10) \times U(1)$ subgroup under which the adjoint representation, which is 78-dimensional, decomposes as follows:

$$78 = 45_0 + 1_0 + 16_{-1/2} + \overline{16}_{1/2}. \tag{25.87}$$

The remaining E_6 gauge bosons are found in the R–NS–NS sector. The left-moving normal ordering constant vanishes. The ground states in this sector are spinors of $O(10)$, the 16 and $\overline{16}$ above. The 248-dimensional representation of the second E_8 is filled out as in the uncompactified theory.

2. *Matter fields* These lie in the fundamental representation of E_6, the 27 under $O(10)$. The 27 decomposes as follows:

$$27 = 1_{-2} + 10_1 + 16_{-1/2}. \tag{25.88}$$

There are nine 10s in the untwisted sectors, corresponding to the states

$$\lambda^a_{-1/2}\lambda^i_{-1/2}\psi^{\bar{j}}_{-1/2}|0\rangle. \tag{25.89}$$

Each of these is one real scalar; we can use the conjugate fields to form nine more real scalars or eight complex scalars. There are nine singlets of charge -2:

$$\lambda^{\bar{i}}_{-1/2}\lambda^{\bar{j}}_{-1/2}\psi^{\bar{k}}_{-1/2}|0\rangle. \tag{25.90}$$

The 16s come from the R–NS–NS sector.

So, we have nine 27s from the twisted sectors, and no $\overline{27}$s; the theory is chiral.

Let us turn now to the twisted sectors. In the Type II case we found moduli in each sector. Here we will find moduli, additional 27s and more. We first need to compute the normal-ordering constants. For the right movers the calculation is exactly as in the Type II theory and gives zero. For the left movers in the NS–NS sector, we have

$$\begin{aligned} a &= -\frac{8}{24} + \frac{6}{4} \times \frac{1}{3} \times \frac{2}{3} - \frac{16}{4} \times \left(-\frac{1}{24} + \frac{1}{4}\right) \\ &\quad + \frac{16}{24} - \frac{10}{4} \times \frac{1}{4} - \frac{6}{4} \times \frac{1}{6} \times \frac{5}{6} \\ &= -1/2, \end{aligned} \tag{25.91}$$

where the first two terms come from the bosons, the next two from the fermions in the unbroken E_8 and the last two from the fermions in the broken E_8. So we can make massless states in a variety of ways:

1. *ten-dimensional represention of* $O(10)$,

$$\lambda^a_{-1/2}|0\rangle_{\text{twist}} \tag{25.92}$$

(note that E_6 invariance requires that this state have $U(1)$ charge $+1$);

2. *a singlet of* $O(10)$ *with* $U(1)$ *charge* -2,

$$\lambda^{\bar{1}}_{-1/6}\lambda^{\bar{2}}_{-1/6}\lambda^{\bar{3}}_{-1/6}|0\rangle_{\text{twist}} \tag{25.93}$$

(together with a set of spinorial states from the R–NS sector, this completes a 27 of E_6);

3. *moduli, other gauge singlets,*

$$\alpha^i_{-1/3}\lambda^{\bar{j}}_{-1/6}|0\rangle \tag{25.94}$$

(if we contract the i and $\bar{j}$ indices, we find the analog of the twisted sector modulus we had in the Type II theory; the other states represent additional singlets).

All together, then, we have found $9 + 27 = 36$ copies of the 27 of E_6, and 36 moduli. Each 27 comfortably accommodates a generation of the Standard Model plus an additional vector-like set of fields. So, while this example is hardly realistic, it is interesting: it predicts a particular number of Standard Model generations, plus additional fields. Whether variants of these ideas can lead to something more realistic is an important question, which we will postpone for the time being.

25.6.1 Discrete symmetries

An unappealing feature of supersymmetric models as theories of nature is the need to postulate discrete symmetries in order to have a sensible phenomenology. This seems rather ad hoc. One aspect of the orbifold construction we have just described is that a variety of discrete symmetries appear naturally. This phenomenon is common in string constructions, as we will see. Here it is particularly easy to exhibit the symmetries.

We have, for simplicity, considered a particular form for the torus – a particular point in the moduli space at which the six-dimensional torus is a product of three two-dimensional tori. But at this point (which is really a surface), there is a large symmetry. First, there is a separate Z_3 symmetry for each plane. (You can check that each plane in fact admits a Z_6 symmetry.) Because of the orbifold projection, one of these symmetries acts trivially on all states but two are non-trivial. If we take the size of each of the three two-dimensional tori to be the same then we also have a permutation symmetry, S_3, among the tori.

The Z_3s are R symmetries. We have already seen that the spinor with index 0 rotates by a phase $e^{2\pi i/6}$ under such a symmetry. By definition, this is an R transformation. This has significant consequences for the low-energy theory, greatly restricting the form of the superpotential.

As an example of the far-reaching consequences of such symmetries, one can show that there are exactly flat directions involving the matter fields. Consider the untwisted moduli. One can give expectation values to the $O(10)$ ten-dimensional and one-dimensional representations in one multiplet in a way which respects the supersymmetry. Specifically, consider the field ϕ given by

$$\phi = \lambda^a_{-1/2}\psi^{\bar{1}}_{-1/2}|0\rangle \tag{25.95}$$

and the corresponding singlet. Both of these are neutral under the rotation in the second plane. So, one cannot construct any superpotential term involving ϕ alone. One can give an expectation value to the singlet and to the 10 in such a way as to cancel the D terms for E_6. The main danger, then, is a superpotential term of the form

$$W = \Psi\phi^2 \tag{25.96}$$

with Ψ some other 27. This is E_6 invariant (in terms of $O(10)$ representations, it involves a product of a singlet and two 10s). But no such term is allowed by the discrete symmetries.

This simple argument shows that the moduli space is even larger than we might have thought. Such symmetries, as they forbid not only certain dimension-four but also certain dimension-five operators, might also be important for understanding the problem of proton stability and other important phenomenological issues.

The model possesses other symmetries as well. There is Z_3 symmetry, under which the twisted sector states transform but the untwisted sector states do not. We will not derive this here but it is plausible, and can be shown readily if one constructs the vertex operators for the twisted states. Many discrete symmetries of the model are subgroups of the Lorentz symmetry of the original higher-dimensional theory. As such they can probably be thought of as gauge symmetries. This is less obvious for other symmetries, but it is generally believed that the discrete symmetries of string theory all have this character. Searches for anomalies in discrete symmetries, for example, have yielded no examples.

One could ask: why would nature choose a point in the moduli space of some string theory at which there is an unbroken discrete symmetry? At the moment our understanding of how to connect string theory to nature is not good enough to give a definite answer to this question but, at the very least, such points are *necessarily* stationary points of the effective potential for the moduli; at the symmetric point, the symmetry forbids linear terms in the action for the charged moduli.

25.6.2 Modular invariance, interactions in orbifold constructions

As in our original string theory constructions, there seems much which is arbitrary in the choices we made above. Also, we have not spelled out what are the appropriate GSO projectors. As for the simple ten-dimensional constructions, the possible GSO projections are constrained by modular invariance. We will leave for the exercises the checking of some particular cases, but the basic result is easy to state. One can project by any transformation, provided that it has a sensible action on fermions and on spinor representations of the gauge group and provided that one has "level matching" in all the twisted sectors. This means that one must be able to construct an infinite tower of states in each sector. To understand the significance of this statement, consider a different choice of group action from that we considered above. Instead of twisting by $(1/3, 1/3, -2/3)$, project by $(1/3, -1/3, 0)$. In this case, for example, in the NS–NS–NS sector, the left-moving normal-ordering constant is $-13/18$. As a result, one cannot construct any states in the twisted sector which satisfy the level-matching condition.

There are other constructions of compactifications with $N=1$ supersymmetry based on free fields. These include models based purely on free fermions. These models are believed to be equivalent to orbifold models in which one *mods out* (performs projections) asymmetrically on the left- and right-moving fields. The latter, "asymmetric orbifold", models are interesting in that they potentially have very few moduli. In order to have sensible, unbroken, discrete symmetries acting on the left and right, typically the original lattice must sit at a self-dual point. So, many moduli are fixed – they are projected out by the orbifold transformation. It is not difficult, in this way, to construct models where there are no moduli that are neutral under space–time symmetries except for the dilaton.

25.7 Effective actions in four dimensions for orbifold models

While string theory provides a very explicit set of computational rules, at least for low orders of perturbation theory, these rules are complicated and rather cumbersome.

Moreover, except in some special circumstances we lack a non-perturbative formulation of the theory. Effective-field-theory methods have proven extremely useful in understanding the dynamics of string theory, both perturbative and non-perturbative. In this section we will work out the effective action for the orbifold models introduced above. More precisely, we work out the Lagrangian for a subset of the fields, up to and including terms with two derivatives. Many features of these Lagrangians will be relevant to the more intricate Calabi–Yau compactifications that we will encounter shortly.

In principle, to calculate the effective action we should calculate the string S-matrix and write down an action for the massless fields which yields the same scattering amplitudes. Alternatively, we can calculate the equations of motion from the beta function and look for an action which reproduces these. But, for low-order terms in the derivative (α') expansion, for the fields in the untwisted sector there is a simpler procedure. We know the form of the ten-dimensional effective action; we can simply truncate the theory to four dimensions. To do this, we start by setting all the charged fields to zero (this includes the gauge fields). We also work at a point with a large discrete symmetry: $Z_3^3/Z_3 \times S_3$. We set all the fields which transform under these symmetries to zero. This includes all the moduli except the one that determines the overall size of the torus and its superpartners. We then write the metric as

$$g_{i\bar{j}}(x^\mu) = g_{\bar{j}i}(x^\mu) = e^{\sigma(x^\mu)}\delta_{i\bar{j}}. \tag{25.97}$$

With this parameterization we are describing the size of the space with respect to a reference metric. We make a similar ansatz for the antisymmetric tensor:

$$b_{i\bar{j}}(x^\mu) = -b_{\bar{j}i}(x^\mu) = b(\sigma(x^\mu))\delta_{i\bar{j}}. \tag{25.98}$$

We must keep also the four-dimensional metric components $g_{\mu\nu}$, the scalar field ϕ and the antisymmetric tensor $B_{\mu\nu}$. We take them all to be functions of x^μ, the uncompactified coordinates, only. Substituting these fields into the ten-dimensional Lagrangian, Eq. (24.8), the integral over the six internal coordinates is easy since all fields are independent of the coordinates. One simply obtains $e^{3\sigma(x)}$ from the $\sqrt{g}$ factor. This is just the volume of the internal space, if σ is constant. There are additional factors $e^{-\sigma}$ coming from the factors of the inverse metric: one from the four-dimensional contribution to the Ricci curvature; one from the kinetic term for ϕ; and three from the $H_{\mu\nu\rho}$ terms. The ten-dimensional curvature term also gives derivative terms in σ. After a short computation we obtain

$$\begin{aligned}\mathcal{L} = &-\frac{1}{2}e^{3\sigma}R^{(4)} - 3e^{3\sigma}\partial_\mu\sigma\partial^\mu\sigma - \frac{9}{16}e^{3\sigma}\frac{\partial_\mu\phi\partial^\mu\phi}{\phi^2} \\ &- \frac{9}{2}e^{\sigma}\phi^{-3/2}\partial_\mu b\partial^\mu b - \frac{3}{4}\phi^{-3/2}H^{\mu\nu\rho}H_{\mu\nu\rho}.\end{aligned} \tag{25.99}$$

It is customary to rescale the metric so that the Einstein term has the standard form

$$g_{\mu\nu} = e^{-3\sigma}g'_{\mu\nu}. \tag{25.100}$$

After this Weyl rescaling, the action becomes

$$\mathcal{L} = -\frac{1}{2}R^{(4)} - 3\partial_\mu\sigma\partial^\mu\sigma - \frac{9}{16}\frac{\partial^\mu\phi\partial_\mu\phi}{\phi^2} - \frac{3}{2}e^{-2\sigma}\phi^{-3/2}\,\partial_\mu b^2 - \frac{3}{4}\phi^{-3/2}e^{6\sigma}H^2_{\mu\nu\rho}. \tag{25.101}$$

It should be possible to cast this Lagrangian as a standard four-dimensional, $N = 1$ supergravity Lagrangian, with a particular Kahler potential. Having set to zero all the fields except for a few moduli, there is no superpotential. To determine the Kahler potential we first note that, in four dimensions, an antisymmetric tensor field is equivalent to a scalar. This follows from counting degrees of freedom; with our usual rules, an antisymmetric tensor in four dimensions has only one degree of freedom. To make this explicit, one performs a "duality transformation" (the term is starting to seem a bit overused!)

$$\phi^{-3/2}e^{6\sigma}H_{\mu\nu\rho}(x) = \epsilon_{\mu\nu\rho\sigma}\partial^{\sigma}a(x). \tag{25.102}$$

The field a is often called the model-independent axion because it couples like an axion and its features do not depend on the details of the compactification. Then we define two chiral superfields, whose scalar components are

$$S = e^{3\sigma}\phi^{-3/4} + 3i\sqrt{2}a \tag{25.103}$$

and

$$T = e^{\sigma}\phi^{3/4} - i\sqrt{2}b. \tag{25.104}$$

Choosing the Kahler potential

$$K = -\ln(S + S^*) - 3\ln(T + T^*) \tag{25.105}$$

reproduces all the terms in Eq. (25.101). The reader may want to check the terms in this equation carefully, but at the least it is a good idea to make sure one understands how the σ^{-1} and ϕ^{-} dependences are reproduced.

Let's now return to the ten-dimensional gauge field terms, Eq. (24.9). This will allow us to include the matter fields as well as the gauge fields. Rather than consider the full set of fields, we can restrict ourselves to the set which is invariant under each separate Z_3 in combination with three separate Z_3s in the gauge group ($\lambda^i \to e^{2\pi k_i i/3}\lambda^i$). This leaves us with three complex scalars C^i corresponding to the states

$$C^i \leftrightarrow \lambda^i_{-1/2}\lambda^a_{-1/2}\psi^{\bar{i}}_{-1/2}|0\rangle \tag{25.106}$$

(here i is not summed). From the point of view of ten dimensions, these are the $A^{ia}_{\bar{i}}$. We also need to include the four-dimensional gauge fields A^{ab}_{μ}. In this way we obtain the additional terms, after Weyl rescaling,

$$\mathcal{L}_{\text{gauge}} = -\frac{1}{4}\phi^{-3/4}e^{3\sigma}F^2_{\mu\nu} - 3e^{-\sigma}\phi^{-3/4}D_{\mu}C^{*\bar{i}}D^{\mu}C^i + \cdots. \tag{25.107}$$

This can still be put into the standard supergravity form. First we need to remember that, in the duality transformation, $H_{\mu\nu\rho}$ now includes the Chern–Simons terms. Then it is necessary to modify the definition of T to include a contribution from the C fields, so that now

$$T = e^{\sigma}\phi^{3/4} - i\sqrt{2}b + C^{*\bar{i}}C^i, \tag{25.108}$$

and to modify the Kahler potential to

$$K = -\ln(S + S^*) - 3\ln(T + T^* - C^*C). \tag{25.109}$$

There is also a coupling of the field S to the gauge fields:

$$\mathcal{L}_S = -\frac{1}{4} S W_\alpha^2. \tag{25.110}$$

This includes a coupling of ϕ and σ to $F_{\mu\nu}^2$, already apparent in Eq. (24.9). The $aF\tilde{F}$ coupling arises from the Chern–Simons term in Eq. (24.10). Recall that

$$H_{\mu\nu\rho} = \partial_{[\mu} B_{\nu\rho]} - \omega_{\mu\nu\rho}. \tag{25.111}$$

So $\int d^4x H^2$, using the definition of a and integrating by parts, gives an $aF\tilde{F}$ coupling. Finally, there is a superpotential that is cubic in the C fields.

25.7.1 Couplings and scales

It is worth pausing to note the connections between the couplings and scales in different dimensions. We will focus first on the heterotic string. We see from Eq. (25.110) that S determines the gauge coupling: $S = 1/g^2$. This is as we would naively expect. The ten-dimensional gauge coupling: is essentially $1/g_s^2$; when we reduce to four dimensions, the four-dimensional gauge fields correspond to modes which are constant on the internal manifold, so that

$$\frac{1}{g_4^2} = \frac{1}{g_s^2} V M_s^6. \tag{25.112}$$

In terms of the fields we defined above, $V = e^{3\sigma}$.

These simple formulas pose a serious problem for the application of weakly coupled heterotic string phenomenology. If we simply identify S with the four-dimensional coupling then the string coupling satisfies

$$g_s^2 = g_4^2 V M_s^6. \tag{25.113}$$

So, we see that at large volume, the limit in which an α' expansion is valid, there is a conflict with small g_s if g_4 is fixed. We can also write a relation between the string scale and the Planck scale in four dimensions:

$$M_{\rm p}^2 = M_s^8 V g_s^{-2}. \tag{25.114}$$

Solving for M_s and substituting in the previous expression gives an expression for g_s which is incompatible with weak coupling, if we assume that $V = M_{\rm gut}^{-6}$.

Later, we will sharpen this strong coupling problem and consider possible solutions.

25.8 Non-supersymmetric compactifications

So far, we have considered compactifications that are supersymmetric. This is not a necessary restriction, but we will see that non-supersymmetric compactifications raise new conceptual and technical problems.

Perhaps the simplest non-supersymmetric compactification is *Scherk–Schwarz compactification*. Here one compactifies the theory (this can be Type I, Type II or heterotic) on a torus. In one direction, say the ninth direction, one imposes the requirement that bosons should obey periodic boundary conditions and fermions anti-periodic ones. One can describe this by taking the radius of the extra dimension to be $2 \times 2\pi R$ and performing a projection

$$P = (-1)^F e^{i(2\pi i)Rp_9}. \tag{25.115}$$

This projection eliminates, for example, the massless gravitinos; there is no supersymmetry and no Bose–Fermi degeneracy in the spectrum. Indeed, in the simplest version there are no massless fermions at all.

As a result, the usual Fermi–Bose cancelation of supersymmetry does not take place and, at one loop, there is a non-zero vacuum energy. More precisely there is a potential for the classical modulus R. The calculation of this potential is just the Casimir calculation we encountered earlier. Only the massless ten-dimensional fields contribute; the massive string states give effects which are exponentially suppressed for large R. To see this one can return to our earlier calculation with a massive state (an oscillator excitation of the string). Replacing the sum over integers by an integral in the complex plane and deforming the contour, as in Eqs. (25.11)–(25.13), yields a term exponentially small in the mass. The detailed results depend on the particular model, but typically the potential is negative and goes to zero at large R. In other words, at one loop the dynamics tends to drive the system to small R. It is not well understood how to study the system beyond one loop.

One can obtain non-supersymmetric theories in four dimensions in many other ways. The Scherk–Schwarz construction can be understood as modding out a supersymmetric compactification by an R symmetry. With this viewpoint, one can simply enumerate the R symmetries of a particular construction and mod out, subject to conditions of modular invariance.

Suggested reading

An introduction to Kaluza–Klein theory prior to the development of string theory is provided in the text *Modern Kaluza–Klein Theories* by Appelquist *et al.* (1985). More thorough discussions of aspects of string compactification are provided by the texts of Green *et al.* (1987) and Polchinski (1998). Some original papers, particularly the orbifold papers, are highly readable; see, for example, Dixon *et al.* (1986). There are many topics here thay we have only touched on in this chapter. We gave an argument that the vanishing of the beta function of the two-dimensional sigma model is equivalent to the equations of motion in space–time, but readers may wish to work through the background field analysis which leads to Einstein's equations. This is described in Polchinski's book and elsewhere. The bosonic formulation of the heterotic string is also well described there, but the original papers are quite readable (Gross *et al.*, 1985, 1986). Bosonization and space–time supersymmetry in the RNS formulation are thoroughly discussed by Polchinski (1998); a clear, but rather brief, introduction, is provided by Peskin's 1996 TASI lectures (Peskin, 1997). The non-supersymmetric compactification described here was introduced by Rohm (1984).

Exercises

(1) Derive the gauge terms in the Lagrangian of Eq. (25.7). You can do this by taking the metric to be flat.

(2) Derive the scalar kinetic terms of Eq. (25.8). You can do this by at first taking the four-dimensional metric to be flat, and allowing only σ to be a function of x.

(3) Verify, by studying the OPEs of the vertex operators for the different massless fields, that the enhanced symmetry of the bosonic string at the point $R = 1/\sqrt{2}$ is $SU(2) \times SU(2)$. Explain why, in the heterotic string, the symmetry is only $SU(2)$. What is the symmetry in the IIA theory?

(4) For the orbifold model, work out the spectrum in the untwisted sectors in greater detail, paying particular attention to spinorial representations of the O groups and to the space–time spinors. In particular, make sure that you are clear that the 27s are chiral, i.e. all the states in the 27s have one four-dimensional chirality and all those in $\overline{27}$s have the opposite chirality.

(5) Derive the term in Eq. (25.99) involving $\partial\sigma^2$.

(6) Verify that the Kahler potential of Eq. (25.109) properly reproduces the kinetic terms of the matter fields.

26 Compactification of string theory II. Calabi–Yau compactifications

Up to now we have focused on rather simple models involving toroidal compactifications and their orbifold generalizations. But, while by far the simplest, these turn out to be only a tiny subset of the possible manifolds on which to compactify string theories. A particularly interesting and rich set of geometries is provided by the Calabi–Yau manifolds. These are manifolds which are Ricci flat, $R_{MN} = 0$. Their interest arises in large part because these compactifications can preserve some subset of the full ten-dimensional supersymmetry. This is significant if one believes that low-energy supersymmetry has something to do with nature. It is also important at a purely theoretical level since, as usual, supersymmetry provides a great deal of control over any analysis; at the same time there is less supersymmetry than in the toroidal case, so a richer set of phenomena is possible.

This chapter is intended to provide an introduction to this subject. In the first section we will develop some mathematical preliminaries. Unlike the toroidal or orbifold compactifications it is not possible, in most instances, to provide explicit formulas for the underlying metric on the manifold and other quantities of interest. The six-dimensional Calabi–Yau spaces, for example, have no continuous isometries (symmetries), so at best one can construct the metrics by numerical methods. But it turns out to be possible from topological considerations to extract much important information without a detailed knowledge of the metric. The machinery required to define these spaces and to extract at least some of this information includes algebraic geometry and cohomology theory, subjects not part of the training of most physicists. The following mathematical interlude provides a brief introduction to the necessary mathematics. There is much more in the suggested reading.

26.1 Mathematical preliminaries

Two notions are very useful for understanding Calabi–Yau spaces: differential forms and vector bundles. Differential forms have already appeared implicitly in our discussion of IIA and IIB string theory. We start with an antisymmetric tensor field $A_{i_1 i_2 \dots i_n}$. Suppose that there is a gauge invariance

$$A_{i_1 \dots i_n} \to A_{i_1 \dots i_n} + \frac{1}{n}\left[\partial_{i_1} \Lambda_{i_2 \dots i_n} - \partial_{i_2} \Lambda_{i_1 i_3 \dots i_n} + \cdots (-1)^r \partial_{i_r} \Lambda_{i_1 \dots i_{r-1} i_{r+1} \dots i_n}\right], \tag{26.1}$$

where Λ is antisymmetric in all its indices. We can write a shorthand for this,

$$\delta A = d\Lambda, \tag{26.2}$$

where $d\Lambda$ is the "exterior derivative." Acting on an antisymmetric tensor of rank p, the exterior derivative produces a rank-$(p+1)$ antisymmetric tensor, dH:

$$dH_{i_1\dots i_{p+1}} = \frac{1}{p+1}\left(\partial_{i_1}H_{i_2\dots i_{p+1}} - \partial_{i_2}H_{i_1 i_3\dots i_p+1} + \cdots\right). \tag{26.3}$$

We can think of this object more abstractly as follows. Antisymmetric tensors with p indices are called p-forms. A "basis" for p-forms is provided by the antisymmetrized products of differentials:

$$dx^{i_1} \wedge dx^{i_2} \wedge \cdots \wedge dx^{i_p}. \tag{26.4}$$

We can then write

$$H = \frac{1}{p!}H_{i_1\dots i_p}dx^{i_1} \wedge \cdots \wedge dx^{i_p}. \tag{26.5}$$

The product of two forms A, B is known as the wedge product, $A \wedge B$. If A is an n-form and B an m-form then

$$(A \wedge B)_{i_1\dots i_{n+m}} = \frac{n!m!}{(n+m)!}A_{i_1\dots i_n}B_{i_{n+1}\dots i_{n+m}} + (-1)^P\text{permutations} \tag{26.6}$$

or, more compactly,

$$A \wedge B = \frac{1}{(n+m)!}A_{i_1\dots i_n}B_{i_{n+1}\dots i_{n+m}}dx^1 \wedge \cdots \wedge dx^{n+m}. \tag{26.7}$$

In this language the exterior derivative can be written as $d \wedge H$ or simply dH, where d is thought of as a one-form with components $d_i = \partial_i$.

It is important to practise with this notation, and some exercises are provided at the end of the chapter. One should check that

$$d^2H = 0. \tag{26.8}$$

It is instructive to write electrodynamics in the language of forms. One should verify that the field strength tensor is a two-form, which can be written as

$$F = dA. \tag{26.9}$$

The homogeneous Maxwell's equations (the Bianchi identities for the field strength) follow from $d^2 = 0$:

$$dF = 0. \tag{26.10}$$

Apart from multiplication and differentiation, there is another important operation, denoted by $*$ and called the Hodge star. In d dimensions, this takes a p-form to a $(d-p)$-form:

$$(*H)_{i_1\dots i_{d-p}} = \frac{1}{p!}\epsilon_{i_1\dots i_{d-p}}{}^{i_{d-p+1}\dots i_d}H_{i_{d-p+1}\dots i_d}. \tag{26.11}$$

A particularly interesting object is $*d$. For example, $*d \wedge d$ is a d-form. But the components of a d-form are necessarily proportional to $\epsilon_{i_1 \dots i_d}$. With a little work, one can show that

$$*(*d \wedge d) = \partial^2. \tag{26.12}$$

Using the $*$ operation, we can write the action for a p-form field as

$$S = \frac{1}{2(p+1)!} \int *F \wedge F \tag{26.13}$$

with $F = dA$. This is clearly gauge invariant. It is easy to check that this reproduces the standard action for electrodynamics.

For physics, we are particularly interested in the zero modes of A, i.e. field configurations that satisfy $dA = 0$ but which are not simply gauge transformations; they cannot everywhere be written as

$$A = d\Lambda. \tag{26.14}$$

A simple example of what is at issue is provided by a gauge field on a circle, $0 \leq y \leq 2\pi R$. The one-form gauge field,

$$A_y = \partial_y \Lambda, \quad \Lambda = cy \tag{26.15}$$

is not a sensible gauge transformation unless $c = n/R$, since a fermion of unit charge will not transform into itself. In electrodynamics, for example, this corresponds to the fact that the Wilson line,

$$U = \exp\left(i \int_0^{2\pi R} dy\, A_y\right) \tag{26.16}$$

is gauge invariant and non-trivial, again, unless $c = n/R$.

This suggests that we want to consider closed p-forms α which satisfy

$$d\alpha = 0, \tag{26.17}$$

but that we are not interested in *exact* forms

$$\alpha = d\beta. \tag{26.18}$$

More generally, we want to define an equivalence class known as the cohomology class of α. We will view α and α' as equivalent if

$$\alpha' = \alpha + d\beta, \tag{26.19}$$

where β is well defined everywhere on the manifold.

In general, for field configurations on a manifold M the number of linearly independent zero modes is known as the *Betti number*, b_p. This number is related to the number of (basis) p-dimensional submanifolds which are not boundaries of $(p+1)$-dimensional surfaces. We will not prove this but will at least make it plausible. Consider the integration of a p-form, α, over a p-dimensional submanifold Σ:

$$\int_\Sigma \alpha_{i_1 \dots i_p}\, d\Sigma^{i_1 i_p}. \tag{26.20}$$

By Stokes' theorem, the integral of the exterior derivative of a $(p-1)$-form β over Σ is related to the integral of β over the boundary of Σ:

$$\int_{\Sigma} d\beta = \int_{\partial\Sigma} \beta. \tag{26.21}$$

If Σ is compact, it has no boundary so the integral of $d\beta = 0$.

Two p-forms are in the same cohomology class if

$$\int_{\Sigma} (\alpha - \alpha') = \int_{\Sigma} d\beta = \int_{\partial\Sigma} \beta = 0. \tag{26.22}$$

Note that, as before, it is important in this expression that β is defined throughout the manifold.

If we consider the structure of a massless chiral multiplet, we note that there are two scalars and a chiral fermion. In compactifications preserving $N = 1$ supersymmetry, modes of antisymmetric tensor fields which are annihilated by d will correspond to massless scalars; supersymmetry guarantees that the other elements of the multiplet are also present. The suggested readings at the end of the chapter contain more detailed discussions of these issues, but it is not too hard to understand how the various states arise in terms of the forms annihilated by d. The other massless *scalar* arises because one can also choose the form in such a way the Laplacian vanishes. The Dirac operator is closely related to differential forms on manifolds. This can be shown using the creation–annihilation operator construction of the Dirac matrices that we used in our discussion of orthogonal groups. One can exhibit in this way the required pairing.

With this machinery we can define an important set of topological invariants of manifolds: characteristic classes. Consider a gauge field F, where $F = dA$. Note that F is closed: $dF = 0$. The gauge field F is said to be an element of $H_1(M, R)$, the second cohomology group of the manifold M with real coefficients. The cohomology class of such two-forms is known as the first Chern class.

When the manifold is topologically non-trivial, if we consider a gauge field then it may not be possible to describe the field everywhere by a single non-singular potential. This problem is familiar to us from the case of the Dirac monopole. Instead, in different regions α and β we have to use different potentials, $A_{(\alpha)}$, $A_{(\beta)}$. In regions where α and β overlap (transition regions), $A_{(\alpha)}$ and $A_{(\beta)}$ will be gauge transforms of one another:

$$A_{(\alpha)} = A_{(\beta)} + \phi_{(\alpha\beta)}. \tag{26.23}$$

Another set of gauge fields is said to be in the same topological class if

$$\tilde{A}_{(\alpha)} = \tilde{A}_{(\beta)} + \phi_{(\alpha\beta)} \tag{26.24}$$

with the *same* transition function ϕ. Now, since the functions A and $\tilde{A}$ are not uniquely defined everywhere, on the one hand $F = dA$ and $\tilde{F} = d\tilde{A}$ are not in the trivial cohomology class in general. On the other hand, $F - \tilde{F}$ is in this class, since the difference $A - \tilde{A} = B$ is well defined. So $F - \tilde{F} = dB$ and F and $\tilde{F}$ are in the same cohomology class. Thus the cohomology class of F, the first Chern class, is a topological invariant.

There is a theorem which states that if the first Chern class is non-zero then one can always find a two-dimensional surface Σ with the property

$$I(\Sigma) = \frac{1}{2\pi} \int_\Sigma F \neq 0. \tag{26.25}$$

Note that this is a kind of magnetic flux. By Dirac's argument (see Chapter 7), $I(\Sigma)$ is an integer. The first Chern class plays an important role in the theory of Calabi–Yau spaces.

These ideas can be generalized to complex spaces. Here we define, as we did for the orbifold, complex coordinates z_i and $\bar{z}_i$. We then define a (p, q)-form ψ to be an object with p z_i-type indices and q $\bar{z}_i$-type indices. Note that ψ is totally antisymmetric in both types of indices. We can define two types of exterior derivatives, ∂ and $\bar{\partial}$, in an obvious way:

$$\partial \psi_{a_1 \dots a_{p+1}, \bar{a}_1 \dots \bar{a}_q} = \frac{1}{p+1} \partial_{a_1} \psi_{a_2 \dots a_{p+2} \bar{a}_1 \dots \bar{a}_q} + (-1)^P \text{ permutations.} \tag{26.26}$$

Note that $\partial^2 = 0$; $\bar{\partial}$ is defined similarly. In terms of these definitions,

$$d = \partial + \bar{\partial}. \tag{26.27}$$

These are known as the Dolbeault operators. We can then consider differential forms annihilated by these operators. The numbers of independent forms annihilated by the ∂ and $\bar{\partial}$ operators are known as the Hodge numbers, $h^{p,q}$. Then, for example, one has the Hodge decomposition

$$b_n = \sum_{p+q=n} h^{p,q}. \tag{26.28}$$

Again, is is possible to choose these forms so that they are annihilated by the Laplacian.

26.2 Calabi–Yau spaces: constructions

We have already constructed a rather rich set of four-dimensional string theories. But they are only a small subset of what appears to be a vast set of possibilities. We saw, for example, that the orbifold compactifications give rise to moduli which describe states which are *not* orbifolds. A rich set of compactifications of string theory, of which the orbifolds we studied in the last chapter are special cases, are provided by the Calabi–Yau spaces. In this section, we introduce these.

Our strategy to construct solutions is to look for solutions of the ten-dimensional field equations. One can ask: why is this sensible? There are two answers. First, if we consider spaces in which the massless ten-dimensional fields are slowly varying, it should be appropriate to integrate out the massive string modes and study the low-energy equations. A more serious question is: why is it that we can simply look at the low-order equations? Even at the classical level, integrating out the massive states will lead to terms with arbitrary numbers of derivatives. This question is far more serious. If we solve the equations, say, involving two derivatives then we can try to find solutions of the terms in up to four derivatives perturbatively. To do this we expand the fields in modes of the lowest-order theory (e.g. eigenfunctions of the Laplace operator on the complex space). These are precisely the Kaluza–Klein modes. Calling these ϕ_n and substituting our lowest-order solution into the next-order terms, we will obtain equations of the form

$$(\nabla^2 + m_n^2)\phi_n = \frac{\alpha'}{R^2}\Gamma_n. \tag{26.29}$$

For $m_n \neq 0$, i.e. for the massive Kaluza–Klein modes, we simply obtain a small shift. But the massless modes are problematic. In the case of Calabi–Yau compactifications it is supersymmetry which will come to our rescue. We will see that, for the massless modes, the tadpoles (Γ_ns) vanish.

We begin with the Type II theory. Rather than examine the equations of motion we look at the supersymmetry variations. In flat-space four-dimensional theories, we are familiar with the idea that we can find minima of the potential by setting the auxiliary fields to zero. We can phrase this in a different, seemingly more obscure, way: we can find static solutions of the classical equations by requiring that the supersymmetry variations of all the fields vanish. That is, we require

$$\delta\psi = \epsilon F = 0, \quad \delta\lambda = \epsilon D = 0. \tag{26.30}$$

We will try the same strategy. In Chapter 17 we introduced the essential elements required to understand spinors in a gravitational background (the reader may want to reread Section 17.6). To make things simple, we will look for solutions where the antisymmetric tensor vanishes and the dilaton is constant, so only the metric is spatially varying. Then the condition that there should be a conserved supersymmetry becomes

$$\delta\psi_M = D_M\eta = 0. \tag{26.31}$$

So η is covariantly constant. This means that, under parallel transport around any closed curve, η returns to itself. As in gauge theories the effect of parallel transport can be described in terms of Wilson lines, where now the Wilson line is written in terms of the spin connection, ω:

$$U = P\exp\left(i\oint \omega\, dx\right). \tag{26.32}$$

The fact that η is unchanged under any such transformation greatly restricts the form of ω. To see how this works, consider that in the ten-dimensional Lorentz group, there is an $O(6)$ subgroup which acts on the compactified coordinates, as well as the four-dimensional Lorentz group acting on the Minkowski coordinates. The 16-component spinor in ten dimensions decomposes under these groups as

$$\eta = (4,2) + (\bar{4}, 2^*). \tag{26.33}$$

By local Lorentz transformations, we can take the $(4,2)$ representation to have the form (suppressing the four-dimensional spinor index)

$$\eta = \begin{pmatrix} 0 \\ 0 \\ 0 \\ \eta_0 \end{pmatrix}. \tag{26.34}$$

In order that this be invariant, we require that the spin connection lie in an $SU(3)$ subgroup of $O(6)$. The space is said to be a space of $SU(3)$ holonomy.

In general ω is an $O(6)$ matrix. Restriction to $SU(3)$ is a strong constraint. Already $U(3)$ holonomy requires that the manifold be complex. We encountered this in the orbifold case, where we introduced three complex coordinates and their conjugates. There is no unique way to introduce the complex coordinates. The continuous set of choices will lead to a set of moduli of our solutions, known as the complex structure moduli. In addition, a manifold of $U(3)$ holonomy is Kahler. This means that the metric can be derived from a function $K(x^i, x^{\bar{i}})$, the Kahler potential, through

$$g_{i\bar{j}} = \partial_i \partial_{\bar{j}} K. \tag{26.35}$$

While proving that a manifold of $U(3)$ holonomy must be Kahler is challenging, it is not hard to check that a Kahler manifold has $U(3)$ holonomy. Some aspects of these manifolds are discussed in the exercises.

The Christoffel symbols (affine connection) and curvature for a Kahler manifold can be written in quite compact forms. (Verification of these formulas is left for the exercises.) The components of the Christoffel symbols are given by

$$\Gamma^a_{bc} = g^{a\bar{d}} \partial_b g_{c\bar{d}}, \quad \Gamma^{\bar{a}}_{\bar{b}\bar{c}} = g^{\bar{a}d} \partial_{\bar{b}} \partial \bar{b} g_{\bar{c}d}. \tag{26.36}$$

As a result, the non-zero components of the Riemann tensor are

$$R^{\bar{a}}_{\bar{b}c\bar{d}} = \partial_c \Gamma^{\bar{a}}_{\bar{b}\bar{d}} \tag{26.37}$$

and the Ricci tensor is

$$R_{\bar{b}c} = -\partial_c \Gamma^{\bar{a}}_{\bar{b}\bar{a}}. \tag{26.38}$$

Using

$$\Gamma^{\bar{a}}_{\bar{b}\bar{a}} = \partial_{\bar{b}} \ln \det g, \tag{26.39}$$

this can be further simplified:

$$R_{\bar{b}c} = -\partial_{\bar{b}} \partial_c \ln \det g. \tag{26.40}$$

Note that our result, Eq. (24.19), for the curvature of a two-dimensional Riemann surface is a special case of this.

The requirement that the metric have $SU(3)$ holonomy has a dramatic consequence for the curvature: the Ricci tensor vanishes. This follows from our discussion of the spin connection as a gauge field for local Lorentz transformations. On a six(real)-dimensional Kahler manifold we have seen that the spin connection is not an $O(6)$ field but, rather, a $U(3)$ field (in four dimensions it is a $U(2)$ field, etc.). The $U(1)$ part of the Riemann tensor is the trace over the Lorentz indices – the group indices, thinking of the Riemann tensor as a non-Abelian field strength. But this object is the Ricci tensor, so $SU(3)$ holonomy requires that the Ricci tensor itself vanish everywhere on the manifold. For such a configuration the lowest-order Einstein equation is automatically satisfied, $R_{i\bar{j}} = 0$. The question which we would like to address is: given a Kahler manifold, is it possible to deform the Kahler potential in such a way that the Ricci tensor vanishes? Clearly a necessary condition for this is that the integral

$$c_1 = \frac{1}{2\pi} \int \operatorname{Tr} R \tag{26.41}$$

vanish. This quantity is the first Chern class, the topological invariant which we discussed earlier. It was Calabi who conjectured that the vanishing of the first Chern class for a manifold was a necessary and sufficient condition that the manifold admit a unique metric of $SU(3)$ holonomy. Yau later proved this conjecture. The spaces constructed in this way are the famous Calabi–Yau spaces. In general, while one can prove that such metrics exist, actually constructing them is a difficult numerical problem. Fortunately, many properties relevant to the low-energy behavior of string theory on these manifolds can be obtained from more limited, topological, information.

It is worthwhile comparing this with our orbifold constructions. The orbifolds are everywhere flat. But the existence of a deficit angle associated with the fixed points means that there is actually a δ-function curvature; this gives precisely the holonomy of these manifolds. If we decompose the spinors as before then, as we transport them about the fixed points, the i-components pick up a phase, $e^{\frac{2\pi i}{3}}$, while the 0-components are invariant. Correspondingly, we find one unbroken supersymmetry.

When we discuss the heterotic theory on a Calabi–Yau space, we will have to choose values for the gauge fields as well. It will not be possible to simply set the gauge fields to zero. From the point of view of four dimensions, gauge fields with indices in the extra dimensions are like scalars, so this will result in the breaking of some or all the gauge symmetry. As we will see in Section 26.6.1, there are many possible choices for these fields, with distinct consequences for the structure of the low-energy theory. In an interesting subclass, some features of the heterotic theory are closely related to those of Type II on Calabi–Yau spaces.

26.3 The spectrum of Calabi–Yau compactifications

In both the Type II and heterotic cases, many features of the low-energy spectrum follow from general topological features of the manifold and do not depend on details of the metric. In the heterotic case the number of generations (minus the number of antigenerations) is a topological invariant. Suppose that we have some number of generations for some choice of metric. If we now make smooth, continuous, changes in the metric then the massless spectrum can change, as generations and antigenerations pair to gain mass or become massless. In other words, a mass term in an effective action can pass through zero but the net number of generations cannot change. In some cases, other features of the spectrum are similarly invariant. So, while it is difficult to write down explicit metrics for manifolds having $SU(3)$ holonomy, it is possible to determine many important features of the low-energy theory from basic topological features of the manifold.

In the Type II theory the numbers of hypermultiplets and vector multiplets are separately topological. They do not pair up as one moves about on the moduli space; the $N=2$ supersymmetry ensures that if a field is massless at one point in the moduli space then it is massless at all points. Even more dramatic is that the massless states found in the lowest order of the α' expansion are in fact massless to all orders α' and in string perturbation theory. So it is enough to study the lowest-order supergravity equations of motion in order to count the massless particles.

The important non-zero Hodge numbers are $h^{2,1}$ and $h^{1,1}$. In the IIA theory there are $h^{1,1}$ vector multiplets and $h^{2,1}$ hypermultiplets. In the IIB theory this is reversed. In the heterotic case, the $(2,1)$-forms will correspond effectively to generations and the $(1,1)$-forms to antigenerations.

The counting of massless fields is not difficult to understand. Since we have taken the antisymmetric tensor fields and fermions to vanish in the background, the equations for these fields are particularly simple. Consider the antisymmetric tensor $B_{\mu\nu}$. On a complex manifold, as we explained earlier, there are $h^{1,1}$ $(1,1)$-forms $b^{(a)}_{i,\bar{j}}$ and $h^{2,1}$ $(2,1)$-forms annihilated by the operators ∂ and $\bar{\partial}$. Since the corresponding three-index field strengths $H = dB$ vanish, there is no energy cost to giving a constant expectation value to the associated four-dimensional fields; they correspond to massless scalars in four dimensions. The fields connected to the $(1,1)$-forms $b_{i,\bar{i}}$, are easy to describe. In addition to the antisymmetric tensor there is also a massless perturbation of the metric:

$$ig_{i\bar{i}}(x,y) = \phi(x)b_{i,\bar{i}}(y). \tag{26.42}$$

Here x refers to the ordinary four-dimensional Minkowski coordinates and y refers to the compactified coordinates. Similarly, in the IIA theory one can find a massless gauge field rounding out the bosonic components of the vector multiplet. This comes from the three-index Ramond field,

$$C_{\mu i,\bar{i}}(x,y) = A_\mu(x)b_{i,\bar{i}}(y). \tag{26.43}$$

We will leave to the reader the problem of working out the structure of the hypermultiplets in terms of the $(2,1)$-forms and also of determining the pairings in the IIB case.

A $(1,1)$-form which is always present is the Kahler form,

$$b^K_{i,\bar{i}} = ig_{i,\bar{i}}, \quad b_{\bar{i},i} = -ig_{i,\bar{i}}. \tag{26.44}$$

This satisfies

$$\partial b^K = \bar{\partial} b^K = 0 \tag{26.45}$$

because $g_{i\bar{i}} = \partial_i \partial_{\bar{i}} K$. The real scalar which sits in the multiplet with b^K is just the metric itself. The corresponding massless field is the radius of the compact space:

$$g_{i,\bar{i}}(x^\mu, z^i) = R^2(x^\mu)g_{i,\bar{i}}(z), \quad B_{i,\bar{i}}(x^\mu, z^i) = b(x^\mu)b_{i,\bar{i}}(z). \tag{26.46}$$

That the field is massless is no surprise; the condition $R_{i\bar{i}} = 0$ is not changed under an overall rescaling of the metric, so the vev is undetermined.

26.4 World-sheet description of Calabi–Yau compactification

Thus far we have described the compactification of string theory in terms of ten-dimensional space–time. This analysis makes sense if the radius of the compactified space is large compared with the string length, ℓ_s. We can also formulate these questions in world-sheet terms. This provides a complementary way to understand many features of the compactified theory and is useful for at least two reasons. First, it provides tools to ask what happens when the compactification radius is of order the string scale or smaller. Second, there are some features of the spectrum and interactions which are more readily accessible in this framework.

In the Type II theory the non-linear sigma model which describes compactification on a Calabi–Yau space has some striking features. First, in the absence of background antisymmetric tensor fields it is left–right symmetric. Second, there are two left-moving and two right-moving supersymmetries on the world sheet as opposed to the one left-moving and one right-moving supersymmetry of a general configuration. This can be usefully understood in a number of ways. In the light cone gauge, one can work with the covariantly constant spinor η and its conjugate $\bar{\eta}$ to construct two left-moving and two right-moving supersymmetry generators, both in the sense of the world sheet and in space–time. We have already seen this in the case of orbifold constructions. There, in the light cone gauge, we have eight left-moving and eight right-moving supersymmetry generators, before the orbifold projection. We can organize these in terms of their transformation properties under the $SU(3) \times U(1)$ holonomy group. For both the left and right movers there are triplets Q_i, antitriplets $\bar{Q}_{\bar{i}}$ and singlets, Q_0 and $\bar{Q}_0$. The triplets and antitriplets are charged under the $U(1)$ symmetry; the singlets are not. The orbifold projection eliminates the triplets. The two singlets survive.

In a purely world-sheet description, non-linear sigma models described by a Kahler metric automatically have two left-moving and two right-moving supersymmetries. To describe these, we can introduce a superspace with four Grassmann coordinates, of which two are left movers and two are right movers: θ_+^A and θ_-^A. This superspace can be thought of as the truncation of $N = 1$ supersymmetry in four dimensions. As in four dimensions we can define, operators D_α and $\bar{D}_\alpha$ and left- and right-moving chiral fields annihilated by the $\bar{D}$s. Correspondingly, we can define chiral left- and right-moving fields

$$X_+^i(z,\theta) = x^i(z) + \theta_+^A \psi_A^i(z) + \text{auxiliary field} \tag{26.47}$$

and similarly for X_-^i. In terms of these fields we can write the action of the conformal field theory as

$$\int d^2\sigma \int d^2\theta_+ d^2\theta_- K(X, \bar{X}). \tag{26.48}$$

Integrating over the θs, the bosonic terms are just $\int d^2\sigma g_{i,\bar{i}}\, \partial_\alpha x^i \partial^\alpha x^{\bar{i}}$, with $g_{i,\bar{i}}$ the Kahler metric.

The superconformal algebra, in these backgrounds, is enlarged to what is referred to as the $N = 2$ superconformal algebra (one such algebra for the left movers, one for the right movers). In addition to the stress tensor and the two supercurrents, this algebra contains a $U(1)$ current. The supersymmetry generators can be constructed by the Noether procedure. They can also be guessed by taking the generators in a flat background and making the expressions covariant:

$$G^+ = g_{i,\bar{i}} DX^i \psi^{\bar{i}}, \quad G^- = g_{i,\bar{i}} DX^{\bar{i}} \psi^i. \tag{26.49}$$

These have opposite charge under the $U(1)$ current (an R current) constructed from the fermions,

$$j(z) = \psi^{\bar{i}}(z)\psi^i(z), \tag{26.50}$$

with a similar current for the left movers. The full algebra is

$$\begin{aligned} T(z)G^{\pm}(0) &\approx \frac{3}{2z^2}G^{\pm}(0) + \frac{1}{z}\partial G^{\pm}(0), \\ T(z)j(0) &\approx \frac{1}{z^2}j(0) + \frac{1}{z}\partial j(0), \\ j(z)G^{\pm}(0) &\approx \pm\frac{1}{z}G^{\pm}(0). \end{aligned} \tag{26.51}$$

These equations say that G has dimension $3/2$ while j has dimension one, and $G^{\pm}$ have $U(1)$ charges plus and minus one. The central charge appears in the relations

$$\begin{aligned} G^+(z)G^-(0) &\approx \frac{2c}{3z^3} + \frac{2}{z^2}j(0) + \frac{2}{b}T(0) + \frac{1}{z}\partial j(0), \\ G^+(z)G^+(0) &\approx 0, \\ j(z)j(0) &\approx \frac{c}{3z^2}. \end{aligned} \tag{26.52}$$

The non-linear sigma models appropriate to *heterotic* compactifications on Calabi–Yau spaces have a number of interesting features. We will see that, for a particular choice of gauge fields, the world-sheet theory which describes the heterotic compactification is identical to that of the Type II theory. Thus again they have two left-moving and two right-moving supersymmetries ((2, 2) supersymmetry). The fact that the world-sheet theories of the two different string theories are the same allows us to argue, as we will below, that Calabi–Yau spaces are solutions of the full, non-perturbative, string equations of motion. But this observation also tells us about interesting features of the spectrum.

To understand the spectrum, it is helpful to ask, first, what is a vertex operator from the perspective of the two-dimensional conformal field theory? The answer is that a vertex operator is a *marginal deformation* of the theory, a perturbation of dimension 2 ((1, 1) in terms of the left- and right-moving Virasoro algebras). The standard way to compute the dimensions of operators is to treat them as perturbations and calculate, for example, the beta function of the perturbation. For marginal operators the beta function vanishes to first order. The moduli correspond to "exactly marginal deformations" of the theory. For these the beta functions vanish to all orders in the perturbation (and non-perturbatively), corresponding to the fact that the theory, even for a finite perturbation, is conformal.

The existence of moduli means that there is a multiparameter set of conformal field theories. Varying the action with respect to the parameters yields operators which are exactly marginal. In this way, we have the two-dimensional version of the correspondence between moduli and massless fields.

An example of a modulus is the radius of the complex space. The lowest-order equation for the metric is invariant under an overall scaling of lengths. But this is not obviously true of the higher-order corrections. For Type II theories the space–time supersymmetry guarantees that there is no potential for the moduli, so the sigma model is a good conformal field theory, suitable for heterotic string compactification. On the heterotic side we can also give a more direct world-sheet argument. Here R^{-2} is the coupling constant of the sigma model. In other words, writing the metric as R^2 times a reference metric of order the string scale, R^2 appears in front of the Lagrangian. We know that the lowest-order beta function equation is the same as the field theory equation. It is trivially independent of R^2, since it is a one-loop effect. For higher orders there is a non-renormalization theorem. This follows from a combined world-sheet and space–time argument. The superpartner of fluctuations in the radius is the fluctuation of the antisymmetric tensor field, $b_{i,\bar{i}} = ig_{i,\bar{i}}$. The associated vertex operator term in the action is a total derivative on the world sheet at zero momentum. It is perhaps easiest to see this by writing the vertex operator at zero momentum in the form

$$\begin{aligned} V_b &= b_{MN}\epsilon^{\alpha\beta}\, \partial_\alpha X^M \partial_\beta X^N \\ &= \partial_M \partial_N K\, \epsilon^{\alpha\beta}\, \partial_\alpha X^M \partial_\beta X^N \\ &= \partial_\alpha (\epsilon^{\alpha\beta}\, \partial_\beta X^M \partial_M K). \end{aligned} \tag{26.53}$$

So b decouples at zero momentum. Because b is in a supermultiplet with R^2 this means that the superpotential, which is a holomorphic function of the superfields, is independent of R^2.

Actually, this statement is not precisely correct because K is not single-valued. In perturbation theory it is true since one is not sensitive to the global structure of the manifold (in perturbation theory, all fluctuations are small). Non-perturbatively, one can encounter instantons in the world-sheet theory. A more detailed analysis is required to determine whether there are corrections to the superpotential. In left–right symmetric compactifications of the heterotic string, i.e. those with two left-moving and two right-moving supersymmetries ((2, 2) models), a study of fermion zero modes in the presence of the instanton shows that no superpotential for the moduli is generated; this is consistent with one's expectations from the Type II theory. For compactifications with two right-moving but no left-moving supersymmetries ((2, 0) models), corrections can be generated though in some cases intricate cancelations still prevent the appearance of a potential for the moduli. These two classes of models are phenomenologically quite distinct, as we will see shortly.

26.5 An example: the quintic in CP^4

It is helpful to have a concrete example of a Kahler manifold with $c_1 = 0$, on which we know that one can construct a metric of $SU(3)$ holonomy. We have previously encountered the complex projective spaces in N dimensions, CP^N. These are defined as spaces with $N+1$ complex coordinates Z_a and with the identification $Z_a \to \lambda Z_a$ for any complex number λ. We have written down a Kahler potential on this space:

$$K = \ln\left(1 + \sum_{a=1}^{N} Z_a \bar{Z}_a\right). \tag{26.54}$$

Any complex submanifold of a Kahler manifold is also a Kahler manifold; one can simply take the Kahler potential to be the Kahler potential of the full manifold, evaluated on the submanifold. To obtain a manifold with three complex dimensions we can start with CP^4 and write down an equation for the vanishing of a polynomial $P(Z)$. The polynomial should be homogeneous in order that it has a sensible action in CP^N. It turns out that it should also satisfy other conditions. Its gradient should at most vanish, at the origin (which is not a point in CP^N). In order that the first Chern class should vanish, it should be quintic. We will give an argument for this shortly.

The simplest (most symmetric) possibility is

$$P = Z_1^5 + Z_2^5 + Z_3^5 + Z_4^5 + Z_5^5 = 0, \tag{26.55}$$

but there are obviously many more. We can deform this polynomial by adding other quintic polynomials. These correspond to varying the complex structure. Since each deformation produces another solution of the string equations, each deformation corresponds to a modulus, one of the complex structure moduli. Associated with each deformation is a form of type (2, 1), which we will not attempt to construct here.

Before listing the deformations, we note that not every deformation corresponds to a change in the physical situation – and thus to a massless particle. Holomorphic changes of the coordinates which are non-singular and invertible do not change the complex structure. The transformation

$$Z_i \to Z_i + \epsilon^{ij} Z_j \tag{26.56}$$

is well defined in CP^4. As a consequence, deformations such as $Z_1^4 Z_2$ are not physical. So we can list the possible deformations:

$$Z_1^3 Z_2^2 \ldots, \quad Z_1^3 Z_2 Z_3 \ldots, \quad Z_1^2 Z_2^2 Z_3, \ldots, \quad Z_1^2 Z_2 Z_3 Z_4, \ldots, \quad Z_1 Z_2 Z_3 Z_4 Z_5. \tag{26.57}$$

All together there are 101 possible deformations of the polynomial, corresponding to $h_{2,1} = 101$. In this example, there is only one Kahler modulus, the overall radius of the compact space.

We can understand heuristically why the first Chern class vanishes, in a way which will help us to understand other features of these manifolds. A characteristic feature of the Calabi–Yau spaces is the existence of a covariantly constant three-form, ω_{ijk}. The existence of this form follows from the existence of a covariantly constant spinor η:

$$\omega_{ijk} = \bar{\eta} \Gamma_{[ijk]} \eta. \tag{26.58}$$

Working in terms of the creation–annihilation operator basis for the Γs, one sees that ω is holomorphic. The Γ_is can be defined in such a way that the $\Gamma_{\bar{i}}$ matrices annihilate η. Then, because of the complete antisymmetrization, only components of ω with indices $1, 2, 3$ are non-vanishing. In the space defined by the vanishing of a quintic polynomial in CP^4, we can show that there exists a holomorphic three-form which is everywhere non-vanishing. Setting $x^i = Z_i/Z_5$, $i = 1, \ldots, 4$,

$$\omega = dx^1 \wedge dx^2 \wedge dx^3 \left(\frac{\partial P}{\partial x^4} \right)^{-1}. \tag{26.59}$$

One can show that this expression does not depend on singling out a particular coordinate and that it is not singular at the points where the derivative vanishes provided that the polynomial P is quintic and that the gradient of P vanishes only at the origin. The existence of such a form can be shown to be equivalent to the vanishing of the first Chern class.

26.6 Calabi–Yau compactification of the heterotic string at weak coupling

Much effort has been devoted to the study of compactifications of the weakly coupled heterotic string on Calabi–Yau spaces. These theories have many features of the Standard Model. They also allow one to consider many questions of Beyond the Standard Model physics. Before beginning an analysis of these models it is worth listing some points that we can address in this framework.

1. *Low-energy supersymmetry* Solutions of the classical equations of the heterotic string theory on Calabi–Yau spaces exist. They have $N = 1$ supersymmetry. Supersymmetry, as in field theory, is unbroken to all orders of perturbation theory but may be broken non-perturbatively.
2. *Low-energy gauge groups* The simplest constructions have gauge group $E_8 \times E_6$, broken perhaps by Wilson lines, which preserve the rank of the gauge group. But many models have a moduli space in which the gauge group is broken to precisely that of the Standard Model.
3. *Generations* The number of generations is typically determined in terms of topological features of the underlying manifold.
4. *Massless particles, not protected by symmetries* Various massless states arise which are not protected by chiral symmetries. This is precisely what we want in order to understand the presence of light Higgs fields in supersymmetric theories. We know that if such fields are present in the low-energy field theory, they are protected from gaining large masses by non-renormalization theorems. In field theory the vanishing of such mass terms appears mysterious; in these string constructions, it is automatic. Such states could play the role of Higgs fields in supersymmetric models. In other words, *the Huggs five-turning problem of ordinary supersymmetric field theories is readily solved in this framework.*
5. *Unification of couplings* The string theories that we are studying are not grand unified theories in the conventional sense. There is no energy scale at which these

compactifications appear as four-dimensional theories with a single unbroken gauge group. Yet, generically, the couplings are unified. These two features, which we will see are easy to understand in terms of the microscopic structure of string theory, are quite surprising from a low-energy point of view. They have sometimes been referred to as "string miracles".

6. *Continuous and discrete symmetries* It is easy to prove that for these compactifications (and for weak-coupling heterotic models in general) there are no continuous global symmetries; all continuous symmetries must be gauge symmetries. Discrete symmetries, however, hand, proliferate and might play the role of R-parity or lead to other interesting phenomena. These discrete symmetries are typically gauge symmetries, in the sense that they are residual symmetries left over after the breaking of continuous gauge symmetries.

We will also see that there are a number of problems with these models, which illustrate some of the basic difficulties in developing a string phenomenology, as follows.

1. *There are too many models* While there are many with three generations, there are also some with hundreds of generations, with non-standard gauge groups and the like.
2. *The problem of moduli* Non-perturbatively, moduli can acquire potentials but they typically vanish in various asymptotic regimes. Simple general arguments indicate that stable supersymmetry-breaking minima, if they exist, must be in regions which are inherently strongly coupled in the sense that *no* weak coupling approximation is available.
3. *The problem of the cosmological constant* This is closely related to the previous one. In many instances moduli potentials can be calculated. For any given value of the moduli the size of these potentials is scaled, as one would expect, by the scale of supersymmetry breaking. As a result, even if strongly coupled stable minima exist it is not clear why the cosmological constant should be small at these points.

We will not offer a solution to these problems in this chapter but will explore at least one proposed answer, known as the "landscape," in Chapter 30.

26.6.1 Features of Calabi–Yau compactifications of the heterotic string

In the previous section we asserted that, in suitable backgrounds, the world-sheet conformal field theory which describes the heterotic string is the same as that which describes the Type II theory. Here, we describe compactifications of the heterotic string theory in more detail.

To construct solutions, we still look for these which preserve a space–time supersymmetry. Again we require the supersymmetry variation of the gravitino to vanish, giving $D_\mu \eta = 0$, so once more we need a covariantly constant spinor. There is now an equation for the variation of the ten-dimensional gaugino, as well:

$$\delta\lambda \propto \Gamma^{ij} F_{ij} \eta. \tag{26.60}$$

One strategy, then, to find solutions which preserve $N = 1$ supersymmetry is to require that $F_{ij}\Gamma^{ij}$ is an $SU(3)$ matrix. There is a simple ansatz which achieves this. Both E_8 and $O(32)$ have $SU(3)$ subgroups:

$$SU(3) \times E_6 \times E_8 \subset E_8 \times E_8, \quad SU(3) \times O(26) \subset O(32). \tag{26.61}$$

On the Calabi–Yau space the spin connection is an $SU(3)$-valued field, so we take the gauge field to be a field in one of these $SU(3)$ subgroups. Then, for gauge generators not in $SU(3)$, expression (26.60) is automatically satisfied. For those in $SU(3)$ the condition is mathematically identical to that for the gravitinos and is again satisfied.

This ansatz satisfies another condition. We have set the antisymmetric tensor field B to zero but, because of the Chern–Simons terms, this does not by itself guarantee that the field strength H is zero. With this ansatz, however, the Chern–Simons terms for the gauge and gravitational fields are identical. As a quick check, note that

$$dH = \text{Tr}\,(R \wedge R) - \text{Tr}\,(F \wedge F), \tag{26.62}$$

and the two terms in this expression clearly cancel. This establishes that here we have a solution of the equations of motion to lowest order in the α' expansion. But there is another way to see this, which will allow us to establish, as we did for the Type II theory, that this is an *exact* solution, perturbatively and non-perturbatively. If we write down the non-linear sigma model which describes the heterotic string in this background, it is identical to that for the Type II theory. To see this, as in the orbifold case, we divide the left-moving gauge fermions into three sets. First, there are the fermions λ^A, $A = 1, \ldots, 16$, in the second E_8 group, which are not affected by the background gauge field and remain free.. In the first E_8, ten fermions, λ^a, $a = 1, \ldots, 10$ (transforming as a vector in the $O(10)$ subgroup of E_6), are also free. The remaining six interacting fermions can be grouped, like the left-moving coordinates, into three complex fermions, λ^i and $\lambda^{\bar{i}}$. These fermions interact in precisely the same way as the left-moving fermions in the Type II theory. This can be seen by writing the action of the Type II fermions in terms of the vierbein and spin structure rather than the metric and the Christoffel connection.

We see from this that the moduli of the Type II theory are also moduli of the heterotic theory. Actually, we knew this had to be so since we know that each of these conformal field theories, on the Type II side, is a good conformal field theory for the heterotic theory. But we can also see this pairing more directly in the language of vertex operators. Here it is somewhat more convenient to work in the RNS picture. The vertex operators correspond to small deformations of the background in the directions of the moduli. In the Type II theory they are built from right-moving fields, ∂X^i and ψ^i, and left-moving fields, $\bar{\partial} X^i$ and $\tilde{\psi}^i$. In the heterotic case we can trade $\tilde{\psi}^i$ with λ^i. Since the action for the λ^is is the same as that for the $\tilde{\psi}^i$s, the dimensions of the vertex operators are exactly the same. This does not preclude the existence of additional moduli on the heterotic side, and we will see that typically there *are* additional moduli in these compactifications.

While all moduli of the Type II theory are moduli of the heterotic theory, not all heterotic moduli correspond to states of the Type II theory. Vertex operators for moduli which preserve only two right-moving supersymmetries ((2, 0)) are not suitable vertex operators for the Type II theory. The moduli we are considering here are distinguished because they preserve the two left-moving world-sheet supersymmetries, and we will refer to these as Type II moduli. Perhaps more interesting, though, than the pairing of moduli is a pairing of the Type II moduli with matter fields. The moduli associated with (2, 1)-forms are paired with 27s of E_6 and (1, 1) moduli with $\overline{27}$s. This is most readily seen in the language of vertex operators, using the world-sheet superconformal symmetry. The vertex operators for the Type II theory are the highest components of the corresponding superconformal multiplets with respect to both left- and right-moving supersymmetries. In superspace they are the $\theta_+^2\theta_-^2$ components of operators of the form

$$f(X^i, \bar{X}^i). \tag{26.63}$$

The $\theta_+\theta_-^2$ component has dimension $(1/2, 1)$. We can form an operator of dimension $(1, 1)$ by multiplying by λ^a, one of the free fermions. This operator does not have the highest weight with respect to the left-moving $N = 2$ algebra, but this is not a problem; this symmetry is not a gauge symmetry on the world sheet but simply an accident of our choice of background field. It is highest-weight with respect to the left-moving Virasoro algebra, which is all that matters.

We have already observed this pairing in the Z_3 orbifold model, which is a special case of the Calabi–Yau construction. In the untwisted sector the vertex operators for the moduli took the form, for the left-movers,

$$\bar{\partial} X^i, \tag{26.64}$$

while for the 27s they took the form

$$\lambda^a \lambda^i. \tag{26.65}$$

The supersymmetry transformation of the latter operator changes λ^i to $\bar{\partial} X^i$.

The distinction between 27s and $\overline{27}$s is readily understood. In the Type II case we can distinguish two types of moduli, depending on their charges under the $U(1)$ symmetry within the left-moving $N = 2$ algebra. In the orbifold context some vertex operators involve $\bar{\partial} X^i$ and some $\bar{\partial} \bar{X}^{\bar{i}}$. In the heterotic case, the world-sheet $U(1)$ symmetry corresponds to the $U(1)$ subgroup of E_6 in the decomposition $O(10) \times U(1) \subset E_6$. This $U(1)$ charge is precisely what distinguishes the 10s, for example, in the 27 and $\overline{27}$. In the Type II case this distinction corresponds to the distinction between (2, 1) and (1, 1) moduli, so we obtain precisely the pairing we described above (note that what one calls a 27 and a $\overline{27}$ is a matter of convention; if one adopts an opposite convention, the identification is reversed).

This result holds everywhere in the moduli space; since the number of moduli of each type does not change as one moves in the moduli space, the number of 27s and $\overline{27}$s does not change. This is a surprising result. One might have thought that, in a complicated construction such as this, 27s and $\overline{27}$s would, whenever possible, pair to gain mass. But this is not the case. This is precisely the sort of phenomena one needs to understand light Higgs particles in supersymmetric theories. We will see shortly how this works in a more detailed model.

26.6.2 Gauge groups: symmetry breaking

The heterotic models we have been considering have group $E_8 \times E_6$. If we are to describe the Standard Model we need to be able to break this symmetry. We have seen in the case of toroidal compactifications that gauge symmetries can be broken by the expectation values of gauge fields with indices in compactified dimensions. Stated in a more gauge-invariant fashion, these are non-trivial expectation values for Wilson lines. In the Calabi–Yau case the same is possible.

We will consider a specific example: the quintic in CP^4, with the vanishing of the polynomial:

$$Z_1^5 + Z_2^5 + Z_3^5 + Z_4^5 + Z_5^5 = 0. \tag{26.66}$$

The corresponding Calabi–Yau manifold, as we saw, has 101 27s and one $\overline{27}$. This polynomial has a variety of symmetries. As in the case of the torus, we can use these to project out states and simplify the spectrum. Consider, for example, the symmetry

$$Z_i \to \alpha^i Z_i, \quad \alpha = e^{2\pi i/5}. \tag{26.67}$$

This is a symmetry of the polynomial. It is somewhat different from the orbifold symmetries we have discussed since, as the reader can check, it *acts without fixed points*. Mathematicians refer to such a symmetry as "freely acting". For the physics it means that if we mod out the Calabi–Yau by this symmetry then, while it is still necessary to include twisted sectors, the twisted strings have mass of order R, the Calabi–Yau radius, and there are no light states in these sectors if R is large.

We can readily classify the states that are invariant under this symmetry. Among the moduli, there are 21 $h_{2,1}$ fields, associated with polynomials such as $Z_1^3 Z_3 Z_4$ and $Z_1 Z_2 Z_3 Z_4 Z_5$. The Kahler modulus (i.e. the overall radius) is also invariant under this transformation, and so survives the projection. The corresponding Euler number is 40, one fifth of the Euler number of the covering space. There are also 21 27s of E_6 and one $\overline{27}$. Further symmetries can be used to reduce the number of generations to as few as four.

But what interests us here is obtaining smaller gauge groups. We can define the Z_5 group to include a transformation in E_6. This is equivalent to the presence of a Wilson line on the manifold. An interesting way to do this is to consider a somewhat different decomposition of E_6 from what we have considered up to now:

$$SU(3) \times SU(3) \times SU(3) \subset E_6. \tag{26.68}$$

An example of a Wilson line in this product of $SU(3)$s is

$$U = \begin{pmatrix} 1 & 0 & 0 \\ 0 & 1 & 0 \\ 0 & 0 & 1 \end{pmatrix} \begin{pmatrix} \alpha & 0 & 0 \\ 0 & \alpha & 0 \\ 0 & 0 & \alpha^3 \end{pmatrix} \begin{pmatrix} \alpha & 0 & 0 \\ 0 & \alpha & 0 \\ 0 & 0 & \alpha^3 \end{pmatrix}. \tag{26.69}$$

This breaks E_6 to $SU(3) \times SU(2) \times SU(2) \times U(1)^2$.

26.6.3 Massless Higgs fields, or the μ problem

When we mod out in such a way as to reduce the gauge symmetry, we also alter the spectrum. We have seen that this greatly reduces the number of moduli and the number of generations. The presence of the Wilson lines also disrupts the left–right symmetry of the model. As a result, the pairing of moduli and matter fields is no longer quite so simple.

In the presence of the Wilson line one still obtains 20 complete E_6 generations. If one thinks, loosely, of some of the massless fields "gaining" mass then elements of the 27 and $\overline{27}$s must pair up to gain mass. More precisely, in this modding out procedure states disappear, but they must disappear in pairs. But one also obtains some incomplete multiplets, where paired states do not disappear. Consider the $\overline{27}$. This is invariant under the original Z_5s, so any state which survives must be invariant under the Wilson line. Using the decomposition of the 27 under $SU(3)^3$, we obtain

$$27 = (3, 1, \bar{3}) + (\bar{3}, 3, 1) + (1, \bar{3}, 3). \tag{26.70}$$

So we obtain Z_5 singlets from only the third multiplet. These form a $(1, 2, 2)$ under $SU(3) \times SU(2) \times SU(2)$, as well as a singlet. There is a corresponding pair of states from the 27s. This is the sort of multiplet we need to help understand the presence of light Higgs particles in supersymmetric models: massless states at tree level which arise, from a low-energy point of view, more or less by accident.

26.6.4 Continuous global symmetries

In the heterotic string theory, there are no continuous global symmetries. We will not give a formal proof here but the basic argument is not hard to understand. If there is a global symmetry, it should be a symmetry of the world-sheet theory. In this way we are guaranteed that vertex operators can be chosen to have well-defined transformation properties and that the S-matrix will transform properly. The global symmetry will be associated with a world-sheet current. This current can be decomposed into left- and right-moving pieces. But, from any left-moving current we can build a gauge boson vertex operator, so the symmetry is necessarily a gauge symmetry. Right-moving currents will not commute with the world sheet supersymmetry generators and will not have a well-defined action on states (in BRST language they do not commute with the BRST operator). So they are not symmetries in space–time.

There are subtleties needed to complete the proof. First, as we have already seen, string theories typically possess, in perturbation theory, symmetries under which a scalar field undergoes a constant shift. These symmetries, as we will discuss further, are only broken non-perturbatively. The space–time version of such symmetries is not a conventional selection rule but rather a statement that scattering amplitudes vanish in the limit that the momenta of certain particles tend to zero. Second are the selection rules associated with the Poincare group. These clearly have a different status. On the one hand, in some sense, these symmetries are connected to the gauge symmetries of general relativity. On the other hand, their world-sheet implementation is different. For example, translations would appear to be non-linearly realized symmetries from a world-sheet point of view, but momentum is still conserved as a consequence of the Mermin–Wagner–Coleman theorem. In any case, these subtleties are readily isolated and resolved.

This argument also indicates that there are no global symmetries in the Type II theories. This is in accord with our expectation that global symmetries are unlikely to arise in a theory of quantum gravity.

26.6.5 Discrete symmetries

When we studied orbifold models, we found that discrete symmetries existed in a subset of vacua on the full moduli space. This is also the case for the Calabi–Yau manifold constructed from the vanishing of a quintic polynomial in CP^4. Such symmetries turn out to be quite common.

The quintic polynomial $P = \sum Z_i^5$ exhibits a set of Z_5 symmetries:

$$Z_i \to \alpha^{k_i} Z_i, \quad \alpha = e^{2\pi i/5}. \tag{26.71}$$

An overall phase rotation of all the Z_is has no effect in CP^4, so the symmetry here is Z_5^4. There is also a permutation symmetry, S_5. This symmetry group is a subgroup of the $O(6)$ symmetry which would act on six non-compact flat dimensions. We can thus think of these symmetries as discrete gauge transformations. So their existence in a theory of gravity is not surprising.

We would like to know whether these symmetries are R symmetries. We can address this by considering their effect on the covariantly constant spinor η. This is more challenging to do than in the orbifold context, since we do not have quite such explicit expressions. It is simplest to look at the covariantly constant three-form. We have already given a construction,

$$\omega = dx^1 \wedge dx^2 \wedge dx^3 \left(\frac{\partial P}{\partial x^4} \right)^{-1}, \tag{26.72}$$

with $x^i = Z_i/Z_5$. This construction treats the coordinates asymmetrically but, as we explained, ω is symmetric among the coordinates. Note that ω transforms essentially like η^2, i.e. like θ^2. So symmetries under which ω transforms non-trivially are R symmetries, and W transforms like ω.

Consider first the Z_5 transformations of the separate Z_is. We can read off immediately how ω transforms under transformations of the first three; the other two follow by symmetry. So, we have

$$\omega \to \alpha^{\sum k_i}. \tag{26.73}$$

Similarly, under those S_5 transformations which permute Z_1, Z_2, Z_3 we can see how ω transforms. If the permutation is odd, ω changes sign. So again the general S_5 transformation is an R symmetry.

Turning on the various complex structure moduli typically breaks some of or all this symmetry. For example, if we turn on the modulus associated with the polynomial

$$z_1 z_2 z_3 z_4 z_5 \tag{26.74}$$

then we break the Z_5^4 symmetry down to a subgroup satisfying $\sum k_i = 0$ mod 5. This group is Z_5^3 but it is a non-R-symmetry, in light of the transformation law of ω. An expectation value for this field clearly preserves the permutation symmetry.

Similarly, turning on say $aZ_1^3 Z_2 + bZ_1^2 Z_2^3$ breaks the symmetries acting on Z_1 and Z_2 as well as some of the permutation symmetry. Turning on enough fields breaks all the symmetry.

One might ask why one should be interested in points or surfaces in the moduli space which preserve a discrete symmetry, when in the bulk of the space there is no symmetry. This question is closely related to the question: what sorts of dynamics might determine the values of the moduli? This is a subject with which we will deal extensively later but for which we can provide no definitive resolution. But, even without understanding this dynamics, there is a simple reason to suspect that points in the moduli space with symmetries might be singled out by the dynamics. Imagine that we somehow manage to compute an effective potential for the moduli, arising, perhaps, due to some non-perturbative string effects. Symmetry points are necessarily *stationary points* of this effective potential. There is, of course, no guarantee that they are *minima* of the potential but they are certainly of interest as candidates for string ground states.

There are, as we have seen, certain facts of nature which suggest that discrete symmetries might play some role in extensions of the Standard Model, including proton decay and dark matter.

26.6.6 Further symmetry breaking: the Standard Model gauge group

The Wilson line mechanism, as we have described it, provides a path to reduce the gauge symmetry from $E_6 \times E_8$ but leaves the rank untouched.[1] We can hope to reduce the gauge symmetry further by giving expectation values to some matter fields. Ideally, these expectation values would be large. The presence of other gauge groups (as well as unwanted matter multiplets) can spoil the prediction of coupling unification and can lead to severe difficulties with proton decay and other rare processes. We are led, then, to ask whether we can consider more general states, in which the spin connection is not equal to the gauge connection and the rank is reduced.

This is a complex subject, which has been only partially explored. At lowest order in the α' expansion there are such flat directions. They are not left–right symmetric and, while in order that they exhibit space–time supersymmetry they have two right-moving supersymmetries, they have no left-moving supersymmetry. So they are not suitable backgrounds for Type II theories and one cannot argue as easily as for the standard embedding that these $(0, 2)$ configurations are solutions of exact classical string equations. They are still subject to perturbative non-renormalization theorems in α'. But a detailed study of instanton amplitudes is required to determine whether these flat directions are lifted non-perturbatively, i.e. by corrections of the form $e^{-R^2\alpha'}$.

[1] Non-Abelian discrete symmetries offer possibilities for reducing the rank but we will not explore these here.

There is, however, a class of vacua with the Standard Model gauge group which can be found by symmetry arguments, much as we found additional flat directions in the Z_3 orbifold model. Consider, again, the quintic in CP^5, with the symmetric polynomial. We can find flat directions of the D terms by taking $27 = \overline{27}$. More precisely, starting with the E_6 decomposition into $O(10)$ representations,

$$27 = 10_1 + 1_{-2} + 16_{-1/2}, \tag{26.75}$$

we can give expectation values to the singlets in the $\overline{27}$ and one of the 27s. These are also flat directions of the F terms. For example, consider the 27 corresponding to the polynomial $Z_1Z_2Z_3Z_4Z_5$. The product $\overline{27}\,27$ is invariant under all the discrete R-symmetries; no terms of the form $(\overline{27}\,27)^n$ can appear in the superpotential. So this direction is exactly flat (terms of the form 27^3, $\overline{27}^3$ cannot lift these directions either). In combination with Wilson lines these flat directions readily break to the $SU(3) \times SU(2) \times U(1)$ group of the Standard Model.

26.6.7 Gauge coupling unification

One of the striking successes of low-energy supersymmetry is its prediction of unification. Within the context of grand unification – where the gauge group of the Standard Model is unified in a simple group at a scale $M_{\rm GUT}$ – the fact that the couplings unify is readily understood. In the context of the compactifications considered here it is not immediately obvious why this should be the case. In the case of symmetry breaking by Wilson lines, for example, the compactification scale and the scale of the symmetry breaking are of the same order. So there is no energy scale where one has a unified, four-dimensional effective theory.

In the weakly coupled heterotic string, however, the couplings *do unify* under rather broad conditions. In the case of Wilson line breaking this can be understood immediately in field-theoretic terms. The effect of the Wilson line is to eliminate states from the E_6 unified theory, but at tree level no couplings are altered. So the couplings of all groups emerging from E_6 are the same. Perhaps more surprising is the fact that the E_6 and E_8 couplings are the same. This can be seen by considering the vertex operators for the gauge bosons in each group. In both cases the vertex operators are constructed in terms of free two-dimensional fields, which obey the same algebra (in the unbroken subgroup) as in the flat-space theory. So, for example, the operator product expansions of these gauge boson vertex operators are unaltered. There are constructions where unification does not hold. They involve replacing the two-dimensional fermions with a current algebra having a different central extension.

In the (2, 1) flat directions considered above we can give an argument, based on the low-energy field theory, that the couplings remain unified as one moves out along the flat direction. A change in the coupling requires that there be a coupling of this modulus to the gauge fields. But, at the classical level, we know that there are no such couplings because any such coupling would violate the axion shift symmetry. This symmetry is unaffected by the expectation value of these moduli.

When we come to consider strongly coupled strings, the problem of coupling unification will be more complicated. It will be less clear in what sense unification is generic. Whether this is a problem for the theory, or a clue to a way forward, is a question for the student to ponder.

26.6.8 Calculating the parameters of the low-energy Lagrangian

As we have explained, on the one hand string theory is a theory without fundamental dimensionless parameters. On the other hand, the structure of the low-energy theory, as we now see, depends on discrete choices: which Calabi–Yau, orbifold etc.?; in how many dimensions?; with how much supersymmetry?; with which Wilson lines and continuous dynamical quantities, the moduli? For any given choice, at least classically it should be a straightforward problem to calculate the parameters of the low-energy theory.

It is easy to calculate the four-dimensional gauge couplings in terms of the ten-dimensional dilaton and the radius. We have already seen how this works for simple compactifications, and this carries over directly to the Calabi–Yau case since the vertex operators for the gauge fields are constructed in terms of two-dimensional fields, as in the orbifold or toroidal case.

The cosmological constant is another interesting quantity in the low-energy theory. At the classical level in the Calabi–Yau compactifications, it vanishes. This can be understood in a variety of ways. First, if we examine the solution of the ten-dimensional equations of motion, we see that since the Ricci tensor vanishes; there is no cosmological term. Second, in the two-dimensional conformal field theory the cosmological constant would give rise to a tadpole for the dilaton but this is forbidden by conformal invariance. Ultimately, the absence of a cosmological constant is inherent in the form of the solution: we have assumed that four dimensions are flat. We will see later that this is not necessary: string theory admits anti-de Sitter (AdS) spaces as well as Minkowski spaces as classical solutions.

From the perspective of trying to understand the Standard Model, a particularly important set of parameters is the set of Yukawa couplings. These can certainly be computed in string theory. In principle we should construct the vertex operators for the appropriate matter fields and then construct the required OPE coefficients or suitable scattering matrices. In practice this can often be short-circuited. In the orbifold models, for example, in the untwisted sectors we can read off the Yukawa couplings by dimensional reduction of the ten-dimensional Lagrangian. The scalar fields are components of the original ten-dimensional gauge fields A_i. Similarly, the fermions are components of the ten-dimensional gauginos. In the orbifold theory, alternatively it is not difficult to construct the vertex operators and to compute the required OPE coefficients.

In the Calabi–Yau case we have seen that, in the α' expansion, the superpotential is independent of R. So one can work at very large radius and pick out the leading contribution. To actually do the computation one can construct the zero modes of the scalar and spinor fields and substitute into the Lagrangian. A priori one might expect that this would be quite difficult, given that one does not have an explicit formula for the metric. But it turns out that the Yukawa couplings have a topological significance, and their values can be inferred by general reasoning. We will not have a particular use for explicit formulas here, but it is important to be aware of their existence.

26.6.9 Other perturbative heterotic string constructions

The quintic is just one of a large class of Calabi–Yau models which can be constructed. The exact number is not actually known. It is not even known, with certainty, whether the number of Calabi–Yau vacua is finite or infinite.

So, while we will not assess here the size of this space, there is clearly a large class of string solutions with gauge group identical to that of the Standard Model. These theories have varying numbers of generations, including both orbifold (or free-fermion) models and Calabi–Yau constructions with three generations. There are many models with groups, numbers of generations, and other features radically different from those of the Standard Model. Still, it is remarkable how easily we have obtained models which accord with some of our speculations for Beyond the Standard Model physics. We have found low-energy supersymmetry, coupling unification, light Higgs particles and discrete symmetries which can potentially suppress proton decay and give rise to a stable dark matter candidate, all in a framework where we can imagine that real calculations are possible.

In subsequent chapters we will turn to the problems of actually turning these observations and discoveries into a real theory which we can confront with experiment.

Suggested reading

Volume 2 of Green *et al.* (1987) provides a comprehensive introduction to Calabi–Yau compactification, and I have borrowed heavily from their presentation. Weakly coupled string models with three generations have been constructed in the context of Calabi–Yau compactification; their phenomenology is considered by Greene *et al.* (1987). Models based on free fermions were been constructed by Faraggi (1999). We will encounter non-perturbative constructions in Chapter 28. At special points in their moduli spaces, some Calabi–Yau spaces can be described in terms of solvable conformal field theories. This program was initiated by Gepner (1987) and is described at some length by Polchinski (1998). A very accessible description, including computations of physically interesting couplings, appears in Distler and Greene (1988).

Exercises

(1) Write down the field strength of electrodynamics as a two-form and express its gauge invariance in the language of forms. Verify that $dF = 0$ is equivalent to the Bianchi identity (the homogeneous Maxwell equations).

(2) Show that, for a Kahler manifold, the non-vanishing components of the affine connection (Christoffel symbols) are given by Eq. (26.36). Then show that the non-zero components of the Riemann tensor are given by Eq. (26.37) and verify Eq. (26.38). Derive Eq. (26.40) by noting that

$$\Gamma^{\bar{a}}_{\bar{b}\bar{a}} = \partial_{\bar{b}} \ln \det g. \tag{26.76}$$

Show that our result for the two-dimensional curvature of a Riemann surface is a special case of this.

(3) For a flat two-dimensional torus, introduce complex coordinates and verify that the bosonic and fermionic terms are just those of the free string action. You can take $K = X^\dagger X$ for this case.

(4) Write out in some detail the action of the heterotic string propagating in the Calabi–Yau background with spin connection equal to the gauge connection. Determine the form of the vertex operators for the 27 and $\overline{27}$ fields, in the RNS formulation (you can limit yourself to the NS–NS sector).

(5) Exhibit a combination of Wilson lines and $SU(5)$ singlet expectation values which breaks the gauge group to that of the Standard Model in the case of the quintic in CP^4.

27 Dynamics of string theory at weak coupling

In previous chapters we have seen that string theory at the classical level shows promise of describing the Standard Model and can realize at least one scenario for the physics beyond: low-energy supersymmetry. But there are many puzzles, most importantly the existence of moduli and the related question of the cosmological constant. At tree level, in the Calabi–Yau solutions the cosmological constant vanishes. But whether this holds in perturbation theory and beyond requires an understanding of the quantum theory.

In studying string theory, we have certain tools:

1. weak coupling expansions;
2. long-wavelength (low-momentum, α') expansions.

We have exploited both these techniques already. In analyzing string spectra we worked in a weak coupling limit. There are corrections to the masses and couplings, for example; in string perturbation theory all but a few states that we have studied have finite lifetimes. At weak coupling these effects are small, but at strong coupling the theories will presumably look dramatically different.

In asserting that Calabi–Yau vacua are solutions of the string equations, we used both the above types of expansion. We wrote down the string equations both in lowest order in the string coupling and also with the fewest number of derivatives (two). Even at weak coupling and in the derivative expansion, we can ask whether Calabi–Yau spaces are actually solutions of the string equations, both classically and quantum mechanically. For example, we have seen that, at lowest order in both expansions, there are typically many massless particles. We might expect tadpoles to appear for these fields, both in the α' expansion and in loops. There is in general no guarantee that we can find a sensible solution by simply perturbing the original one.

Yet there are many cases where we can make exact statements. In both Type II and heterotic string theories, we can often show that Calabi–Yau vacua correspond to exact solutions of the classical string equations. We can also show that they are good vacua – there are no tadpoles for massless fields – to all orders of the string perturbation expansion. More dramatically, we can sometimes show that these vacua are good, non-perturbative, states of the theory. This is perhaps surprising since we lack a suitable non-perturbative formulation in which to address this question directly. The key to this magic is supersymmetry. In the framework of quantum field theory we have already seen that supersymmetry gives a great deal of control over the dynamics, both perturbative and non-perturbative. We were able to prove a variety of non-renormalization theorems from very simple starting points. The more that supersymmetry is involved, the stronger the results we could establish simply from symmetry considerations, without a detailed understanding of the dynamics. The same is true in string theory. We can easily prove a variety of non-renormalization theorems for string perturbation theory. We can show that with $N = 1$ supersymmetry in four dimensions the superpotential is not renormalized from its tree level form in perturbation theory; the gauge coupling functions are not renormalized beyond one loop. These same considerations indicate the sorts of non-perturbative corrections which can (and do) arise. In theories with more supersymmetries one can prove stronger statements: that the superpotential is not renormalized at all and that there are strong constraints on the kinetic terms. These sorts of results will be important when we try to understand weak–strong coupling dualities.

27.1 Non-renormalization theorems

In each superstring theory one can prove a variety of non-renormalization theorems. Consider, first, the case of ten dimensions. At the level of two derivative terms the actions with $N = 1$ or $N = 2$ supersymmetry (16 or 32 supercharges) are unique. So, *both perturbatively and non-perturbatively, there is no renormalization.* This is a variant of our discussion in four-dimensional field theories. If we dimensionally reduce the Type II theories on a six-dimensional torus, we obtain a four-dimensional theory with 32 supercharges ($N = 8$ in four dimensions); if we reduce the heterotic theory we obtain a theory with $N = 4$ supersymmetry in four dimensions (16 supercharges). In either case the supersymmetry is enough to prevent corrections to either the potential or the kinetic terms, not only perturbatively but non-perturbatively.

These are quite striking results. From this we learn that the question of whether the universe is four-dimensional or whether it has, say, four or eight supersymmetries, or none, is not simply a dynamical question (at least in the naive sense of comparing the energies of different states or their relative stability). Other issues, perhaps cosmological, must come into play. We will save speculations on these questions for later.

27.1.1 Non-renormalization theorems for world-sheet perturbation theory

Let us turn now to compactified theories. Consider first a Type II theory compactified on a Calabi–Yau space. In this case the low-energy theory has $N = 2$ supersymmetry. Again, this is enough to guarantee that there is no potential generated for the moduli, perturbatively or non-perturbatively. In other words, starting with a solution of the equations of the low-energy effective field theory, at lowest order in g_s and R^2/α', we are guaranteed that we have an exact solution to all orders – and non-perturbatively – in both parameters.

Now consider the compactification of the heterotic string theory on the same Calabi–Yau space, with spin connection equal to the gauge connection. Then the world-sheet theory, as we saw, has two left-moving and two right-moving supersymmetries. It is identical to the theory which describes the corresponding Type II background. But we have just established that the Calabi–Yau space is a solution of the classical string equations, which means that there is a corresponding superconformal field theory with central charge

$c = 9$. This is an exact statement, so the background corresponds to an exact solution of the classical string equations. This does *not* establish that the Calabi–Yau space corresponds quantum mechanically to an exact vacuum, as it does in the Type II case. For example, the intermediate states in quantum loops in the two theories are different.

We can establish this result in a different way. Consider the $h_{1,1}$ (1, 1)-forms $b^{(a)}_{i\bar{i}}$; one of these is the Kahler form, where $b_{i,\bar{i}} = g_{i,\bar{i}}$. In world-sheet perturbation theory we have seen that these fields decouple at zero momentum. The fact that all scattering amplitudes involving external b particles vanish at zero momentum has consequences for the structure of the low-energy effective Lagrangian: only derivatives of b appear in the Lagrangian. This is reminiscent of the couplings of Goldstone bosons; the Lagrangian, in world-sheet perturbation theory, is symmetric under

$$b(x) \to b(x) + \alpha \tag{27.1}$$

for constant α. We will refer to fields exhibiting such perturbative shift symmetries quite generally as *axions*.

This result implies a non-renormalization theorem for sigma-model perturbation theory; b lies in a supermultiplet with r^2, the modulus which describes the size of the Calabi–Yau space. This is apparent from the fact that they are both Kaluza–Klein modes associated with the metric $g_{i,\bar{i}}$; r^2 is the symmetric part and b is the antisymmetric part. So this is similar to the situation in which we could prove non-renormalization theorems in field theory. Different orders of sigma model perturbation theory are associated with different powers of r^{-2}. But, in holomorphic quantities such as the superpotential and gauge coupling function, additional powers of r^{-2} are accompanied by powers of b. So only terms which are independent of r^{-2} are permitted by the shift symmetry. As a result, the superpotential computed at lowest order is not corrected in sigma model perturbation theory. This means that particles which are moduli at the leading order in α' are moduli to all orders of sigma model perturbation theory.

This non-renormalization theorem does not quite establish that these are good solutions of the classical string theory; there is still the possibility that non-perturbative effects in the sigma model will give rise to potentials for the lowest-order moduli. Indeed, our argument for the vanishing of the b couplings is not complete. At zero momentum the vertex operator for b, V_b, is topological; while it is the integral of a total divergence, it does not necessarily vanish. There generally exist classical Euclidean solutions of the two-dimensional field theory – instantons – for which the vertex operator is non-zero. These world-sheet instantons raise the possibility that non-perturbative effects on the world sheet will lift some of or all the vacuum degeneracy. For the (2, 2) theories, however, we already know that this does not occur. Earlier we argued, by considering the compactification of the related Type II theories, that the corresponding sigma models are exactly conformally invariant. It is possible (and not terribly difficult), by examining the structure of the two-dimensional instanton calculation (i.e. for the "world-sheet instanton") to show that no superpotential is generated. While we will not review this analysis here, the techniques involved are familiar from our discussion of four-dimensional instantons. One wants to determine whether instantons can generate a superpotential. It is necessary, as in four dimensions, to count the fermion zero modes and see whether they can lead to a non-vanishing correlation function at zero momentum for an appropriate set of fields. In the (2, 2) case one finds that they cannot. One can then ask whether quantum corrections (i.e. due to small fluctuations) to the instanton result can yield such a correction. Here one notes that, as in perturbation theory, holomorphy fixes uniquely the dependence on the coupling. So if the lowest-order contribution vanishes, higher orders will vanish as well.

In the case of (2, 0) compactifications of the heterotic string the situation is more complicated. Perturbatively, we can argue, as before, that solutions of the string equations at lowest order are solutions to all orders in the α' expansion. Non-perturbatively, however, the situation is less clear. For such compactifications there is no corresponding Type II compactification, so we can not rely on the magic of $N = 2$ supersymmetry; it is necessary to examine in detail the effects of world-sheet instantons. In general, if one does the sort of zero-mode counting described above then one finds that it is possible to generate a superpotential. But in many cases one can argue that there are cancelations, and the superpotential vanishes.

It is important to understand that the non-renormalization theorems do not imply that the Calabi–Yau manifold is itself an exact solution to the classical string equations; rather, the point is that a solution is guaranteed to exist nearby. There can be – and are – tadpoles for massive particles in sigma model perturbation theory. A tadpole corresponds to a correction of the equations of motion as follows:

$$\nabla^2 h + m^2 h = \Gamma. \tag{27.2}$$

This is solved by a perturbatively small shift in, the h field;

$$h = -\frac{\Gamma}{m^2}. \tag{27.3}$$

For the massless fields, however, one cannot find a solution in this way, and in general, if there is a tadpole, there is no nearby (static) solution of the equations. This is why the low-energy effective action is such a useful tool in addressing such questions: it is precisely the tadpoles for the massless fields which are important.

27.1.2 Non-renormalization theorems for string perturbation theory

In field theory we proved non-renormalization theorems by treating couplings as background chiral fields and exploring the consequences of the holomorphy of the effective action as a function of these fields. In string theory we have no coupling constants, but the moduli determine the effective couplings and, since they are themselves fields, they are restricted by the symmetries of the theory. We exploited this connection in the previous subsection to prove non-renormalization theorems for sigma model perturbation theory. In this subsection we prove similar statements for string perturbation theory.

We begin with the heterotic string theory, on a Calabi–Yau manifold or an orbifold. In this case we have seen that there is a field S which we called the dilaton (it is sometimes called the *four-dimensional dilaton*). The vertex operator for the imaginary part of this field, $a(x)$, at $k = 0$ is simply

$$V_a = \int d^2\sigma\, \epsilon^{ab}\, \partial_a X^\mu \partial_b X^\nu b_{\mu\nu}. \tag{27.4}$$

This is, again, a total derivative on the world sheet. So this particle, which we saw earlier is an axion, decouples at zero momentum. Again there is a shift symmetry – this is just the axion shift symmetry. Again, this means that the superpotential must be independent of S. But, since powers of perturbation theory come with powers of S, this establishes that the superpotential is not renormalized to all orders of perturbation theory!

As in the world-sheet case there can be non-perturbative corrections to the superpotential, and this raises the possibility that potentials will be generated for the moduli. We will see shortly that gluino condensation, as in supersymmetric field theories, is one such effect.

First, we consider other string theories. In the case of Type II compactified on a Calabi–Yau space, the $N=2$ supersymmetry is enough to ensure that no superpotential is generated perturbatively or non-perturbatively: Calabi–Yau spaces correspond to exact ground states of the theory, and the degeneracies are exact as well. As in field theories with $N = 2$ supersymmetry, corrections to the metric (the Kahler potential (26.44)) are possible. Theories with more supersymmetry (heterotic on tori, or Type II theories on $K3$ with $N = 4$ supersymmetry or Type II theories on tori with eight supersymmetries) are even more restricted.

27.2 Fayet–Iliopoulos *D* terms

In deriving non-renormalization theorems for string perturbation theory, we established that there is no renormalization of the superpotential or of the gauge coupling function beyond one loop. But this is not quite enough to establish that there is no renormalization of the *potential*. We must also check whether Fayet–Iliopoulos terms are generated. From field-theoretic reasoning we might guess that any renormalization would occur only at one loop. In globally supersymmetric theories in superspace, a Fayet–Iliopoulos term has the form

$$\zeta^2 D = \int d^4\theta\, V. \tag{27.5}$$

This term is just barely gauge invariant: under $V \to V + \Lambda + \Lambda^\dagger$ it is invariant because $\int d^4\theta \Lambda = 0$ since Λ is chiral. If we treat the gauge coupling (or any other couplings) as a background field, any would-be corrections to D would have the form

$$\int d^4\theta\, g(S, S^\dagger) V, \tag{27.6}$$

which is only invariant if g is a constant. Thus any D term is independent of the coupling, in the normalization where $1/g^2$ appears in front of the gauge terms. So at most there is a one-loop correction.

Before going on to string theory, it is interesting to look at the structure of any one-loop term. Call the associated $U(1)$ generator Y. If the supersymmetry is unbroken then massive

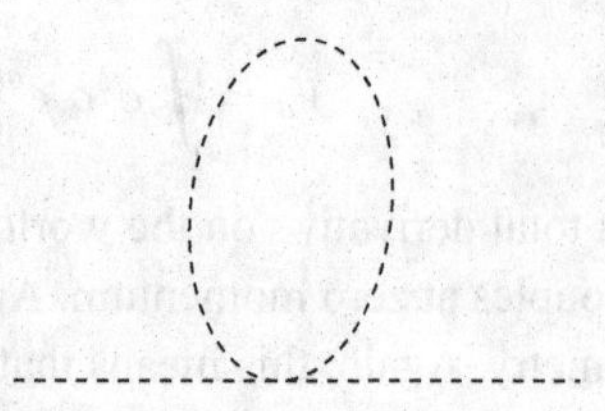

Fig. 27.1 The Feynman diagram which contributes to the D term.

fields come in pairs with opposite values of Y, so only massless fields contribute. The Feynman diagram which contributes to the D term is shown in Fig. 27.1. It is given by:

$$\zeta^2 = \operatorname{Tr} Y \int \frac{d^4k}{(2\pi)^4} \frac{1}{k^2}. \tag{27.7}$$

So a vanishing D term requires that the trace of the $U(1)$ generator vanish. The one-loop diagram is quadratically divergent, but let us rewrite Eq. (27.7) in a way which resembles expressions we have seen in string theory. We can introduce a "Schwinger parameter," which we will call τ_2. Then

$$\begin{aligned} \zeta^2 &= 2\pi \operatorname{Tr} Y \int_0^\infty d\tau_1 \int \frac{d^4k}{(2\pi)^4} e^{-2\pi\tau_2 k^2} \\ &= \frac{1}{32\pi^3} \operatorname{Tr} Y \int_0^\infty \frac{d\tau_2}{\tau_2^2} \int_{-1/2}^{1/2} d\tau_1. \end{aligned} \tag{27.8}$$

We have written the expression in this way because we want to consider it as an integral over the modular parameter of the torus. At this stage the integral is still quadratically divergent. But, under modular transformations, the complex τ plane is mapped into itself an infinite number of times. We can define a *fundamental domain*,

$$-\frac{1}{2} \leq \tau_1 \leq \frac{1}{2}, \quad |\tau| \geq 1. \tag{27.9}$$

If we restrict the integration to the fundamental domain, the result is finite. In string theories, the correct answer terms out to be

$$\zeta^2 = \frac{1}{192\pi^2} \operatorname{Tr} Y. \tag{27.10}$$

This result can be derived by a straightforward string computation. However, in string models where $\operatorname{Tr} Y$ is non-zero we can give a low-energy field theory argument which completely fixes the coefficient of the D term and also sheds light on possible perturbative corrections. If $\operatorname{Tr} Y \neq 0$, the low-energy theory has a gravitational anomaly. This anomaly is rather similar to the gauge anomalies we discussed in the context of field theory. It arises from a diagram with one external gauge boson and an external graviton. String models with such anomalies typically have gauge anomalies as well, which we can readily evaluate. As an example, consider the compactification of the $O(32)$ heterotic string on a Calabi–Yau space, with spin connection equal to the gauge connection. In this case the low-energy gauge group is $SO(26) \times U(1)$. There are $h_{1,1}$ 26s with $U(1)$ charge 1, and $h_{2,1}$ 26s with $U(1)$ charge -1. There are also corresponding singlets, with charges $+2$ and -2

respectively. These are in precise correspondence with the fields we found in E_6; the 26s arise in parallel to the $O(10)$ 10s and the singlets arise in parallel to the $O(10)$ singlets. But now it is clear that there are anomalies in the gauge symmetries. For example, there is a $U(1) \times O(26)^2$ anomaly proportional to

$$A = h_{2,1} - h_{1,1} \tag{27.11}$$

and a $U(1)^3$ anomaly given by

$$A' = (h_{2,1} - h_{1,1})(26 - 8). \tag{27.12}$$

This is, however, a modular invariant configuration of string theory, so there should not be any inconsistency, at least in perturbation theory. Therefore something must cancel the anomaly. The cancelation is actually a variant of the mechanism discussed originally by Green and Schwarz in ten dimensions, now specialized to four dimensions. We know that there is a coupling:

$$\int d^2\theta \, S W_\alpha^2. \tag{27.13}$$

This gives rise to a coupling of the axion to the $F\tilde{F}$ terms of each group. The anomaly calculation in the low-energy theory implies a variation of the action proportional to the anomaly coefficient and $F\tilde{F}$. So, if the axion transforms under the gauge symmetry as

$$a(x) \rightarrow a(x) + c\omega(x) \tag{27.14}$$

then this can cancel the anomaly. It is crucial that the anomaly coefficients are the same for each group.

We can check whether this hypothesis is correct. If $a(x)$ transforms as above then it must couple to the gauge field. The required covariant derivative is

$$D_\mu a = \partial_\mu a - \frac{1}{c} A_\mu. \tag{27.15}$$

So, from the kinetic term in the action there is a coupling of A_μ to a. One can compute this coupling without great difficulty and verify that it has the required magnitude.

More interesting, however, is to consider the implications of supersymmetry. We can generalize the coupling above to superspace. The transformation law for a now becomes a transformation law for S:

$$S \rightarrow S + \Lambda + \Lambda^\dagger, \tag{27.16}$$

where Λ is the chiral gauge transformation parameter. The gauge-invariant action for S is

$$-\int d^4\theta \, \ln\left(S + S^\dagger - \frac{1}{c} V\right). \tag{27.17}$$

If we expand this Lagrangian in a Taylor series, we see that, in addition to the $A_\mu \partial^\mu a$ coupling, we generate a Fayet–Iliopoulos D term,

$$\int d^4\theta \frac{1}{c(S + S^\dagger)} V. \tag{27.18}$$

One can verify that this term – and the other terms implied by this analysis – are present. First, we can ask: at what order in perturbation theory should each of these terms appear? To establish this we need to remember that the standard supergravity Lagrangian is written in a frame where M_p^2 appears in front of the Einstein term in the effective action. In the string frame, it is the dilaton – essentially S – which appears in front. If we rescale the four-dimensional metric according to

$$g_{\mu\nu} \to S g_{\mu\nu} \tag{27.19}$$

then S appears in front of the Lagrangian. With this same rescaling, the "kinetic" term, which had an S in front, has S^3. The Fayet–Iliopoulos D term, originally had a $1/S$ in front. Correspondingly, the resulting scalar mass term would be proportional to $1/S^2$. After the metric rescaling this would be independent of S; that is, in the heterotic string theory the D term should appear at one loop, in accord with our field theory intuition. Similarly, the coupling $A_\mu \partial^\mu a$ should appear at one loop, while there should be a contribution to the cosmological constant at two loops. All these results can be found by straightforward string computations (some of them are described in the suggested reading).

In essentially all the known examples this one-loop D term does not lead to supersymmetry breaking. There always seem to be fields which cancel the D term. Consider, again, the $O(32)$ theory. Here we can try to cancel the D term by giving an expectation value to one of the singlets, 1_{-2}. The question is whether this gives a non-zero contribution to the potential when we consider the superpotential. The most dangerous coupling is a term $1_{-2}1_{+2}$ involving some other singlet. But such terms are absent at lowest order, and their absence to higher orders is guaranteed by the non-renormalization theorems. Charge conservation forbids terms of the form 1^n_{-2}; there are no other dangerous terms. So this corresponds to an exact "F-flat" direction of the theory, in which all F vevs vanish. So, in perturbation theory there exists a good vacuum. While a general argument is not known, empirically this possibility for cancelation appears to arise in every known example.

What does the theory look like in this new vacuum?

1. Supersymmetry is restored and the vacuum energy vanishes.
2. The $U(1)$ gauge boson has a mass-squared of order g_s^2 times the string scale.
3. The longitudinal mode of the gauge boson is principally the imaginary part of the charged scalar field whose vev canceled the D term. There is still a light axion.

From the perspective of a very low energy observer, the D term is not a dramatic development. It plays some role in determining the physics at a very high energy scale (albeit not quite as high as the string scale). What is perhaps most impressive is the utility of effective-field-theory arguments in sorting out a microscopic string problem. Prior to the discovery of the D term, for example, there had been many papers "proving" a strict non-renormalization theorem for the *potential*; this, we see, is not correct (it is not hard to determine, in retrospect, what went wrong in the original proofs). The effective-field-theory arguments make clear when the potential is renormalized in perturbation theory and when it is not. They also permit one to easily find the "new vacuum" in cases where a Fayet–Iliopoulos term appears. It is possible, in principle, to find this vacuum by string methods, but this is distinctly more difficult. Finally, these arguments give insight into the non-perturbative fate of the non-renormalization theorems.

27.3 Gaugino condensation: breakdown of axion shift symmetries beyond perturbation theory

We have seen that, in string theory, if supersymmetry is unbroken at tree level in some particular constructions then it is unbroken to all orders of perturbation theory. The argument, as in field theory, allows exponential dependence on the coupling. In the case of a heterotic string compactified on a Calabi–Yau space, gaugino condensation, as in supersymmetric field theories, generates a superpotential on the moduli space.

Consider the $E_8 \times E_8$ theory compactified on a Calabi–Yau space, with spin connection equal to the gauge potential and without Wilson lines. In this case there is an $E_6 \times E_8$ gauge symmetry. There are typically several fields in the 27 of E_6, but there are no chiral fields transforming in the E_8. One has a pure E_8 supersymmetric gauge theory. The couplings of the E_6 and E_8 are equal at the high scale, so the E_8 coupling becomes strong first. This leads, as we have seen, to gaugino condensation. We have also seen that at tree level there is a coupling

$$SW_\alpha^2. \tag{27.20}$$

Just as before, this leads to a superpotential for S,

$$W(S) = Ae^{-3S/b_0}. \tag{27.21}$$

One often hears this described as a "field theory analysis," as if it is not necessarily a feature of the string theory. But string theory obeys all the principles of quantum field theory. If we correctly integrate out the high-energy string effects then *the low-energy analysis is necessarily reliable.* So the only question is: are there terms in the low-energy effective action that lead to larger effects? One might worry that, since we understand so little about non-perturbative string theory, it would be hard to address this. But, with some very mild assumptions, we can establish that *the low-energy effects are parametrically larger than any high-energy string effects.*

The basic assumption is that, as in field theory, non-perturbatively the theory obeys a discrete shift symmetry (for a suitable normalization of a):

$$a(x) \to a(x) + 2\pi. \tag{27.22}$$

When we discuss non-perturbative string theory, we will give some evidence for this assumption; it will turn out to be one of the milder assertions on the subject of string duality. For now, we note that if we accept this assumption then, any superpotential for S arising from high-energy string effects will be of the form

$$W_{\text{np}} = C_n e^{-nS} \tag{27.23}$$

for integer n. So, such effects are exponentially smaller than gaugino condensation.

What does the low-energy theory look like? The dilaton potential goes rapidly to zero for large S, i.e. in the weak coupling limit. We might have hoped that somehow we would find that supersymmetry is broken and the moduli fixed. But, instead, gaugino condensation leads to a runaway potential. At large S we have just argued that no additional string effects can stabilize this behavior.

We can imagine more elaborate versions of this phenomenon, involving matter fields as well, in some sort of hidden sector. But it is difficult to construct models where the moduli are stabilized in any controlled fashion along these lines.

27.4 Obstacles to a weakly coupled string phenomenology

We have seen that string theory is a theory without dimensionless parameters. This is an exciting prospect, but it also raises the question: how are the parameters of low-energy physics then determined? We have argued that the answer to this question lies in the dynamics of the moduli: the expectation values of these fields determine the couplings in the low-energy Lagrangian.

In non-supersymmetric string configurations, perturbative effects already lift the degeneracy among different vacua, giving rise to a potential for the moduli. In the previous section we have learned that in supersymmetric compactifications non-perturbative effects generically lift the flat directions of the potential. In other words, the moduli are not truly moduli at the quantum level. At best, we can speak of approximate moduli in regions of the field space where the couplings are weak. The potentials, both perturbative and non-perturbative, all tend to zero at zero coupling. This is not surprising; with a little thought it becomes clear that this behavior is not specific to perturbation theory or some particular non-perturbative phenomenon such as gaugino condensation; at very weak coupling, we expect that the potential always tends rapidly to zero. This means that if the potential has a minimum, this occurs when the coupling is not small. This is troubling, for it means that it is likely to be hard – if possible at all – to do computations which will reveal detailed features of the state of string theory which describes the world we see around us.

In the next chapter we will see that much is known about non-perturbative string physics. Most striking is a set of dualities which relate regimes of very strong coupling in one string theory to weak coupling in another. While impressive, these by themselves do not help with the strong coupling problem we have elucidated above; if, at very strong coupling, the theory is equivalent to a weakly coupled theory then the potential will again tend to zero. In other words, it is likely that stable ground states of string theory exist only in regions where no approximation scheme is available.

Perhaps just as troubling is the problem of the cosmological constant. Neither perturbative nor non-perturbative string theory seems to have much to say. The potentials are more or less of the size one would guess from dimensional analysis (and the expected dependence on the coupling). Perhaps most importantly they are, up to powers of the coupling, as large as the scale set by supersymmetry breaking.

There are, however, some reasons for optimism. Perhaps the most important is provided by nature itself: the gauge and Yukawa couplings of the Standard Model are small. Another is provided by string theory. As we will discuss later, there are ways in which large pure numbers can arise dynamically in the theory. These might provide mechanisms to understand the smallness of couplings, even in situations where asymptotically the potential vanishes. Finally, we will see that there is, at present, only one proposal to understand the smallness of the cosmological constant, and string theory may provide a realization of this suggestion.

Suggested reading

The result that there are no continuous global symmetries in string theory is fundamental. For the heterotic theory, it appears in Banks and Dixon (1988). Non-renormalization theorems for world-sheet perturbation theory and issues in the construction of $(0, 2)$ models were described by Witten (1986) and by Green *et al.* (1987). The non-renormalization theorem for string perturbation theory is described in Dine and Seiberg (1986). The space–time argument for the Fayet–Iliopoulos D term appears in Dine *et al.* (1987c); world-sheet computations appear in Atick *et al.* (1987) and Dine *et al.* (1987a). World-sheet instantons are discussed in Dine *et al.* (1986, 1987b); cancelations of instanton effects relevant to $(0, 2)$ theories were studied by Silverstein and Witten (1995).

28 Beyond weak coupling: non-perturbative string theory

In the previous chapter we were forced to face the fact that on the one hand string theory, if it describes nature, is not weakly coupled. On the other hand, the very formulation of the theory that we have put forward is perturbative. We have described the quantum mechanics of single strings and given a prescription for calculating their interactions order by order in perturbation theory in a parameter g_s. There is a parallel here to Feynman's early work on relativistic quantum theory: Feynman guessed a set of rules for computing the perturbative amplitudes of electrons. In that case, however, one already had a candidate for an underlying description: quantum electrodynamics. It was Dyson who clarified the connection. For Abelian theories a non-perturbative approach probably does not exist, but in the case of non-Abelian gauge theories it does. The field-theoretic formulation provides an understanding of the underlying symmetry principles and access to a treasure trove of theoretical information.

A string field theory would be a complicated object. The string fields themselves would be functionals of the classical two-dimensional fields which describe the string. The quantization of such fields is sometimes called the "third quantization." Much effort has been devoted to writing down such a field theory. For open strings one can obtain relatively manageable expressions which reproduce string perturbation theory. For closed strings, infinite sets of contact interactions are required. But, quite apart from their cumbersome structure, there are reasons to suspect that this is not a useful formulation. There would seem to be, for example, vastly too many degrees of freedom. At one loop we have seen that the string amplitudes are to be integrated only over the fundamental region the moduli space. Naively, a field theory which simply describes all of the states of the string would have amplitudes integrated over the whole region, and the cosmological constant would be extremely divergent. The contact interaction terms mentioned above solve this problem but not in a very satisfying way.

Despite this, there has been great progress in understanding the non-perturbative aspects of the known string theories. Most strikingly, it is now known that all theories with 16 or more supersymmetries are the same. Many tools have been developed to study phenomena beyond string perturbation theory, especially *D*-branes and supersymmetry. There exist some cases where non-perturbative formulations of string theory are possible, and we will discuss them briefly in this chapter. They are technically and conceptually much simpler than string field theory. They have a puzzling, perhaps disturbing feature, however: they are special to strings propagating in particular backgrounds. It is as if, in Einstein's theory, for each possible geometry one had to give a different Hamiltonian. All these results are "empirical." They have been developed by collecting circumstantial evidence on a case-by-case basis. There is still much that is not understood. In Chapter 31, we will discuss how this developing understanding might lead to a closer connection of string theory and nature.

28.1 Perturbative dualities

Before considering examples of weak–strong coupling dualities, we return to the large-radius–small-radius duality (T-duality) we studied in Section 25.3; many dualities that we will study have a similar flavor to this, even though they cannot be demonstrated so directly. Thus we saw that there is an equivalence of the heterotic string theory at small radius to the theory at large radius. By examining the action of these transformations at their fixed points, we saw that these duality symmetries are gauge symmetries. We could ask, as well, the significance of duality transformations in the IIA and IIB theories. As with other closed strings, in addition to transforming the radii, the duality transformation is as follows:

$$\partial X^9 \to -\partial X^9, \quad \bar{\partial} X^9 \to \bar{\partial} X^9. \tag{28.1}$$

Because of the world-sheet supersymmetry, the transformation has the same action on the fermions: $\psi^9 \to -\psi^9$; $\tilde{\psi}^9 \to \tilde{\psi}^9$. But, under this, the chirality operator appearing in the GSO projector is reversed in sign, i.e. duality interchanges the IIA and IIB theories: *the small-radius IIA theory is equivalent to the large-radius IIB theory* and vice versa. There are other weak coupling connections between string theories. For example, the compactified $O(32)$ heterotic string theory is equivalent to the $E_8 \times E_8$ theory.

28.2 Strings at strong coupling: duality

Duality is a term used in physics to label different descriptions of the same physical situation. At the level of perturbation theory we have learned about five apparently different string theories. On the basis of on the perturbative dualities discussed above, we see that there are at most three inequivalent string theories, the Type I, Type II and heterotic theories. But it is tempting to ask whether there are more connections between the theories. In this chapter we will see that all the known string theories are equivalent in a similar way, but these equivalences relate small and large coupling. For example, the strong coupling limit of the $O(32)$ heterotic string theory is the weak coupling limit of the Type I string theory; the strongly coupled limit of the $E_8 \times E_8$, compactified to six dimensions on a torus, is the weakly coupled limit of the Type II theory compactified on a K3 manifold (K3 manifolds are essentially four-dimensional Calabi–Yau spaces); the ten-dimensional Type II theory is self-dual and, perhaps most intriguingly of all, the strong coupling limit of the Type IIA theory in ten dimensions is described, at low energies, by a theory whose low-energy limit is eleven-dimensional supergravity.

Lacking a non-perturbative formulation of the theory, the evidence for these connections is necessarily circumstantial. While circumstantial, however, it is compelling. All the evidence relies on supersymmetry. We will not be able to review it all here but will try to give the flavor of some of the arguments. Supersymmetry, especially supersymmetry with 16 or 32 supercharges, allows one to write down a variety of exact formulas, for Lagrangians (based on strong non-renormalization theorems) and for spectra (based on BPS formulas), which can be trusted in both weak and strong coupling limits. This allows detailed tests of the various dualities.

28.3 *D*-branes

When we discussed strong–weak (electric–magnetic) dualities in field theory, topological objects played a crucial role. The same is true in string theory, where the solitons are various types of branes. In general, a *p-brane* is a soliton with a $(p+1)$-dimensional world volume, so a 0-brane is a particle, a 1-brane is a string, a 2-brane is a membrane and so on. One might construct these by solving complicated non-linear differential equations. But a large and important class of topological objects can be uncovered in string theory in a different – and much simpler – way. These are the *D*-branes. These branes fill an important gap in our understanding of the Type I and Type II theories. In these theories we encountered gauge fields in the Ramond–Ramond sectors: two-forms in Type I, one-forms and three-forms in Type IIA, zero-forms, two-forms, and four-forms in Type IIB. One natural question is: what are the charged objects that couple to these fields? They are not within the perturbative string spectrum. The vertex operators for these fields involve the gauge-invariant field strengths only, so in perturbation theory there are no objects with minimal coupling. The answer is that they are *D*-branes. Their masses (*tensions*) are proportional to $1/g_s$, so at weak coupling they are very heavy. This is why they are not encountered in the string perturbation expansion.

When we discussed open strings we noted that there are two possible choices of boundary condition: Neumann and Dirichlet. At first sight, Neumann boundary conditions appear more sensible; Dirichlet boundary conditions would violate translational invariance, implying that strings end at a particular point or points. But we have already encountered violations of translational invariance within translationally invariant theories: solitons, for example magnetic monopoles or higher-dimensional objects such as cosmic strings or domain walls. Admitting the possibility of Dirichlet boundary conditions for some of or all the coordinates leads to a class of topological objects known as *D*-branes (for Dirichlet branes). If $d-p-1$ of the boundary conditions are Dirichlet while $p+1$ are Neumann, the system is said to describe a *Dp*-brane.

We can be quite explicit. We start with the bosonic string. For the Neumann directions we have our previous open-string mode expansion of Eq. (21.16). For the Dirichlet directions we have:

$$X^I = x_0^I + i\sum_{n\neq 0} \frac{1}{n}\alpha_n^I e^{-in\tau} \sin n\sigma, \quad I = 1, d-p-1. \tag{28.2}$$

Note that there are no momenta associated with the Dirichlet directions. The x_0^Is should be thought of as collective coordinates. We will argue shortly that the tension of the branes is proportional to M_s^{p+1}/g_s.

Consider an extreme case, that of a $D0$-brane. There are 25 collective coordinates and no momenta, so this object is a conventional soliton. In field theory the excitations near the soliton, which describe the scattering of mesons (field theory excitations) from the soliton must be found by studying the eigenfunctions of the quadratic fluctuation operator. But here they are very simple: they are just the excitations of the open string. As a second example, consider a $D3$-brane. Now the momentum has four components, so the excitations which propagate on the brane are four-dimensional fields. These break up into two types. The Neumann fields X^μ give rise to a massless gauge boson, the state $\alpha_{-1}^\mu|0\rangle$; the Dirichlet fields X^I give rise to massless scalars on the brane $\alpha_{-1}^I|0\rangle$. In the superstring version of this construction there are six scalars, a gauge boson and their superpartners. In $N=1$ language this amounts to a vector multiplet and three chiral multiplets, the content of $N=4$ Yang–Mills theory with gauge group $U(1)$.

Before considering some of these statements in greater detail, let us explore a further aspect of this construction. Suppose that we have several branes, say $D3$-branes, parallel to each other; here, "parallel" just means that the strings which end on these branes have Dirichlet or Neumann boundary conditions for the same coordinate. Now, however, we have the possibility that the strings end on different branes. Take the simplest case of two branes. If the branes are separated by a distance r, in addition to the modes above, labeled by the collective coordinate $x_i^I, i=1,2$, we have to allow for expansions of the form

$$X^I(\sigma,\tau)=x_i^I+\sigma\frac{r}{\pi}\left(x_j^I-x_i^I\right)+i\sum_{n\neq 0}\frac{1}{n}\alpha_n^I e^{-in\tau}\sin n\sigma,\quad I=1,\ldots,d-p-1. \tag{28.3}$$

There are two such configurations, one starting on the first brane and ending on the second and one starting on the second brane and ending on the first. The ground states in these sectors have mass-squared proportional to r^2. For $r\neq 0$, all these states are massive. The massless bosons consist of a $U(1)$ gauge boson on each brane, as well as scalars. As $r\to 0$, we have two additional massless gauge bosons. If we generalize to n branes, we have n massless gauge bosons and $6n$ scalars; as we bring the branes close together, we have n^2 gauge bosons and $6n^2$ scalars.

There is a natural conjecture as to what is going on here. When all the branes coincide we have a $U(n)$ gauge symmetry, with three complex scalars transforming in the adjoint representation of the group. As the branes are separated, the adjoint scalars acquire (commuting) expectation values; these break the gauge symmetry to $U(1)^n$, giving mass to the other gauge bosons. In principle we would like to check that these n^2 gauge bosons interact as required for Yang–Mills theories, as we did for the gauge bosons of the heterotic string. This is more challenging here, since we need vertex operators which connect strings ending on different branes, and we will not attempt this. We will provide further evidence for the correctness of this picture shortly.

The branes break some of the supersymmetry of the Type II theory in infinite space; instead of 32 conserved supercharges there are 16. A simple way to understand this uses the light cone gauge construction. There are now open strings ending on the brane. For the world-sheet fermions, the boundary conditions relate the left and right movers on the string. Calling these S_a and $\tilde{S}_a$, we have

$$S^a(\sigma,\tau) = \sum_n S_n^a e^{-in(\tau+\sigma)}, \quad \tilde{S}^a(\sigma,\tau) = \sum_n S_n^a e^{-in(\tau-\sigma)}. \tag{28.4}$$

Recall that half the supercharges have the very simple form

$$Q^a = \int d\sigma\, S^a, \quad \tilde{Q}^a = \int d\sigma\, \tilde{S}^a, \tag{28.5}$$

so $Q^a = \tilde{Q}^a$. This is the structure of a broken supersymmetry generator, with S the goldstino. The other set supercharges is linearly realized. Other configurations, such as non-parallel sets, preserve less supersymmetry. Brane–anti-brane configurations preserve no supersymmetry at all.

We can imagine other sets of branes, which would respect different amounts of supersymmetry. If we have branes which are not parallel, for example, different sets of supersymmetries will be preserved. In order to count supersymmetries we need to compare the supersymmetries on different branes at different angles relative to one another.

28.3.1 Brane charges

We have seen that the simplest D-brane configurations preserve half the supersymmetries. In other words, they are BPS states. Typically BPS states are associated with conserved charges. In the case of the IIA and IIB theories, in the Ramond–Ramond sectors there are gauge fields but, in perturbation theory, no charged objects. Polchinski guessed – and showed – that the objects which carry Ramond–Ramond charges are D-branes. In the IIA case the gauge fields are a one-form and a three-form; in the IIB case they are a zero-form a two-form, and a (self-dual) four-form. In relativistic mechanics, a gauge field couples to a particle – a zero-brane. We have seen that a two-index tensor couples naturally to a string – a one-brane. So, this suggests that, in the IIA theory, there should be Dp-branes with p even, coupling to the corresponding R–R gauge fields, while in the IIB theory there should be Dp-branes with p odd. Polchinski verified this by direct calculation. He computed the one-loop amplitude for two separated branes. For large separations he found the poles associated with exchange of the massless gauge fields (more precisely, for fixed separation r one should see a falloff with powers of $1/r$). His calculation not only yields the brane charges, it also gives the brane tensions.

Consider the case of two branes, separated by a distance y. In empty flat space, the trace over states in the one-loop amplitude for open strings gives a generic scattering amplitude of the form

$$\mathcal{A} = C \int_0^\infty \frac{dt}{t^2}. \tag{28.6}$$

The power of t arises from the momentum integral $\int d^8k \exp(-k^2)$, as well as from manipulation of the oscillator traces. The main difference in the case of two separated branes is that the mass-squared has a contribution y^2, from the brane separation y, and $9-p$ coordinates of the brane are fixed, so they do not have associated momenta. So, the result has the form

$$\mathcal{A} = C\int_0^\infty \frac{dt}{t^2}(8\pi^2\alpha' t)^{(9-p)/2}\exp\left(-\frac{ty^2}{2\pi\alpha'}\right)$$
$$\sim y^{-(7-p)} \sim G_{9-p}(y). \tag{28.7}$$

Here $G_d(y)$ is the scalar Green's function in d dimensions. So, one can think of a potential between the branes associated with the exchange of massless states. These massless states include antisymmetric tensor fields and their superpartners as well as gravitons and gravitinos. These contributions can be isolated, and the tensions and charges of the D-branes determined. In the case a the superstring, the full potential vanishes due to boson and fermion cancelations.

28.3.2 Brane actions

We are familiar with the actions for zero-branes and one-branes. The action for a general p-brane is a generalization of these:

$$S_p = -T_p\xi\int d^{p+1}\xi\ \det\left(\frac{\partial X^\mu}{\partial\xi^a}\frac{\partial X^\nu}{\partial\xi^b}\eta_{\mu\nu}\right)^{1/2}, \tag{28.8}$$

where T_p is the brane tension. In the zero-brane case this is the action for a particle; $X^\mu(\tau)$ is the collective coordinate which describes the position of the soliton and T_0 is its mass. For a general background with a bulk metric, a dilaton and an antisymmetric tensor field this generalizes to

$$S_p = -T_p\int d^{p+1}\xi\ e^{-\Phi}[-\det(G_{ab}+B_{ab}+2\pi\alpha' F_{ab})]^{1/2}. \tag{28.9}$$

The terms involving the metric and antisymmetric tensor are similar to those we have encountered elsewhere in string theory, and their form is not surprising. The factor $e^{-\Phi}$ arises because in the open-string sector the coupling constant is the square root of that for the closed-string sector.

28.4 Branes from the T-duality of Type I strings

There is another way to think about D-branes, which provides further insight. We have seen that closed-string theories exhibit a duality between large and small radius. In the heterotic theory there is an exact equivalence of the theories at large and small radius, which can be understood as a gauge symmetry. In Type II theories, T-duality relates two apparently different theories. Therefore, is natural to ask what is the connection between large and small radius in theories with open strings. Open strings have momentum states but no winding states. So, there cannot be a self-duality. Instead we look for an equivalence between the open-string theory at one radius and some other theory at the inverse radius. Here we uncover *D*-branes.

Consider the boundary conditions on the strings in the compactified direction. For the closed-string fields, the effect of the duality transformation in the compactified direction X is:

$$X_L \to X_L, \quad X_R \to -X_R. \tag{28.10}$$

In terms of left- and right-moving bosons in open-string theories, the Neumann boundary conditions are

$$\partial_\tau X = (\partial_{\sigma_+} + \partial_{\sigma_-})X = 0. \tag{28.11}$$

So, after a *T*-duality transformation we would expect that

$$(\partial_{\sigma_+} - \partial_{\sigma_-})X = \partial_\sigma X = 0, \tag{28.12}$$

i.e. we have traded Neumann for Dirichlet boundary conditions. While this follows from simple calculus manipulations, it is instructive to formulate it in terms of the mode expansion for an open string. Prior to the duality transformation, we have

$$X^9 = x_i^9 + \frac{1}{2}p(\tau+\sigma) + \frac{1}{2}p(\tau-\sigma) + i\sum_{n\neq 0}\frac{1}{n}\left(\alpha_n^9 e^{-in(\tau+\sigma)} + \alpha_n^9 e^{-in(\tau-\sigma)}\right). \tag{28.13}$$

The effect of the duality transformation is to change the sign of the terms which depend on $\tau - \sigma$. So, instead of an expansion in terms of cosines we have an expansion in terms of sines:

$$X^9 = x_0^9 + p\sigma + i\sum_{n\neq 0}\frac{1}{n}\alpha_n^9 e^{-in\tau}\sin n\sigma. \tag{28.14}$$

These are precisely the Dirichlet branes. Note the role of p: in the *T*-dual picture it is a sort of winding: it describes strings which start on the brane, wind around the compact dimension some number of times and then end on the brane.

This *T*-duality of open strings also allows us to understand better the appearance of gauge interactions associated with stacks of branes. In the original open-string picture, gauge degrees of freedom are described by Chan–Paton factors, i.e. charges on the ends of the string. In the case of Type I strings these are described by states of the form $|AB\rangle$, $A, B = 1, \ldots, 32$. Now consider a $U(16)$ subgroup of $O(32)$. The string ends carry labels i,j, within $U(16)$. Taking the diagonal generators of $U(N)$ to be the matrices

$$T_1 = \mathrm{diag}(1,0,0,\ldots), \quad T_2 = \mathrm{diag}(0,1,0,\ldots) \tag{28.15}$$

etc., the state $(\bar{i},j)$ carries charge -1 under T_i, $+1$ under T_j and zero under the other generators.

We can consider constant background gauge fields in the 9 direction. We can write these as

$$A = \mathrm{diag}(a_1, a_2, \ldots, a_{16}). \tag{28.16}$$

This has a gauge-invariant description in terms of the Wilson line:

$$U = \exp\left(i \oint d\vec{x} \cdot \vec{A}\right), \tag{28.17}$$

where the integral is taken in the periodic directions. Such a background gauge field in general breaks the gauge symmetry to $U(1)^{16}$; the other gauge bosons should gain mass. In field theory the corresponding mass terms are proportional to

$$[A^\mu, A^9]^2, \tag{28.18}$$

so the diagonal gauge bosons are massless and those corresponding to the non-Hermitian generator

$$T^{kl}_{ij} = \delta^k_i \delta^l_j \tag{28.19}$$

have mass-squared

$$m^2 = (a_i - a_j)^2. \tag{28.20}$$

This is similar to the calculations we made of symmetry breaking in grand unified theories.

We would like to understand how this result arises directly in string theory. It is simplest to consider the case of a string which is constant in σ, the space-like world-sheet coordinates. The coupling of the string depends on the Chan–Paton factors; see Section 21.1. In the light cone frame the action in the presence of a gauge field is like that of a particle:

$$\frac{1}{2}\int d\tau \left[\left(\frac{\partial \dot{X}^9}{\partial \tau}\right)^2 + (a_i - a_j)\frac{\partial X^9}{\partial \tau}\right]. \tag{28.21}$$

For a non-constant string the situation is somewhat more complicated, since the gauge fields couple at the string's end points.

The extra term modifies the canonical momenta. These are now

$$P = \frac{n}{R} = \frac{\partial \dot{X}^9}{\partial \tau} + a_i - a_j. \tag{28.22}$$

This means that the leading term in the string mode expansion is

$$X^9 = \left[\frac{n}{R} - (a_i - a_j)\right]\tau. \tag{28.23}$$

This gives an extra contribution to the mass. If $n = 0$, this is *exactly* what we expect from field-theoretic reasoning.

Now we will consider the T-dual picture. Under T-duality the zero-mode part of X transforms into

$$X^9 = x_0 + \left[\frac{n}{R} - (a_i - a_j)\right]\sigma. \tag{28.24}$$

For $i = j$ this corresponds to a string that begins and ends on the same D-brane. For $i \neq j$ the string ends at different points, i.e. on separated D-branes. At least for the Type I theory we have *derived* the picture we conjectured earlier: a stack of N coincident branes describes a $U(N)$ gauge symmetry; as the branes are separated, the gauge symmetry is broken by a field in the adjoint representation.

28.4.1 Orientifolds

We have seen that we can understand the appearance of D-branes by considering T-duality transformations of open strings. The Type I theory is a theory of *oriented* strings. In the closed-string sector the action has a parity symmetry which interchanges left and right on the world sheet. Calling the corresponding operator Ω, one keeps only states which are invariant under the action of Ω. This is necessary for the consistency of interactions of open and closed strings. This means that closed-string states like

$$\alpha_{-2}\tilde{\alpha}_{-1}\tilde{\alpha}_{-1}|0\rangle \tag{28.25}$$

are not allowed, but symmetrized combinations such as

$$(\alpha_{-2}\tilde{\alpha}_{-1}\tilde{\alpha}_{-1} + \tilde{\alpha}_{-2}\alpha_{-1}\alpha_{-1})|0\rangle \tag{28.26}$$

are allowed. This projection is similar to the orbifold projections that we have encountered earlier.

Consider the action of Ω in the T-dual theory. We have seen that, in terms of the original fields,

$$(X^9)' = -X^9_{\rm L} + X^9_{\rm R}. \tag{28.27}$$

So, the effect of interchanging left and right is to change the sign of X^9, i.e. Ω is a combination of a world-sheet parity transformation and a reflection in space–time.

The effect of this projection on states is similar to a Z_2 orbifold projection. We can combine momentum states to form states with definite transformation properties under the reflection

$$|\tilde{F}\rangle = |p\rangle \pm |-p\rangle. \tag{28.28}$$

Gravitons $G_{\mu\nu}$, for example, with indices in the non-compact directions, must have momentum states which are even; in coordinate space this means that graviton states must be even functions of x. The fields $G_{\mu 9}$ must be odd functions, and so on. It is as if there is an entity, the *orientifold*, sitting at the origin, the fixed point of the reflection. This object in fact has a negative tension. One way to see this is simply to note that the effect of the T-duality transformation is to produce a set of D-branes. These branes have a positive tension. From the point of view of the non-compact dimensions this is a cosmological constant. But the original theory had no such cosmological constant – this must be canceled by the orientifold.

Just as it is not necessary to start from the Type I theory and its dualities in order to encounter D-branes, it is not necessary to start from the Type I theory to consider orientifolds. Starting from Type II theories, in particular, we can perform a projection by world-sheet parity times some Z_2 space–time symmetry. For example, consider a Type II theory with a single compact dimension. On this theory, we can make a projection which is a combination of world-sheet parity Ω and a reflection in the compact dimension.

28.5 Strong–weak coupling dualities: the equivalence of different string theories

We have seen that, at weak coupling, there are a variety of connections between different string theories which are surprising from a field-theoretic perspective. The heterotic string, compactified on a circle of very large radius, is equivalent to a string theory compactified at very small radius (with a different coupling). The Type IIA theory at large radius is equivalent to the IIB theory at small radius. The $O(32)$ heterotic string is equivalent to the $E_8 \times E_8$ theory. All of these equivalences involve significant rearrangement of the degrees of freedom. Typically, Kaluza–Klein modes, which are readily understood from a space–time field-theory point of view, must be exchanged with winding modes, which seem inherently stringy-like. So, perhaps it is not surprising that there are other equivalences, involving weak and strong coupling. Again, we have had some inkling of this in field theory, when we studied $N = 4$ Yang–Mills theory. There, the theory at weak coupling is equivalent to a theory at strong coupling. To see this equivalence one needs to significantly rearrange the degrees of freedom. States with different electric and magnetic charges exchange roles as the coupling is changed from strong to weak.

In string theory there is a complex web of dualities. The IIB theory in ten dimensions exhibits a strong–weak coupling duality very similar to that of $N = 4$ Yang–Mills theories; weak and strong coupling are completely equivalent. The $O(32)$ heterotic string theory, in ten dimensions, is equivalent at strong coupling to the weakly coupled Type I theory. These relations are surprising, in that these theories appear to involve totally different degrees of freedom at weak coupling. But there are more surprises still. The strong coupling limit of the IIA theory in ten dimensions is a theory whose low-energy limit is eleven-dimensional supergravity. If we allow for compactifications of the theory, this set of dualities is already enough to establish an equivalence of all string theories as well as some as yet not fully understood theory whose low-energy limit is eleven-dimensional supergravity. But, as we compactify, we find further intricate relations. For example, the Type IIA theory on K3 is equivalent to $E_8 \times E_8$ on T^4. Given that all the sensible theories of quantum gravity we know are equivalent, it is plausible that, in some sense, there is a unique theory of quantum gravity. As we will see, however, we only know this reliably for theories with at least 16 supercharges. For theories with four or fever, the situation is less clear; it is by no means obvious that the statement is even meaningful.

In the sections that follow, we will explore some of these dualities and the evidence for them. We will also discuss two particularly surprising equivalences. We will argue that certain string theories are equivalent to *quantum field theories* – even to quantum mechanical systems. The very notion of space–time in this framework will be a derived concept.

28.6 Strong–weak coupling dualities: some evidence

In the case of T-dualities, i.e. dualities which relate the behavior of string theories at weak coupling and different radii, it is straightforward to understand the precise mappings between the different descriptions. Lacking a general non-perturbative definition of string theory, it is not possible to do something similar in the case of strong–weak coupling dualities. Instead, one can try to put together compelling circumstantial evidence. Without supersymmetry even this is essentially impossible. But, in the presence of sufficient supersymmetry, one has a high degree of control over the dynamics. Evidence for equivalence can be provided by studying the following.

1. *The effective action* In ten or eleven dimensions the terms in the action with up to two derivatives are uniquely determined by supersymmetry, so they do not receive corrections either perturbatively or non-perturbatively. A similar statement holds for $N \geq 4$ actions in four dimensions (and actions with varying degrees of supersymmetry in between). In some cases one can check higher-derivative terms in the effective action as well.
2. *The spectrum of BPS objects* In many cases the low-lying states are BPS objects. They cannot disappear from the spectrum as the coupling or other parameters are varied. With 16 or more supercharges, they obey *exact* mass formulas. The identity of the BPS states for different theories provides non-trivial evidence for these equivalences.

We will explore only some of the simplest connections here, but it is important to stress that these identifications are often subtle and intricate. In many instances where one might have thought the dualities mentioned above would fail, they do not.

28.6.1 From IIA to eleven-dimensional supergravity (M theory)

We will start with the IIA theory, where we can readily access both aspects of the duality. Comparing the actions of eleven-dimensional supergravity and the IIA theory is particularly straightforward, as the Lagrangian of the IIA theory is often obtained by compactifying eleven-dimensional supergravity on a circle, keeping only the zero modes. The basic degrees of freedom in eleven dimensions are the graviton g_{MN}, the antisymmetric tensor gauge field C_{MNO} and the gravitino ψ_M. We are not going to work out the detailed properties of this theory, but it is a useful exercise to check that the numbers of bosonic and fermionic degrees of freedom are the same. As usual, we can count degrees of freedom by going to the light cone (or using the "little group," the group of rotations in $D = 11 - 2 = 9$). The metric is a symmetric traceless tensor; for the gravitino, we need also to impose the constraint $\gamma^I \psi_i = 0$. For the metric, then, we have $((9 \times 10)/2) - 1 = 44$ degrees of freedom while from the three-index antisymmetric tensor we have $(9 \times 8 \times 7)/3! = 84$, giving a total of 128 bosonic degrees of freedom. From the gravitino we have $9 \times 16 - 16 = 128$ degrees of freedom.

If we compactify x^{10} on a circle of radius R, we obtain the following bosonic degrees of freedom in ten dimensions:

1. the ten-dimensional metric $g_{\mu\nu}$ $(\mu, \nu = 0, \ldots, 9)$;
2. from $g_{10\mu}$ we obtain a vector gauge field, which is identified with the Ramond–Ramond vector field of the IIA theory;
3. from $C_{10\mu\nu}$ we obtain an antisymmetric tensor field, which is identified with the antisymmetric tensor $B_{\mu\nu}$ of the NS–NS sector of the IIA theory;
4. from $C_{\mu\nu\rho}$ we obtain the three-index antisymmetric tensor field of the R–R sector of the IIA theory;
5. from $g_{10,10}$ we obtain a scalar field in ten dimensions, the dilaton of the IIA theory; note that this mode corresponds to the radius R of the eleventh dimension.

Now consider the action. We will examine just the bosonic terms. These are constructed in terms of the curvature tensor, the three-index antisymmetric tensor and its corresponding four-index field strength F:

$$\mathcal{L} = \frac{-1}{2\kappa^2}\sqrt{g}\mathcal{R} - \frac{1}{48}\sqrt{g}F^2_{MNPQ} - \frac{\sqrt{2}\kappa}{3456}\epsilon^{M_1\ldots M_{11}}F_{M_1\ldots M_4}F_{M_5\ldots M_8}C_{M_9\,M_{10}\,M_{11}}. \tag{28.29}$$

As we have indicated, the dimensional reduction of this theory gives the Lagrangian of the IIA theory in ten dimensions. It is convenient to parameterize the fields in terms of the vielbein e^A_M. Then

$$e^A_M = \begin{pmatrix} e^A_\mu & A_\mu \\ 0 & R_{11} \end{pmatrix}. \tag{28.30}$$

Correspondingly, the metric has the structure

$$g_{MN} = e^A_M e^B_N \eta_{AB} = \begin{pmatrix} g_{\mu\nu} & R_{11}A_\mu \\ R_{11}A_\nu & R^2_{11} \end{pmatrix}. \tag{28.31}$$

If we simply substitute these expressions into the Lagrangian, the coefficient of the Einstein $\mathcal{R}$ term, will be proportional to R. In order to bring this Lagrangian to the canonical, Einstein, form, it is necessary to perform a Weyl rescaling of the metric. Instead, through we will perform the rescaling in such a way as to bring the action to the "string frame". In this frame, all the NS–NS fields have a factor $e^{-2\phi}$ at the front, where $e^{-2\phi}$ is the string coupling (the dilaton). In ten dimensions, $\sqrt{g} = e$ transforms like $(g_{\mu\nu})^5$ under an overall rescaling of the metric; $\mathcal{R}$ transforms like $(g_{\mu\nu})^{-1}$. So we need to make the rescaling:

$$g_{\mu\nu} \rightarrow R_{11}^{-2/3} g_{\mu\nu}. \tag{28.32}$$

The three-form C in Eq. (28.29), upon reduction, leads to various fields in ten dimensions. The components $C_{10\mu\nu}$ give the NS–NS two-form. The fields $C_{\mu\nu\rho}$ give the R–R three-form. The R–R one-form field arises from the $g_{10,\mu}$ components of the metric. The ten-dimensional action becomes

$$S = S_{\mathrm{NS}} + S_{\mathrm{R}}, \tag{28.33}$$

with

$$S_{\text{NS}} = \frac{1}{2}\int d^{10}x\,\sqrt{g}e^{-2\phi}\left(\mathcal{R} + (\nabla\phi)^2 - \frac{1}{2}H^2\right), \tag{28.34}$$

$$I_{\text{R}} = -\int d^{10}x\,\sqrt{g}\left(\frac{1}{4}F^2 + \frac{1}{2\times 4!}F_4^2\right) - \frac{1}{4}F_4\wedge F_4\wedge B. \tag{28.35}$$

We have seen that, when the action is written in this way, R is related to the coupling of the ten-dimensional string theory. The Weyl rescaling $g_{\mu\nu} \to R_{11}^{-3/4} g_{\mu\nu}$ gives an action with R^3 at the front, i.e.

$$\mathcal{L} = R_{11}^{-3}\left(-\frac{1}{2}\mathcal{R} - \frac{3}{4}R^{-3/2}H_{\mu\nu\rho}^2 - \frac{9}{16}\left(\frac{\partial_\mu R_{11}}{R_{11}}\right)^2\right). \tag{28.36}$$

In this form the unit of length is the string scale ℓ_s. So, loops come with a factor R_{11}^3 (the ultraviolet cutoff is ℓ_s^{-1}). We see that

$$g_s^2 = \frac{R_{11}^3}{\ell_{11}^3}. \tag{28.37}$$

We can derive this relation in another way (we will ignore the factor $2\pi s$), which makes a more direct connection between eleven-dimensional supergravity and strings. The eleven-dimensional theory has membrane solutions. We will not exhibit then here, but this fact should not be too surprising since the three-form C_{MNO} couples naturally to membranes. The eleven-dimensional theory has only one scale, ℓ_{11}, so the tension of the membranes is of order ℓ_{11}^{-3}. We can wrap one coordinate of the membrane around the eleventh dimension. If the eleventh dimension is very small, the result is a string propagating in ten dimensions, with tension

$$T = \ell_{11}^{-3}R = \ell_s^{-2}. \tag{28.38}$$

Now, again the ten-dimensional gravitational coupling is related to ℓ_{11} by

$$G_{10} = \frac{\ell_{11}^9}{R_{11}}. \tag{28.39}$$

So we find, once more,

$$g_s^2 = \frac{R_{11}^3}{\ell_{11}^3}. \tag{28.40}$$

Here we have our first piece of circumstantial evidence for the connection. Let us turn now to the BPS spectrum. Consider, first, the eleven-dimensional supersymmetry algebra. Eleven-dimensional spinors can be decomposed into ten-dimensional spinors of definite chirality, with indices α and $\dot{\alpha}$. In this basis,

$$\Gamma_{11} = \begin{pmatrix} 0 & 1 \\ 1 & 0 \end{pmatrix}. \tag{28.41}$$

The eleven-dimensional momenta decompose into ten-dimensional momenta and p_{11} in an obvious way:

$$\{Q_\alpha, Q_{\dot\alpha}\} = \not{p}_{\alpha,\dot\alpha} + p_{11}\delta_{\alpha,\dot\alpha}. \tag{28.42}$$

From a ten-dimensional point of view, the last term is a central charge. In the presence of such a central charge, we can prove a BPS bound as we did for the monopole. This bound is saturated by the Kaluza–Klein modes of the graviton and the antisymmetric tensor field. To what charge does this central charge correspond in the IIA theory, and to which states do the momentum states correspond? It is natural to guess that this is an R–R charge. The simplest possibility is the charge associated with the one-form gauge field. The carriers of the one-form charge are $D0$-branes. The $D0$-branes are BPS states – they preserve half the ten-dimensional supersymmetry. So *states of definite eleven-dimensional momentum are states of definite D-brane charge.* More precisely, localized states with N units of Kaluza–Klein momentum correspond to the zero-energy bound states (so-called threshold bound states) of N D-branes.

There are numerous further tests of this duality. For example, if one compactifies the theory further, there are connections to IIB theory. There are also connections involving $M5$-branes. This short discussion should gives some flavor of the duality, and the evidence, for it, however.

28.6.2 IIB self-duality

The IIB theory exhibits an interesting self-duality. We can understand this, first, from the Lagrangian. The Lagrangian for the NS–NS fields is the same as for the IIA theory. For the R–R fields we have now zero- two- and four-form fields. The Lagrangian for these is similar, with appropriate indices, to that for the R–R fields of the IIA case. A careful examination shows that, under the transformation $\phi \to -\phi$, the Lagrangian goes into itself. At the classical level, the action is also invariant under shifts of the axion.

Grouping the dilaton e^ϕ and the Ramond–Ramond scalar θ into a complex field

$$\tau = \frac{4\pi i}{g_s} + \frac{\theta}{2\pi}, \tag{28.43}$$

it is then natural to conjecture that the underlying theory has an $SL(2,Z)$ symmetry similar to that of $N = 4$ Yang–Mills theory:

$$\tau \to \frac{a\tau + b}{c\tau + d}, \quad ad - bc = 1. \tag{28.44}$$

Further evidence for this symmetry is obtained by studying BPS objects: the various branes of the theory. In the IIB theory we have fundamental strings and $D1$-branes; we also have $D5$-branes. Under this duality the fundamental strings are mapped into $D1$-branes by the $SL(2,Z)$ transformations. Correspondingly, the H_3-form (which couples to fundamental strings) should be mapped into the F_3-form (which couples to $D1$ strings). The $D3$-branes are associated with the gauge-invariant five-form field strength, which is self-dual, so we might expect the $D3$-branes to be invariant. A study of the BPS formulas for these states lends support to these conjectures.

This leaves the $D5$-branes. These couple to the Ramond–Ramond six-form gauge field, which is associated with a seven-form field strength that is in turn dual to the three-form R–R field strength. In other words the $D5$-brane is a magnetic source for F_3. So, we might expect these to be dual to something which is a magnetic source for the NS three-form. This would be an NS five-brane. Such an object can be constructed as a soliton of the ten-dimensional IIB supergravity theory. It plays an important role in understanding the duality of these theories and also appears in other contexts. For example, in M theory, it is associated with a seven-form field strength, which is dual to the four-form field strength that we have already encountered. The $M5$ solution is

$$g_{mn} = e^{2\phi}\delta_{mn}, \quad g_{\mu\nu} = \eta_{\mu\nu}, \tag{28.45}$$

$$H_{mno} = -\epsilon_{mno}{}^{p}\partial_p\phi, \tag{28.46}$$

$$e^{2\phi} = e^{2\phi(\infty)} + \frac{Q}{2\pi^2 r^2}. \tag{28.47}$$

Here μ, ν are the coordinates tangent to the brane (they are the world-volume coordinates) and $m, n\ldots$ are the coordinates transverse to the brane. The $SL(2,Z)$ duality of the IIB theory is quite intricate and beautiful. There are many subtle and interesting checks.

28.6.3 Duality of Type I and $O(32)$

The duality between the Type I and $O(32)$ theories is particularly intriguing, as it is a duality between a theory with open and closed strings and a theory with closed strings only. It is also puzzling, since the perturbative spectra of these theories, at the level of massive states, are quite different. The $O(32)$ heterotic theory contains towers of massive states in spinor representations; there is nothing like this in the perturbative spectrum of the Type I theory. By way of evidence we can begin, again, with the effective Lagrangian. For the heterotic theory this can be written

$$\int d^{10}x\, e^{-2\phi}(\mathcal{R} + |\nabla\phi|^2 + F^2 + dB^2). \tag{28.48}$$

Here $e^{-2\phi}$ is the dilaton field, and we have written the action in the string frame. Consider, now, the transformation

$$g = e^{\phi}g', \quad \phi = -\phi'. \tag{28.49}$$

This takes the action to

$$\int d^{10}x\sqrt{g}[e^{-2\phi'}(R + |\nabla\phi'|^2) + e^{-\phi'}F^2 + dB^2]. \tag{28.50}$$

This is the action for the bosonic fields of the Type I theory. The closed-string fields couple with g^2 while the open-string fields couple with g. In the Type I theory the antisymmetric tensor is an R–R field and, as a result, no factor equal to the coupling (the dilaton) appears out front of its kinetic term.

Now we can ask: how do the hetorotic strings appear in the open-string theory? Here, we might guess that these strings would appear as solitons. More precisely, they are just the

$D1$-branes of the Type I theory. At weak coupling the tension of these strings will behave as $1/g$, i.e. it will be quite large. In this sector one can find states in spinorial representations of $O(32)$ arising from configurations of ($D1$–$D9$)-branes. Most importantly, the $D1$-branes are BPS. As a result they persist to strong coupling, and in this regime their tension is small. We will not explore the various subtle tests of this correspondence, but other features that one can investigate include the identification of the winding strings of the heterotic theory.

Many other dualities among different string theories have been explored. These include an equivalence between heterotic string theory on a four-torus and Type IIA on K3 and equivalences of Calabi–Yau compactifications of the Type II theory and heterotic theory on K3 $\times$ $T2$.

28.7 Strongly coupled heterotic string

In ten dimensions we have seen that the strong coupling limit of the IIA theory is a theory whose low-energy limit is eleven-dimensional supergravity. The strong coupling limit of the IIB theory is again the IIB theory. The strong coupling limit of the $O(32)$ heterotic string is the Type I string. This still leaves the question: what is the strong coupling limit of the $E_8 \times E_8$ heterotic string? The answer is intriguing. It has some tantalizing connections to facts we see in nature. It also suggests different ways of thinking about compactifications – giving an inkling of the large extra dimension and warped-space pictures which we will discuss in the next chapter.

Horava and Witten recognized that the strong coupling limit of the heterotic string, like the IIA theory, is an eleven-dimensional theory. The theory is defined on an interval of radius R_{11}. The relation of R_{11} to the string tension and coupling are exactly as in the IIA case. This means that as the coupling becomes large the interval becomes large. We will refer to the full eleven-dimensional space as the "bulk." The fields propagating in the bulk are a full eleven-dimensional supergravity multiplet: graviton, gravitino and three-form field. At the end of the interval there are two walls (Fig. 28.1). These walls are similar to orientifolds in that they are not dynamical (there are no degrees of freedom corresponding to motion of the walls). The low-lying degrees of freedom on each wall are those of a

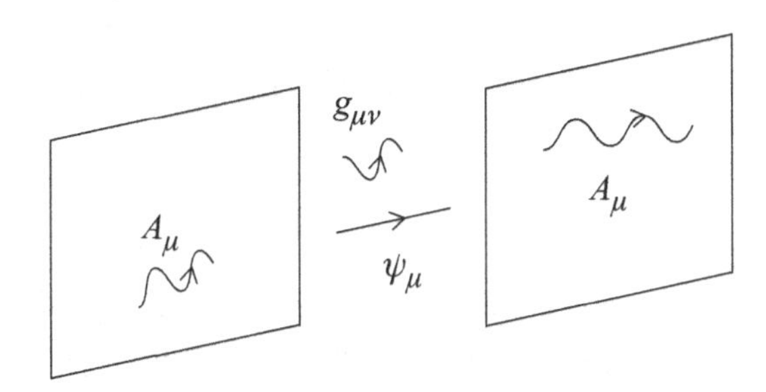

Fig. 28.1 The strongly coupled heterotic string is described by an eleven-dimensional bulk theory and two segregated walls, on which gauge degrees of freedom propagate.

supersymmetric E_8 gauge theory: gauge bosons and gauginos in the adjoint representation. The Lagrangian has the structure of a bulk plus a boundary term:

$$S = -\frac{1}{2\kappa^2}\int d^{11}x\sqrt{g}\mathcal{R} - \sum_{i=1}^{2}\frac{1}{8\pi}(4\pi\kappa^2)^{2/3}\int d^{10}x\sqrt{g}\,\mathrm{Tr}\,F_i^2 + \cdots. \tag{28.51}$$

Note that the gauge coupling is simply proportional to the sixth power of the eleven-dimensional Planck length.

Support for this picture comes from a variety of sources. First, there is a subtle cancelation of gauge and gravitational anomalies. Second, the long-wavelength limit of this theory is ten-dimensional gravity plus Yang–Mills theory, with a relation between the gauge and gravitational couplings appropriate to the heterotic string (this is one way to determine the relations between the coupling constants). Further compactifications provide further checks.

28.7.1 Compactification of the strongly coupled heterotic string

One puzzle in the phenomenology of the weakly coupled heterotic string concerns the value of the gauge coupling and the unification scale. In the MSSM the unification scale is two orders of magnitude below the Planck scale. If we imagine that the unification scale corresponds to a scale of compactification then

$$\alpha_{\mathrm{gut}} \propto \frac{g_s^2}{V}. \tag{28.52}$$

If we treat the left-hand side as fixed then as V becomes large so does g_s. Substituting in the observed values, we see that g_s is quite large. As we will now show, the situation in the strong coupling limit is quite different – and much more promising.

Now consider the compactification of the strongly coupled theory on a Calabi–Yau space. The full compact manifold, from the point of view of an eleven-dimensional observer, is the product of the interval times a Calabi–Yau space X. Such a configuration is an approximate solution of the lowest-order equations of motion. Even at the level of the classical equations of this theory, there are corrections arising from the coupling of bulk and boundary fields. These corrections can be constructed in a power series expansion. Terms in the expansion grow with R_{11}, owing to the one-dimensional geometry in the eleventh dimension. They are proportional to $\kappa^{2/3}$, from the bulk–brane coupling in Eq. (28.51). On dimensional grounds there is a factor R^{-4}, where R is the Calabi–Yau radius. The expansion parameter is thus

$$\epsilon = \kappa^{2/3}\frac{R_{11}}{R^4}. \tag{28.53}$$

We can readily obtain the relation between the four-dimensional and eleven-dimensional quantities. Using the string relations (here we need to be careful about factors of 2 and π)

$$G_N = \frac{e^{2\phi}(\alpha')^4}{64\pi V}, \quad \alpha_{\mathrm{gut}} = \frac{e^{2\phi}(\alpha')^3}{16\pi V}, \tag{28.54}$$

where V is the volume of the compact space X, and the eleven-dimensional relations

$$G_N = \frac{\kappa^2}{16\pi^2 V R_{11}}, \quad \alpha_{\text{gut}} = \frac{(4\pi\kappa^2)^{2/3}}{2V}, \tag{28.55}$$

we have

$$R_{11}^2 = \frac{\alpha_{\text{gut}}^3 V}{512\pi^4 G_N^2}, \quad M_{11} = R^{-1}\left[2(4\pi)^{-2/3}\alpha_{\text{gut}}\right]^{-1/6}. \tag{28.56}$$

where $R = V^{1/6}$. Substituting value of α_{gut} obtaining from running the couplings as in the MSSM (Chapter 11) and the four-dimensional Planck mass gives:

$$R_{11}M_{11} = 18, \quad R = 2\ell_{11} = 3 \times 10^{16}\ \text{GeV}. \tag{28.57}$$

The regime of validity of the strongly coupled description is the regime where V and R_{11} are large compared with ℓ_{11}. We see that nature might well be in such a regime. When we evaluate the expansion parameter ϵ, we find $\epsilon \sim 1$. Adopting the viewpoint that the ground state of string theory which describes nature should be strongly coupled, this, again, seems promising: the parameters of grand unification correspond to the point where the eleven-dimensional expansion is just breaking down, $\epsilon \approx 1$. This is in contrast with the weak coupling picture, which seems far from its range of validity.

Apart from this phenomenological application of string theory ideas, there are two new possibilities which this analysis suggests. First, some compact dimensions might be large compared with the Planck scale (or any fundamental scale). Second, in a case with a one-dimensional geometry, this dimension can be significantly warped, i.e. the metric need not be a constant. These ideas underlie the large-extra-dimension and Randall–Sundrum models of compactification which we will encounter in the next chapter.

28.8 Non-perturbative formulations of string theory

We have seen that, at least in cases with a great deal of supersymmetry, there is a surprisingly large access to non-perturbative dynamics. But much of the evidence for the various phenomena we have described is circumstantial, matching actions and spectra in various regions of a given string moduli space. We lack a general non-perturbative formulation of the theory, analogous to, say, the lattice formulations of Yang–Mills theories which we encountered in Part 1. One might have hoped that there would be a *string field theory* that would be analogous to ordinary quantum field theories, but such a possibility is fraught with conceptual and technical difficulties. We have mentioned some of these. In this section we will describe situations where one can give a complete non-perturbative description. These descriptions are specific to particular backgrounds: flat space in higher dimensions and certain AdS spaces. In eleven dimensions, the flat-space supersymmetric theory can be described as an ordinary quantum mechanical system, while the theory compactified on an n-dimensional torus is described by a field theory in $n + 1$ space–time dimensions, up to $n = 3$. Quite generally, string theory (gravity) in AdS spaces is described by *conformal* field theories (CFTs); this is known as the AdS–CFT correspondence. Both formulations exhibit what is believed to be a fundamental feature of any quantum theory of gravity: holography. The holographic principle asserts that the number of degrees of freedom of a quantum theory of gravity grows, not as the volume of the system, but as its area.

28.8.1 Matrix theory

We have seen that the strong coupling limit of the IIA theory is an eleven-dimensional theory, whose low-energy limit is eleven-dimensional supergravity; $D0$-branes were crucial in making the correspondence. The Kaluza–Klein states of the eleven-dimensional theory are bound states of $D0$-branes; states with momentum N/R_{11} correspond to zero-energy ("threshold") bound states of N $D0$-branes. The world-line theory of N $D0$-branes is ten-dimensional $U(N)$ Yang–Mills theory reduced to zero dimensions. The action which describes this system is

$$S = \int dt \left[\frac{1}{g} \mathrm{Tr}(D_t X^i D_t X^i) + \frac{1}{2g} M^6 R_{11}^2 \,\mathrm{Tr}([X^i, X^j][X^i, X^j]) \right.$$

$$\left. + \frac{1}{g} \mathrm{Tr}(i\theta^T D_t \theta + M^3 R_{11} \theta^T \gamma^i [X^i, \theta]) \right], \qquad (28.58)$$

where R_{11} is the eleven-dimensional radius, M is the eleven-dimensional Planck mass and $g = 2R_{11}$. The Xs are the bosonic variables $X_I, I = 1, \ldots, 9$; the θs are the fermionic coordinates. It is necessary to impose Gauss's law as a constraint on states.

Classically and quantum mechanically this system has a large moduli space, corresponding to configurations with commuting X^Is. For large X^I, the spectrum in these directions consists, in the language of quantum mechanics, of $9N$ free particles and a set of oscillators with frequencies of order $|\vec{X}|$. We can integrate out the fast degrees of freedom, obtaining an effective action for the low-energy degrees of freedom, the X^Is and their superpartners. The bosonic states are just momentum states for these particles. They are the states corresponding to the collective modes of the D-branes.

Banks, Fischler, Shenker and Susskind made the bold hypothesis of identifying these degrees of freedom and the Lagrangian of Eq. (28.58), as a complete description of the eleven-dimensional theory, in the limit that $N \to \infty$. They called this the *matrix model.* The Hamiltonian following from the action of (28.58) is identified with the light cone Hamiltonian, and N is identified with the light cone momentum, $P^+ = N/R$. In the large-N limit this becomes a continuous variable; it is necessary to take $R \to \infty$ at a suitable rate. The first step in this identification is to note that the spectrum of low-lying states of the matrix model is precisely that of the light cone supergravity theory. We have already noted that the states are labeled by a momentum nine-vector $\vec{p}$. In addition, there are 16 fermionic variables, the partners of the bosons. As in other contexts we can define eight fermionic creation operators and eight fermionic destruction operators. From these we can construct a Fock space with 256 states, of which half are space–time bosons (i.e. they have integer spin) and half are fermions. This is just the correct number to describe a graviton and an antisymmetric tensor in eleven dimensions and their superpartners. The states transform correctly under the little group.

A more convincing piece of evidence comes from studying the S-matrix of the matrix theory. Consider, for example, graviton–graviton scattering. Integrating out the massive states of the theory gives an action involving derivatives of x. We will not reproduce the detailed calculation here but the basic behavior is easy to understand. One can compute the action from Feynman graphs, just as in field theory. With four external Xs, simple power counting gives an action, in coordinate space, behaving as

$$\mathcal{L}_I \approx \dot{X}^4 \int \frac{dk}{(k^2+M^2)^4} \cong c\frac{v^4}{M^7}. \tag{28.59}$$

Here $M \propto |X| = R$, the separation of the gravitons. The four factors of v correspond to the four derivatives in the graviton–graviton amplitude; $1/R^7$ is precisely the form of the graviton propagator in coordinate space. With a little more work one can show that one obtains precisely the four-graviton amplitude in eleven dimensions, for suitable kinematics.

The M theory compactified on an n-torus is described by an $(n+1)$-dimensional field theory. We won't argue this through but will just note that in this case the power counting gives the right graviton–graviton scattering amplitude. If $n > 3$, however, the theory is non-renormalizable and the description does not make sense. An alternative description can be formulated for dimensions down to six. The matrix model has been subjected to a variety of other tests. It turns out that the large-N limit is not necessary; for fixed N one can describe a discretized version of the light cone theory (DLCQ). One can actually derive this result from with the assumed duality between IIA theory and eleven-dimensional supergravity.

All this is quite remarkable. Without even postulating the existence of ordinary space–time we have actually uncovered space–time and general relativity in a simple quantum mechanics model. One interesting feature of these constructions is the crucial role played by supersymmetry. Without it, quantum effects would lift the flat directions and one would not have space–time – though one would still have a sensible quantum system. One might speculate that what we think of as space–time is not fundamental but almost an accident associated with the dynamics of particular systems. Lacking, however, a formulation for a realistic non-supersymmetric system, this remains as speculation.

28.8.2 The AdS–CFT correspondence

An equally remarkable equivalence arises in the case of string theory on anti-de Sitter spaces. This connection was first conjectured by Maldacena and is referred to as the AdS–CFT correspondence. It asserts that gravitational theories in AdS spaces have a description in terms of conformal field theories on the boundary of the space.

28.8.2.1 A little more general relativity: AdS space

We could construct anti-de Sitter space by solving the Friedmann equation with a negative cosmological constant. Instead we will adopt a more geometrical viewpoint. Starting with a flat $(p+3)$-dimensional space, with metric

$$ds^2 = -dx_0^2 - dx_{p+2}^2 + \sum_{i=1}^{p+1} dx_i^2, \tag{28.60}$$

we consider the hyperboloid

$$x_0^2 + x_{p+2}^2 - \sum_{i=1}^{p+1} x_i^2 = R^2. \tag{28.61}$$

These coordinates can be parameterized in various ways. For example, one can take

$$\begin{aligned} x_0 &= R\cosh\rho\cos\tau, \quad x_{p+2} = R\cosh\rho\sin\tau, \\ x_i &= R\sinh\rho\,\Omega_i, \quad i = 1,\dots p+1, \quad \Omega_i^2 = 1. \end{aligned} \tag{28.62}$$

This automatically satisfies (28.61) and yields the metric

$$ds^2 = R^2(-\cosh^2\rho d\tau^2 + d\rho^2 + \sinh^2\rho\, d\Omega^2). \tag{28.63}$$

In making the AdS–CFT correspondence, another parameterization is helpful. This covers half the hyperboloid

$$\begin{aligned} x_0 &= \frac{1}{2u}[1 + u^2(R^2 + \vec{x}^2 - t^2)], \quad x_{p+2} = Rut, \\ x^i &= Rux^i, \quad i = 1,\dots p, \\ x^{p+1} &= \frac{1}{2u}[1 - u^2(R^2 - \vec{x}^2 + t^2)]. \end{aligned} \tag{28.64}$$

The metric is then

$$ds^2 = R^2\left[\frac{du^2}{u^2} + u^2(-dt^2 + d\vec{x}^2)\right]. \tag{28.65}$$

Anti-de Sitter space has interesting features, which we will not fully explore here. There is a boundary at spatial infinity ($u = \infty$). Light can reach the boundary in finite time, but massive particles cannot do so. In a cosmological context, the negative cosmological constant leads not to an eternal AdS space but to a singularity. The last form of the metric will be useful in making the AdS–CFT correspondence in a moment. The metric has isometries (symmetries); the group of isometries can be seen from the form of the hyperboloid and the underlying metric of the $(p+3)$-dimensional space; it is $SO(2, p+1)$. This turns out to be the same symmetry as conformal symmetry in $p+1$ dimensions; this, again, is a crucial aspect of the AdS–CFT correspondence.

28.8.2.2 Maldacena's conjecture

Maldacena originally discovered this connection for the case of string theory on $AdS_5 \times S_5$. One suggestive argument starts by considering a set of N parallel $D3$-branes. We have discussed such configurations as open-string configurations but they can also be uncovered as solitonic solutions of the supergravity equations, here of the IIB theory. For these, the metric has the form

$$ds^2 = H(y)^{-1/2}dx^\mu dx_\mu + H(y)^{1/2}\left(dy^2 + y^2 d\Omega_5^2\right)$$
$$F_{\mu\nu\rho\sigma\tau} = \epsilon_{\mu\nu\rho\sigma\tau\alpha}\partial^\alpha H. \tag{28.66}$$

Here the x^μs are the coordinates tangent to the branes, while the ys (and their associated angles) are the transverse coordinates. The dilaton in this configuration is a constant; the other antisymmetric tensors vanish. The function H, for N parallel branes, is

$$H(\vec{y}) = 1 + \sum_{i=1}^{N} \frac{4\pi g_s(\alpha')^2}{|\vec{y} - \vec{y}_i|^4}. \tag{28.67}$$

This can be rewritten as

$$ds^2 = \left(1 + \frac{L^2}{y^4}\right)^{-1/2} \eta_{\mu\nu}dx^\mu dx^\nu + \left(1 + \frac{L^2}{y^4}\right)^{1/2} \left(dy^2 + y^2 d\Omega_5^2\right). \tag{28.68}$$

The parameter L is related to the string coupling g_s, the brane charge (the number of branes) N and the string tension α' by:

$$L^4 = 4\pi g_s N(\alpha')^2. \tag{28.69}$$

It is convenient to introduce a coordinate $u = L^2/y$ and to take a limit where N and g_s are fixed while $\alpha' \to 0$. The metric then becomes:

$$ds^2 = L^2\left(\frac{1}{u^2}\eta_{\mu\nu}dx^\mu dx^\nu + \frac{du^2}{u^2} + d\Omega_5^2\right). \tag{28.70}$$

We have seen the terms involving u and x previously; this is the geometry of AdS_5. The remaining terms describe a five-sphere of radius L.

Now, from a string point of view the low-energy limit of the system of N $D3$-branes is described by $N = 4$ Yang–Mills theory. So we might, with Maldacena, conjecture that there is just such an equivalence between the brane configuration of the string theory (a gravity theory in AdS space) and the field theory. Not surprisingly, demonstrating this equivalence is not simple. One needs to argue that on the string side the bulk modes (graviton, antisymmetric tensors and so on) decouple, as do the massive excitations of the open strings ending on the branes. One cannot argue this at weak coupling, and it would be surprising if one could since in that case one could calculate any quantity in the gravity theory in a weak coupling perturbation expansion in the Yang–Mills theory. This is similar to the situation in the matrix model. There are, however (as in the matrix model), many quantities which are protected by supersymmetry, and these permit quite detailed, consistency checks both in this case and for many other examples of the correspondence.

Suggested reading

Non-perturbative string dualities are discussed extensively in the second volume of Polchinski's (1998) book. This provides an excellent introduction to D-branes. They are treated at length in the text by Johnson (2003), as well. The reader may want to consult earlier papers on duality, especially Witten (1995). Matrix theory and the AdS–CFT correspondence are treated in several excellent pedagogical reviews (Bigatti and Susskind, 1997; Aharony *et al.*, 2000; D'Hoker and Freedman, 2002), but the original papers are very instructive; see, for example, Banks *et al.* (1997); Seiberg (1997), Maldacena (1997) and Witten (1998).

Exercises

(1) *D-branes* For a stack of N *D*-branes, write the open-string mode expansions. Show that, for small separations, the spectrum looks like that of a Higgs $U(N)$ field theory, with the Higgs in the adjoint representation. In the light cone gauge, check the counting of supersymmetries for open strings and *D*-branes.

(2) Verify the construction of the bosonic terms in the ten-dimensional action from the dimensional reduction of the eleven-dimensional action.

(3) Verify that the NS5-brane is a solution of the ten-dimensional supergravity equations.

(4) Take the long-wavelength limit of the Horava–Witten theory (see Section 28.7). Write down the Lagrangian in the ten-dimensional Einstein frame and verify that the gauge and gravitational couplings obey a relation appropriate to the heterotic string theory,

$$g_{\mathrm{ym}}^2 = 4\kappa^2\alpha'^{-1}. \tag{28.71}$$

(5) Calculate the effective action of the matrix model at one loop Eq. (28.58) in more detail. Verify that, treated in the Born approximation, this yields the correct graviton–graviton scattering matrix element for the eleven-dimensional theory. You may find the background-field method helpful for this computation.

(6) Check that the configuration of Eq. (28.66) solves the field equations of IIB supergravity in the case of a single brane. You may want to use some available programs for evaluating the curvature. Verify that in the Maldacena limit, the metric can be recast as in Eq. (28.30). If one requires that the curvature of the AdS space is small, it needs to be checked that the *D*-brane theory is strongly coupled. Discuss the problem of decoupling.

29 Large and warped extra dimensions

Considerations of the sort we encountered in the previous chapter have inspired two approaches to Beyond the Standard Model physics: large extra dimensions (LED or ADD) and warped spaces (Randall–Sundrum). In this chapter we will provide a brief introduction to each.

29.1 Large extra dimensions: the ADD proposal

In string theory it is natural to imagine that the compactification scale is not much different from the Planck scale. The size of the compact space is typically a modulus, and if it is stabilized then one might this to happen at a value not much different from one, in string (and therefore Planck) units. In terms of our general discussion of moduli stabilization we have seen that, once the radius becomes very large, any potential, perturbative or non-perturbative, tends to zero.

But if we are willing to discard this natural prejudice, an extraordinary possibility opens up. Perhaps the extra dimensions are not Planck size but much larger, even macroscopic? Arkani-Hamed, Dimopoulos and Dvali (ADD) realized that, from an experimental point of view, the limits on the size of such large compact dimensions are surprisingly weak. Allowing the extra dimensions to be large totally reorients our thinking about the nature of couplings and scales in string theory (or any underlying fundamental theory). Such a viewpoint places the hierarchy problem in a whole different light, perhaps allowing solutions entirely different from technicolor or supersymmetry.

Branes are crucial to this picture. The observed gauge couplings are small, but not extremely small. In Kaluza–Klein theory and in weakly coupled string theories, however, they are related to the underlying scales in a clear way. For example, in the heterotic string,

$$g_4^{-2} \cong g_s^{-2} M_s^6 R^6. \tag{29.1}$$

So, if g_4 is fixed then as $R \to \infty$, $g_s \to \infty$, but even in a compactified theory the gauge coupling on $D3$-branes is insensitive to the large volume. With more general branes one has more intricate possibilities, depending on how the branes wrap the internal space. However, gravity becomes weak as R becomes large:

$$G_N = \frac{1}{M_p^2} = \frac{1}{M_p^8 R^6} = \frac{g_s^2}{\ell_s^8 R^6}. \tag{29.2}$$

Here M_p is the Planck mass. Now, if g_s is fixed and of order one, as $R \to \infty$, the Planck length tends to zero.

How large might we imagine R could be? If we assume that R is macroscopic, or nearly so, then on distance scales smaller than R the force of gravity will be that appropriate to a higher-dimensional theory. In d space–time dimensions,

$$\text{force}_g \sim \frac{1}{r^{d-2}}. \tag{29.3}$$

If there are a large extra dimensions, any others will be comparable in size with the fundamental scale, $M_p^2 = M_{\text{fund}}^{2+a} R^a$, or

$$M_{\text{fund}} = \left(M_p^2 R^{-a}\right)^{1/(2+a)}, \quad R = M_p^{-1}(M_p/M_{\text{fund}})^{-(2+a)/a}. \tag{29.4}$$

A new viewpoint on the hierarchy problem arises by supposing that M_{fund} is close to the scale of weak interactions, say $M_{\text{fund}} \sim 1$ TeV.Then we can use Eq. (29.4) to relate R to the Planck scale and the weak scale. For example, if $a = 2$, $R \approx 0.01$ cm! For larger a, R is smaller, but still dramatically large; for $a = 3$, for example, it is about 10^{-7} cm. But the value of R for $a = 1$ would be, quite literally, astronomical in size and is clearly ruled out by observations.

Subsequently to the ADD proposal there has been a serious campaign to improve the experimental limits on gravity at mm and smaller scales. With

$$V(r) = -G_N \frac{m_1 m_2}{r}\left(1 + \alpha e^{-r/R}\right), \tag{29.5}$$

one now knows that $R < 37$ μm.

The possibility of large extra dimensions offers a different perspective on the hierarchy problem. The weak scale is fundamental; the issue is to understand why the radius of the large dimensions is so large. One possibility which has been seriously considered is that there are some very large fluxes. For example, if H_{MN} is a two-form associated with a $U(1)$ gauge field and Σ is some closed two-dimensional surface, we could have

$$\int_\Sigma H_{MN} dx^M \wedge dx^N = N. \tag{29.6}$$

If the radius of the dimensions associated with Σ were large then

$$H \sim \frac{N}{R^2}. \tag{29.7}$$

The potential, in turn, would receive a contribution behaving as N^2/R^2. If there were also a (positive) cosmological constant then

$$V = \Lambda R^2 + \frac{N^2}{R^2} \tag{29.8}$$

and, assuming that Λ were of order the fundamental scale,

$$R^4 \sim N^2 \ell_{\text{fund}}^4. \tag{29.9}$$

To obtain a sufficiently large radius in this way, then, requires an extremely large flux. There are some circumstances where such large pure numbers may not be required; supersymmetry and low dimensionality ($a = 2$) would help.

For now we will assume that somehow a large radius arises, for dynamical reasons, and consider some other questions which, ultimately, such a picture raises.

1. *Proton decay* With no further assumptions about the theory we would expect that baryon-number-violating operators would arise, suppressed only by the TeV scale. It would then be necessary to suppress operators of very high dimension. One possible resolution of this problem is elaborate discrete symmetries. Another suggestion has been that the modes responsible for the different low-energy fermions might be very nearly orthogonal.
2. *Other flavor-changing processes* For the same reason, flavor changing processes in weak interactions, processes such as $\mu \to e + \gamma$ and the like pose a danger. One possible solution is that there is a fundamental scale a few orders of magnitude larger than the weak scale. This raises the question of why the weak scale is small – the hierarchy problem again. The orthogonality of fermions, again, can help with many of these difficulties.

We turn, finally, to the phenomenology of large extra dimensions. Here there are exciting possibilities. If R is large then the Kaluza–Klein modes are very light. They are very weakly coupled, but there are many of them and little energy is required for their production. So, let us consider the inclusive production of Kaluza–Klein particles in an accelerator. In terms of $G_{\rm N} = \kappa^2/(8\pi)$, the amplitude for the emission of a Kaluza–Klein particle is proportional to κ. For any given mode, then, the cross section behaves as $\sigma_n \sim G_{\rm N}E^2$, where the E^2 factor follows from dimensional analysis. We need to sum over n or, equivalently, to integrate over a-dimensional phase space. As a crude estimate that we can treat the amplitude as constant and cut off the integration at E, so

$$\sigma_{\rm tot} = R^a \int d^a k \, \sigma_k = G_{\rm N} R^a E^{2+a}. \tag{29.10}$$

Recalling that $G_{\rm N} = G_{\rm fund} R^{-a}$, we see that the tower of Kaluza–Klein particles couples like a $(4 + a)$-dimensional particle: at high energies the extra dimensions are manifest! The cross section exhibits exactly the behavior with energy that one expects in $4 + a$ dimensions.

The actual processes which might be observed in accelerators are quite distinctive. One would expect to see, for example, the production of high-energy photons accompanied by missing energy, with the cross section showing a dramatic rise with energy. Such signatures have already been used (as of the time of writing) to set limits on such couplings.

The production of Kaluza–Klein particles in astrophysical environments can be used to set limits on extra dimensions as well. For example, in the case of two large dimensions and a fundamental scale of order 1 TeV, we saw that the scale of the Kaluza–Klein excitations – the inverse of the radius of the extra dimensions – is of order 10^{-12} GeV, so such particles are easy to produce. Like axions, they might be readily produced in stars.

29.2 Warped spaces: the Randall–Sundrum proposal

Having entertained the possibility that some compact dimensions of space might be very large, one might wonder why the extra dimensions should be flat. In fact, in the Horava–Witten theory the extra dimensions are not that. Taking the formulas of this theory literally, we have seen that if it describes nature then the eleventh dimension is quite large in fundamental units. The metric of this dimension is significantly distorted; we might say that it is warped. This is not surprising; the geometry is essentially one-dimensional. The Green's functions for the fields grow linearly with distance. One of the appealing features of the Horava–Witten proposal is that the dimensions are just large enough that the distortion of the geometry is of order one.

Randall and Sundrum made a more radical proposal: they argued that the warping might be enormous and might account for the large hierarchy between the weak scale and the Planck scale. In the simplest version of their model there is again one extra dimension; call its coordinate ϕ, $0 < \phi < \pi$. The model contains two branes, one at $\phi = 0$, one at $\phi = \pi$. The tensions of the two branes are taken to be equal and opposite. One imagines that the Standard Model fields propagate on one brane, the "visible sector" brane, while some other, hidden, sector fields propagate on the other. The action is then

$$S = S_{\text{grav}} + S_{\text{vis}} + S_{\text{hid}}. \tag{29.11}$$

The bulk gravitational action S_{grav} includes a cosmological constant term:

$$S_{\text{grav}} = \int d^4x \int d\phi \sqrt{-G}(-\Lambda + 2M^3\mathcal{R}), \tag{29.12}$$

where M is the five-dimensional Planck mass. The brane actions are

$$S_{\text{vis}} = \int d^4x \sqrt{-g_{\text{vis}}}(\mathcal{L}_{\text{vis}} - \Lambda_{\text{vis}}), \quad S_{\text{hid}} = \int d^4x \sqrt{-g_{\text{hid}}}(\mathcal{L}_{\text{hid}} - \Lambda_{\text{hid}}). \tag{29.13}$$

Here we have separated off a brane tension term on each brane; we have also distinguished the bulk five-dimensional metric G_{MN} from the metrics on each of the branes, $g_{\mu\nu}$. This has the structure of a gravitational problem in five dimensions, with δ-function sources at $\phi = 0, \pi$. Einstein's equations are

$$\sqrt{-G}\left(R_{MN} - \frac{1}{2}G_{MN}R\right) = -\frac{1}{4M^3}\Big[\Lambda\sqrt{-G}G_{MN} + \Lambda_{\text{vis}}\sqrt{-g_{\text{vis}}}g^{\text{vis}}_{\mu\nu}\delta^\mu_M\delta^\nu_N\delta(\phi-\pi)$$
$$+ \Lambda_{\text{hid}}\sqrt{-g_{\text{hid}}}g^{\text{hid}}_{\mu\nu}\delta^\mu_M\delta^\nu_N\delta(\phi)\Big]. \tag{29.14}$$

Now one makes an ansatz for the metric which leads to warping:

$$ds^2 = e^{-2\sigma(\phi)}\eta_{\mu\nu}dx^\mu dx^\nu + r_c^2 d\phi^2. \tag{29.15}$$

Here r_c is the radius of the compact dimension. Substituting the ansatz Eq. (29.15) into the five-dimensional Einstein equation (29.14) one obtains equations for σ:

$$\frac{6\sigma'^2}{r_c^2} = \frac{-\Lambda}{4M^3}, \quad \frac{3\sigma''}{r_c^2} = \frac{\Lambda_{\text{hid}}}{4M^3 r_c}\delta(\phi) + \frac{\Lambda_{\text{vis}}}{4M^3 r_c}\delta(\phi - \pi). \tag{29.16}$$

This is solved by

$$\sigma = r_c|\phi|\sqrt{-\frac{\Lambda}{24M^3}}, \tag{29.17}$$

provided that the following conditions on the Λs hold:

$$\Lambda_{\text{hid}} = \Lambda_{\text{vis}} = 24M^3 k, \quad \Lambda = -24M^3 k^3. \tag{29.18}$$

In this case the metric varies exponentially rapidly. Note that r_c does not need to be extremely large in order that one obtain an enormous hierarchy. One might worry, though, about the identification of the graviton. It turns out that the metric has zero modes:

$$ds^2 = e^{-2kr_c|\phi|}[\eta_{\mu\nu} + \tilde{h}_{\mu\nu}(x)dx^\mu dx^\nu + T^2(x)d\phi^2], \tag{29.19}$$

where T^2 represents a variation on r_c, usually referred to as the *radion*, and $\tilde{h}_{\mu\nu}$ is the four-dimensional metric. If one substitutes into the action, one finds

$$S = \int d^4x \int d\phi \, 2M^3 r_c e^{-2kr_c|\phi|}\sqrt{-\tilde{g}}\tilde{R}. \tag{29.20}$$

From this we can read off the effective Planck mass:

$$M_p^2 = M^3 r_c \int d\phi \, e^{-2kr_c|\phi|} = \frac{M^3}{k}\left(1 - e^{-2kr_c}\right). \tag{29.21}$$

So, the four-dimensional Planck scale is comparable with the fundamental five-dimensional scale.

To see that the physical masses on the visible brane are small, consider the visible sector action for a scalar particle:

$$S_{\text{vis}} = \int d^4x \sqrt{-g} e^{-4kr_c\pi}\left[\tilde{g}_{\mu\nu}e^{2kr_c\pi}|D_\mu\phi|^2 - \lambda\left(|\phi|^2 - v_0^2\right)^2\right]. \tag{29.22}$$

Rescaling ϕ to $e^{kr_c\pi}\phi$, we have

$$S_{\text{vis}} = \int d^4x \sqrt{-g}\left[\tilde{g}_{\mu\nu}|D_\mu\phi|^2 - \lambda\left(|\phi|^2 - e^{-2kr_c\pi}v_0^2\right)^2\right], \tag{29.23}$$

so the scale is indeed exponentially smaller than the scale on the other brane.

There are many questions one can ask about this structure.

1. How robust is this type of localization of gravity?
2. How do higher excitations, e.g. bulk fields, interact with the fields on the brane? Is the hierarchy stable? (The answer is yes.)
3. Does this sort of warping arise in string theory? Again, the answer is yes, though the details look different.
4. As in the case of large extra dimensions, if this picture makes sense then there are many excitations on the branes; higher-dimension operators are suppressed only by the TeV scale. As there, one has to ask: how does one understand the conservation of baryon number? Other flavor-changing processes? Neutrino masses? Precision electroweak physics? Answers have been put forward to all these questions, but they remain suitable subjects for research. Precision electroweak corrections typically require that the lightest Kaluza–Klein K modes be more massive than 3 TeV.

5. Again as in the case of large extra dimensions, for experimental searches one wants to focus on the additional degrees of freedom associated with bulk fields and the brane. In this case, *unlike* the case of large extra dimensions, the Kaluza–Klein states are not dense. Instead, the low-lying states have masses and spacings of order the TeV scale. Their couplings are not of gravitational strength but, rathers, scaled by inverse powers of the scale of the visible sector brane. The limits are model-dependent (e.g. they depend on which are the bulk fields residing on one brane) but, from LHC result, are in many cases larger than 2 TeV.
6. Given the relatively large scales, how does one understand a comparatively light Higgs? Obtaining a custodial $SU(2)$ symmetry (see Section 8.1) tends to require a large gauge group in the bulk. One might suspect that tuning, similar to that of supersymmetric theories, is also required to obtain a light Higgs.

Finally, there are other variants of the Randall–Sundrum proposal which have been put forward. Perhaps the most interesting is one in which space is not compactified at all but simply warped, with gravity localized on the visible brane. These ideas suggest a rich set of possibilities for what might underlie a quantum theory of gravity. Some features – the exponential warping of the metric, in particular – have been observed in string theory but many, at least to date, have not. This is a potentially important area for further research.

Suggested reading

The original paper of Arkani-Hamed *et al.* (1999) is quite clear and comprehensive, as is the paper of Randall and Sundrum (1999). Good reviews of the Randall–Sundrum proposal are provided by the lecture notes of Sundrum (2005), Csaki *et al.* (2005) and Kribs (2006). The Particle Data Group website provides an up-to-date summary of experimental limits on both large and warped extra dimensions.

Exercise

(1) Verify the Randall–Sundrum solution of Eq. (29.14).

30 The landscape: a challenge to the naturalness principle

We have focused in this text on several questions of naturalness, and have used them to motivate searches for possible new physics. It is fair to say that most physicists find this principle compelling and are reluctant to accept extreme (or even modest!) fine tunings in theories of natural phenomena. But, during the past decade, a plausible, if highly speculative, alternative picture has gained currency, known as the *landscape*. If correct it provides a picture for the emergence of the laws of nature in which fine tunings are not surprising and provide few or no clues as to new degrees of freedom that might lie at higher energy scales.

We will divide our discussion into two parts. First we will explain, in very general terms, what is meant by a landscape and how it might address some naturalness problems in our current understanding of particle physics. Then we consider models for how a landscape might arise in string theory. These models are at best plausible; the existence of any non-supersymmetric states in string theory (apart, possibly, from certain special AdS vacua), much less vast numbers of them, is hardly established.

30.1 The cosmological constant revisited

We have stressed that the cosmological constant (i.e. the dark energy) presents potentially the most striking failure of naturalness. One might hope to solve this problem by introducing new degrees of freedom. Supersymmetry helps to some extent. In global supersymmetry the ground state energy is well defined and of order the scale of supersymmetry breaking raised to the fourth power. In local supersymmetry there is also the term $-3|W|^2$ in the potential. The problem is that this last term must very nearly cancel the positive contributions from supersymmetry breaking. The superpotential W can naturally be small as a result of R symmetries, but no one has proposed a mechanism, based on either dynamics or symmetries, which would lock W onto its required value. Many physicists have searched for an analog of the axion solution of the strong CP problem, in which some light field would adjust in such a way as to cancel the c.c. Without reviewing the various proposals, one might expect that the basic obstacle is in fact illustrated by the Peccei–Quinn mechanism. The axion solution to the strong CP problem relies critically on the existence of an approximate CP symmetry of QCD at $\theta = 0$; small θ is singled out within the Standard Model. There is no clear analog of this (approximate) enhanced symmetry for the cosmological constant. More strikingly, the *measured* value of the dark energy is itself quite peculiar, being nearly coincident with the density of dark matter (and baryonic matter), *at this moment in the history of the universe.*

Weinberg, following suggestions of Banks and Linde, put forward a very different sort of proposal to understand why there could be a small cosmological constant value. At the time he made this proposal, the dark energy had not been observed and there was a prejudice among many theorists that the cosmological constant was rendered exactly zero by some mechanism. Weinberg asked how, in the presence of a cosmological constant, the universe would differ from what we observe. He assumed that other important cosmological quantities, and particularly the spectrum of the initial density perturbations remained unchanged and that matter–radiation equality is obtained at a time of order 10^5 years as in the standard big bang theory. He noted that in that case galaxy formation began when these fluctuations became non-linear, about 10^9 years after the big bang. If the universe was dominated by a cosmological constant at that time, the galaxies would not have formed. This limits the cosmological constant to be less than about 100 times its observed value.

By itself this is an interesting observation, a statement that certain facts about the universe and the underlying laws are consistent. But Weinberg went further. As had been stressed by Linde, in a universe which has undergone inflation, our observable universe is typically only a small part of some larger *metaverse*. Suppose that in different regions of this metaverse, the constants of nature and in particular the cosmological constant, differ: in most regions the cosmological constant is large, but there are observers only in that fraction in which the cosmological constant is extremely small. This is much like the situation of fish and water. Only a very tiny fraction of the universe contains water, but fish inevitably find themselves in that tiny fraction. He dubbed this principle the *weak anthropic principle*.

Now, the most likely value of the cosmological constant would then, be expected to be that value which was most common in a landscape consistent with this anthropic constraint. More precisely, we might imagine that there is a distribution function $f(\Lambda)$, for cosmological constants and a function $\mathcal{E}(\Lambda)$ which describes the likelihood of there being observers in a particular environment and that the probability of a given value of Λ would be obtained by integrating over the product of these. Weinberg reasoned that since a small value of Λ is not favored by any symmetry, one would expect $f(\Lambda)$ to be roughly flat; as a crude model one might then take $\mathcal{E}(\Lambda)$ to be a step function. Then one could predict that the most common value of Λ is close to the maximum allowed by the anthropic constraint.

This argument can be viewed as a *prediction* of the dark energy. The result is somewhat large compared with observation but not too bad on a log scale. One could contemplate refinements which would do better. In particular, $\mathcal{E}$ might well not be a θ function. One could also consider the consequences of allowing other parameters to vary, or "scan", significantly complicating the question of prediction.

There has been much discussion about the use of the arthropic principle and whether it has scientific validity. On the one hand, it is the only explanation so far offered which is at all compelling. On the other hand, to be really persuasive one should have, at the very least, some sort of underlying theory which gives rise to a landscape.

30.2 Candidates for an underlying landscape

Weinberg's argument is interesting, but how might a *metaverse* or *landscape* of this type arise? One proposal was put forth by Bousso and Polchinski. They noted that, as we have seen, string theories possess different types of flux. These can sometimes be thought of as electric, sometimes as magnetic. They are typically quantized, by Dirac's argument. In particular, on compact spaces, fluxes with indices in the compact space will take discrete values and can be labeled by integers n_i, in some units appropriate. Here $i = 1, \ldots, N$ runs over the different types of flux; n_i is often itself constrained by various consistency conditions, e.g.

$$\sum_{i=1}^{N} n_i^2 \leq \chi. \tag{30.1}$$

If N is large and χ is a large integer then the number of possible flux choices will be very large, of order the volume of a sphere in N dimensions (a computation familiar from dimensional regularization in quantum field theory) of radius $\sqrt{\chi}$:

$$\chi^{N/2} \frac{2\pi^{N/2}}{\Gamma(N/2)}. \tag{30.2}$$

Bousso and Polchinski wrote down toy models involving four-form flux, but it was subsequently recognized that other types of flux might dominate, such as three-form fluxes in the case of Type II string theories compactified on Calabi–Yau manifolds.

It turns out also that fluxes can stabilize, even classically, many moduli of the Type II theories, and furthermore there exist scenarios for how the remaining moduli might be stabilized. These are, at the moment, merely scenarios but they provide models for how Weinberg's proposal might be implemented in a microscopic theory.

30.3 The nature of physical law in a landscape

In flux landscapes the features of whatever low-energy theories emerge depend on which vacuum, or ground state, the system occupies. This includes the low-energy degrees of freedom (the light fields) and the parameters of the underlying Lagrangian. For the cosmological constant, in particular, one might expect more or less random values to emerge, at least if there are no symmetry considerations such as supersymmetry. The resulting distribution of parameters was dubbed a *discretuum* by Bousso and Polchinski. In order to obtain the value of the cosmological constant, in a theory where the typical energy scale is the Planck scale, one would need more than 10^{120} such states, so one should certainly be able to think of the distribution as approximately continuous. If random, with zero not a special value, one will inevitably obtain Weinberg's flat distribution.

But, having opened up this possibility, that the parameters in a landscape could be scanned for the cosmological constant, there is no obvious reason why other parameters might not scan as well. Among the parameters of the Standard Model, we would include the Higgs mass and quartic coupling, the gauge couplings and the quark and lepton Yukawa couplings as well as the QCD scale and the θ parameter.

We could well imagine that on the one hand there is some anthropic selection for some of these parameters. If we hold the others fixed, the rates for important stellar processes, relevant to the creation of heavy elements, depend on the value of the weak scale. The proton–neutron mass difference, and thus the values of the u and d quark masses, might also be importance for the existence of observers. On the other hand our existence is not contingent, at least in any obvious way, on the masses of the heavier quarks and leptons or on the mixing angles, and so one might expect them to be random numbers, picked from some underlying distribution. These distributions might not be uniform; the theory *is* found to be more symmetric as these couplings become small, for example. Various possibilities have been considered.

Particularly puzzling from this viewpoint is the θ parameter. While we have seen that experimentally θ must be extremely small, for quantities such as nuclear reaction rates θ has the potential to play only a minor role. It is hard to imagine an anthropic constraint which would require θ even as small as 0.01, much less 10^{-10}. So, something more is required if the anthropic principle is to be viable. Conceivably axion dark matter is important for the formation of structure in the universe, and this somehow leads to a small θ. But it is probably fair to say that no convincing case for this has yet been made.

30.4 Physics beyond the Standard Model in a landscape

One might argue that that if one adopts an anthropic viewpoint then there is no need for physics beyond the Standard Model, at least until one reaches scales such as those associated with the right-handed neutrino mass. In particular, there need not be new phenomena associated with electroweak symmetry breaking. This viewpoint might be correct, and the experimental situation at the LHC in late 2015 might give some limited support for this possibility, but there are reasons to question it.

For definiteness, let us focus on supersymmetry. In a landscape one would expect that there are states with no supersymmetry, with some approximate supersymmetry and with unbroken supersymmetry. The class of states with approximate supersymmetry might well provide a realization of conventional notions of naturalness. One might expect that, among these, states with a low value of the weak scale (compared with $M_{\rm p}$) typically have a low value of the supersymmetry breaking scale. So, if the supersymmetric states are somehow more numerous, or otherwise favored, one would predict low-scale supersymmetry breaking. It could be, however, that the non-supersymmetric states are far more numerous than the supersymmetric ones and that low-energy supersymmetry is extremely rare. One might then obtain a low-energy theory which appears extremely tuned. Detailed studies of model landscapes lead to refinements of these considerations.

Flux models with and without supersymmetry have been extensively studied. In these studies, "without supersymmetry" typically means that one starts with a locally supersymmetric action and studies the stationary points of an effective action computed in a crude (i.e. not systematic) approximation. At some of these stationary points the supersymmetry is badly broken but at others it is not. These models lead, in many cases, to distributions of low-energy parameters which appear potentially robust. For example, superpotential parameters are often uniformly distributed, for small values of the parameters, as complex numbers. From these sorts of studies, at least three branches of the landscapes are suggested:

1. a non-supersymmetric branch;
2. a supersymmetric branch with spontaneous (non-dynamical) supersymmetry breaking;
3. a supersymmetric branch with dynamical supersymmetry breaking.

On the second branch the distribution of supersymmetry breaking scales, for a fixed value of the weak scale and a small cosmological constant, favors *very high* scales of supersymmetry breaking. This runs counter to the intuition which generates much of the interest in low-energy supersymmetry. It results from very simple considerations, however, such as assuming the uniformity of superpotential parameters. Roughly speaking, if one has a field Z which contains the goldstino (the longitudinal mode of the gravitino), then there are three renormalizable parameters in its superpotenitial, two of which must be small for low-scale breaking; there is also the parameter W_0, the expectation value of the superpotential. One assumes that one pays a price of m_H^2/M_p^2 for the tuning of the Higgs mass. If one also requires a small μ parameter for the Higgs, and this is also uniformly distributed, high scale breaking is even more strongly favored.

On the third branch, things can be better. In this case the supersymmetry-breaking scale is distributed uniformly on a log scale. If W_0 is uniform as a complex variable then supersymmetry breaking is distributed uniformly on a log scale. So, while this does not particularly favor very high scale breaking, it also does not point to TeV breaking scales. To account for scales of order TeV or perhaps slightly higher, one would need to introduce other considerations (perhaps the cosmology of moduli or the density of dark matter). A non-dynamical μ term again pushes towards higher scales.

We returning to the question: are there more or fewer states on the supersymmetric than on the non-supersymmetric branches. One's first guess would be that supersymmetry is special and that non-supersymmetric states might be far more common. Against this are two arguments, both based on questions of *stability*. The first is perturbative. In landscape models (Type II with fluxes in particular) there are many fields. At the stationary points it is important that the curvature be positive in all directions. For a random potential for N fields, one might expect that only $1/2^N$ of the non-supersymmetric stationary points would be stable; it turns out that the suppression is even larger. But this only addresses the question of perturbative stability. Among the remaining states, only an exponentially small fraction are long lived. Supersymmetric states that have a small cosmological constant, are in fact generically stable in both senses. So this *might* indicate that the supersymmetric branch is more heavily populated than the non-supersymmetric branch.

30.5 't Hooft's naturalness priciple challenged

Finally, we can return to 't Hooft's principle of naturalness itself. Why, in fact, would we expect that states with symmetries are favored? One argument has to do, again, with the stationary points of potentials: symmetric points are always stationary. Another argument, in a landscape framework, is the possibility that symmetric points, being special, might be singular points in the distributions of parameters and thus favored.

In a flux landscape one can give a tentative answer: symmetries are *highly disfavored*. Consider, for example, a discrete symmetry. Some fluxes will be invariant under the symmetry, but typically most will not. Since the number of states goes as a power of the number of fluxes, symmetric states will be an exponentially small fraction of the total. It could be that some other model for landscapes would favor symmetric states. It is also possible that adding, for example, cosmological considerations would make the distribution singular at symmetric points. Still, from a landscape perspective, 't Hooft's principle is not self evident. We have given arguments why states with greater *supersymmetry* might be favored, but these are at best tentative and it is not clear how they might extend to more conventional bosonic symmetries.

We are left, then, with a great deal of uncertainty. The very existence of a landscape remains purely a matter of conjecture. If it does exist, the manner in which one should enforce anthropic constraints (or even just experimental priors) is not completely clear. Finally, the features of the putative landscape will determine questions such as: is there supersymmetry at scales well below the Planck scale? For the moment, it would seem that we least have to at admit such questions, especially until we have experimental evidence that more traditional notions of naturalness are operative at least for the understanding the scale of weak interactions.

30.6 Small and medium size hierarchies: split supersymmetry

If a landscape picture is operative, it raises the possibility that there are simply large hierarchies. This might be understood anthropically but, whether or not one likes such an approach, the picture raises the possibility that there is no low-energy explanation of these surprising failures of dimensional analysis. But such a picture also raises the possibility of more modest hierarchies. One might imagine that there is some tension between the anthropic requirements for, say, dark matter and the weak scale and that this might account for a somewhat large scale of supersymmetry breaking. Alternatively, simply imposing certain facts – that matter–radiation equality occurs at a temperature of approximately 1 eV, on underlying theories, say, with moduli, implies a supersymmetry-breaking scale of about 30 TeV, compatible with the observed Higgs mass. One proposal is known as "split supersymmetry". Here it is assumed that the dark matter is a *wino* in an underlying theory with an anomaly-mediated spectrum. To account for the dark matter, the wino mass must be of order several hundred GeV, and the gravitino and squarks and leptons must be more massive by factors of order π/α. In such a picture it is conceivable that we could find gluinos and some other supersymmetric particles in an accelerator with energies somewhat higher than those of the LHC. Alternatively, however, one could imagine that all the new supersymmetric states are rather heavy, with dark matter in, say, the form of axions.

Suggested reading

The cosmological constant problem, and Weinberg's proposal, are discussed in Weinberg's review (1989). A good review of the issues in landscape statistics is provided in Denef *et al.* (2007). Ideas surrounding split supersymmetry are discussed in Arkani-Hamed *et al.* (2005).

31 Coda: Where are we heading?

The LHC, in its first years of running, has been a remarkable success. The discovery of the Higgs boson in an extremely complex environment is an extraordinary achievement, both experimentally and also in the application of our understanding of many facets of the Standard Model. This particle appears, at the 10%–20% level, in several channels, to be the Higgs field of the simplest version of the Standard Model. Over the next few years these measurements will improve and additional channels will be studied. In Chapter 4 of this text we studied the Standard Model as an effective-field theory. In that discussion our treatment of the Higgs sector was somewhat tentative; we entertained the possibility that the Standard Model might fail at scales of order 1 TeV. It is quite possible, however, that over the next few years we will establish that the Standard Model, *including only a single Higgs doublet*, provides a complete description of nature up to a scale of a few TeV. This would represent an extraordinary achievement.

Yet we have many unanswered questions. As this book goes to press the LHC is beginning to run at close to its design energy of 14 TeV. It is quite possible that, as we explore this new energy frontier, we will see one or more major discoveries – a candidate for dark matter, evidence for supersymmetry, additional Higgs fields, a new $U(1)$ gauge boson Z' or something totally unanticipated. Experiments at the *cosmic frontier* searching for dark matter, CMB polarization or non-Gaussianity and other phenomena are coming on line and/or improving their reach, and major discoveries might be made over the next few years. Alternatively we have seen that the LHC has already excluded many possibilities for new physics. It is conceivable that the answers to many questions do not lie at energies which will be accessible in the next few years.

To conclude this book, an assessment of some of the ideas for Beyond the Standard Model physics, and their prospects, is appropriate.

31.1 The hierarchy or naturalness problem

The hierarchy problem is strongly suggestive of new physics at TeV energy scales. Supersymmetry, broken at around one TeV, is a possible solution which we have explored extensively in this book. But the mass of the Higgs and LHC exclusions strongly suggest that, if supersymmetry is present at all, it is broken at scales of order tens of TeV or even higher. This raises significant experimental challenges. Even a collider with center of mass energy in the 100 TeV range does not have a 10 TeV reach, much less 30 TeV or more. At a theoretical level there is the question, what might account for such a scale? We have discussed some possible explanations for varying degrees of tuning but certainly have not established a compelling criterion; basically once one has admitted tuning, it is hard to decide how much is too much.

Strongly interacting Higgs and other (non-supersymmetric) dynamical explanations of the hierarchy problem have to confront different challenges. For technicolor and its variants, there are the long-standing questions of flavor-changing neutral currents and precision electroweak physics; now there is the additional puzzle of why there should be a particle behaving like an elementary Higgs field. This latter question is confronted more directly in "little Higgs theories" (and their variants), where the Higgs emerges as a pseudogoldstone boson of other interactions. Here perhaps the biggest challenge is the complex set of constraints on these theories, not least of which is simply: how do the required non-Abelian global symmetries emerge? Many ideas are being explored and, at the same time, the possibility of composite Higgs particles provide another target for experimental searches.

31.2 Dark matter, the baryon asymmetry and dark energy

Dark matter remains the subject of extensive search efforts. Apart from the hierarchy problem, the "wimp miracle" is another pointer to possible new physics at the electroweak scale. Much of the parameter space for supersymmetric wimps has been ruled out by direct and indirect detection searches, but some remains, and there are tantalizing hints for dark matter with more interesting properties. At the same time we have seen that axions provide a plausible candidate for the dark matter. The ADMX experiment is, as of the time of writing, probing an interesting part of the axion parameter space. There are ideas under consideration to search for far lighter axions. It is quite possible that in the next few years we will see a discovery; alternatively, there will be important exclusions.

The origin of the baryon asymmetry is an interesting question about which we don't have sharp clues or evidence. Electroweak baryogenesis in the Standard Model itself has long since been ruled out. Within supersymmetric theories there remain corners of the parameter space where it might yet be allowed but, given the present tunings involved in most supersymmetric theories, this seems a bit of a long shot. Affleck–Dine baryogenesis, discussed in Chapter 19, could still be operative even if supersymmetry is broken at some high energy scale but, without the discovery of supersymmetric particles, it is unclear how one might accumulate evidence for this mechanism. Leptogenesis as a possibility receives support from the discovery of the neutrino mass. But, if right-handed neutrinos are at scales of order 10^{14}–10^{16} GeV, the reheating temperature after inflation needs to be of this order, which seems unlikely. Put differently, a determination of the scale of inflation, and the development of a theory of reheating, would significantly constrain models of leptogenesis (and neutrinos mass).

For dark energy, the principle experimental question appears to be whether the dark energy is indeed a cosmological constant, or whether $w = -1$ (see Eq. (18.15)) in its equation of state. This is already established at the 10% level; upcoming experiments, such as the Dark Energy Survey, will reduce the errors further.

31.3 Inflationary cosmology

Here perhaps the largest question to which we may obtain experimental access over the next few years is the energy scale of inflation. If this scale is in the range $10^{15}-10^{16}$ GeV, it is likely to be established by observation of the tensor polarization of the CMB, known as *B mode* polarization. This would be remarkable, first, in that it would represent our earliest observation of the universe, a time of order 10^{-25} seconds after the big bang. Second, it would be a guide to thinking about other energy scales in physics. Is this scale, perhaps, related to a scale of unification or string theory, for example? It would certainly be a guide to modeling inflation.

31.4 String theory and other approaches to foundational questions

String theory, thought of literally as a quantum theory of strings, is remarkable in many ways. It is a consistent theory of quantum gravity. It incorporates gauge interactions like those of the Standard Model. It can exhibit other striking features of the Standard Model, such as repetitive generations. String theory also includes many elements which have appeared in our speculations about Beyond the Standard Model physics, including:

1. axions: axions appear with approximate Peccei–Quinn symmetries which are potentially good enough to solve the strong CP problem;
2. low-energy supersymmetry;
3. new strong interactions;
4. multiple generations of quarks and leptons;
5. unification of the known forces;
6. possibly large extra dimensions, warped spaces and the like;
7. discrete symmetries of sorts interesting for model building but, as would be expected of theories of quantum gravity, no global continuous symmetries.

At the same time string theories appear robust as quantum theories of gravity. But, as currently understood, it is hard to see how weakly coupled strings could provide a complete description of nature.

There are several issues, as we have seen. Principal among these are understanding the fixing of moduli and supersymmetry breaking. These problems are intimately connected. For string theories (compactifications) without supersymmetry, even at one loop there is a potential for the moduli; this tends to zero for large radius and small couplings, the regions where the calculations are reliable. Supersymmetric models either respect supersymmetry exactly (due to the presence of more than four supersymmetries or due to discrete symmetries) or they break supersymmetry non-perturbatively and are subject to the same difficulties.

So we face the problem that superstring theories, in the realms in which we understand them, almost certainly cannot describe nature. Instead, we can retreat a bit and take from string theory the lesson that sensible theories of quantum gravity exist and can account for many features of the low-energy world (gauge theories, chiral fermions). But there is almost certainly some other structure needed to describe the world around us. Whether string theory is a part of this larger structure, or whether such a structure describing nature is a distinct entity, we do not know. The landscape hypothesis is tied to the former view. Efforts to escape it would seem tied to the latter. Clues for exploring these questions include the web of dualities and questions such as the presence or absence of quantum tunneling between different vacua.

The success of the Standard Models of particle physics and of cosmology mean that we can formulate very precise questions about how nature might be structured. But these questions are challenging. The author, for one, hopes for discoveries over the next decade which provide direction to our speculations. It is to be hoped that this book has laid out a range of theoretical tools of value to those who seek an understanding of the universe at a deeper level.

Suggested reading

An enumeration of the conceptual problems of the landscape and an approach to thinking about a reformulation of quantum general relativity appears in Banks (2014).

PART 4

APPENDICES

Appendix A Two-component spinors

The Dirac equation simplifies dramatically in the case where the fermion mass is zero. The equation

$$\not{D}\psi = 0 \tag{A1}$$

has the feature that if ψ is a solution then so is $\gamma_5\psi$:

$$\not{D}(\gamma_5\psi) = 0. \tag{A2}$$

The matrices

$$P_\pm = \frac{1}{2}(1 \pm \gamma_5) \tag{A3}$$

are projectors:

$$P_\pm^2 = P_\pm, \quad P_+P_- = P_-P_+ = 0. \tag{A4}$$

To understand the physical significance of these projectors it is convenient to use a particular basis for the Dirac matrices γ^μ, often called the chiral or Weyl basis:

$$\gamma^\mu = \begin{pmatrix} 0 & \sigma^\mu \\ \bar{\sigma}^\mu & 0 \end{pmatrix}, \tag{A5}$$

where

$$\sigma^\mu = (1, \vec{\sigma}), \quad \bar{\sigma}^\mu = (1, -\vec{\sigma}). \tag{A6}$$

In this basis,

$$\gamma_5 = i\gamma^0\gamma^1\gamma^2\gamma^3 = \begin{pmatrix} -1 & 0 \\ 0 & 1 \end{pmatrix}, \tag{A7}$$

so that

$$P_+ = \begin{pmatrix} 0 & 0 \\ 0 & 1 \end{pmatrix}, \quad P_- = \begin{pmatrix} 1 & 0 \\ 0 & 0 \end{pmatrix}. \tag{A8}$$

We will adopt certain notation that follows the text of Wess and Bagger:

$$\psi = \begin{pmatrix} \chi_\alpha \\ \phi^{*\dot{\alpha}} \end{pmatrix}. \tag{A9}$$

Correspondingly, we label the indices on the matrices σ^μ and $\bar{\sigma}^\mu$ as

$$\sigma^\mu = \sigma^\mu_{\alpha\dot{\alpha}}, \quad \bar{\sigma}^\mu = \bar{\sigma}^{\mu\dot{\beta}\beta}. \tag{A10}$$

This allows us to match the "upstairs" and "downstairs" indices and will prove quite useful. The Dirac equation now becomes

$$i\sigma^{\mu}_{\alpha\dot{\alpha}}\partial_\mu\phi^{*\dot{\alpha}} = 0, \quad i\bar{\sigma}^{\mu\dot{\alpha}\alpha}\partial_\mu\chi_\alpha = 0. \tag{A11}$$

Note that χ and ϕ^* are equivalent representations of the Lorentz group; χ and ϕ obey identical equations. We may proceed by complex-conjugating the second of Eqs. (A11) and noting that $\sigma_2\sigma^{\mu *}\sigma_2 = \bar{\sigma}^\mu$.

Before discussing this identification in terms of representations of the Lorentz group, it is helpful to introduce some further notation. First, we define the action of complex conjugation as that of changing dotted to undotted indices. So, for example,

$$\phi^{*\dot{\alpha}} = (\phi^\alpha)^*. \tag{A12}$$

Then we define the antisymmetric matrices $\epsilon_{\alpha\beta}$ and $\epsilon^{\alpha\beta}$ by

$$\epsilon^{12} = 1 = -\epsilon^{21}, \quad \epsilon_{\alpha\beta} = -\epsilon^{\alpha\beta}. \tag{A13}$$

The matrices with dotted indices are defined identically. Note that, for the upstairs indices, $\epsilon = i\sigma_2$ and $\epsilon_{\alpha\beta}\epsilon^{\beta\gamma} = \delta^\gamma_\alpha$. We can use these matrices to raise and lower indices on spinors. Define $\phi_\alpha = \epsilon_{\alpha\beta}\phi^\beta$, and similarly for the dotted indices. So

$$\phi_\alpha = \epsilon_{\alpha\beta}(\phi^{*\dot{\beta}})^*. \tag{A14}$$

Finally, we will define the complex conjugation of a product of spinors as inverting the order of factors; so, for example, $(\chi_\alpha\phi_\beta)^* = \phi^*_{\dot{\beta}}\chi^*_{\dot{\alpha}}$.

With this in hand, the reader should check that the action for our original four-component spinor is:

$$\begin{aligned} S = \int d^4x\, \mathcal{L} &= \int d^4x \left(i\chi^*_{\dot{\alpha}}\bar{\sigma}^{\mu\dot{\alpha}\alpha}\partial_\mu\chi_\alpha + i\phi^\alpha\sigma^\mu_{\alpha\dot{\alpha}}\partial_\mu\phi^{*\dot{\alpha}} \right) \\ &= \int d^4x \left(i\chi^\alpha\sigma^\mu_{\alpha\dot{\alpha}}\partial_\mu\chi^{*\dot{\alpha}} + i\phi^\alpha\sigma^\mu_{\alpha\dot{\alpha}}\partial_\mu\phi^{*\dot{\alpha}} \right). \end{aligned} \tag{A15}$$

At the level of Lorentz-invariant Lagrangians or equations of motion, there is *only one* irreducible representation of the Lorentz algebra for massless fermions.

Two-component fermions have definite helicity. For a single-particle state with momentum $\vec{p} = p\hat{z}$, the Dirac equation reads

$$p(1 \pm \sigma_z)\phi = 0. \tag{A16}$$

Similarly, the reader should check that the antiparticle has the opposite helicity.

It is instructive to describe quantum electrodynamics with a massive electron in two-component language. Write

$$\psi = \begin{pmatrix} e \\ \bar{e}^* \end{pmatrix}. \tag{A17}$$

In the Lagrangian we need to replace ∂_μ with the covariant derivative, D_μ. Note that e contains annihilation operators for the left-handed electron and creation operators for the corresponding antiparticle. Note also that $\bar{e}$ contains annihilation operators for particles with the opposite helicity and charge to e and $\bar{e}^*$ and creation operators for the corresponding antiparticle.

The mass term $m\bar{\psi}\psi$ becomes:

$$m\bar{\psi}\psi = me^\alpha \bar{e}_\alpha + me^*_{\dot{\alpha}}\bar{e}^{*\dot{\alpha}}. \tag{A18}$$

Again, note that both terms preserve electric charge. Note also that the equations of motion now couple e and $\bar{e}$.

It is helpful to introduce one last piece of notation. Set

$$\psi\chi = \psi^\alpha \chi_\alpha = -\psi_\alpha \chi^\alpha = \chi^\alpha \psi_\alpha = \chi\psi. \tag{A19}$$

Similarly,

$$\psi^*\chi^* = \psi^*_{\dot{\alpha}}\chi^{*\dot{\alpha}} = -\psi^{*\dot{\alpha}}\chi^*_{\dot{\alpha}} = \chi^*_{\dot{\alpha}}\psi^{*\dot{\alpha}} = \chi^*\psi^*. \tag{A20}$$

Finally, note that, with these definitions,

$$(\chi\psi)^* = \chi^*\psi^*. \tag{A21}$$

Appendix B Goldstone's theorem and the pi mesons

It is easy to prove Goldstone's theorem for theories with fundamental scalar fields. But the theorem is more general than that, and some of its most interesting applications are in theories without fundamental scalars. We can illustrate this with QCD. In the limit where there are two massless quarks (i.e. in the limit where we neglect the masses of the u and d quarks), we can write the QCD Lagrangian in terms of spinors

$$\Psi = \begin{pmatrix} u \\ d \end{pmatrix} \tag{B1}$$

as

$$\mathcal{L} = \bar{\Psi} i \gamma^\mu D_\mu \Psi - \frac{1}{4} F_{\mu\nu}^2. \tag{B2}$$

This Lagrangian has symmetries

$$\Psi \rightarrow e^{i\omega^a \tau^a/2} \Psi, \quad \Psi \rightarrow e^{i\omega^a \tau^a - \gamma^5/2} \Psi \tag{B3}$$

(the τ_s^a are the Pauli matrices). In the limit where the two quarks are massless, QCD is thus said to have the symmetry $SU(2)_{\mathrm{L}} \times SU(2)_{\mathrm{R}}$.

So, writing a general four-component fermion as

$$\Psi = \begin{pmatrix} q \\ \bar{q}^* \end{pmatrix}, \tag{B4}$$

the Lagrangian has the form:

$$\mathcal{L} = i\Psi \sigma^\mu D_\mu \Psi^* + i\bar{\Psi} \sigma^\mu D_\mu \bar{\Psi}^*. \tag{B5}$$

In this form, we have two separate symmetries:

$$\Psi \rightarrow \exp\left(i\omega_{\mathrm{L}}^a \frac{\tau^a}{2} \right) \Psi, \quad \bar{\Psi} \rightarrow \exp\left(i\omega_{\mathrm{R}}^a \frac{\tau^a}{2} \right) \bar{\Psi}. \tag{B6}$$

Written in this way, it is clear why the symmetry is called $SU(2)_{\mathrm{L}} \times SU(2)_{\mathrm{R}}$.

Now, it is believed that in QCD the operator $\overline{\Psi}\Psi$ has a non-zero vacuum expectation value, i.e.

$$\langle \overline{\Psi}\Psi \rangle \approx (0.3\,\mathrm{GeV})^3 \delta_{ff'}. \tag{B7}$$

This is in four-component language; in two-component language this becomes:

$$\langle \bar{\Psi}_f \Psi_{f'} + \bar{\Psi}_f^* \Psi_{f'}^* \rangle \neq 0. \tag{B8}$$

This leaves ordinary isospin, the transformation without the γ_5 in four-component language, or with $\omega_L^a = -\omega_R^a$, unbroken, in two-component language.

But there are three broken symmetries. Correspondingly, we expect that there are three Goldstone bosons. To prove this, write

$$\mathcal{O} = \bar{\Psi}\Psi, \quad \mathcal{O}^a = \bar{\Psi}\gamma^5 \frac{\tau^a}{2}\Psi. \tag{B9}$$

Under an infinitesimal transformation,

$$\delta\mathcal{O} = 2i\omega^a\mathcal{O}^a, \quad \delta\mathcal{O}^a = i\omega^a\mathcal{O}. \tag{B10}$$

In the quantum theory these give the commutation relations

$$[Q^a, \mathcal{O}] = 2i\mathcal{O}^a, \quad [Q^a, \mathcal{O}^b] = i\delta^{ab}\mathcal{O}. \tag{B11}$$

Here Q^a is the integral of the time component of a current. To see that there must be a massless particle, we study

$$0 = \int d^4x\, \partial_\mu \left[\langle\Omega|T(j^{\mu a}(x)\mathcal{O}^b(0))|\Omega\rangle e^{-ip\cdot x} \right] \tag{B12}$$

(this follows because the integral of a total derivative is zero). We can evaluate the right-hand side, carefully writing out the time-ordered product in terms of ϑ-functions and noting that the action of ∂_0 on the ϑ-functions gives δ-functions:

$$0 = \int d^4x\, \langle\Omega|\, [j^{oa}(x), \mathcal{O}^b(0)]\delta(x^0)|\Omega\rangle e^{-ip\cdot x} - ip_\mu \int d^4x\, \langle\Omega|T(j^{\mu a}(x)\mathcal{O}^b(0))|\Omega\rangle. \tag{B13}$$

Now consider the limit $p^\mu = 0$. The first term on the right-hand side becomes the matrix element of $[Q^a, \mathcal{O}^b(0)] = \mathcal{O}(0)$. This is non-zero. The second term must be singular, then, if the equation is to hold. This singularity, as we will now show, requires the presence of a massless particle. For this we use the spectral representation of the Green's function. In general a pole can arise at zero momentum only from a massless particle. To understand this singularity we introduce a complete set of statesand, say for $x^0 > 0$, write it as

$$\sum_\lambda \int \frac{d^3p}{2E_p(\lambda)} \langle\Omega|j^{\mu a}(x)|\lambda_p\rangle\langle\lambda_p|\mathcal{O}^b(0)|\Omega\rangle. \tag{B14}$$

In the sum we can separate the term coming from the massless particle. Call this particle π^b. On Lorentz-invariance grounds,

$$\langle\Omega|j^{\mu a}|\pi^b(p)\rangle = f_\pi p^\mu \delta^{ab}. \tag{B15}$$

Set

$$\langle\lambda_q|\mathcal{O}^a(x)|\pi^b(p)\rangle = Z\delta^{ab}e^{-ip\cdot x} \tag{B16}$$

Adding the contribution from the time ordering $x_0 < 0$, we obtain for the left-hand side a massless scalar propagator i/p^2 multiplied by $Zf_\pi p^\mu$, so the equation is now consistent:

$$\langle\bar{\Psi}\Psi\rangle = \frac{p^2}{p^2} f_\pi Z. \tag{B17}$$

It is easy to see that, in this form, Goldstone's theorem generalizes to any theory without fundamental scalars in which a global symmetry is spontaneously broken.

Returning to QCD, what about the fact that the quarks are massive? The quark mass terms break the symmetries explicitly. But if these masses are small, we should be able to think of the potential as "tilted", i.e. almost, but not quite, symmetric as in Section 5.3.1. This gives rise to small masses for the pions. We could compute these by studying, again, correlation functions of derivatives of currents. A simpler procedure is to consider the symmetry-breaking terms in the Lagrangian:

$$\mathcal{L}_{\rm sb} = \bar{\Psi} M \Psi, \tag{B18}$$

where M is the quark mass matrix,

$$M = \begin{pmatrix} m_u & 0 \\ 0 & m_d \end{pmatrix}. \tag{B19}$$

Since the π mesons are, by assumption, light, we can focus on these. If we have a non-zero pion field, we can think of the fermions as being given by:

$$\Psi = \exp\left(i\frac{\pi^a}{f_\pi}\gamma^5\frac{\tau^a}{2}\right)\Psi. \tag{B20}$$

In other words, the pion fields behave like symmetry transformations of the vacuum (and everything else).

Now assume that there is an "effective interaction" for the pions, containing kinetic terms $(1/2)(\partial_\mu \pi^a)^2$. Taking the form above for Ψ, the pions obtain a potential from the fermion mass terms. To work out this potential one substitutes this form for the fermions into the Lagrangian and replaces the fermion bilinear form by its vacuum expectation value. This gives

$$V(\pi) = \langle \bar{q}q \rangle \,{\rm Tr}\big[\exp\left(i\omega^a \gamma_5 \tau^a\right) M\big]; \tag{B21}$$

one can now expand to second order in the pion fields, obtaining:

$$m_\pi^2 f_\pi^2 = (m_u + m_d)\langle \bar{q}q \rangle. \tag{B22}$$

Exercises

(1) Verify Eq. (B13).
(2) Derive Eq. (B22), known as the Gell-Mann–Oakes–Renner formula.

Appendix C

Some practice with the path integral in field theory

The path integral is extremely useful, both in field theory and in string theory. This appendix provides a brief review of path integration, and some applications. Many of the examples are drawn from finite-temperature field theory. These are instructive since one can easily write explicit expressions. They are also useful for understanding the high-temperature universe and are closely connected to the computations which arise in compactified theories.

C.1 Path integral review

Feynman gave an alternative formulation of quantum mechanics in which one calculates amplitudes by summing over the possible trajectories of a system, weighting by $e^{iS/\hbar}$, where S is th classical action of the trajectory. For a particle, the path integral is

$$Z = \int [dx]\, e^{iS/\hbar}. \tag{C1}$$

Here $\int [dx]$ implies an instruction to sum over all possible paths of the particle.

This generalizes immediately to field theory, where surprisingly it is often more useful than in the case of quantum systems with a small number of degrees of freedom:

$$Z = \int [d\phi]\, e^{iS}. \tag{C2}$$

For a single field ϕ it is useful to introduce sources $J(x)$ and to define

$$Z[J] = \int [d\phi] \exp\left\{ i \int d^4x \left[\frac{1}{2}(\partial\phi)^2 - V(\phi) + J\phi \right] \right\}. \tag{C3}$$

Green's functions for ϕ can then be obtained by the functional differentiation of Z with respect to J:

$$T\langle \phi(x_1)\cdots\phi(x_n)\rangle = \frac{\delta}{i\delta J(x_1)} \cdots \frac{\delta}{i\delta J(x_n)} Z[J]. \tag{C4}$$

For free fields the integral can be performed by completing the squares. Writing the action as

$$S_{\text{free}} = \int d^4x \left[\frac{1}{2}\phi(x) D^{-1}\phi(x) + \phi(x)\, J(x) \right], \tag{C5}$$

with

$$D^{-1} = \partial^2 - m^2 = p^2 - m^2, \tag{C6}$$

we can complete the squares in the action:

$$S_{\rm free} = \int d^4x \left[\frac{1}{2}\phi(x) + \int d^4y J(y) D(y,x)\right] D^{-1} \left[\phi(x) + \int d^4z J(z) D(z,x)\right] - \int d^4x d^4y J(x) D(x,y) J(y). \tag{C7}$$

Now, in the free field functional integral one can shift the ϕ integral, obtaining

$$Z_0[J] = \Delta \exp\left[\frac{-i}{2}\int d^4x d^4y J(x) D(x,y) J(y)\right]. \tag{C8}$$

Here Δ is the free field functional integral at $J = 0$. It is the square root of the functional determinant of the operator D; D itself is the propagator of the scalar. This expression can then be used to develop perturbation theory. For example, with a $(\lambda/4!)\phi^4$ interaction we can write

$$Z[J] = \exp\left[i\int d^4x \frac{\lambda}{4!}\left(\frac{\delta}{i\delta J(x)}\right)^4\right] Z_0[J]. \tag{C9}$$

Working out the terms in the power series reproduces precisely the Feynman diagram expansion.

This has generalizations to non-Abelian gauge theories, both those with unbroken and those with broken symmetries, which we discuss in Section 2.3. We will also find it useful for addressing other questions.

C.2 Finite-temperature field theory

As an application of path integral methods and because of its importance in cosmology, we consider at some length the problem of field theory at finite temperatures.

In statistical mechanics one is interested in the partition function,

$$Z[\beta] = {\rm Tr}\, e^{-\beta H}. \tag{C10}$$

For a quantum mechanical system in contact with a heat bath, we have

$$Z[\beta] = \sum_n \langle n| e^{-\beta E_n} |n\rangle, \tag{C11}$$

where n labels the energy eigenstates.

For a harmonic oscillator of unit mass, $H = [(p^2/2) + (\omega^2/2)]x^2$ and the partition function is:

$$e^{-\beta F} = \sum_n e^{-\beta\omega(n+1/2)}$$
$$= e^{-\omega\beta/2}\frac{1}{1-e^{-\beta\omega}}. \tag{C12}$$

Now, we can think of

$$\langle x|e^{-\beta H}|x\rangle \tag{C13}$$

as the amplitude for starting at x and ending up at x after propagating through an imaginary time $-i\beta$. This can be represented as a path integral:

$$\langle x|e^{-\beta H}|x\rangle = \int_{x(0)=x(\beta)=x}[dx]\exp\left(-\int_0^\beta dt\,L_{\rm E}\right), \tag{C14}$$

where $L_{\rm E}$ is the Euclidean Lagrangian,

$$L_{\rm E} = \left(\frac{dx}{dt}\right)^2 + \frac{1}{2}\omega^2x^2 \tag{C15}$$

(note the signs here!). The partition function is now

$$Z[\beta] = \int_{x(0)=x(\beta)=x_0}^{dx_0}[dx]\exp\left(-\int_0^\beta dt\,L_{\rm E}\right), \tag{C16}$$

i.e. we integrate over the possible values of x at $t = 0$ in order to take the trace. This is the problem of a box periodic in the time direction. For this simple system with one degree of freedom, we can write:

$$x(t) = \sum_n \frac{1}{\sqrt{T}}a_n e^{-2\pi int/\beta}. \tag{C17}$$

We will simplify the problem slightly by taking $x(t)$ to be complex (you can think of this simply as corresponding to an isotropic harmonic oscillator in two dimensions). The action of this configuration is

$$S = \sum_{n=-\infty}^{\infty}\frac{1}{2}(\omega_n^2+\omega^2)|a_n|^2. \tag{C18}$$

The path integral is now

$$Z[\beta] = \prod\int da_n da_n^*\,e^{-S_{\rm E}}. \tag{C19}$$

The integrals are just Gaussian integrals. For a complex variable z we have

$$\int d^2z\,e^{-a|z|^2} = \frac{\pi}{a}, \tag{C20}$$

so we have the following result for Z:

$$Z[\beta] = \prod \frac{1}{\omega^2 + \omega_n^2}, \tag{C21}$$

where $\omega_n = 2\pi n/T$.

Now, before trying to evaluate this product, it is useful to pause and note that it can be expressed in terms of the determinant of a matrix. Quite generally, Gaussian path integrals take the form of (inverse) determinants. In this case, if we write $\mathcal{M}$ as the differential operator

$$\mathcal{M} = \frac{1}{2}\left(-\frac{d^2}{dt^2} + \omega^2\right), \tag{C22}$$

its eigenfunctions are just $e^{i\omega_n t}$, with eigenvalues $\omega_n^2 + \omega^2$. So Z is just the inverse determinant of $\mathcal{M}$. Had we worked with only one real coordinate, we would have obtained the square root of the inverse determinant.

The determinant of an infinite matrix may seem a daunting object, but there are some tricks that permit evaluation in many cases. The first thing is to write the determinant as a sum, by taking logarithms. In general,

$$\det M = \exp(\mathrm{Tr} \ln M) \tag{C23}$$

(to see this, diagonalize M). It is easier to evaluate derivatives of the determinant rather than the determinant itself. We can obtain a very useful formula for the derivative of a determinant by writing

$$\begin{aligned} \det(M + \delta M) &= \exp\left[\mathrm{Tr}\ln(M + \delta M)\right] = \exp\left[\mathrm{Tr}\ln M + \ln(1 + M^{-1}\delta M)\right] \\ &= \exp\left(\mathrm{Tr}\ln M\right)\exp\left(\mathrm{Tr}\, M^{-1}\delta M\right) \approx \det M(1 + \mathrm{Tr}\, M^{-1}\,\delta M). \end{aligned} \tag{C24}$$

Dividing by δM gives the derivative.

In our case, it is convenient to study

$$\frac{1}{Z}\frac{d}{d\omega^2}Z = \sum_n \frac{1}{\omega^2 + \omega_n^2}. \tag{C25}$$

This is progress. Our infinite product is now an infinite sum. The question is: how do we do the sum? The trick is to look for a periodic function which is well-behaved at infinity but has poles at the integers. A suitable choice is

$$\frac{1}{e^{iz\beta} - 1}. \tag{C26}$$

We can then replace any sum of the form $\sum f(n)$ by a contour integral,

$$\frac{1}{2\pi}\int dz\, f(z)\frac{1}{e^{iz\beta} - 1}. \tag{C27}$$

Here the contour is a line running just above the real z axis and back again just below it. The residues of the (infinite number of) poles give back the original sum.

Now one can deform the contour, taking one line into the upper half plane and the other into the lower, picking up the poles at $z = \pm i\omega$. This leaves us with

$$\frac{dF}{d\omega^2} = \left(\frac{1}{e^{-\omega\beta} - 1} - \frac{1}{e^{\omega\beta} - 1}\right)\frac{1}{2\omega}. \tag{C28}$$

We could analyze this problem further, but let us jump instead to free-field theory. Then

$$Z[\beta] = \int_{\phi(\beta)=\phi(0)} [d\phi] \exp\left\{-\int d^4x[(\partial_\mu\phi)^2 + m^2\phi^2]\right\}. \tag{C29}$$

In a finite box, with periodic boundary conditions, we can make the following expression:

$$\phi(\vec{x}, t) = \sum_{\vec{k},m} \exp\left(i\vec{k}_n \cdot \vec{x} + i\omega_m t\right)\phi_{\vec{k},m}, \tag{C30}$$

where $\omega_m = 2\pi mT$.

In this form we have that

$$Z[\beta] = \det(-\partial^2 + m^2)^{-1/2}. \tag{C31}$$

Again, this is somewhat awkward to work with. It is easier to differentiate it:

$$\frac{1}{Z}\frac{\partial Z}{\partial m^2} = \frac{1}{Z}\int [d\phi] \exp\left(-\int d^4x\, \mathcal{L}_\mathcal{E}\right)\int d^4z\, \frac{1}{2}\phi^2(z). \tag{C32}$$

This is just the propagator, with periodic boundary conditions in the time direction:

$$\int d^4z\, \langle\phi(z)\phi(z)\rangle = \beta V\langle\phi(0)\phi(0)\rangle. \tag{C33}$$

The propagator is given by

$$\langle\phi(0)\phi(0)\rangle = \sum_m \sum_k \frac{1}{\omega_m^2 + \vec{k}^2 + m^2}. \tag{C34}$$

We can convert this into a more recognizable form by means of the same trick as above. The propagator is given by the expression below:

$$\langle\phi(0)\phi(0)\rangle = \int \frac{d^3k}{(2\pi)^3}\frac{1}{2\pi}\int \frac{dz}{e^{iz\beta} - 1}\frac{1}{(2\pi nT)^2 + \vec{k}^2 + m^2}. \tag{C35}$$

Now deform the contour as before, picking up the poles at $\pm i\sqrt{\vec{k}^2 + m^2}$. Both poles make the same contribution, yielding

$$\begin{aligned} &\frac{1}{2\sqrt{\vec{k}^2 + m^2}}\left(\frac{1}{\exp\left(-\beta\sqrt{\vec{k}^2 + m^2}\right) - 1} - \frac{1}{\exp\left(\beta\sqrt{\vec{k}^2 + m^2}\right) - 1}\right) \\ &= \frac{1}{2\sqrt{\vec{k}^2 + m^2}}\left(1 + \frac{2}{\exp\left(\beta\sqrt{\vec{k}^2 + m^2}\right) - 1}\right). \end{aligned} \tag{C36}$$

Note the appearance of the Bose–Einstein factors here. Note also that the first term has the structure of the zero-temperature expression for the energy; the second is the finite-temperature expression. This is what we find on differentiating Eq. (C36):

$$\beta F = V \int \frac{d^3k}{(2\pi)^3} \left[\frac{1}{2} E_k + \beta^{-1} \ln(1 - e^{-\beta E_k}) \right]. \tag{C37}$$

Note the connection with the result for the single oscillator. So far our discussion has been for free-field theory but we can extend it immediately to interacting theories by developing a perturbation order-by-order in the couplings, just as at zero temperature.

C.3 QCD at high temperatures

Two particularly important cases are QCD and the weak-interaction theory. At low energies QCD is a complicated theory but, at high temperatures, things simplify drastically. In perturbation theory, if we are studying the free energy, for example, over above path integral analysis instructs us to study a Euclidean problem with discrete energies which are multiples of T. So, provided that we do not encounter infrared problems, the free energy should be a power series in $g^2(T)$, calculable in perturbation theory.

One can argue that there is actually a phase transition between a confined phase and a deconfined phase. To find an order parameter for this transition, we start by considering a Wilson line, running between imaginary times $t = 0$ and $t = \beta$,

$$U_T(\vec{x}) = P \exp \left[i \int_0^\beta A^0(\vec{x}, t) dt \right]. \tag{C38}$$

Because of the periodic boundary conditions, this expression is gauge invariant. The correlation of two such operators is related to the potential of two static quarks:

$$P(R) = \langle U_T(\vec{R}) U_T(0) \rangle = C \exp[-\beta V(R)]. \tag{C39}$$

In a confining phase, with a linear potential between the quarks, $P(R)$ vanishes exponentially with R. In a Coulomb phase (nearly free quarks), it will tend to a constant. At very high temperatures we would expect that we could compute P in a power series in $g^2(T)$ and that we will find free-quark behavior. Numerical studies show that there is indeed a phase transition at a particular temperature between confined and unconfined phases. The order of the transition depends on the group.

Finite-temperature perturbation theory suffers from infrared divergences, even at very high temperatures. The problem is the zero-frequency modes in the sum over frequencies. If we simply set all the frequencies to zero, we have the Feynman diagrams of a three-dimensional field theory. At four loops the divergence is logarithmic. At higher loops it is power law.

We can understand this directly in the path integral. Consider a massless scalar field. The exponent in the path integral is

$$\int_0^\beta dt d^3x\, (\partial_\mu \phi)^2. \tag{C40}$$

For small β, assuming it makes sense to treat fields as constant in β, the path integral thus becomes

$$\int [d\phi(\vec{x})]\, e^{-\beta H}, \tag{C41}$$

which is the classical partition function for the three-dimensional system.

Thought of in this way, there is a natural guess for how the infrared divergences are cut off. A three-dimensional gauge theory has a dimensionful coupling λ^2. One might expect that such a theory has a mass gap proportional to λ^2 (in three dimensions, the gauge coupling has the dimensions of $\sqrt{M}$). In the present case the coupling is $\lambda = g^2 T$. This scale then would cut off the infrared divergence. This suggests that the theory at finite temperature makes sense but does not help a great deal with computations. The problem is that in four loops we obtain a contribution $g^8 \ln g^2$ but, at higher orders, we obtain a power series in g^2/g^2, i.e. we can at best compute the leading logarithmic term at four loops. It is possible to study some of these issues numerically in lattice gauge theory, which provides some support for this picture.

Instanton effects at high temperatures

In QCD at zero temperature we saw that instanton calculations were plagued by infrared divergences. At high temperatures this is not the case. The scale invariance of the zero-energy theory is lost and the instanton solution has a definite scale, of order the temperature. As a result, instanton effects behave as $\exp[-8\pi^2/g^2(T)]$ and are calculable. Thus it is possible to compute the θ-dependence systematically. This is particularly relevant to the understanding of the axion in the early universe.

C.4 Weak interactions at high temperatures

The weak interactions exhibit different phenomena at high temperatures. Most strikingly, there is a transition between a phase in which the gauge bosons are massive and one in which they are massless. This transition can be uncovered in perturbation theory. By analogy with the phase transition in the Landau–Ginzburg model of superconductivity, one might expect that the value of $\langle \Phi \rangle$ will change as the temperature increases. To determine the value of Φ one must compute the free energy as a function of Φ. The leading temperature-dependent corrections are obtained by simply noting that the masses of the various fields in the theory (the W and Z bosons and the Higgs field, in particular) depend on Φ. So the contributions of each species to the free energy are Φ-dependent:

$$\mathcal{F}(\Phi) V_T(\Phi) = \pm \sum_i \int \frac{d^3p}{2\pi^3} \ln \left\{ 1 \mp \exp\left[-\beta \sqrt{p^2 + m_i^2(\Phi)} \right] \right\}, \tag{C42}$$

where $\beta = 1/T$, T is the temperature, the sum is over all particle species (physical helicity states) and the plus sign is for bosons, the minus for fermions. In the Standard Model, for temperature $T \sim 10^2$ GeV, one can treat all the quarks as massless except for the top quark. The effective potential (C42) then depends on the top quark mass m_t, the vector boson masses M_Z and m_W and the Higgs mass m_H. Performing the integral in the equation yields

$$V(\Phi, T) = D\big(T^2 - T_0^2\big)\Phi^2 - ET\Phi^3 + \frac{\lambda}{4}\Phi^4 + \cdots . \tag{C43}$$

The parameters T_0, D and E are given in terms of the gauge boson masses and the gauge couplings. For the moment, though, it is useful to note certain features of this expression. The quantity E turns out to be a rather small dimensionless number, of order 10^{-2}. If we ignore the ϕ^3 term then we have a second-order transition, at temperature T_0, between a phase with $\phi \neq 0$ and a phase with $\phi = 0$. Because the W and Z masses are proportional to ϕ, this is a transition between states with massive and massless gauge bosons.

Because of the ϕ^3 term in the potential, the phase transition is potentially at least weakly first order. A second, distinct, minimum appears at a critical temperature. A first-order transition is not, in general, an adiabatic process. As we lower the temperature to the transition temperature, the transition proceeds by the formation of bubbles; inside the bubble the system is in the true equilibrium state (the state which minimizes the free energy) while outside it tends to the original state. These bubbles form through thermal fluctuations at different points in the system and grow until they collide, completing the phase transition. The moving bubble walls are regions where the Higgs fields are changing and all Sakharov's conditions are satisfied.

C.5 Electroweak baryon number violation

We have seen that, at low temperatures, violations of baryon and lepton number are extremely small. This is not the case at high temperatures, where baryon number violation is a rapid process which can come to thermal equilibrium. This has at least two possible implications. First, it is conceivable that these sphaleron (see below) processes can themselves be responsible for generating a baryon asymmetry. This is called electroweak baryogenesis. Second, sphaleron processes can change an existing lepton number, producing a net lepton and baryon number. This is the process called leptogenesis. In this section, we summarize the main arguments showing that the electroweak interactions violate baryon number at high temperature.

Recall that, classically, the ground states are field configurations for which the energy vanishes. The trivial solution of this condition is $\vec{A} = 0$, where $\vec{A}$ is the vector potential. More generally, one can consider an $\vec{A}$ which is a "pure gauge",

$$\vec{A} = \frac{1}{i} g^{-1} \vec{\nabla} g, \tag{C44}$$

where g is a gauge transformation matrix. In an Abelian ($U(1)$) gauge theory, fixing the gauge eliminates all but the trivial solution, $\vec{A} = 0$.[1] This is not the case for non-Abelian gauge theories. There is a class of gauge transformations, labeled by a discrete index n, which do not tend to unity as $|\vec{x}| \to \infty$ and which therefore must be considered to be distinct states. These have the form:

$$g_n(\vec{x}) = e^{inf(\vec{x})\hat{x}\cdot\tau/2}, \tag{C45}$$

where $f(x) \to 2\pi$ as $\vec{x} \to \infty$ and $f(\vec{x}) \to 0$ as $\vec{x} \to 0$.

So, the ground states of the gauge theory are labeled by an integer n. Now if we evaluate the integral of the current K^0, we obtain a quantity known as the *Chern–Simons number*:

$$n_{\rm CS} = \frac{1}{16\pi^2}\int d^3x K^0 = \frac{2/3}{16\pi^2}\int d^3x\, \epsilon_{ijk}{\rm Tr}(g^{-1}\partial_i g g^{-1}\partial_j g g^{-1}\partial_k g). \tag{C46}$$

For $g = g_n$, $n_{\rm CS} = n$. The reader can also check that for $g' = g_n(x)h(x)$, where h is a gauge transformation which tends to unity at infinity (a so-called "small gauge transformation"), this quantity is unchanged. The Chern–Simons number $n_{\rm CS}$, is topological in this sense (for $\vec{A}$s which are not pure gauge, $n_{\rm CS}$ is in no sense quantized).

Schematically, we can thus think of the vacuum structure of a Yang–Mills theory as indicated in Fig. C.1. We have, at weak coupling, an infinite set of states, labeled by integers, and separated by barriers from one another. In tunneling processes which change the Chern–Simons number, because of the anomaly the baryon and lepton numbers will change. The exponential suppression found in the instanton calculation is typical of tunneling processes, and in fact the instanton calculation which leads to the result for the amplitude is nothing other than a field-theoretic WKB calculation.

One can determine the height of the barrier separating configurations having different $n_{\rm CS}$ by looking for the field configuration which corresponds to a particle top of the barrier. This is a solution of the static equations of motion with finite energy. It is known as a *sphaleron*. When one studies the small fluctuations about this solution, one finds that there is a single negative mode, corresponding to the possibility that the system will roll downhill into one or the other well. The sphaleron energy is of order

$$E_{\rm sp} = \frac{c}{g^2}M_W. \tag{C47}$$

This can be seen by using scaling arguments on the classical equations; determining the coefficient c requires a more detailed analysis. The rate for thermal fluctuations to cross the barrier per unit time per unit volume should be of order the Boltzmann factor for this configuration, multiplied by a suitable prefactor:

$$\Gamma_{\rm sp} = T^4 e^{-E_{\rm sp}/T}. \tag{C48}$$

Note that the rate becomes large as the temperature approaches the W boson mass. The W boson mass itself goes to zero as one approaches the electroweak phase transition.

[1] More precisely, this is true in axial gauge. In the gauge $A_0 = 0$, it is necessary to sum over all time-independent transformations in order to construct a state which obeys Gauss's law.

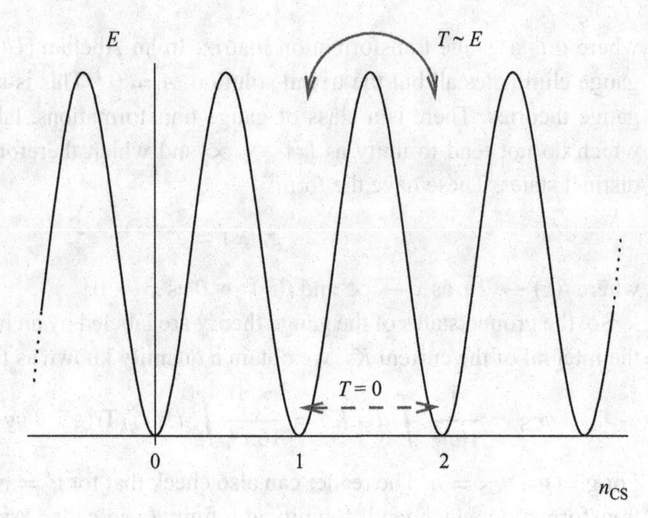

Fig. C.1 Schematic Yang–Mills vacuum structure. At zero temperature instanton transitions between vacua with different Chern–Simons numbers are suppressed. At finite temperature these transitions can proceed via sphalerons.

At this point the computation of the transition rate is a difficult problem – there is no small parameter – but general scaling arguments show that the transition rate is of the form:[2]

$$\Gamma_{\text{bv}} = \alpha_W^4 T^4. \tag{C49}$$

Suggested reading

The path integral is well treated in most modern field theory textbooks. Peskin and Schroder (1995) provide a concise introduction. High-temperature field theory is developed in a number of textbooks, such as that of Kapusta (1989).

Exercises

(1) Go through the calculation of the free energy of a free scalar field, being careful about factors of 2 and π.

(2) Compute the constants appearing in Eq. (C43). Plot the free energy, and show that the transition is weakly first order.

(3) Show, by power counting, that infrared divergences first appear in the free energy of a gauge theory at three loops. To do this you can look at the zero-frequency terms in the sums over frequency. Show that the divergences become more severe at higher orders.

[2] More detailed considerations alter slightly the parametric form of the rate.

Appendix D

The beta function in supersymmetric Yang–Mills theory

We have seen that holomorphy is a powerful tool with which to understand the dynamics of supersymmetric field theories. But one can easily run into puzzles and paradoxes. One source of confusion is the holomorphy of the gauge coupling. At tree level, the gauge coupling arises from a term in the action of the form

$$\int d^2\theta S W_\alpha^2, \tag{D1}$$

where $S = -1/4g^2 + ia$. This action, in perturbation theory, has a symmetry

$$S \to S + i\alpha. \tag{D2}$$

This is just an axion shift symmetry. Combined with holomorphy, it greatly restricts the form of the effective action. The only allowed terms are:

$$\mathcal{L}_{\text{eff}} = \int d^2\theta\, (S + \text{constant}) W_\alpha^2. \tag{D3}$$

The constant term corresponds to a one-loop correction. But higher-loop corrections are forbidden.

However, it is well known that there are two-loop corrections to the beta function in supersymmetric Yang–Mills theories (higher-loop corrections have also been computed). Does this represent an inconsistency? This puzzle can be stated – and has been stated – in other ways. For example, the axial anomaly lies in a supermultiplet with the conformal anomaly – the anomaly in the trace of the stress tensor. One usually says that the axial anomaly is not renormalized but that the trace anomaly is proportional to the beta function.

The resolution to this puzzle was provided by Shifman and Vainshtein; we will present it in a form developed by Arkani-Hamed and Murayama and updated by Dine and Festuccia. The idea is to exploit the finiteness of $N = 4$ supersymmetric Yang–Mills to use it as a regulator for the pure $N = 1$ gauge theory (this can be generalized to a variety of other theories as well). We take the $N = 4$ theory and add masses for the adjoint chiral fields. Calling these masses M, the low-energy theory is just pure Yang–Mills. The ultraviolet divergences of the theory necessarily become logarithms of M.

In the language of $N = 1$, the $N = 4$ theory is often presented in a way which makes the $SU(4)$ R-symmetry manifest:

$$\mathcal{L} = \int d^4\theta \frac{1}{g^2}\Phi_i^\dagger \Phi_i - \frac{1}{32\pi^2}\int d^2\theta \left(\frac{8\pi^2}{g^2} + i\theta\right) W_\alpha^2 + \int d^2\theta \frac{1}{g^2}\Phi_1\Phi_2\Phi_3. \tag{D4}$$

It is helpful to present the theory in a fashion which is holomorphic in the gauge coupling, $\tau = 8\pi^2/g^2 + i\theta$. This is achieved by the rescaling $\Phi_i \to g^{2/3}\Phi_i$. Then, including a holomorphic mass term for the Φ_is,

$$\mathcal{L} = \int d^4\theta \frac{1}{g^{2/3}} \Phi_i^\dagger \Phi_i - \frac{1}{32\pi^2} \int d^2\theta \left(\frac{8\pi^2}{g^2} + i\theta \right) W_\alpha^2 + \int d^2\theta \left(\Phi_1 \Phi_2 \Phi_3 + m_{\text{hol}} \Phi_i \Phi_i \right). \tag{D5}$$

Now, consider integrating out the physics between two scales, m_1 and m_2. Since there are no infrared divergences and we have written the Lagrangian in a manifestly holomorphic form, the coupling renormalization is necessarily holomorphic:

$$\frac{8\pi^2}{g^2(m_2)} = \frac{8\pi^2}{g^2(m_1)} + b_0 \log(m_{\text{hol}}^{(2)}/m_{\text{hol}}^{(1)}). \tag{D6}$$

From the Lagrangian, Eq. (D5) we see that, at the classical level, the physical mass (i.e. the actual mass of the Φ_i particles) and the holomorphic mass are related by

$$m_{\text{hol}} = g^{-2/3} m. \tag{D7}$$

So, defining the beta function by

$$\beta(g) = \frac{\partial g}{\partial \log m} \tag{D8}$$

and differentiating Eq. (D6) yields

$$\beta(g) = -\frac{g^3}{16\pi^2} \frac{3N}{1 - 2Ng^2/(16\pi^2)}. \tag{D9}$$

This expression is known as the Novikov–Shifman–Vainstein–Zakharov (NSVZ) beta function. It is, in some sense, exact since the holomorphic expression is exact. However, if we insist, for example, that m should be the physical ("pole") mass of the Φ_i particles then the relation between m and m_{hol} is corrected in each order of perturbation theory. Indeed, the scheme in which the NSVZ beta function is exact is precisely that in which one insists that m and g are related as in Eq. (D7). So, such exact relations must be used with care. In any case this analysis is readily extended to gauge theories with matter, which can be embedded in finite $N = 2$ theories.

Suggested reading

The use of finite theories as regulators was developed in Arkani-Hamed and Murayama (2000); the presentation described here appears in Dine *et al.* (2011).

Exercise

Starting with the finite $N-2$ theories discussed in Chapter 16, proceed as we did for the $N-4$ theories (Eqs. (D4)–(D6)) to derive the beta function for $N-1$ theories with matter, Eq. (16.35). Note that the analysis is valid for only a restricted number of flavors.

References

Affleck, I. (1980). Testing the instanton method. *Phys. Lett.* **B, 92**, 149.

Affleck, I., Dine, M. and Seiberg, N. (1984). Dynamical supersymmetry breaking in supersymmetric QCD. *Nucl. Phys. B*, **241**, 493.

Aharony, O., Gubser, S. S., Maldacena, J. M., Ooguri, H. and Oz, Y. (2000). Large *N* field theories, string theory and gravity. *Phys. Rept*, **323**, 183 [arXiv:hep-th/9905111].

Albrecht, A. and Steinhardt, P. J. (1982). Cosmology for grand unified theories with radiatively induced symmetry breaking. *Phys. Rev. Lett.*, **48**, 1220.

Almheiri, A., Marolf, D., Polchinski J., and Sully, J. (2013). Black holes: complementarity or firewalls?, *JHEP* **1302**, 062 [hep-th/arXiv:1207.3123].

Appelquist, T., Chodos, A. and Freund, P. G. O. (1985). *Modern Kaluza–Klein Theories*. Menlo Park: Benjamin/Cummings.

Arkani-Hamed, N., Dimopoulos, S., Giudice G. F., and Romanino, A. (2005). Aspects of split supersymmetry, *Nucl. Phys. B*, **709**, 3 [arXiv:hep-ph/0409232].

Arkani-Hamed, N., Dimopoulos, S. and Dvali, G. R. (1999). Phenomenology, astrophysics and cosmology of theories with sub-millimeter dimensions and TeV scale quantum gravity. *Phys. Rev. D*, **59**, 086004 [arXiv:hep-ph/9807344].

Arkani-Hamed N., and Murayama, H. (2000). Holomorphy, rescaling anomalies and exact beta functions in supersymmetric gauge theories, *JHEP*, **0006**, 030 [arXiv:hep-th/9707133].

Ashok, S. and Douglas, M. R. (2004). Counting flux vacua. *JHEP*, **0401**, 060 [arXiv:hep-th/0307049].

Atick, J. J., Dixon, L. J. and Sen, A. (1987). String calculation of Fayet–Iliopoulos D terms in arbitrary supersymmetric compactifications. *Nucl. Phys. B*, **292**, 109.

Bagger, J. A., Moroi, T. and Poppitz, E. (2000). Anomaly mediation in supergravity theories. *JHEP*, **0004**, 009 [arXiv:hep-th/9911029].

Bailin, D. and Love, A. (1993). *Introduction to Gauge Field Theory*. London: Institute of Physics.

Banks, T., Kaplan, D. B. and Nelson, A. E. (1994). Cosmological implications of dynamical supersymmetry breaking. *Phys. Rev. D*, **49**, 779 [hep-ph/9308292].

Banks, T., Fischler, W., Shenker, S. H. and Susskind, L. (1997). M theory as a matrix model: A conjecture. *Phys. Rev. D*, **55**, 5112 [arXiv:hep-th/9610043].

Banks, T., Dine, M. and Graesser, M. (2003). Supersymmetry, axions and cosmology. *Phys. Rev. D*, **68**, 075011 [arXiv:hep-ph/0210256].

Berger, V., Marfatia, D. and Whisnant, K. (2012). *The Physics of Neutrinos,* Princeton, NJ: Princeton University Press.

Becker, K., Becker, M. and Schwarz, J. H. (2007). *String Theory and M-Theory: A Modern Introduction*, Cambridge: Cambridge. University Press.

Berkooz, M., Dine, M. and Volansky, T. (2004). Hybrid inflation and the moduli problem. *Phys. Rev. D*, **71**, 103 502 [arXiv:hep-ph/0409226].

Bigatti, D. and Susskind, L. (1997). Review of matrix theory. In *Proc. Conf. on Strings, Branes and Dualities*, eds. L. Balieu, P. Di Francesco, M. Douglas *et al.* Amsterdam: Kluwer [arXiv:hep-th/9712072].

Bousso, R. and Polchinski, J. (2000). Quantization of four-form fluxes and dynamical neutralization of the cosmological constant. *JHEP*, **0006**, 006 [arXiv:hep-th/0004134].

Brown, L. S., Carlitz, R. D., Creamer, D. B. and Lee, C. K. (1978). Propagation functions in pseudoparticle fields. *Phys. Rev. D*, **17**, 1583.

Buchmuller, W., Di Bari, P. and Plumacher, M. (2005). Leptogenesis for pedestrians. *Ann. Phys.*, **315**, 305 [arXiv:hep-ph/0401240].

Buican, M., Meade, P., Seiberg, N. and Shih, D. (2009). Exploring general gauge mediation, *JHEP*, **0903**, 016 [arXiv:hep-ph/0812.3668].

Carena, M., Quiros, M., Seco, M. and Wagner, C. E. M. (2003). Improved results in supersymmetric electroweak baryogenesis. *Nucl. Phys. B*, **650**, 24 [arXiv:hep-ph/0208043].

Carpenter, L. M. Dine, M., Festuccia, G. and Mason, J. D. (2009). Implementing general gauge mediation, *Phys. Rev. D*, **79**, 035002. [arXiv:hep-ph 0805.2944/hep-ph].

Carroll, S. (2004). *Spacetime and Geometry: An Introduction to General Relativity*. San Francisco: Addison-Wesley.

Casher, A., Kogut, J. B. and Susskind, L. (1974). Vacuum polarization and the absence of free quarks. *Phys. Rev. D*, **10**, 732.

Cheng, T. and Li, L. (1984). *Gauge Theory of Elementary Particle Physics*. Oxford: Clarendon Press.

Chivukula, R. S. (2000). Technicolor and compositeness, arXiv:hep-ph/0011264.

Cohen, A. G., Kaplan, D. B. and Nelson, A. E. (1993). Progress in electroweak baryogenesis. *Ann. Rev. Nucl. Part. Sci.*, **43**, 27 [arXiv: hep-ph/9302210].

Coleman, S. (1983). The magnetic monopole fifty years later. In ed. A. Zichichi. *The Unity of the Fundamental Interactions*. New York: Plenum Press.

Coleman, S. (1985). The uses of instantons. In *Aspects of Symmetry*. Cambridge: Cambridge University Press.

Cottingham, W. N. and Greenwood, D. A. (1998). *An Introduction to the Standard Model of Particle Physics*. Cambridge: Cambridge University Press.

Cremmer, E., Julia, B. and Scherk, J. *et al.* (1979). Spontaneous symmetry breaking and Higgs effect in supergravity without cosmological constant. *Nuclear Physics B*, **147**, 105.

Creutz, M. (1983). *Quarks, Gluons and Lattices*. Cambridge: Cambridge University Press.

Crewther, R. J., Di Vecchia, P., Veneziano, G. and Witten, E. (1979). *Phys. Lett.*, **88B**, 123.

Csaki, C., Hubisz, J. and Meade, P. (2005). TASI lectures on electroweak symmetry breaking from extra dimensions, arXiv:hep-ph/0510275.

Davoudiasl, H., Hewett, J. L. and Rizzo, T. G. (2000). Phenomenology of the Randall–Sundrum gauge hierarchy model. *Phys. Rev. Lett.*, **84**, 2080 [arXiv:hep-ph/9909255].

de Carlos, B., Casas, J. A., Quevedo, F. and Roulet, E. (1993). Model independent properties and cosmological implications of the dilaton and moduli sectors of 4-d strings. *Phys. Lett. B*, **318**, 447 [arXiv:hep-ph/9308325].

Denef, F., Douglas, M. R. and Kachru, S. (2007). Physics of string flux compactifications, *Ann. Rev. Nucl. Part. Sci.*, **57**, 119 [arXiv:hep-th/0701050].

D'Hoker, E. and Freedman, D. Z. (2002). Supersymmetric gauge theories and the AdS/CFT correspondence [arXiv:hep-th/0201253].

Di Pietro, L., Dine, M. and Komargodski, Z. (2014). (Non-)decoupled supersymmetric field theories, *JHEP*, **1404**, 073 (2014) [arXiv:hep-th/1402.3385].

Dimopoulos, S. and Georgi, H. (1981). Softly broken supersymmetry and *SU*(5). *Nucl. Phys. B*, **193**, 150.

Dine, M. and Kusenko, A. (2003). The origin of the matter–antimatter asymmetry. *Rev. Mod. Phys.*, **76**, 1 [arXiv:arXiv:hep-ph/0303065].

Dine, M. and Seiberg, N. (1986). Nonrenormalization theorems in superstring theory. *Phys. Rev. Lett.*, **57**, 2625.

Dine, M. and Seiberg, N. (2007). Comments on quantum effects in supergravity theories, *JHEP*, **0703**, 040 [arXiv:hep-th/0701023].

Dine, M., Seiberg, N., Wen, X. G. and Witten, E. (1986). Nonperturbative effects on the string world sheet. *Nucl. Phys. B*, **278**, 769.

Dine, M., Ichinose, I. and Seiberg, N. (1987a). F terms and D terms in string theory. *Nucl. Phys. B*, **293**, 253.

Dine, M., Seiberg, N., Wen, X. G. and Witten, E. (1987b). Nonperturbative effects on the string world sheet. 2. *Nucl. Phys. B*, **289**, 319.

Dine, M., Seiberg, N. and Witten, E. (1987c). Fayet–Iliopoulos terms in string theory. *Nucl. Phys. B*, **289**, 589.

Dine, M., Festuccia, G., Pack, L., Park, C. S., Ubaldi L. and W. Wu, (2011). Supersymmetric QCD: exact results and strong coupling. *JHEP*, **1105**, 061 [arXiv:hep-th/1104.0461].

Dine, M. and Draper, P. (2014). Anomaly mediation in local effective theories. *JHEP*, **1402**, 069 [arXiv:hep-ph/1310.2196].

Distler, J. and Greene, B. R. (1988). Some exact results on the superpotential from Calabi–Yau compactifications. *Nucl. Phys. B*, **309**, 295.

Dixon, L. J. (2013). A brief introduction to modern amplitude methods, arXiv:hep-ph/1310.5353.

Dixon, L. J., Harvey, J. A., Vafa, C. and Witten, E. (1986). Strings on orbifolds. 2. *Nucl. Phys. B*, **274**, 285.

Dodelson, S. (2004). *Modern Cosmology*. Burlington: Academic Press.

Donoghue, J. F., Golowich, E. and Holstein, B. R. (1992). *Dynamics of the Standard Model*. Cambridge: Cambridge University Press.

Englert, F. and Brout, R. (1964). Broken symmetry and the mass of gauge vector mesons, *Phys. Rev. Lett.*, **13**, 321.

Eidelman, S. *et al.* [Particle Data Group] (2004). Review of particle physics. *Phys. Lett. B*, **592**, 1 (This article – and others – can be found on the Particle Data Group website).

Enqvist, K. and Mazumdar, A. (2003). Cosmological consequences of MSSM flat directions. *Phys. Rept*, **380**, 99 [arXiv:hep-ph/0209244].

Faraggi, A. E. (1999). Toward the classification of the realistic free fermionic models. *Int. J. Mod. Phys. A*, **14**, 1663 [arXiv:hep-th/9708112].

Feng., J. L., March-Russell, J., Sethi, S. and Wilczek, F. (2001). Saltatory relaxation of the cosmological constant. *Nucl. Phys. B*, **602**, 307 [arXiv:hep-th/0005276].

Fradkin, E. and Shenker, S. (1979). Phase diagrams of lattice gauge theories with Higgs fields. *Phys. Rev. D*, **19**, 3602.

Gates, S. J., Grisaru, M. R. and Siegel, W. (1983). *Superspace: or One Thousand and One Lessons in Supersymmetry*. San Francisco: Benjamin/Cummings.

Gepner, D. (1987). Exactly solvable string compactifications on manifolds of $SU(N)$ holonomy. *Phys. Lett. B*, **199**, 380.

Giudice, G. F. and Rattazzi, R. (1999). Theories with gauge-mediated supersymmetry breaking. *Phys. Rept.*, **322**, 419 [arXiv:hep-ph/9801271].

Glashow, S. L., Iliopoulos, J. and Maiani, L. (1970). Weak interactions with lepton–hadron symmetry. *Phys. Rev. D*, **2**, 1285.

Green, M. B., Schwarz, J. H. and Witten, E. (1987). *Superstring Theory,* Cambridge: Cambridge University Press.

Greene, B. R., Kirklin, K. H., Miron, P. J. and Ross, G. G. (1987). A three generation superstring model. 2. Symmetry breaking and the low-energy theory, *Nucl. Phys. B*, **292**, 606.

Gross, D. J. and Wilczek, F. (1973). Ultraviolet behavior of non-abelian gauge theories. *Phys. Rev. Lett.*, **30**, 1343.

Gross, D. J., Harvey, J. A., Martinec, E. J. and Rohm, R. (1985). Heterotic string theory. 1. The free heterotic string. *Nucl. Phys. B*, **256**, 253.

Guralnik, G. S. Hagen, C. R. and Kibble, T. W. B. (1964) Global conservation laws and massless particles, *Phys. Rev. Lett.* **13**, 585.

Gross, D. J., Harvey, J. A., Martinec, E. J. and Rohm, R. (1986). Heterotic string theory. 2. The interacting heterotic string. *Nucl. Phys. B*, **267**, 75.

Hartle, J. B. (2003). *Gravity, an Introduction to Einstein's General Relativity*. San Francisco: Addison-Wesley.

Harvey, J. (1996). Magnetic monopoles, duality, and supersymmetry, arXiv:hep-th/-9603086.

Hall, L. J., Pinner, D. and Ruderman, J. T. (2012). A natural SUSY Higgs near 126 GeV, JHEP **1204**, 131. [arXiv:hep-ph/1112.2703].

Hawking, S. W. (1976). Breakdown of predictability in gravitational collapse, *Phys. Rev. D*, **14**, 2460.

Higgs, P. W. (1964). Broken symmetries and the masses of gauge bosons. *Phys. Rev. Lett.* **13**, 508.

Intriligator, K. A. and Seiberg, N. (1996). Lectures on supersymmetric gauge theories and electric–magnetic duality. *Nucl. Phys. Proc. Suppl.*, **45BC**, 1 [arXiv:hep-th/9509066].

Intriligator, K. A., Seiberg, N. and D. Shih (2006). Dynamical SUSY breaking in meta-stable vacua. *JHEP*, **0604**, 021 [arXiv:hep-th/0602239].

Intriligator, K. and Thomas, S. (1996). Dynamical supersymmetry breaking on quantum moduli spaces. *Nucl. Phys.*, **B473**, 121, arXiv:hep-th/9603158.

Jackson, J. D. (1999). *Classical Electrodynamics*. Hoboken: Wiley.

Johnson, C. (2003). *D-Branes*. Cambridge: Cambridge University Press.

Kachru, S., Kallosh, R., Linde, A. and Trivedi, S. (2003). De Sitter vacua in string theory. *Phys. Rev. D*, **68**, 046005 [arXiv:hep-th/0301240].

Kapusta, J. I. (1989). *Finite-Temperature Field Theory*. Cambridge: Cambridge University Press.

Kolb, E. W. and Turner, M. S. (1990). *The Early Universe*. Redwood City: Addison-Wesley.

Kribs, G. D. (2004). TASI 2004 lectures on the phenomenology of extra dimensions, arXiv:hep-ph/0605325.

Linde, A. D. (1982). A new inflationary universe scenario: a possible solution of the horizon, flatness, homogeneity, isotropy and primordial monopole problems. *Phys. Lett.*, **108B**, 389.

Linde, A. (1990). *Particle Physics and Inflationary Cosmology*. Reading: Harwood Academic.

Linde, A. D. (1994). Hybrid inflation. *Phys. Rev. D*, **49**, 748 [arXiv:astro-ph/9307002].

Lykken, J. D. (1996). Introduction to supersymmetry, arXiv:hep-th/9612114.

Maldacena, J. M. (1997). The large *N* limit of superconformal field theories and supergravity. *Adv. Theor. Math. Phys.*, **2**, 231 [arXiv:hep-th/9711200].

Manohar, A. V. and Wise, M. B. (2000). *Heavy Quark Physics*. Cambridge: Cambridge University Press.

Martin, S. P. and Vaughn, M. T. (1994). Two loop renormalization group equations for soft supersymmetry breaking couplings. *Phys. Rev. D*, **50**, 2282 [arXiv:hep-ph/9311340].

Masiero, A. and Silvestrini, L. (1997). Two lectures on FCNC and CP violation in supersymmetry, arXiv:hep-ph/9711401.

Meade, P. Seiberg, N. and Shih, D. (2009). General gauge mediation, *Prog. Theor. Phys. Suppl.* **177**, 143. [arXiv:hep-ph/0801.3278].

Mohapatra, R. N. (2003). *Unification and Supersymmetry: The Frontiers of Quark – Lepton Physics*. Berlin: Springer-Verlag.

Murayama, H. and Pierce, A. (2002). Not even decoupling can save minimal supersymmetric *SU*(5). *Phys. Rev. D* **65**, 055009. [arXiv:hep-ph/0108104].

Nilles, H. P. (1984). Supersymmetry and supergravity. *Phys. Rept*, **110**, 1.

Olive, D. I. and Witten, E. (1978). Supersymmetry algebras that include topological charges. *Phys. Lett. B*, **78**, 97.

Pais, A. (1986). *Inward Bound*. Oxford: Clarendon Press.

Peet, A. (2000). TASI lectures on black holes in string theory [arXiv:hep-th/0008241].

Perelstein, M. (2007). Little Higgs models and their phenomenology. *Prog. Part. Nucl. Phys.* **58**, 247. [arXiv:hep-ph/0512128].

Peskin, M. E. (1985). An introduction to the theory of strings. In *Proc. Yale Summer School on High Energy Physics*, eds. M. I. Bowick and F. Gussey. World Scientific.

Peskin, M. E. (1987). In *From the Planck Scale to the Weak Scale: Towards a Theory of the Universe*, ed. H. E. Haber. Singapore: World Scientific.

Peskin, M. E. (1990). *Theory of Precision Electroweak Measurements*. Lectures given at 17th SLAC Summer Inst.: Physics at the 100 GeV Mass Scale, Stanford, CA.

Peskin, M. E. (1997). In *Fields, Strings and Duality: TASI 96*, ed. C. Efthimiouo and B. Greene. Singapore: World Scientific.

Peskin, M. E. and Schroeder, D. V. (1995). *An Introduction to Quantum Field Theory*. Menlo Park: Addison Wesley.

Peskin, M. E. and Takeuchi, T. (1990). A new constraint on a strongly interacting Higgs sector. *Phys. Rev. Lett.*, **65**, 964.

Pokorski, S. (2000). *Gauge Field Theories*. Cambridge: Cambridge University Press.

Polchinski, J. (1998). *String Theory*. Cambridge: Cambridge University Press.

Politzer, H. D. (1973). Reliable perturbative results for strong interactions? *Phys. Rev. Lett.*, **30**, 1346.

Ramond, P. (1999). *Journeys Beyond the Standard Model*. New York: Perseus Books.

Randall, L., Soljacic, M. and Guth, A. H. (1996). Supernatural inflation: inflation from supersymmetry with no (very) small parameters. *Nucl. Phys. B*, **472**, 377 [arXiv:hep-ph/9512439].

Randall, L. and Sundrum, R. (1999). A large mass hierarchy from a small extra dimension. *Phys. Rev. Lett.*, **83**, 3370 [arXiv:hep-ph/9905221].

Rohm, R. (1984). Spontaneous supersymmetry breaking in supersymmetric string theories. *Nucl. Phys. B*, **237**, 553.

Ross, G. G. (1984). *Grand Unified Theories*. Boulder: Westview Press.

Salam, A. and Ward, J. C. (1964). Electromagnetic and weak interactions. *Phys. Lett.*, **13**, 168.

Sannan, S. (1986). Gravity as the limit of the type II superstring theory. *Phys. Rev. D*, **34**, 1749.

Schmaltz, M. and D. Tucker-Smith (2005). Little Higgs review. *Ann. Rev. Nucl. Part. Sci.*, **55**, 229. [arXiv:hep-ph/0502182].

Schwartz, M. D. (2014). *Quantum Field Theory and the Standard Model*: Cambridge, Cambridge University Press.

Seiberg, N. (1993). Naturalness versus supersymmetric nonrenormalization theorems. *Phys. Lett. B*, **318**, 469 [arXiv:hep-ph/9309335].

Seiberg, N. (1994a). The power of holomorphy: exact results in 4-D SUSY field theories. In *Int. Symp. on Particles, Strings and Cosmology (PASCOS94)*, ed. K. C. Wali. Singapore: World Scientific [arXiv:hep-th/9408013].

Seiberg, N. (1994b). Exact results on the space of vacua of four-dimensional SUSY gauge theories. *Phys. Rev. D*, **49**, 6857 [arXiv:hep-th/9402044].

Seiberg, N. (1995a). The power of duality: exact results in 4D SUSY field theory. *Int. J. Mod. Phys. A*, **16**, 4365 [arXiv:hep-th/9506077].

Seiberg, N. (1995b). Electric–magnetic duality in supersymmetric nonAbelian gauge theories. *Nucl. Phys. B*, **435**, 129 [arXiv:hep-th/9411149].

Seiberg, N. (1997). Why is the matrix model correct? *Phys. Rev. Lett.*, **79**, 3577 [arXiv:hep-th/9710009].

Seiberg, N. and Witten, E. (1994). Electric–magnetic duality, monopole condensation, and confinement in $N = 2$ supersymmetric Yang–Mills theory. *Nucl. Phys. B*, **426**, 19 [Erratum: *ibid. B*, **430**, 485 (1994)] [arXiv:hep-th/9407087].

Seiden, A. (2005). *Particle Physics: A Comprehensive Introduction*. San Francisco: Addison-Wesley.

Shadmi, Y. and Shirman, Y. (2000). Dynamical supersymmetry breaking. *Rev. Mod. Phys.*, **72**, 25 [arXiv:hep-th/9907225].

Silverstein, E. and Witten, E. (1995). Criteria for conformal invariance of (0, 2) models. *Nucl. Phys. B*, **444**, 161 [arXiv:hep-th/9503212].

Srednicki, M. (2007). *Quantum Field Theory*. Cambridge: Cambridge University Press.

Strominger, A. and C. Vafa (1996). Microscopic origin of the Bekenstein–Hawking entropy. *Phys. Lett. B*, **379**, 99. [arXiv:hep-th/9601029].

Sundrum, R. (2005). To the fifth dimension and back. In *Tasi 2004 lectures* arXiv:hep-th/0508134.

Susskind, L. (1977). Coarse grained quantum chromodynamics. In *Weak and Electromagnetic Interactions at High Energy*, eds. R. Balian and C. H. Llewelleyn Smith. Amsterdam: North-Holland.

Terning, J. (2003). Non-perturbative supersymmetry, arXiv:hep-th/0306119.

't Hooft, G. (1971). Renormalization of massless Yang–Mills fields. *Nucl. Phys. B*, **33**, 173.

't Hooft, G. (1976). *Phys. Rev. D*, **14**, 3432; erratum: *ibid. D*, **18**, 2199 (1978).

't Hooft, G. (1980). In *Recent Developments in Gauge Theories*, eds. G. 't Hooft *et al.* New York: Plenum Press.

Turner, M. S. (1990). Windows on the axion. *Phys. Rept*, **197**, 67.

Vilenkin, A. (1995). Predictions from quantum cosmology. *Phys. Rev. Lett.*, **74**, 846 [arXiv:gr-qc/9406010].

Wald, R. M. (1984). *General Relativity*. Chicago: University of Chicago Press.

Weinberg, S. (1967). A model of leptons. *Phys. Rev. Lett.*, **19**, 1264.

Weinberg, S. (1972). *Gravitation and Cosmology, Principles and Applications of the General Theory of Relativity*. New York: John Wiley and Sons.

Weinberg, S. (1989). The cosmological constant problem. *Rev. Mod. Phys.*, **61**, 1.

Weinberg, S. (1995). *The Quantum Theory of Fields*. Cambridge: Cambridge University Press.

Weinberg, S. (2000). The cosmological constant problems [arXiv:astro-ph/0005265].

Weinberg, S. (2008). *Cosmology*. Oxford: Oxford University Press.

Wess, J. and Bagger, J. (1992). *Supersymmetry and Supergravity*, Princeton: Princeton University Press.

Wilson, K. G. (1974). Confinement of quarks. *Phys. Rev. D*, **10**, 2445.

Witten, E. (1981). Dynamical breaking of supersymmetry. *Nucl. Phys. B*, **188**, 513.

Witten, E. (1986). New issues in manifolds of *SU*(3) holonomy. *Nucl. Phys. B*, **268**, 79.

Witten, E. (1995). String theory dynamics in various dimensions. *Nucl. Phys. B*, **443**, 85 [arXiv:hep-th/9503124].

Witten, E. (1998). Anti-de Sitter space and holography. *Adv. Theor. Math. Phys.*, **2**, 253 [arXiv:hep-th/9802150].

Yang, C. N. and Mills, R. L. (1954). Conservation of isotopic spin and isotopic gauge invariance. *Phys. Rev.*, **96**, 191.

Index

Printed in the United States
by Baker & Taylor Publisher Services